The secret of KOTA
now at your Doorstep

DPP

Daily Practice Problem Sheets

for NEET/AIIMS

BIOLOGY

DPP BOOKLET

60 Topic-wise + **38** Chapter-wise

Tests for Concept Checking & Speed Building

Improves your learning by at least 20%

- Collection of 3000 + MCQ's of all variety of questions
- Unique & innovative way of learning
- Detailed solutions to Topic-wise & Chapter-wise practice sheets
- Covers all important concepts of each topic
- As per latest pattern & syllabus

• **Corporate Office** : 45, 2nd Floor, Maharishi Dayanand Marg, Corner Market,
Malviya Nagar, New Delhi-110017
Tel. : 011-49842349 / 49842350

Typeset by Disha DTP Team

For further information about books from DISHA,
Log on to **www.dishapublication.com** or email to **info@dishapublication.com**

Daily Practice Problem (DPP) Sheets BIOLOGY for NEET (Kota's formula to Success)

PREPARE ASSESS IMPROVE

Assessment is the most integral part of a student's preparation but still most of them avoid it. Only assessment can tell where you stand and how you can improve from that point. So it is very important that you take the right assessment, which is on the correct pattern, has the same level of difficulty as the actual exam and covers all the important concepts of the subject.

Disha Publication launches a first of its kind product which changed the way coaching was conducted in KOTA - the hub of Engineering and Medical Entrance education in India.

The book "Daily Practice Problem (DPP) Sheets for NEET/AIIMS" is precise, apt and tuned to all the requirements of a NEET/AIIMS aspirant.

KEY DIFFERENTIATING FEATURES OF THE DPP SHEETS

- Division of the complete NEET/AIIMS syllabus of Biology into 60 most important Topics. Each chapter has been broken into 2 or more topics.
- Time Limit, Maximum Marks, Cut-off, Qualifying Scores have been provided for each DPP Sheet/ topic. You must attempt each Sheet in test like conditions following the time limits. The Cut-off for each Sheet indicate that students must at least score this much whereas if a student scores more or equal to Qualifying Score it means his/ her preparation is perfect in that topic.
- Ultimate tool for Concept Checking & Speed Building.
- Collection of 3000 Standardised MCQ's of all variety of NEW pattern questions – MCQ only one correct and Assertion-Reason.
- Unique & innovative way of learning. Whenever you have prepared a topic just attempt that worksheet.
- 2 Separate Booklets - DPP Booklet & Solution Booklet. Do not refer the Solution Booklet until and unless you have made all the efforts to solve the DPP Sheets.
- Covers all important Concepts of each Topic in the form of different Questions in the DPP Sheets.
- As per latest pattern & syllabus of NEET/AIIMS exam.
- Compliant to all boards of education.

No matter where you PREPARE from – a coaching or NCERT books or any other textbook/ Guide - Daily Practice Problem Sheets provides you the right ASSESSMENT on each topic. Your performance provides you the right cues to IMPROVE your concepts so as to perform better in the final examination.

It is to be noted here that these are not tests but act as a checklist of student's learning and ability to apply concepts to different problems. Do proper analysis after you attempt each DPP sheet and try to locate your weak areas.

It is our strong belief that if an aspirant works hard on the cues provided through each of the DPP sheets he/ she can **improve his/ her learning and finally the SCORE by at least 20%.**

• •••• •• ••••••

The book comprise of following two parts

TOPIC-WISE DPP SHEETS WITH SOLUTIONS

INDEX/SYLLABUS

Some Question Updations & Errata

We have made changes in certain questions for them to be more meaningful for the aspirants. The following must be referred while answering the DPPs:

DPP 1, Q 1: Read the question as:

Which of the following is not the feature of living systems?

(a) Produce gametes
(b) Take in energy and use it for maintenance and growth
(c) Adapt to the environment
(d) Respond to their environment

DPP 2, Q 5: Read the question as:

Which of the following does not evolve oxygen?

(a) Green sulfur bacteria
(b) Blue green algae
(c) Green algae
(d) Autotrophic plants

DPP 2, Q 22: Read the question as:

Which of the following can be used for differentiating Gram positive bacteria from Gram negative?

(a) Negative staining
(b) Carbol fuschin staining
(c) Acid fast staining
(d) Gram's staining

DPP 5, Q 17: Read the question as:

The pigment phycoerythrin is characteristic of

(a) green algae
(b) brown algae
(c) *Spirogyra*
(d) Rhodophyta

DPP 12, Q 14: Read the question as:

Which one of the following is the characteristic for birds?

(a) They are ammonotelic organisms
(b) They have scales
(c) They are warm blooded animals
(d) They are quadruped

DPP 20, Q 1: Read the question as:

Identify the correct match between types of chromosomes and their description.

Chromosomes		Position of centromere
A. Metacentric	1.	At the tip
B. Sub metacentric	2.	Almost near tip
C. Acrocentric	3.	At the middle
D. Telocentric	4.	Slightly away from the middle

Codes

(a) A-1, B-3, C-2, D-4
(b) A-3, B-4, C-2, D-1
(c) A-1, B-2, C-3, D-4
(d) A-4, B-3, C-1, D-2

DPP 37, Q 22: Read the question as:

Electron microscopic studies of the sarcomeres have revealed that during muscle contraction

(a) the width of A-band decreases.
(b) the width of H-zone decreases.
(c) the width of I-band increases.
(d) the diameter of fibre increases.

DPP 44, Q 28: Read the question as:

Epiboly is the process of

(a) mass migration of cells from the animal hemisphere so that upper micromeres begin to migrate over the edge of the dorsal lip and roll inside and are tucked beneath the outer layer.
(b) the expansion of one cell sheet over another cells.
(c) rotation of gastrula within the vitelline membrane so that the animal pole becomes anterior.
(d) formation of a small slit like invagination occurring on the grey crescent.

DPP 48, Q 25: Read the question as:

Operon is

(a) a set of closely linked genes regulating a metabolic pathway in prokaryotes
(b) the sequence of three nitrogen bases determining a single amino acid
(c) the sequence of nitrogen bases in mRNA which codes for a single amino acid
(d) a gene responsible for switching on or off other genes

DPP 50, Q 20: Read the question as:

Which of the disease is transmitted by houseflies?

(a) Typhoid
(b) Yellow fever
(c) Cholera
(d) Dysentery

DPP - Daily Practice Problems

Name :

Date :

Start Time :

End Time :

BIOLOGY 01

SYLLABUS : The Living World

Max. Marks : 120

Time : 60 min.

GENERAL INSTRUCTIONS

- The Daily Practice Problem Sheet contains 30 MCQ's. For each question only one option is correct. Darken the correct circle/ bubble in the Response Grid provided on each page.
- You have to evaluate your Response Grids yourself with the help of solution booklet.
- Each correct answer will get you 4 marks and 1 mark shall be deduced for each incorrect answer. No mark will be given/ deducted if no bubble is filled. Keep a timer in front of you and stop immediately at the end of 60 min.
- The sheet follows a particular syllabus. Do not attempt the sheet before you have completed your preparation for that syllabus. Refer syllabus sheet in the starting of the book for the syllabus of all the DPP sheets.
- After completing the sheet check your answers with the solution booklet and complete the Result Grid. Finally spend time to analyse your performance and revise the areas which emerge out as weak in your evaluation.

DIRECTIONS (Q.1-Q.30) : There are 30 multiple choice questions. Each question has 4 choices (a), (b), (c) and (d), out of which ONLY ONE choice is correct.

1. The most important feature of all living systems is to
   (a) utilize oxygen to generate energy.
   (b) replicate the genetic information.
   (c) produce gametes.
   (d) utilize solar energy for metabolic activities.

2. First life on earth was
   (a) cyanobacteria (b) chemoheterotrophs
   (c) autotrophs (d) photoautotrophs

3. Which of the following is not true for a species?
   (a) Members of a species can interbreed.
   (b) Gene flow does not occur between the populations of a species.
   (c) Each species is reproductively isolated from every other species.
   (d) Variations occur among members of a species.

4. Classification given by Bentham and Hooker is
   (a) artificial (b) natural
   (c) phylogenetic (d) numerical

RESPONSE GRID 1. ⓐⓑⓒⓓ 2. ⓐⓑⓒⓓ 3. ⓐⓑⓒⓓ 4. ⓐⓑⓒⓓ

Space for Rough Work

5. Basic unit or smallest taxon of taxonomy/classification is
(a) species (b) kingdom
(c) family (d) variety

6. A taxon is a
(a) group of related families.
(b) group of related species.
(c) type of living organisms.
(d) taxonomic group of any ranking.

7. An important criterion for modern day classification is
(a) resemblances in morphology.
(b) anatomical and physiological traits.
(c) breeding habits.
(d) presence or absence of notochord.

8. The term phylum was given by
(a) Cuvier
(b) Haeckel
(c) Theophrastus
(d) Linnaeus

9. Sequence of taxonomic categories is
(a) Class – Phylum – Tribe – Order – Family – Genus – Species
(b) Division – Class – Family – Tribe – Order – Genus – Species
(c) Division – Class – Order – Family – Tribe – Genus – Species
(d) Phylum – Order – Class – Tribe – Family – Genus – Species

10. Binomial nomenclature means
(a) one name given by two scientists.
(b) one scientific name consisting of a generic and specific epithet.
(c) two names, one latinised, other of a person.
(d) two names of same plant.

11. Linnaeus is credited with
(a) binomial nomenclature
(b) theory of biogenesis
(c) discovery of microscope
(d) discovery of blood circulation

12. Species is
(a) unit of classification.
(b) unit in the evolutionary history of a tree.
(c) specific class of evolution.
(d) not related to evolution.

13. "Taxonomy without phylogeny is similar to bones without flesh" is the statement of
(a) Oswald Tippo
(b) John Hutchinson
(c) Takhtajan
(d) Bentham and Hooker

14. The practical purpose of classification of living organisms is to
(a) explain the origin of living organisms.
(b) trace the evolution of living organisms.
(c) name of the living organisms.
(d) facilitate identification of unknown organisms.

RESPONSE GRID					
	5. ⓐⓑⓒⓓ	6. 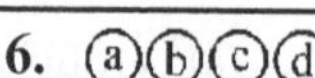	7. 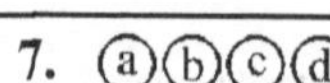	8. 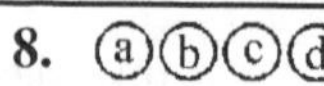	9. ⓐⓑⓒⓓ
	10. ⓐⓑⓒⓓ	11. ⓐⓑⓒⓓ	12. ⓐⓑⓒⓓ	13. ⓐⓑⓒⓓ	14. ⓐⓑⓒⓓ

Space for Rough Work

15. The book '*Genera Plantarum*' was written by
(a) Bessy
(b) Hutchinson
(c) Engler and Prantl
(d) Bentham & Hooker

16. Which of the following is less general in characters as compared to genus?
(a) Species
(b) Division
(c) Class
(d) Family

17. Genus is a group of similar and related
(a) order
(b) genera
(c) families
(d) species

18. Founder of "*Taxonomy*" is
(a) Aristotle
(b) John Ray
(c) Haeckel
(d) Linnaeus

19. Which one of the following is the species?
(a) *Carnivora*
(b) *Canis*
(c) *Familiaris*
(d) *Canis familiaris*

20. Naming of which of the following is incorrect?
(a) *Apis indica*
(b) *Trypanosoma gambiense*
(c) *Ficus Bengalensis*
(d) *Mangifera indica*

21. Which of the following covers the largest number of organisms?
(a) Genus (b) Family
(c) Phylum (d) Class

22. Which one of the following statement is correct regarding 'L' in the scientific name of *Mangifera indica* L. ?
(a) Letter L. signifies latin language.
(b) The name is reverse with *Indica* preceeding *mangifera.*
(c) Letter L. signifies taxonomist Linnaeus.
(d) Letter L. is superfluous.

23. A species is defined as "the group of actually or potentially inter-breeding natural population producing fertile offspring and reproductively isolated from such other groups". The above statement is given by
(a) Carolus Linnaeus
(b) Mayr
(c) J.B. Lamarck
(d) Charles Darwin

24. Specimen used for original publication by the author is
(a) holotype (b) isotype
(c) syntype (d) lactotype

RESPONSE GRID					
	15. ⓐⓑⓒⓓ	16. ⓐⓑⓒⓓ	17. ⓐⓑⓒⓓ	18. ⓐⓑⓒⓓ	19. ⓐⓑⓒⓓ
	20. ⓐⓑⓒⓓ	21. ⓐⓑⓒⓓ	22. ⓐⓑⓒⓓ	23. ⓐⓑⓒⓓ	24. ⓐⓑⓒⓓ

Space for Rough Work

25. The complete name of a plant in binomial nomenclature has

(a) three parts (words) (b) two parts (words)

(c) five parts (words) (d) one part (word)

26. The term '*New systematics*' was introduced by

(a) Adolf Engler (b) Karl Prantl

(c) George Bentham (d) Julian Huxley

27. Biosystematics aims at

(a) identification and arrangement of organisms on the basis of their cytological characteristics.

(b) the classification of organisms based on broad morphological characters.

(c) delimiting various taxa of organisms and establishing their relationships.

(d) the classification of organisms based on their evolutionary history and establishing their phylogeny on the totality of various parameters from all fields of studies.

28. In a hierarchical system of plant classification, which one of the following taxonomic ranks generally ends in 'ceae'?

(a) Family

(b) Genus

(c) Order

(d) Class

29. Branch connected with nomenclature, identification and classification is called

(a) ecology

(b) taxonomy

(c) morphology

(d) physiology

30. The science of naming the plant is known as

(a) classification

(b) identity

(c) nomenclature

(d) taxonomy

RESPONSE GRID					
	25. ⓐⓑⓒⓓ	26. ⓐⓑⓒⓓ	27. ⓐⓑⓒⓓ	28. ⓐⓑⓒⓓ	29. ⓐⓑⓒⓓ
	30. ⓐⓑⓒⓓ				

DAILY PRACTICE PROBLEM SHEET 1 - BIOLOGY			
Total Questions	30	Total Marks	120
Attempted		Correct	
Incorrect		Net Score	
Cut-off Score	40	Qualifying Score	64
Success Gap = Net Score – Qualifying Score			
Net Score = (Correct × 4) – (Incorrect × 1)			

Space for Rough Work

DPP - Daily Practice Problems

Name : | Date :

Start Time : | End Time :

BIOLOGY 02

SYLLABUS : Biological Classification: Monera/Virus

Max. Marks : 120 | **Time : 60 min.**

GENERAL INSTRUCTIONS

- The Daily Practice Problem Sheet contains 30 MCQ's. For each question only one option is correct. Darken the correct circle/ bubble in the Response Grid provided on each page.
- You have to evaluate your Response Grids yourself with the help of solution booklet.
- Each correct answer will get you 4 marks and 1 mark shall be deduced for each incorrect answer. No mark will be given/ deducted if no bubble is filled. Keep a timer in front of you and stop immediately at the end of 60 min.
- The sheet follows a particular syllabus. Do not attempt the sheet before you have completed your preparation for that syllabus. Refer syllabus sheet in the starting of the book for the syllabus of all the DPP sheets.
- After completing the sheet check your answers with the solution booklet and complete the Result Grid. Finally spend time to analyse your performance and revise the areas which emerge out as weak in your evaluation.

DIRECTIONS (Q.1-Q.30) : There are 30 multiple choice questions. Each question has 4 choices (a), (b), (c) and (d), out of which ONLY ONE choice is correct.

1. In *Escherichia coli*
   (a) an organised nucleus is present.
   (b) one chromosome is present.
   (c) one DNA molecule is present.
   (d) one RNA molecule is present.
2. Bacteria are considered as plants, because
   (a) these have a rigid cell wall.
   (b) these have a green colour.
   (c) they can reproduce.
   (d) they are present everywhere.
3. 'Peptidoglycan' is a characteristic constituent of the cell wall of
   (a) eubacteria and unicellular eukaryotes.
   (b) bacteria and cyanobacteria.
   (c) archaebacteria and eukaryotes.
   (d) all members of 'monera' and 'protista'.
4. Which one of the following is correct for structure of cell wall of fungi and bacteria ?
   (a) Both have glycopeptide.
   (b) Both are made up of N-acetylglucosamine.
   (c) Both are made up of N-acetylglucosamine and cellulose.
   (d) Both are made up of chitin.

RESPONSE GRID	1. ⓐⓑⓒⓓ	2. ⓐⓑⓒⓓ	3. ⓐⓑⓒⓓ	4. ⓐⓑⓒⓓ

Space for Rough Work

5. Which of the following does not evolve oxygen?
(a) Photosynthetic bacteria
(b) Blue-green algae
(c) Green algae
(d) Autotrophic plants

6. Bacterial cell divides in every minute and takes one hour to fill up a cup. How much time will be taken to fill half the cup?
(a) 59 minutes (b) 30 minutes
(c) 60 minutes (d) 29 minutes

7. How many bacteria are produced in four hours if a bacterium divides once in half an hour ?
(a) 8 (b) 64
(c) 16 (d) 256

8. Nitrifying bacteria, *Nitrosomonas* and *Nitrobactor*
(a) convert (oxidize) ammonia or ammonium compounds into nitrates.
(b) convert nitrate into nitrogen.
(c) convert nitrogen into nitrates.
(d) convert carbon dioxide into carbohydrates.

9. Mycoplasma is a
(a) eukaryotic and multicellular.
(b) prokaryotic and multicellular.
(c) prokaryotic and unicellular.
(d) eukaryotic and unicellular.

10. What is incorrect for mycoplasma?
(a) It shows osmotic response.
(b) It shows absence of cell wall.
(c) They are sensitive to modern antibiotics.
(d) They are obligate intracellular parasites.

11. In which of the following there is no sexual reproduction?
(a) *Ulothrix* (b) *Nostoc*
(c) *Aspergillus* (d) *Volvox*

12. Which is not true for the heterocyst of *Nostoc*?
(a) Intercalary position.
(b) Site of nitrogen fixation.
(c) Presence of polar nodules.
(d) CO_2 fixation ability.

13. Cyanophyceae has got
(a) definite nucleus and plastid.
(b) no definite nucleus but plastid.
(c) neither definte nucleus nor plastid.
(d) definite nucleus but no plastid.

14. Which was the first photosynthetic organism?
(a) Green algae
(b) Red algae
(c) Cyanobacteria
(d) Brown algae

RESPONSE GRID					
	5. ⓐⓑⓒⓓ	6. ⓐⓑⓒⓓ	7. ⓐⓑⓒⓓ	8. ⓐⓑⓒⓓ	9. ⓐⓑⓒⓓ
	10. ⓐⓑⓒⓓ	11. ⓐⓑⓒⓓ	12. ⓐⓑⓒⓓ	13. ⓐⓑⓒⓓ	14. ⓐⓑⓒⓓ

Space for Rough Work

15. Nitrogen fixation by *Nostoc* takes place in

(a) heterocysts (b) vegetative cells

(c) akinetes (d) hormogonia

16. Auxospores and hormocysts are fomed respectively, by

(a) some diatoms and several cyanobacteria.

(b) some cyanobacteria and many diatoms.

(c) several cyanobacteria and several diatoms.

(d) several diatoms and a few cyanobacteria.

17. Which of the following plants is used as biofertilizer?

(a) *Nostoc* (b) *Funaria*

(c) *Volvox* (d) *Rhizopus*

18. The site of respiration in bacteria is

(a) episome

(b) mesosome (cytoplasmic membrane)

(c) ribosome

(d) microsome

19. The chief component of bacterial cell wall is

(a) cellulose and chitin

(b) cellulose and pectin

(c) amino acids and polysaccharides

(d) cellulose and carbohydrates

20. Chemosynthetic bacteria are those bacteria which

(a) carry out photosynthesis.

(b) synthesize food without light.

(c) are unable to carry out chemical process.

(d) carry out photosynthesis in absence of light.

21. Bacteria bearing flagella all over the body are called

(a) peritrichous (b) atrichous

(c) monotrichous (d) cephalotrichous

22. Under which of the following condition a Gram positive bacteria turns Gram negative?

(a) Defective staining technique.

(b) Changed *pH* of medium.

(c) Aged bacterial culture.

(d) All of the above.

23. Which one of the following is also called halophiles?

(a) Eubacteria

(b) Actinomycetes

(c) Archaebacteria

(d) Cyanobacteria

24. Bacteria reproduces sexually by

(a) endospores

(b) transformation

(c) conidia

(d) exospores

RESPONSE GRID					
	15. ⓐⓑⓒⓓ	16. ⓐⓑⓒⓓ	17. ⓐⓑⓒⓓ	18. ⓐⓑⓒⓓ	19. ⓐⓑⓒⓓ
	20. ⓐⓑⓒⓓ	21. ⓐⓑⓒⓓ	22. ⓐⓑⓒⓓ	23. ⓐⓑⓒⓓ	24. ⓐⓑⓒⓓ

Space for Rough Work

25. The symbiotic nitrogen fixing bacteria present in root nodules of legumes belongs to genus
(a) *Xanthomonas* (b) *Pseudomonas*
(c) *Rhizobium* (d) *Acetobacter*

26. Mycoplasma are not sensitive to
(a) streptomycin (b) penicillin
(c) erythromycin (d) neomycin

27. Which of the following shows the absence of chlorophyll 'b'?
(a) green algae
(b) red algae
(c) blue-green algae
(d) brown algae

28. Sexual reproduction is absent in
(a) *Cyanobacteria* (b) bacteria
(c) eukaryote (d) all of these

29. Heterocysts are found in certain
(a) viruses (b) bacteria
(c) *Cyanobacteria* (d) *Mycoplasmas*

30. Single filament of *Nostoc* without mucilage sheath is known as
(a) hyphae
(b) colony
(c) trichome
(d) mycelium

RESPONSE GRID	
	25. ⓐⓑⓒⓓ 26. ⓐⓑⓒⓓ 27. ⓐⓑⓒⓓ 28. ⓐⓑⓒⓓ 29. ⓐⓑⓒⓓ
	30. ⓐⓑⓒⓓ

DAILY PRACTICE PROBLEM SHEET 2 - BIOLOGY			
Total Questions	30	Total Marks	120
Attempted		Correct	
Incorrect		Net Score	
Cut-off Score	44	Qualifying Score	68
Success Gap = Net Score – Qualifying Score			
Net Score = (Correct × 4) – (Incorrect × 1)			

Space for Rough Work

DPP - Daily Practice Problems

Name :

Date :

Start Time :

End Time :

BIOLOGY 03

SYLLABUS : Biological Classification : Protista

Max. Marks : 120 **Time : 60 min.**

GENERAL INSTRUCTIONS

- The Daily Practice Problem Sheet contains 30 MCQ's. For each question only one option is correct. Darken the correct circle/ bubble in the Response Grid provided on each page.
- You have to evaluate your Response Grids yourself with the help of solution booklet.
- Each correct answer will get you 4 marks and 1 mark shall be deduced for each incorrect answer. No mark will be given/ deducted if no bubble is filled. Keep a timer in front of you and stop immediately at the end of 60 min.
- The sheet follows a particular syllabus. Do not attempt the sheet before you have completed your preparation for that syllabus. Refer syllabus sheet in the starting of the book for the syllabus of all the DPP sheets.
- After completing the sheet check your answers with the solution booklet and complete the Result Grid. Finally spend time to analyse your performance and revise the areas which emerge out as weak in your evaluation.

DIRECTIONS (Q.1-Q.30) : There are 30 multiple choice questions. Each question has 4 choices (a), (b), (c) and (d), out of which ONLY ONE choice is correct.

1. Protista includes
   (a) protozoa, algae and fungi
   (b) algae, bryophyta, bacteria and fungi
   (c) fungi, slime moulds and vascular plants
   (d) protozoa, bacteria, algae and bryophyta
2. Which protist reproduces both by binary fission and conjugation?
   (a) *Amoeba* (b) *Paramecium*
   (c) *Euglena* (d) *Monocystis*
3. Protista contains
   (a) *Euglena*, dinoflagellates and *Yeast*
   (b) *Amoeba, Paramecium, Hydra*
   (c) *Euglena*, *Paramecium*, mushroom
   (d) *Amoeba, Paramecium* and Dinoflagellates
4. Characteristic spores of diatoms are
   (a) ascospores
   (b) basidiospores
   (c) auxospores
   (d) zoospores

RESPONSE GRID	1. ⓐⓑⓒⓓ	2. ⓐⓑⓒⓓ	3. ⓐⓑⓒⓓ	4. ⓐⓑⓒⓓ

Space for Rough Work

5. Protists are
   1. Unicellular and prokaryote.
   2. Unicellular and eukaryote.
   3. Multicellular and eukaryote.
   4. Autotroph and heterotroph.

   (a) 1 + 2 + 3 (b) 2 + 3 + 4
   (c) 3 + 4 (d) 2 + 4

6. Two species of *Amoeba* X and Y were kept in fresh water and got adapted. Species X developed contractile vacuole. When both were transferred to sea water and got adapted, both X and Y lost their contractile vacuole. From these observation we conclude that
   (a) both X and Y are marine species.
   (b) species Y is marine and X is fresh water.
   (c) species X is marine and Y is fresh water.
   (d) both X and Y are fresh water.

7. Which organism behaves like plants in the presence of light and absence of organic food, but in reverse conditions behaves like animals?
   (a) *Archaebacteria* (b) *Euglena*
   (c) *Nostoc* (d) *Paramecium*

8. Spore producing body of a cellular slime mould is
   (a) pseudoplasmodium (b) plasmodium
   (c) sporangium (d) sporophore

9. The slime moulds are characterized by the presence of
   (a) elaters (b) pseudoelaters
   (c) capillitium (d) capitulum

10. *Amoeba* reacts
   (a) negatively to strong light and positively to weak light.
   (b) positively to strong light and negatively to weak light.
   (c) unaffected by light intensity.
   (d) positive to both strong and weak light.

11. Periodic appearance of malaria symptoms occurs due to periodic
   (a) entry of merozoites into erythrocytes.
   (b) attack of liver cells by merozoites.
   (c) formation of signet ring.
   (d) release of pyrogen in blood.

12. Microfossil often present in petroleum producing formation are those of
   (a) *Radiolarians* (b) *Diatoms*
   (c) *Helizoans* (d) *Foraminiferans*

13. During conjugation in *Paramecium*
   (a) out of the four micronuclei formed, three nuclei degenerate.
   (b) out of the twelve macronuclei formed, four nuclei degenerate.
   (c) zygote nucleus undergoes eight successive divisions in each conjugant.
   (d) out of the sixteen nuclei formed from zygote, 12 become macronuclei and 4 micronuclei.

14. In Whittaker's five kingdom classification, unicellular eukaryotes primarily aquatic and having various cell organelles constitute
   (a) monera (b) protista
   (c) animalia (d) plantae

RESPONSE GRID					
	5. ⓐⓑⓒⓓ	6. 	7. ⓐⓑⓒⓓ	8. ⓐⓑⓒⓓ	9. ⓐⓑⓒⓓ
	10. ⓐⓑⓒⓓ	11. ⓐⓑⓒⓓ	12. ⓐⓑⓒⓓ	13. ⓐⓑⓒⓓ	14. ⓐⓑⓒⓓ

Space for Rough Work

15. The scientist who coined the term '*Protista*' to include both plant and animal like unicellular organism was
(a) Robert Koch (b) E. F. Haeckel
(c) L. Pasteur (d) Joseph Lister

16. Which one of the following can photosynthesize its food?
(a) *Hydra* (b) *Paramecium*
(c) *Monocystis* (d) *Euglena*

17. Bacteria lack alternation of generation because there is
(a) neither syngamy nor reduction division.
(b) distinct chromosomes which are absent.
(c) no conjugation.
(d) no exchange of genetic material.

18. Pseudopodia is a characteristic feature of which class ?
(a) Mastigophora
(b) Sarcodina
(c) Sporozoa
(d) Ciliata

19. Which of the following classes of protozoa is exclusively endoparasitic?
(a) Mastigophora
(b) Sarcodina
(c) Opalinata
(d) Sporozoa

20. If a pond dries, *Amoeba*
(a) reproduces
(b) encysts
(c) degenerates
(d) behaviour is uncertain

21. Which of the following unicellular organism has a macronucleus for trophic function and one or more micronuclei for reproduction?
(a) *Euglena* (b) *Amoeba*
(c) *Paramecium* (d) *Trypanosoma*

22. .... is not true for *Euglena*
(a) Presence of chl. – 'a' and 'b'.
(b) Presence of protoplasmic capsule.
(c) Presence of cellulose cell wall.
(d) Presence of proteinaceous pellicle.

23. In which of the following binary fission is not seen ?
(a) *Plasmodium*
(b) *Amoeba*
(c) *Euglena*
(d) *Paramecium*

24. The class of *Trichonympha* is
(a) calcaria
(b) scyphozoa
(c) sporozoa
(d) mastigophora

RESPONSE GRID					
	15. ⓐⓑⓒⓓ	16. ⓐⓑⓒⓓ	17. ⓐⓑⓒⓓ	18. ⓐⓑⓒⓓ	19. ⓐⓑⓒⓓ
	20. ⓐⓑⓒⓓ	21. ⓐⓑⓒⓓ	22. ⓐⓑⓒⓓ	23. ⓐⓑⓒⓓ	24. ⓐⓑⓒⓓ

Space for Rough Work

25. Who discovered malarial parasite ?
   (a) Sir Ronald Ross (b) Charles Laveran
   (c) Patrick Manson (d) Grassi
26. *Trypanosoma gambiense* is spread by
   (a) *Aedes aegyptii* (b) *Culex fatigans*
   (c) *Stomoaxys calcitrans* (d) *Glossina palpalis*
27. Just as *Xenopsylla* is to *Yersinia pestis,* so is.
   (a) *Glossina palpalis* to *Wuchereria bancrofti.*
   (b) *Culex to Plasmodium falciparum.*
   (c) *Homo sapiens* to *Taenia solium.*
   (d) *Phlebotomus to Leishmania denovani.*
28. Nutrition in *Amoeba* is
   (a) holophytic (b) parasitic
   (c) holozoic (d) saprobic
29. Unlike other algae, diatoms do not readily decay due to
   (a) siliceous wall (b) mucilaginous wall
   (c) water proof cell wall (d) non-living cells
30. Sporogony of malarial parasite occurs in
   (a) liver of man.
   (b) RBCs of man.
   (c) stomach wall of mosquito.
   (d) salivary glands of mosquito.

RESPONSE GRID	
	25. ⓐⓑⓒⓓ 26. ⓐⓑⓒⓓ 27. ⓐⓑⓒⓓ 28. ⓐⓑⓒⓓ 29. ⓐⓑⓒⓓ
	30. ⓐⓑⓒⓓ

DAILY PRACTICE PROBLEM SHEET 3 - BIOLOGY			
Total Questions	30	Total Marks	120
Attempted		Correct	
Incorrect		Net Score	
Cut-off Score	32	Qualifying Score	56
Success Gap = Net Score – Qualifying Score			
Net Score = (Correct × 4) – (Incorrect × 1)			

Space for Rough Work

DPP - Daily Practice Problems

Name :   Date :

Start Time :   End Time :

BIOLOGY 04

SYLLABUS : Biological Classification : Fungi

Max. Marks : 120   **Time : 60 min.**

GENERAL INSTRUCTIONS

- The Daily Practice Problem Sheet contains 30 MCQ's. For each question only one option is correct. Darken the correct circle/ bubble in the Response Grid provided on each page.
- You have to evaluate your Response Grids yourself with the help of solution booklet.
- Each correct answer will get you 4 marks and 1 mark shall be deduced for each incorrect answer. No mark will be given/ deducted if no bubble is filled. Keep a timer in front of you and stop immediately at the end of 60 min.
- The sheet follows a particular syllabus. Do not attempt the sheet before you have completed your preparation for that syllabus. Refer syllabus sheet in the starting of the book for the syllabus of all the DPP sheets.
- After completing the sheet check your answers with the solution booklet and complete the Result Grid. Finally spend time to analyse your performance and revise the areas which emerge out as weak in your evaluation.

DIRECTIONS (Q.1-Q.30) : There are 30 multiple choice questions. Each question has 4 choices (a), (b), (c) and (d), out of which ONLY ONE choice is correct.

1. Mycology (mycetology) is a branch which deals with the study of

   (a) viruses   (b) algae

   (c) bacteria   (d) fungi

2. Which one is saprophytic in nutrition?

   (a) *Penicillium*   (b) *Aspergillus*

   (c) *Neurospora*   (d) *Saccharomyces*

3. A group of fungi with septate mycelium in which sexual reproduction is either unknown or lacking are classified under

   (a) phycomycetes   (b) deuteromycetes

   (c) ascomycetes   (d) basidiomycetes

4. Fungal spores produced asexually at the tips of hyphae are called

   (a) sporangiophores

   (b) anthospores

   (c) conidiophores

   (d) meiospores

RESPONSE GRID	1. ⓐⓑⓒⓓ	2. ⓐⓑⓒⓓ	3. ⓐⓑⓒⓓ	4. ⓐⓑⓒⓓ

Space for Rough Work

5. Torula stage of *Mucor* is in _______rich medium.
   (a) pesticides (b) sugar
   (c) antibiotic (d) none of these
6. Facultative saprophyte is
   (a) a parasite which can live as saprophyte.
   (b) always a parasite.
   (c) always a saprophyte.
   (d) may acquire more of life.
7. Which of the following is not correctly matched?
   (a) Root knot disease – *Meloidogyne javanica*
   (b) Smut of bajra – *Tolysporium penicillariae*
   (c) Covered smut of barley – *Ustilage nuda*
   (d) Late blight of potato – *Phytophotora infestans*
8. The hyphae of *Aspergillus* are
   (a) aseptate and multinucleate.
   (b) septate and multinucleate.
   (c) aseptate and uninucleate.
   (d) septate and uninucleate.
9. Dikaryon formation is characteristic of
   (a) ascomycetes and basidiomycetes.
   (b) phycomycetes and basidiomycetes.
   (c) ascomycetes and phycomycetes.
   (d) phycomycetes and zygomycetes.
10. Plasmogamy is fusion of
    (a) two haploid cell including their nuclei.
    (b) two haploid cells without nuclear fusion.
    (c) sperm and egg.
    (d) sperm and two polar nuclei.
11. Algal fungi are placed in
    (a) ascomycetes (b) basidiomycetes
    (c) phycomycetes (d) deuteromycetes
12. A fungus contains cells with two nuclei from different genomes. The nuclei do not fuse but divide independently and simultaneously as new cells are formed. It belongs to
    (a) phycomycetes (b) zygomycetes
    (c) deuteromycetes (d) basidiomycetes
13. Zygospore is formed from
    (a) gametangial copulation
    (b) gametic union
    (c) zygote
    (d) oogamy
14. In *Mucor*, asexual reproduction takes place by
    (a) motile zoospores
    (b) spores
    (c) zoogametes
    (d) zygospores

RESPONSE GRID					
	5. ⓐⓑⓒⓓ	6. ⓐⓑⓒⓓ	7. ⓐⓑⓒⓓ	8. ⓐⓑⓒⓓ	9. ⓐⓑⓒⓓ
	10. ⓐⓑⓒⓓ	11. ⓐⓑⓒⓓ	12. ⓐⓑⓒⓓ	13. ⓐⓑⓒⓓ	14. ⓐⓑⓒⓓ

Space for Rough Work

15. Which of the following types of heterothallism is found in *Mucor*?
(a) Morphological heterothallism
(b) Physiological 2–allele heterothallism
(c) Physiological bipolar multiple allele heterothallism
(d) Physiological tetrapolar multiple allele heterothallism

16. *Mucor* is classified under
(a) basidiomycetes (b) phyeomycetes
(c) ascomycetes (d) Deuteromycetes

17. In *Mucor* and *Rhizopus* there occurs a phenomenon known as heterothallism which means
(a) fusion of two gametes from two thalli of opposite strains.
(b) fusion of two gametes from thalli of similar strain.
(c) formation of a zygospore parthenogenetically.
(d) torula stage.

18. Which one of the following fungus shows heterothallism?
(a) *Erisyphe* (b) *Peziza*
(c) *Rhizopus* (d) *Perongospora*

19. *Rhizopus* reproduces asexually through
(a) sporangiospores (b) aplanospores
(c) chlamydospores (d) all of these

20. Oldest cultivated organism *Yeasts* was discovered by
(a) Aristotle (b) Linnaeus
(c) Leeuwenhoek (d) Theophrastus

21. Yeasts cell wall is rich in
(a) protein (b) cellulose
(c) carbohydrates (d) lipids

22. In endomycorrhiza, the fungus is commonly
(a) zygomycetes
(b) deuteromycetes
(c) ascomycete
(d) basidiomycetes

23. Lichens are ecologically important as they
(a) purify air
(b) are pioneers of barren rocks
(c) are symbionts of algae and fungi
(d) are associated with mycorrhizal roots

24. Which of the following symbiotic association is not useful as biofertilizer?
(a) *Rhizobium* and leguminous roots
(b) Algae and fungi
(c) *Azolla* and *Anabaena*
(d) Roots of higher plants and fungi

RESPONSE GRID	15. ⓐⓑⓒⓓ	16. ⓐⓑⓒⓓ	17. ⓐⓑⓒⓓ	18. ⓐⓑⓒⓓ	19. ⓐⓑⓒⓓ
	20. ⓐⓑⓒⓓ	21. ⓐⓑⓒⓓ	22. ⓐⓑⓒⓓ	23. ⓐⓑⓒⓓ	24. ⓐⓑⓒⓓ

Space for Rough Work

25. Heleotism is a term used for the symbiosis of
(a) algae and bacteria
(b) algae and bryophytes
(c) algae and pteridophytes
(d) algae and fungi

26. A fungus which requires only one single host for completion of its life cycle is called
(a) heteroecious (b) autoecious
(c) heterothallism (d) heterosporous

27. In *Yeast*, cell wall contains
(a) amylose and glucose
(b) glucose and mannose
(c) glucose and muramic acid
(d) sucrose and mannose

28. VAM represents
(a) saprophytic fungi
(b) symbiotic fungi
(c) saprophytic bacteria
(d) symbiotic bacteria

29. Which one of the following group of fungi is sometimes known as "fungal waste basket"?
(a) Phycomycetes
(b) Ascomycetes
(c) Basidiomycetes
(d) Deuteromycetes

30. The scientific name of red bread mould is
(a) *Neurospora* (b) *Mucor*
(c) *Rhizopus* (d) *Aspergillus*

RESPONSE GRID					
	25. ⓐⓑⓒⓓ	26. ⓐⓑⓒⓓ	27. ⓐⓑⓒⓓ	28. ⓐⓑⓒⓓ	29. ⓐⓑⓒⓓ
	30. ⓐⓑⓒⓓ				

DAILY PRACTICE PROBLEM SHEET 4 - BIOLOGY			
Total Questions	30	Total Marks	120
Attempted		Correct	
Incorrect		Net Score	
Cut-off Score	40	Qualifying Score	64
Success Gap = Net Score – Qualifying Score			
Net Score = (Correct × 4) – (Incorrect × 1)			

Space for Rough Work

DPP - Daily Practice Problems

Name : Date :

Start Time : End Time :

BIOLOGY 05

SYLLABUS : Biological Classification : Algae

Max. Marks : 120 **Time : 60 min.**

GENERAL INSTRUCTIONS

- The Daily Practice Problem Sheet contains 30 MCQ's. For each question only one option is correct. Darken the correct circle/ bubble in the Response Grid provided on each page.
- You have to evaluate your Response Grids yourself with the help of solution booklet.
- Each correct answer will get you 4 marks and 1 mark shall be deduced for each incorrect answer. No mark will be given/ deducted if no bubble is filled. Keep a timer in front of you and stop immediately at the end of 60 min.
- The sheet follows a particular syllabus. Do not attempt the sheet before you have completed your preparation for that syllabus. Refer syllabus sheet in the starting of the book for the syllabus of all the DPP sheets.
- After completing the sheet check your answers with the solution booklet and complete the Result Grid. Finally spend time to analyse your performance and revise the areas which emerge out as weak in your evaluation.

DIRECTIONS (Q.1-Q.30) : There are 30 multiple choice questions. Each question has 4 choices (a), (b), (c) and (d), out of which ONLY ONE choice is correct.

1. Phycology is the study of

(a) algae (b) fungi

(c) bacteria (d) all of these

2. Sexual reproduction and mobile cells are absent in

(a) chlorophyceae (b) myxophyceae

(c) rhodophyceae (d) phaeophyceae

3. *Agar-Agar* is obtained from

(a) *Gigartina*

(b) *Gelidium*

(c) *Gracillaria*

(d) all of these

4. Example of coenocytic algae is

(a) *Vaucheria* (b) *Chara*

(c) *Nostoc* (d) *Polysiphonia*

RESPONSE GRID 1. ⓐⓑⓒⓓ 2. ⓐⓑⓒⓓ 3. ⓐⓑⓒⓓ 4. ⓐⓑⓒⓓ

Space for Rough Work

5. Coenobium means
   (a) a hollow spherical colony.
   (b) a group of filaments.
   (c) palmelloid form.
   (d) none of the above.

6. In physiological anisogamy
   (a) gametes are morphologically similar but physiologically dissimilar.
   (b) gametes are morphologically dissimilar and physiologically similar.
   (c) gametes are morphologically and physiologically similar.
   (d) none of the above.

7. Fusion of mature individuals which directly act as gametes, is called
   (a) isogamy (b) anisogamy
   (c) hologamy (d) autogamy

8. The non-motile, greatly thickened asexual spores are called
   (a) hypnospores (b) aplanospores
   (c) macrozoospores (d) microzoospores

9. Which one of the following statements regarding the algae is incorrect ?
   (a) Most algae are photosynthetic.
   (b) Algae can be classified according to their pigments.
   (c) All algae are filamentous.
   (d) *Spirogyra* does not produce zoospores.

10. Rhodophyceae is red coloured due to
    (a) xanthophyll
    (b) carotenoids
    (c) γ-phycoerythrin
    (d) γ-phycocyanin

11. Which of the following alga is rich in protein?
    (a) *Chlorella*
    (b) *Spirogyra*
    (c) *Oscillatoria*
    (d) *Ulothrix*

12. Pyrenoids are made up of
    (a) core of starch surrounded by sheath of protein.
    (b) core of protein surrounded by fatty sheath.
    (c) proteinaceous centre and starchy sheath.
    (d) core of nucleic acid surrounded by protein sheath.

13. Which of the following pigments is presents in all algae?
    (a) Chlorophyll–*a*
    (b) Chlorophyll–*b*
    (c) Chlorophyll–*c*
    (d) Chlorophyll–*d*

14. Sexual reproduction in algae results in the formation of
    (a) oospore
    (b) zoospore
    (c) zygote
    (d) zygospore

RESPONSE GRID	5. ⓐⓑⓒⓓ	6.  ⓐⓑⓒⓓ	7. ⓐⓑⓒⓓ	8. ⓐⓑⓒⓓ	9. ⓐⓑⓒⓓ
	10. ⓐⓑⓒⓓ	11. ⓐⓑⓒⓓ	12. ⓐⓑⓒⓓ	13. ⓐⓑⓒⓓ	14. ⓐⓑⓒⓓ

Space for Rough Work

15. In oogamy, fertilization involves
(a) a large non-motile female gamete and small non-motile male gamete.
(b) a large motile female gamete and small non-motile male gamete.
(c) a small non-motile female gamete and large motile male gamete.
(d) a large non-motile female gamete and small motile male gamete.

16. Pyrenoids in green algal cells are related to
(a) starch formation
(b) protein storage
(c) general metabolism
(d) enzyme secretion

17. The pigment phycoerythrin is characteristic of
(a) green algae (b) brown algae
(c) *Spirogyra* (d) blue green algae

18. Which of the following is a flagellated alga?
(a) *Chlamydomonas* (b) *Ulothrix*
(c) *Spirogyra* (d) *Acetabularia*

19. Filaments of *Spirogyra* are
(a) uniseriate and unbranched
(b) uniseriate and branched
(c) multiseriate and unbranched
(d) multiseriate and branched

20. In *Spirogyra,* ladder like structure is formed in
(a) lateral conjugation (b) gragmentation
(c) palmella stage (d) scalariform conjugation

21. In *Spirogyra* sometimes the gametes behave directly as zygospores without fusion. Such reproductive bodies are called
(a) azygospores
(b) hypnospores
(b) zygospores
(d) aplanospores

22. The cell wall of *Spirogyra* is made up of
(a) cellulose (b) pectin
(c) lignin (d) chitin

23. *Spirogyra* differs from *Mucor* in having
(a) uninuclear gametangia
(b) multicellular gametes
(c) anisogamete
(d) sexual reproduction

24. Sexual reproduction in *Spirogyra* is morphologically characterized by
(a) oogamy
(b) anisogamy
(c) isogamy
(d) isogamy and oogamy both

RESPONSE GRID	15. ⓐⓑⓒⓓ	16. ⓐⓑⓒⓓ	17. ⓐⓑⓒⓓ	18. ⓐⓑⓒⓓ	19. ⓐⓑⓒⓓ
	20. ⓐⓑⓒⓓ	21. ⓐⓑⓒⓓ	22. ⓐⓑⓒⓓ	23. ⓐⓑⓒⓓ	24. ⓐⓑⓒⓓ

Space for Rough Work

25. *Spirogyra* has a

(a) haplontic life cycle (b) diplontic life cycle

(c) haplobiontic life cycle (d) diplobiontic life cycle

26. In *Ulothrix* meiosis (reduction division) occurs in

(a) gamete (b) zygospore

(c) zoospore (d) thallus

27. Which of the following statement is correct for *Ulothrix*?

(a) Filamentous alga with flagellated reproductive stages.

(b) Filamentous alga with non-flagellated reproductive stages.

(c) Membranous alga producing zoospores.

(d) Non-motile colonial alga lacking reproductive stages.

28. Sexual reproduction in *Spirogyra* can be described as

(a) morphological anisogamy and physiological isogamy.

(b) morphological as well as physiological isogamy.

(c) morphological as well as physiological anisogamy.

(d) morphological isogamy and physiological anisogamy.

29. Meiosis is *Ulothrix* takes place during

(a) zoospore formation (b) gamete formation

(c) zygote germination (d) zoospore germination

30. Physiological anisogamous sexual reproduction is found in

(a) *Mucor* (b) *Selaginella*

(c) *Pinus* (d) *Spirogyra*

RESPONSE GRID	
	25. ⓐⓑⓒⓓ 26. ⓐⓑⓒⓓ 27. ⓐⓑⓒⓓ 28. ⓐⓑⓒⓓ 29. ⓐⓑⓒⓓ
	30. ⓐⓑⓒⓓ

DAILY PRACTICE PROBLEM SHEET 5 - BIOLOGY			
Total Questions	30	Total Marks	120
Attempted		Correct	
Incorrect		Net Score	
Cut-off Score	44	Qualifying Score	72
Success Gap = Net Score – Qualifying Score			
Net Score = (Correct × 4) – (Incorrect × 1)			

Space for Rough Work

DPP - Daily Practice Problems

Name : | Date :

Start Time : | End Time :

BIOLOGY 06

SYLLABUS : Biological Classification : Bryophyta

Max. Marks : 120 | Time : 60 min.

GENERAL INSTRUCTIONS

- The Daily Practice Problem Sheet contains 30 MCQ's. For each question only one option is correct. Darken the correct circle/ bubble in the Response Grid provided on each page.
- You have to evaluate your Response Grids yourself with the help of solution booklet.
- Each correct answer will get you 4 marks and 1 mark shall be deduced for each incorrect answer. No mark will be given/ deducted if no bubble is filled. Keep a timer in front of you and stop immediately at the end of 60 min.
- The sheet follows a particular syllabus. Do not attempt the sheet before you have completed your preparation for that syllabus. Refer syllabus sheet in the starting of the book for the syllabus of all the DPP sheets.
- After completing the sheet check your answers with the solution booklet and complete the Result Grid. Finally spend time to analyse your performance and revise the areas which emerge out as weak in your evaluation.

DIRECTIONS (Q.1-Q.30) : There are 30 multiple choice questions. Each question has 4 choices (a), (b), (c) and (d), out of which ONLY ONE choice is correct.

1. The largest bryophyte is

   (a) *Funaria* (b) *Marchantia*

   (c) *Megaceros* (d) *Dowsonia*

2. Heteromorphic alternation of generation is commonly found in

   (a) algae (b) fungi

   (c) bryophyta (d) all of these

3. In which of the following groups would you place a plant which produces spores and embryos but lacks seeds and vascular tissue?

   (a) Fungi (b) Bryophytes

   (c) Pteridophytes (d) Gymnosperms

4. Among the following which is not characteristic feature of bryophyta?

   (a) Motile sperms

   (b) Presence of archaegonium

   (c) Water essential for fertilization

   (d) Photosynthetically independent sporophyte

RESPONSE GRID	1. ⓐⓑⓒⓓ	2. ⓐⓑⓒⓓ	3. ⓐⓑⓒⓓ	4. ⓐⓑⓒⓓ

Space for Rough Work

5. Which one of the following is true moss?
(a) Club moss (b) Reindeer moss
(c) Irish moss (d) Bog moss

6. Elaters are present in sporogonium of
(a) *Riccia* (b) *Marchantia*
(c) *Selaginella* (d) *Sphagnum*

7. Gametophytic generation is dominant in
(a) pteridophyta (b) bryophyta
(c) angiosperms (d) gymnosperms

8. First land inhabiting plants are
(a) angiosperms (b) gymnosperms
(c) bryophytes (d) pteridophytes

9. Bryophytes can be separated from algae, because they
(a) are thalloid forms.
(b) have no conducting tissue.
(c) possess archaegonia.
(d) contain chloroplast.

10. Bryophytes comprise
(a) sporophyte which has longer duration.
(b) dominant phase of gametophyte which is parasitic.
(c) dominant phase of gametophyte which produces spores.
(d) small sporophyte phase and generally parasitic on gametophyte.

11. Bryophytes are dependent on water because
(a) archegonium has to remain filled with water for fertilization.
(b) water is essential for fertilization for their homosporous nature.
(c) water is essential for their vegetative propagation.
(d) the sperms can easily reach upto egg in the archaegonium.

12. What is incorrect for bryophytes?
(a) Vascular tissue is lacking.
(b) Independent sporophyte is absent.
(c) Gametophyte reduced and dependent.
(d) Asexual reproduction by zoospores is absent.

13. Which of the following have capacity of absorbing water, used to replace cotton and used as a fuel ?
(a) *Marchantia*
(b) *Riccia*
(c) *Sphagnum*
(d) *Funaria*

14. Leafy gametophytes is formed from protonema in
(a) *Marchantia*
(b) *Riccia*
(c) *Funaria*
(d) *Anthoceros*

RESPONSE GRID					
	5. ⓐⓑⓒⓓ	6. ⓐⓑⓒⓓ	7. ⓐⓑⓒⓓ	8. ⓐⓑⓒⓓ	9. ⓐⓑⓒⓓ
	10. ⓐⓑⓒⓓ	11. ⓐⓑⓒⓓ	12. ⓐⓑⓒⓓ	13. ⓐⓑⓒⓓ	14. ⓐⓑⓒⓓ

Space for Rough Work

15. Protonema is found in the life cycle of
(a) *Spirogyra* (b) *Rhizopus*
(c) *Funaria* (d) *Riccia*

16. In *Funaria,* calyptra is derived from
(a) antheridium (b) columella
(c) capsule (d) archegonium

17. The dehiscene of moss capsule takes place by rupture of
(a) operculum
(b) peristome
(c) annulus
(d) calyptra

18. *Funaria* gametophyte is
(a) dioecious and autoecious.
(b) monoecious and autoecious.
(c) dioecious and heteroecious.
(d) monoecious and heteroecious.

19. Which of the following is a diploid?
(a) Capsule of moss.
(b) Gametophyte, archegonia and egg of moss.
(c) Gametophyte, antheridia and sperm of moss.
(d) Gametophyte and spore of moss.

20. Femalesex organ in *Funaria* is called
(a) paraphysis (b) oospores
(c) archaegonium (d) antheridium

21. Spore mother cells in *Funaria* are
(a) haploid
(b) tetraploid
(c) diploid
(d) triploid

22. Calyptra is seen in
(a) antheridiophore (b) archegoniophore
(c) sporophyte (d) gametophyte

23. The sporophyte of *Funaria* begins development within
(a) antheridia
(b) capsule
(c) protonema
(d) archaegonium

24. In *Funaria* (moss) spore germinates to produce
(a) protonema
(b) prothallus
(c) proembryo
(d) embryo

RESPONSE GRID					
	15. ⓐⓑⓒⓓ	16. ⓐⓑⓒⓓ	17. ⓐⓑⓒⓓ	18. ⓐⓑⓒⓓ	19. ⓐⓑⓒⓓ
	20. ⓐⓑⓒⓓ	21. ⓐⓑⓒⓓ	22. ⓐⓑⓒⓓ	23. ⓐⓑⓒⓓ	24. ⓐⓑⓒⓓ

Space for Rough Work

25. Thallus of *Ricca* is
   (a) triploid
   (b) diploid
   (c) haploid
   (d) tetraploid

26. In *Ricca / Marchantia*, the rhizoids are
   (a) branched unicellular
   (b) branched multicellular
   (c) unbranched multicellular
   (d) unbranched unicellular

27. The sporophyte of *Riccia* is
   (a) chemoautotrophs
   (b) partial parasite
   (c) photosynthetically independent
   (d) total parasite

28. An archegonium of *Riccia* has
   (a) 4 neck canal cells, 1 venter canal and one oosphere.
   (b) 4 neck canal cells, 2 venter canal cells and one oosphere.
   (c) 4 neck canal cells, one venter canal and two oosphere.
   (d) 6 neck canal cells, 2 venter canal cells and one oosphere.

29. Life cycle of *Funaria* is not completed without water. Choose the correct statement.
   (a) As *Funaria* is a bryophyte plant.
   (b) As branches will not develop.
   (c) As fertilization takes place in presence of water only.
   (d) As plant is delicate and will become dry and die without water.

30. *Funaria* is attached to substratum by
   (a) roots (b) rhizoids
   (c) haustoria (d) stem

RESPONSE GRID	
	25. ⓐⓑⓒⓓ 26. ⓐⓑⓒⓓ 27. ⓐⓑⓒⓓ 28. ⓐⓑⓒⓓ 29. ⓐⓑⓒⓓ 30. ⓐⓑⓒⓓ

DAILY PRACTICE PROBLEM SHEET 6 - BIOLOGY

Total Questions	30	Total Marks	120
Attempted		Correct	
Incorrect		Net Score	
Cut-off Score	36	Qualifying Score	64
Success Gap = Net Score – Qualifying Score			
Net Score = (Correct × 4) – (Incorrect × 1)			

Space for Rough Work

DPP - Daily Practice Problems

Name : 　　　　　　　　　　Date :

Start Time : 　　　　　　　　　　End Time :

BIOLOGY 07

SYLLABUS : Biological Classification : Pteridophyta

Max. Marks : 120 　　　　　　　　　　**Time : 60 min.**

GENERAL INSTRUCTIONS

- The Daily Practice Problem Sheet contains 30 MCQ's. For each question only one option is correct. Darken the correct circle/ bubble in the Response Grid provided on each page.
- You have to evaluate your Response Grids yourself with the help of solution booklet.
- Each correct answer will get you 4 marks and 1 mark shall be deduced for each incorrect answer. No mark will be given/ deducted if no bubble is filled. Keep a timer in front of you and stop immediately at the end of 60 min.
- The sheet follows a particular syllabus. Do not attempt the sheet before you have completed your preparation for that syllabus. Refer syllabus sheet in the starting of the book for the syllabus of all the DPP sheets.
- After completing the sheet check your answers with the solution booklet and complete the Result Grid. Finally spend time to analyse your performance and revise the areas which emerge out as weak in your evaluation.

DIRECTIONS (Q.1-Q.30) : There are 30 multiple choice questions. Each question has 4 choices (a), (b), (c) and (d), out of which ONLY ONE choice is correct.

1. Which one of the following belongs to vascular cryptogams?
   (a) Bryophyta　　(b) Pteridophyta
   (c) Gymnosperms　　(d) Angiosperms
2. Heterospory is production of
   (a) sexual and asexual spores.
   (b) large and small spores.
   (c) haploid and diploid spores.
   (d) diploid and tetraploid spores.
3. Heterothallism refers to
   (a) fusion which is not accompanied with zygote formation.
   (b) fusion between morphologically similar strain.
   (c) fusion between the strains of structurally similar and physiologically different.
   (d) all of the above
4. An advantage which made seed bearing plants so greatly surpass the ferns in colonzing the land, is the evolution of the
   (a) ability to photosynthesize.
   (b) fibrous root system.
   (c) woody stem.
   (d) non-swimming sperms.

RESPONSE GRID 1. ⓐⓑⓒⓓ 2. ⓐⓑⓒⓓ 3. ⓐⓑⓒⓓ 4. ⓐⓑⓒⓓ

Space for Rough Work

5. In which of the following plant diploid stage generally dominates?
(a) Algae (b) Ferns
(c) Mosses (d) Fungi

6. Pteridophytes are characterized by
(a) presence of vascular system.
(b) absence of flowers.
(c) absence of fruits.
(d) all of the above

7. Which of the following is not a pteridophyte?
(a) *Ginkgo* (b) *Selaginella*
(c) *Polypodium* (d) *Azolla*

8. Seed habit originated firstly in some
(a) pteridophytes
(b) pines
(c) monocots
(d) dicots

9. Which of the following helps in coal formation?
(a) Bacteria
(b) Gymnosperm
(c) Pteridophytes
(d) *Archaebacteria*

10. Which of the following is not involved in the fertilization of fern?
(a) Pollen tube
(b) Water
(c) Archaegonia
(d) Flagellated sperms

11. *Azolla* is used as biofertilizer because it
(a) multiplies very fast to produce massive biomass.
(b) has association of nitrogen-fixing *Rhizobium.*
(c) has association of nitrogen fixing *Cyanobacteria.*
(d) has association of mycorrhiza.

12. Most primitive living vascular plants are
(a) brown algae (b) *Sphagnum*
(c) ferns (d) *Cycads*

13. Meristeles are found in
(a) cycas stem (b) pinus needle
(c) fern leaf (d) fern rhizome

14. In *Selaginella* tapetum orginates from
(a) wall of sporangium
(b) sporogenous tissue
(c) sporophyll
(d) basal part of cone

RESPONSE GRID					
	5. ⓐⓑⓒⓓ	6. ⓐⓑⓒⓓ	7. ⓐⓑⓒⓓ	8. ⓐⓑⓒⓓ	9. ⓐⓑⓒⓓ
	10. ⓐⓑⓒⓓ	11. ⓐⓑⓒⓓ	12. ⓐⓑⓒⓓ	13. ⓐⓑⓒⓓ	14. ⓐⓑⓒⓓ

Space for Rough Work

15. Which of the following is not common in *Funaria* and *Selaginella*?
(a) Roots (b) Archaegonium
(c) Embryo (d) Motile sperms

16. Which of the following is known as 'resurrection plant'?
(a) *Selaginella* (b) *Welwitschia*
(c) *Rafflesia* (d) *Chlorella*

17. *Selaginella* has the character of evolutionary importance. That character is
(a) ligule (b) seed
(c) heterospory (d) strobilus

18. Antherozoids of *Selaginella* has
(a) elongated body with many flagella.
(b) an elongated body with two flagella at one end.
(c) top shaped body with many flagella.
(d) oval body with two flagella at one end.

19. Stele of *Selaginella* is
(a) protosteleic
(b) eusteleic
(c) dictyosteleic
(d) solenosteleic

20. Which one character is common for *Selaginella* and fern ?
(a) Heterospory
(b) Protostele
(c) Development of sporangium
(d) Absence of ovule

21. Heteromorphic alternation of generation is found in
(a) *Spirogyra* (b) *Mucor*
(c) *Selaginella* (d) *Pinus*

22. Spores with chloroplast is present in
(a) *Selaginella* (b) *Equisetum*
(c) *Puccinia* (d) *Rhizopus*

23. In Pteridophytes / *Dryopteris* meiosis occurs at the time of
(a) gamete formation
(b) spore formation
(c) formation of prothallus
(d) formation of sex organs

24. Fern prothallus is
(a) homothallic (b) heterothallic
(c) heterotrophic (d) heteromorphic

RESPONSE GRID					
	15. ⓐⓑⓒⓓ	16. ⓐⓑⓒⓓ	17. ⓐⓑⓒⓓ	18. ⓐⓑⓒⓓ	19. ⓐⓑⓒⓓ
	20. ⓐⓑⓒⓓ	21. ⓐⓑⓒⓓ	22. ⓐⓑⓒⓓ	23. ⓐⓑⓒⓓ	24. ⓐⓑⓒⓓ

Space for Rough Work

25. What is not true for *Dryopteris* ?
   (a) Mature archaegonia secrete malic acid.
   (b) Antherozoids are sickle shaped and multiflagellate.
   (c) Archaegonia has one binuleate neck canal cell.
   (d) Antheridia develop on ventral surface of prothallus.

26. In *Dryopteris* the opening mechanism of sporangiun is effectively operated by
   (a) stalk (b) stomium
   (c) annulus (d) none of these

27. What is correct?
   (a) Protonema of Moss and prothallus of *Dryopteris* are sporophyt.e
   (b) Protonema of Moss and prothallus of *Dryopteris* are gametophytic.
   (c) Protonema of Moss is sporophyte, *Pteris* prothallus is gametophytic but plant body of *Pteris* and *Funaria* are gametophytic.
   (d) Plant body of Moss is gametophyte while that of *Dryopteris* is both gametophytic and sporophytic.

28. Which of the following is not correct with reference to *Selaginella*?
   (a) *Selaginella* is commonly distributed on hills and plains.
   (b) Some species of *Selaginella* are truely xerophytic.
   (c) The vascular cylinder is protostelic.
   (d) The endodermis is trabeculated.

29. *Selaginella* is heterosporous. It produces two types of spores. The microspore develops into microgametophyte and the megaspore develops into megagametophyte. With respect to gametophyte, which one statement is correct?
   (a) The microgametophyte is devoid of chlorophyll at maturity.
   (b) The megagametophyte possesses chlorophyll at maturity.
   (c) The megagametophyte develops within the wall of megaspore.
   (d) The microgametophyte develops chlorophyll at maturity.

30 Which of the following does not belong to *Selaginella*?
   (a) Ramenta (b) Trabeculae
   (c) Rhizophore (d) Ligule

RESPONSE GRID	
	25. ⓐⓑⓒⓓ 26. ⓐⓑⓒⓓ 27. ⓐⓑⓒⓓ 28. ⓐⓑⓒⓓ 29. ⓐⓑⓒⓓ
	30. ⓐⓑⓒⓓ

DAILY PRACTICE PROBLEM SHEET 7 - BIOLOGY			
Total Questions	30	Total Marks	120
Attempted		Correct	
Incorrect		Net Score	
Cut-off Score	32	Qualifying Score	52
Success Gap = Net Score – Qualifying Score			
Net Score = (Correct × 4) – (Incorrect × 1)			

Space for Rough Work

DPP - Daily Practice Problems

Name : | Date :

Start Time : | End Time :

BIOLOGY 08

SYLLABUS : Biological Classification : Gymnosperm / Angiosperm

Max. Marks : 120 **Time : 60 min.**

GENERAL INSTRUCTIONS

- The Daily Practice Problem Sheet contains 30 MCQ's. For each question only one option is correct. Darken the correct circle/ bubble in the Response Grid provided on each page.
- You have to evaluate your Response Grids yourself with the help of solution booklet.
- Each correct answer will get you 4 marks and 1 mark shall be deduced for each incorrect answer. No mark will be given/ deducted if no bubble is filled. Keep a timer in front of you and stop immediately at the end of 60 min.
- The sheet follows a particular syllabus. Do not attempt the sheet before you have completed your preparation for that syllabus. Refer syllabus sheet in the starting of the book for the syllabus of all the DPP sheets.
- After completing the sheet check your answers with the solution booklet and complete the Result Grid. Finally spend time to analyse your performance and revise the areas which emerge out as weak in your evaluation.

DIRECTIONS (Q.1-Q.30) : There are 30 multiple choice questions. Each question has 4 choices (a), (b), (c) and (d), out of which ONLY ONE choice is correct.

1. Which one of the following is not a cryptogam?
   (a) Algae (b) Bryophyta
   (c) Pteridophyta (d) Gymnosperms
2. In which of the following groups would you placed a plant which produces seeds but lacks flower and fruits?
   (a) Fungi (b) Bryophytes
   (c) Pteridophyta (d) Gymnosperms
3. Gymnosperms are called naked seed plants because these lack
   (a) cotyledon (b) endosperm
   (c) ovary wall (d) testa
4. What is the similarity between gymnosperms and angiosperms?
   (a) Phloem of both have companian cells.
   (b) Endosperm is formed before fertilization in both.
   (c) Origin of ovule and seed is similar in both.
   (d) Both have leaves, stem and roots.

RESPONSE GRID	1. ⓐⓑⓒⓓ	2. ⓐⓑⓒⓓ	3. ⓐⓑⓒⓓ	4. ⓐⓑⓒⓓ

Space for Rough Work

5. Which of the following is living fossil?
(a) *Pinus* (b) *Ginkgo*
(c) *Thuja* (d) *Deodar*

6. Which of the following is not the feature of gymnosperms?
(a) Parallel venation
(b) Perennial plants
(c) Distinct branches (long and short branches)
(d) Xylem with vessels

7. A typical gymnospermous plant has 8 chromosomes in leaf cells. What will be number of chromosomes in the cells of its gametophyte?
(a) 16 (b) 8
(c) 4 (d) 24

8. Pollination of pollen grains in *Pinus* takes place at
(a) one celled stage (b) two celled stage
(c) three celled stage (d) four celled stage

9. Non-motile male gametes are formed in
(a) *Funaria* (b) *Selaginella*
(c) *Fern* (d) *Pinus*

10. Which of the following structures in *Pinus* are haploid?
(a) Megaspore, integument, root
(b) Endosperm, megaspore, pollen grain
(c) Pollen grain, leaf, root
(d) Megaspore, endosperm, embryo

11. In *Pinus* seeds, there are
(a) two cotyledons
(b) one cotyledon
(c) fleshy cotyledons
(d) many cotyledons

12. In the embryo of *Pinus*, rosette cells lie
(a) above suspensor cells.
(b) between suspensor and embryonal cells.
(c) between primary and secondary suspensor cells.
(d) between embryonal and apical cells.

13. Which of the following is the best explanation for the excurrent (pyramidal) shape of *Pinus* (the branches giving the plant the appearance of a tall cone)?
(a) Competition among adjoining trees for sunlight.
(b) The effect of auxins on growth of stem tips.
(c) Adaptation for wind pollination.
(d) Efficiency of water transport from ground to leaves.

14. Which statement is correct with reference to *Pinus*?
(a) It is of much economic value.
(b) It is cosmopolitan in distribution.
(c) It grows in deserts and exhibits xerophytic character.
(d) It form deciduous trees in temperate region.

RESPONSE GRID					
	5. ⓐⓑⓒⓓ	6. ⓐⓑⓒⓓ	7. ⓐⓑⓒⓓ	8. ⓐⓑⓒⓓ	9. ⓐⓑⓒⓓ
	10. ⓐⓑⓒⓓ	11. ⓐⓑⓒⓓ	12. ⓐⓑⓒⓓ	13. ⓐⓑⓒⓓ	14. ⓐⓑⓒⓓ

Space for Rough Work

15. The secondary wood of *Pinus* is characterized by
    (a) presence of resin cells.
    (b) presence of resin ducts.
    (c) absence of resin ducts.
    (d) presence of vessels.

16. Ovules in *Pinus* are generally
    (a) anatropous with four integuments.
    (b) anatropous with one integuments.
    (c) amphitropous with three integuments.
    (d) orthotropous with two integuments.

17. In *Pinus* megaspore on germination forms
    (a) pollen tube with motile sperms.
    (b) endosperm with archaegonia.
    (c) endosperm with sperms.
    (d) endosperm with embryo sac.

18. The wood of *Pinus* is
    (a) pycnoxylic and monoxylic.
    (b) pycnoxylic and heteroxylous.
    (c) manoxylic and homoxylous.
    (d) manoxylic and heteroxylous.

19. Top-shaped multiciliate male gametes, and the mature seed which bears only one embryo with two cotyledons are characteristic features of
    (a) polypetalous angiosperms
    (b) gamopetalous angiosperms
    (c) conifers
    (d) cycads

20. Wood of *Cycas* is
    (a) monoxylic and manoxylic
    (b) manoxylic and polyxylic
    (c) diploxylic
    (d) monoxylic

21. Which is incorrect about *Cycas*?
    (a) Its xylem has vessels.
    (b) It has circinate venation.
    (c) It does not have well organised female flower.
    (d) Its roots posses some blue-green algae.

22. Diploxylic condition occurs in *Cycas* in
    (a) root
    (b) stem
    (c) coralloid root
    (d) leaflet

23. Vascular bundles of *Cycas* stem are
    (a) conjoint.
    (b) collateral and closed.
    (c) conjoint, collateral and open.
    (d) conjoint, collateral and closed.

24. In *Cycas*
    (a) male strobilus and megasporophylls occurs on the same individual.
    (b) the same cone consists of both micro and megasporophylls.
    (c) the same sporophyll bears microsporangia and ovules.
    (d) male strobilus (cone) and megasporophylls occur on separate individuals.

RESPONSE GRID	15. ⓐⓑⓒⓓ	16. ⓐⓑⓒⓓ	17. ⓐⓑⓒⓓ	18. ⓐⓑⓒⓓ	19. ⓐⓑⓒⓓ
	20. ⓐⓑⓒⓓ	21. ⓐⓑⓒⓓ	22. ⓐⓑⓒⓓ	23. ⓐⓑⓒⓓ	24. ⓐⓑⓒⓓ

Space for Rough Work

25. *Cycas* has two cotyledons but is not included under angiosperms because it has
(a) circinate ptyxis (b) compound leaves
(c) monocot like stem (d) naked seeds

26. Siphonogamy occurs in
(a) *Pinus* (b) *Pteris*
(c) *Funaria* (d) *Ulothrix*

27. Characters of both conifers and cycads are found in
(a) *Ginkgo* (b) *Ephedra*
(c) *Cupressus* (d) *Thuja*

28. Which Gymnospermic order includes all fossil plants?
(a) Cycadales (b) Cycadofilicales
(c) Cycadeoidales (d) Coniferales.

29. Heteromorphic alternation of generation occurs in
(a) *Funaria, Spirogyra, Selaginella*
(b) *Funaria, Selaginella, Cycas*
(c) *Spirogyra, Rhizopus, Selaginella*
(d) *Rhizopus, Funaria, Spirogyra*

30. Number of cotyledons in *Zea*, *Cycas* and *Pinus* respectively are
(a) 1, 1, many (b) 1, 2, 1
(c) 1, 1, 1 (d) 1, 2, many

RESPONSE GRID	25. ⓐⓑⓒⓓ	26. ⓐⓑⓒⓓ	27. ⓐⓑⓒⓓ	28. ⓐⓑⓒⓓ	29. ⓐⓑⓒⓓ
	30. ⓐⓑⓒⓓ				

DAILY PRACTICE PROBLEM SHEET 8 - BIOLOGY			
Total Questions	30	Total Marks	120
Attempted		Correct	
Incorrect		Net Score	
Cut-off Score	32	Qualifying Score	52
Success Gap = Net Score – Qualifying Score			
Net Score = (Correct × 4) – (Incorrect × 1)			

Space for Rough Work

DPP - Daily Practice Problems

Name : 　　　　　　　　　　　　　　　　　Date :

Start Time : 　　　　　　　　　　　　　　　End Time :

BIOLOGY 09

SYLLABUS : Animal Kingdom: Non-Chordata – Porifera, Coelenterata, Platyhelminthes, Nemathelminthes

Max. Marks : 120 　　　　　　　　　　　　　　　　**Time : 60 min.**

GENERAL INSTRUCTIONS

- The Daily Practice Problem Sheet contains 30 MCQ's. For each question only one option is correct. Darken the correct circle/ bubble in the Response Grid provided on each page.
- You have to evaluate your Response Grids yourself with the help of solution booklet.
- Each correct answer will get you 4 marks and 1 mark shall be deduced for each incorrect answer. No mark will be given/ deducted if no bubble is filled. Keep a timer in front of you and stop immediately at the end of 60 min.
- The sheet follows a particular syllabus. Do not attempt the sheet before you have completed your preparation for that syllabus. Refer syllabus sheet in the starting of the book for the syllabus of all the DPP sheets.
- After completing the sheet check your answers with the solution booklet and complete the Result Grid. Finally spend time to analyse your performance and revise the areas which emerge out as weak in your evaluation.

DIRECTIONS (Q.1-Q.30) : There are 30 multiple choice questions. Each question has 4 choices (a), (b), (c) and (d), out of which ONLY ONE choice is correct.

1. Radial symmetry occurs in
   (a) fishes 　　(b) molluscs
   (c) coelenterata 　　(d) sponges

2. Level of organisation in sponges is
   (a) cellular level 　　(b) acellular level
   (c) tissue level 　　(d) organ-system level

3. Sponges and coelenterates resemble each other in being
   (a) monoblastic and acoelomate
   (b) diploblastic and acoelomate
   (c) triploblastic and acoelomate
   (d) triploblastic and pseudocoelomate

4. Evolution of porifera from protozoan is evidenced by the animals like
   (a) *Euglena*
   (b) *Chlamydomonas*
   (c) *Volvox*
   (d) *Paramecium*

RESPONSE GRID 1. ⓐⓑⓒⓓ 2. ⓐⓑⓒⓓ 3. ⓐⓑⓒⓓ 4. ⓐⓑⓒⓓ

Space for Rough Work

5. In sponges, free swimming larva is called as
   (a) veliger (b) trochophore
   (c) parenchymula (d) bipinnaria
6. Water currents in *Leucosolenia* are produced by
   (a) choanocytes (b) pinacocytes
   (c) archaeocytes (d) thesocytes
7. The common bath sponge belongs to the genus
   (a) *Sycon* (b) *Leucosolenia*
   (c) *Euspongia* (d) *Spongilla*
8. Path of water in a sponge is
   (a) Dermal ostia → Gastral ostia → Osculum.
   (b) Dermal ostia → Gastral ostia → Spongocoel → Osculum.
   (c) Osculum → Spongocoel → Choanocytes → Ostia.
   (d) Pinacocytes → Choanocytes → Enteron → Osculum.
9. Choanocytes in ascon-type of canal system form lining of
   (a) spongocoel (b) porocyte
   (c) apopyle (d) incurrent canal
10. Which of the following are 'multicellular grade' organisms?
   (a) Sponges (b) Coelenterates
   (c) Prokaryotes (d) Vertebrates
11. *Sycon* belongs to a group of animals which are best described as
   (a) Unicellular or acellular.
   (b) Multicellular without any tissue organisation.
   (c) Multicellular with a gastrovascular cavity.
   (d) Multicellular having tissue organisation, but no body cavity.
12. Which is not applicable to coelenterata?
   (a) Coelenteron
   (b) Choanocytes
   (c) Nematocyst
   (d) Radial symmetry
13. Budding is a normal mode of asexual reproduction in
   (a) starfish and *Hydra*
   (b) *Hydra* and sponges
   (c) tapeworm and *Hydra*
   (d) sponges and starfish
14. Nematocysts take part in
   (a) locomotion (b) offence and defence
   (c) food capture (d) all of the above

RESPONSE GRID					
	5. ⓐⓑⓒⓓ	6. ⓐⓑⓒⓓ 	7. ⓐⓑⓒⓓ	8. ⓐⓑⓒⓓ	9. ⓐⓑⓒⓓ
	10. ⓐⓑⓒⓓ	11. ⓐⓑⓒⓓ	12. ⓐⓑⓒⓓ	13. ⓐⓑⓒⓓ	14. ⓐⓑⓒⓓ

Space for Rough Work

15. *Hydra is*
(a) triploblastic, radial symmetry and acoelomate.
(b) triploblastic, radial symmetry and coelomate.
(c) diploblastic, radial symmetry and acoelomate.
(d) diploblastic, radial symmetry and coelomate.

16. Which one of the following is an example of platyhelminthes?
(a) *Trypanosoma* (b) *Schistosoma*
(c) *Plasmodium* (d) *Wuchereria*

17. The excretory organs of flatworms are
(a) flame cells (b) nephridia
(c) malpighian tubules (d) rennate glands

18. What is correct about *Taenia?*
(a) Male organs occur in posterior proglottides.
(b) Male organs occur in anterior proglottides.
(c) Female organs occur in anterior proglottides.
(d) Mature proglottides contain both male and female organs.

19. Secondary host of *Taenia soluim* is
(a) cow (b) sheep
(c) pig (d) man

20. *Fasciola hepatica* lives in
(a) liver of sheep (b) blood of sheep
(c) intestine of sheep (d) spleen of sheep

21. Bilaterally symmetrical but acoelomate animal is
(a) liver fluke (b) jelly fish
(c) round worms (d) crab

22. "Triploblastic, unsegmented, acoelomate exhibiting bilateral symmetry and reproducing both asexually and sexually with parasitic forms." The above description is the characteristic of phylum
(a) platyhelminthes (b) annelida
(c) ctenophora (d) cnidaria

23. Which of the following groups have one or more animals which are not pseudocoelomate?
(a) *Bipalium, Taenia*
(b) *Enterobius, Wuchereria*
(c) *Ancylostoma, Dracunculus*
(d) *Ascaris, Ancylostoma*

24. The adult *Wuchereria bancrofti* lives in
(a) human sub dermal spaces
(b) muscles of *Culex*
(c) salivary glands of *Culex*
(d) human lymph glands

RESPONSE GRID					
	15. ⓐⓑⓒⓓ	16. ⓐⓑⓒⓓ	17. ⓐⓑⓒⓓ	18. ⓐⓑⓒⓓ	19. ⓐⓑⓒⓓ
	20. ⓐⓑⓒⓓ	21. ⓐⓑⓒⓓ	22. ⓐⓑⓒⓓ	23. ⓐⓑⓒⓓ	24. ⓐⓑⓒⓓ

Space for Rough Work

25. Coelom in *Ascaris lumbricoides* is called as

(a) pseudocoelom (b) true coelom

(c) schizocoelom (d) none of these

26. In humans, elephantiasis is caused by

(a) *Ascaris lumbricoides*

(b) *Dracunculus medinensis*

(c) *Wuchereria bancrofti*

(d) *Ancylostoma duodenale*

27. Differentiating trait of *Ascaris* is

(a) sexual dimorphism and rhabditiform larva

(b) unisexual and digenetic parasite

(c) pseudocoelom and metameric segmentation

(d) hermaphrodite and pseudocoelom

28. Pseudocoelom develop from

(a) blastopore lip (b) archenteron

(c) embryonic mesoderm (d) blastocoel

29. In nemathelminthes which coelom is not lined by peritoneum ?

(a) Acoelom (b) Pseudocoelom

(c) Enterocoelom (d) Haemocoel

30. Choose the correct statement with reference to *Ascaris?*

(a) Hatching of embryos takes place in the stomach due to lytic enzyme.

(b) Adulthood is reached inside the body of the host in ten days time.

(c) Development and moulting takes place in the alveoli of lungs.

(d) Hatching of embryo takes places within ten hours.

RESPONSE GRID	
	25. ⓐⓑⓒⓓ 26. ⓐⓑⓒⓓ 27. ⓐⓑⓒⓓ 28. ⓐⓑⓒⓓ 29. ⓐⓑⓒⓓ
	30. ⓐⓑⓒⓓ

DAILY PRACTICE PROBLEM SHEET 9 - BIOLOGY			
Total Questions	30	Total Marks	120
Attempted		Correct	
Incorrect		Net Score	
Cut-off Score	44	Qualifying Score	68
Success Gap = Net Score – Qualifying Score			
Net Score = (Correct × 4) – (Incorrect × 1)			

Space for Rough Work

DPP - Daily Practice Problems

Name :

Date :

Start Time :

End Time :

BIOLOGY 10

SYLLABUS : Animal Kingdom: Non-Chordata – Porifera, Coelenterata, Platyhelminthes, Nemathelminthes

Max. Marks : 120 **Time : 60 min.**

GENERAL INSTRUCTIONS

- The Daily Practice Problem Sheet contains 30 MCQ's. For each question only one option is correct. Darken the correct circle/ bubble in the Response Grid provided on each page.
- You have to evaluate your Response Grids yourself with the help of solution booklet.
- Each correct answer will get you 4 marks and 1 mark shall be deduced for each incorrect answer. No mark will be given/ deducted if no bubble is filled. Keep a timer in front of you and stop immediately at the end of 60 min.
- The sheet follows a particular syllabus. Do not attempt the sheet before you have completed your preparation for that syllabus. Refer syllabus sheet in the starting of the book for the syllabus of all the DPP sheets.
- After completing the sheet check your answers with the solution booklet and complete the Result Grid. Finally spend time to analyse your performance and revise the areas which emerge out as weak in your evaluation.

DIRECTIONS (Q.1-Q.30) : There are 30 multiple choice questions. Each question has 4 choices (a), (b), (c) and (d), out of which ONLY ONE choice is correct.

1. Annelids are

   (a) radially symmetrical (b) externally segmented

   (c) triploblastic (d) pseudocoelomate

2. Body is segmented in

   (a) coelenterate (b) annelida

   (c) porifera (d) mollusca

3. In which of the following, clitellum is absent?

   (a) Polychaeta (b) Oligochaeta

   (c) *Hirudinea* (d) All of these

4. In *Pheretima*, there are red coloured round bodies, in 4th, 5th and 6th segments above the alimentary canal. They are believed to be involved in

   (a) excretion

   (b) digestion

   (c) reproduction

   (d) leucocyte production

RESPONSE GRID 1. ⓐⓑⓒⓓ 2. ⓐⓑⓒⓓ 3. ⓐⓑⓒⓓ 4. ⓐⓑⓒⓓ

Space for Rough Work

5. The colour of the body in earthworm is brown due to the presence of
   (a) porphyrin (b) haemoglobin
   (c) blood (d) haemocyanin

6. Region of earthworm which is forest of nephridia is called
   (a) clitellar region (b) pharyngeal region
   (c) typhlosolar region (d) intestinal region

7. Hearts of *Pheretima* are situated in the segments
   (a) 10, 13, 16 and 17 (b) 7, 9, 12 and 13
   (c) 4, 5, 10 and 13 (d) 11, 14, 17 and 18.

8. *Pheretima posthuma* is highly useful as
   (a) their burrows make the soil loose.
   (b) they make the soil porous, leave their castings and take organic debris in the soil.
   (c) they are used as fish meal.
   (d) they kill the birds due to biomagnification of chlorinated hydrocarbons.

9. Blood of *Pheretima* is
   (a) blue with haemocyanin in corpuscles.
   (b) blue with haemocyanin in plasma.
   (c) red with haemglobin in corpuscles.
   (d) red with haemoglobin in plasma.

10. Chloragogen cells are involved in
   (a) digestion (b) excretion of water
   (c) respiration (d) fat storage

11. In *Pheretima* nephridia occur in
   (a) All segment except 1 – 4 and 10 – 14.
   (b) 1 – 2, 4 – 6, 15 to last segments.
   (c) Meganephridia in pre-clitellar and micronephridia in post-clitellar segments.
   (d) Micronephridia in all segments and meganephridia from clitellar region to end.

12. Which one of the following groups of structures/organs have similar function?
   (a) Typhlosole in earthworm, intestinal villi in rat and contractile vacuole in *Amoeba.*
   (b) Nephridia in earthworm, Malpighian tubules in cockroach and urinary tubules in rat.
   (c) Antennae of cockroach, typmanum of frog and clitellum of earthworm.
   (d) Incisors of rat, gizzard (proventriculus) of cockroach and tube feet of starfish.

13. Insects are
   (a) amminotelic (b) ammonotelic
   (c) ureotelic (d) uricotelic

14. The process of conversion of a small cockroach into an adult cockroach is called as
   (a) moulting (b) metamorphosis
   (c) ecdysis (d) transformation

RESPONSE GRID					
	5. ⓐⓑⓒⓓ	6. 	7. ⓐⓑⓒⓓ	8. ⓐⓑⓒⓓ	9. ⓐⓑⓒⓓ
	10. ⓐⓑⓒⓓ	11. ⓐⓑⓒⓓ	12. ⓐⓑⓒⓓ	13. ⓐⓑⓒⓓ	14. ⓐⓑⓒⓓ

Space for Rough Work

15. Which one of the following belongs to phylum arthropoda?
(a) *Octopus* (b) *Termite*
(c) *Nereis* (d) Leech

16. In cockroach, the excretory organs are
(a) malpighian tubules
(b) nephridia
(c) malpighian corpuscles
(d) solenocytes (flame cells)

17. What is the similarity between cockroach, *Anopheles* and housefly?
(a) Cuticle covering the body.
(b) Two pairs of wings.
(c) Three pairs of legs.
(d) Presence of cephalothorax.

18. An insect differs from spider in having
(a) 3 pairs of legs and 2 pairs of wings instead of 4 pairs of legs and no wings.
(b) 3 pairs of legs and 2 pairs of wings instead of 3 pairs of legs and no wings.
(c) 3 pairs of legs and no wings instead of 5 pairs of legs and generally one pair of wings.
(d) 6 pairs of legs and one pairs of wings instead of 4 pairs of legs and no wings.

19. Which is a matching set in taxonomy?
(a) Leech, locust, sea urchin, lobster.
(b) Star fish, jelly fish, cuttle fish, octopus.
(c) Millipede, crab, centipede, cockroach.
(d) *Nereis*, *Planaria*, round worm, earthworm.

20. Set of useful insects is
(a) wasp, *Anopheles* and bed bug.
(b) bed bug, silk moth and *Anopheles*.
(c) honey bee, bed bug and locust.
(d) cochineal insect, lac insect and honey bee.

21. Hormone produced by corpora allata in insects is
(a) growth hormone (b) moulting hormone
(c) inhibiting hormone (d) juvenile hormone

22. *Periplaneta* shows
(a) complete metamorphosis
(b) incomplete metamorphosis
(c) no metamorphosis
(d) gradual metamorphosis

23. Which one possess larval stages?
(a) Cockroach and Housefly
(b) Housefly and Butterfly
(c) Cockroach and Honey Bee
(d) Grasshopper and Dragonfly

24. Which one of the following is a matching pair of an animal and a certain phenomenon it exhibits?
(a) *Taenia* - Polymorphism
(b) *Pheretima* - Sexual dimorphism
(c) *Musca* - Complete metamorphosis
(d) *Chamaeleon* – Parthenogenesis

RESPONSE GRID					
	15. ⓐⓑⓒⓓ	16. ⓐⓑⓒⓓ	17. ⓐⓑⓒⓓ	18. ⓐⓑⓒⓓ	19. ⓐⓑⓒⓓ
	20. ⓐⓑⓒⓓ	21. ⓐⓑⓒⓓ	22. ⓐⓑⓒⓓ	23. ⓐⓑⓒⓓ	24. ⓐⓑⓒⓓ

Space for Rough Work

25. *Octopus* belongs to class
(a) mollusca
(b) pelecypoda
(c) arthropoda
(d) cephalopoda

26. Mollusca is differentiated from other by
(a) bilateral symmetry and external skeleton.
(b) mantle and gill.
(c) shell and non-segmented body.
(d) mantle and non-segmented body.

27. Aristotle's lantern is found in
(a) jelly fish (b) sea anemone
(c) sea lily (d) sea urchin

28. An animal phylum having radially symmetrical adults but bilateral symmetrical larvae is
(a) porifera (b) coelenterata
(c) echinodermata (d) annelida

29. Which one of the the following pairs of animals is correctly matched with the kind of their body symmetry?
(a) *Hydra* and shark - Bilateral symmetry
(b) Tapeworm and octopus - Radial symmetry
(c) *Amoeba* and sea urchin - Asymmetry
(d) Jellyfish and starfish - Radial symmetry

30. Mark the correct one?

	Phylum	Class	Example
(a)	Annelida	Oligochaeta	*Nereis*
(b)	Mollusca	Pelecypoda	Cuttle fish
(c)	Reptilia	Ophidia	Lizard
(d)	Echinodermata	Holothuroidea	*Cucumaria*

RESPONSE GRID	
	25. ⓐⓑⓒⓓ 26. ⓐⓑⓒⓓ 27. ⓐⓑⓒⓓ 28. ⓐⓑⓒⓓ 29. ⓐⓑⓒⓓ 30. ⓐⓑⓒⓓ

DAILY PRACTICE PROBLEM SHEET 10 - BIOLOGY

Total Questions	30	Total Marks	120
Attempted		Correct	
Incorrect		Net Score	
Cut-off Score	36	Qualifying Score	64
Success Gap = Net Score – Qualifying Score			
Net Score = (Correct × 4) – (Incorrect × 1)			

Space for Rough Work

DPP - Daily Practice Problems

Name : Date :

Start Time : End Time :

BIOLOGY 11

SYLLABUS : Animal Kingdom: Chordata – Feature of Chordata, Pisces, Amphibia

Max. Marks : 120 **Time : 60 min.**

GENERAL INSTRUCTIONS

- The Daily Practice Problem Sheet contains 30 MCQ's. For each question only one option is correct. Darken the correct circle/ bubble in the Response Grid provided on each page.
- You have to evaluate your Response Grids yourself with the help of solution booklet.
- Each correct answer will get you 4 marks and 1 mark shall be deduced for each incorrect answer. No mark will be given/ deducted if no bubble is filled. Keep a timer in front of you and stop immediately at the end of 60 min.
- The sheet follows a particular syllabus. Do not attempt the sheet before you have completed your preparation for that syllabus. Refer syllabus sheet in the starting of the book for the syllabus of all the DPP sheets.
- After completing the sheet check your answers with the solution booklet and complete the Result Grid. Finally spend time to analyse your performance and revise the areas which emerge out as weak in your evaluation.

DIRECTIONS (Q.1-Q.30) : There are 30 multiple choice questions. Each question has 4 choices (a), (b), (c) and (d), out of which ONLY ONE choice is correct.

1. The lamprey (*Petromyzon*) is included in the same taxonomic class as the
   (a) Chamaeleon (*Anolis*)
   (b) Hag fish (*Myxine*)
   (c) Salamander (*Ambystoma*)
   (d) Lung fish (*Neoceratodus*)
2. Cold blooded animals are those having
   (a) cold blood
   (b) variable body temperature according to the temperature of atmosphere
   (c) always constant temperature
   (d) blood, which can flow even below 4°C
3. Which is correct matching set?
   (a) Annelida, Insecta, Mammalia are classes.
   (b) Arthropoda, Mollusca, Pisces are phyla.
   (c) Protozoa, Rhizopoda, Chordata are phyla.
   (d) Calcarea, Oligochaeta, Aves are class.
4. *Petromyzon* is a connecting link between
   (a) *Balanoglossus* and *Amphioxus*
   (b) *Amphioxus* and cyclostomata
   (c) cyclostomata and pisces
   (d) pisces and amphibians

RESPONSE GRID	1. ⓐⓑⓒⓓ	2. ⓐⓑⓒⓓ	3. ⓐⓑⓒⓓ	4. ⓐⓑⓒⓓ

Space for Rough Work

5. Which of the following is the smallest taxonomic group of animals having a cranium, vertebral column, ventral heart, pulmonary respiration and two pairs of limbs?
   (a) Gnathostomata (b) Tetrapoda
   (c) Vertebrata (d) Chordata
6. In which of the following notochord is absent?
   (a) Adult *Herdmania* and *Balanoglossus*
   (b) Adult *Herdmania* and adult *Branchiostoma*
   (c) Larva of *Herdmania* and *Branchiostoma*
   (d) Larva of *Herdmania* and *Balanoglossus*
7. The correct classification of *Balanoglossus* is
   (a) Chordata → Vertebrata → Enteropneusta
   (b) Chordata → Vertebrata → Pterobranchia
   (c) Chordata → Hemichordata → Pterobranchia
   (d) Chordata → Hemichordata → Enteropneusta
8. Which of the following staements is true?
   (a) All chordates are vertebrates.
   (b) All vertebrates are chordates.
   (c) Non chordates have a vertebral column.
   (d) Invertebrates possess a tubular nerve cord.
9. Which is a true fish?
   (a) Cat fish
   (b) Jelly fish
   (c) Cuttle fish
   (d) Silver fish
10. Electric organs are found in
   (a) sharks (b) porpoises
   (c) goldfish (d) rays (*Torpedo*)
11. The fishes possess gills and fins. Which of the following is not a true fish?
   (a) Silver fish (*Lepisma*)
   (b) Gold fish (*Carassius*)
   (c) Silver carp (*Hythalamictyes*)
   (d) Sea horse (*Hippocampus*)
12. Which of the following is characteristic feature of fishes?
   (a) Tail and venous heart
   (b) Venous heart and gills
   (c) Epidermal scales and tail
   (d) Epidermal scales and gills
13. Anadromous fishes move
   (a) from seas to freshwater
   (b) from sea to estuary
   (c) from river to sea
   (d) from estuary to sea
14. Freshwater bony fishes maintain water balance by
   (a) excreting hypotonic urine.
   (b) excreting salt across their gills.
   (c) drinking small amount of water.
   (d) excreting waste in the form of uric acid.

RESPONSE GRID					
	5. ⓐⓑⓒⓓ	6. ⓐⓑⓒⓓ	7. ⓐⓑⓒⓓ	8. ⓐⓑⓒⓓ	9. ⓐⓑⓒⓓ
	10. ⓐⓑⓒⓓ	11. ⓐⓑⓒⓓ	12. ⓐⓑⓒⓓ	13. ⓐⓑⓒⓓ	14. ⓐⓑⓒⓓ

Space for Rough Work

15. Which fins are paired in fishes ?
(a) Dorsal fin and anal fin
(b) Pelvic fin and ventral fin
(c) Pectoral fin and pelvic fin
(d) Caudal fin and dorsal fin

16. *Salamander* belongs to the class
(a) reptilia (b) amphibia
(c) aves (d) mammalia

17. A hibernating frog respires with
(a) lung
(b) diaphragm
(c) buccal epithelium
(d) skin

18. Absence of thumb is characterstic of
(a) rabbit
(b) man
(c) frog
(d) monkey

19. A frog lives in water or near water because
(a) it can get its food easily in water.
(b) its hindlimbs are webbed and helps in swimming.
(c) it can see through its transparent eyelids while swimming.
(d) it respires through the skin.

20. Frog has
(a) five fingers and four toes
(b) four fingers and five toes
(c) five fingers and five toes
(d) four fingers and four toes

21. Neoteny is found in
(a) tadpole (b) *Salamander*
(c) *hyla* (d) *Axolotl*

22. Limbless amphibians belongs to the order
(a) anura
(b) urodela
(c) gymnophiona
(d) squamata

23. Frog is called amphibian because
(a) it has no tail.
(b) it has got lungs.
(c) its young one, tadople is aquatic in its habitat.
(d) it lives both on land and water.

24. Characteristic feature of Cyclostomata are
(a) elongated cylindrical body.
(b) round mouth.
(c) round mouth without jaws and pair appendages.
(d) round mouth with jaw and unpaired appendages.

RESPONSE GRID					
	15. ⓐⓑⓒⓓ	16. ⓐⓑⓒⓓ	17. ⓐⓑⓒⓓ	18. ⓐⓑⓒⓓ	19. ⓐⓑⓒⓓ
	20. ⓐⓑⓒⓓ	21. ⓐⓑⓒⓓ	22. ⓐⓑⓒⓓ	23. ⓐⓑⓒⓓ	24. ⓐⓑⓒⓓ

Space for Rough Work

25. Gnathostomata is divided into 2 super classes, these are
    (a) acraniata and craniata
    (b) pisces and tetrapoda
    (c) amniota and bipedal
    (d) tetrapoda and gnathosomata
26. Heterocercal tail is found in
    (a) cartilaginous fishes (b) bony fishes
    (c) whale (d) amphibians
27. Fishes having swim bladder, which do not have direct communication with the exterior and where resorbent and secretory part is not sharply separated from one another, are called
    (a) physostomes (b) physoclists
    (c) euphysoclits (d) paraphysoclists
28. Which of the following class is called tailed amphibians ?
    (a) Apoda (b) Urodela
    (c) Anura (d) None of these
29. Which of the following is a lving fossil and connecting link between fishes and amphibians ?
    (a) Sea horse
    (b) *Gambusia*
    (c) *Latimeria*
    (d) *Torpedo*
30. Metanephros are present in
    (a) birds (b) reptilia
    (c) mammals (d) all of these

RESPONSE GRID	25. ⓐⓑⓒⓓ	26. ⓐⓑⓒⓓ	27. ⓐⓑⓒⓓ	28. ⓐⓑⓒⓓ	29. ⓐⓑⓒⓓ
	30. ⓐⓑⓒⓓ				

DAILY PRACTICE PROBLEM SHEET 11 - BIOLOGY			
Total Questions	30	Total Marks	120
Attempted		Correct	
Incorrect		Net Score	
Cut-off Score	40	Qualifying Score	64
Success Gap = Net Score – Qualifying Score			
Net Score = (Correct × 4) – (Incorrect × 1)			

Space for Rough Work

DPP - Daily Practice Problems

Name : Date :

Start Time : End Time :

BIOLOGY 12

SYLLABUS : Animal Kingdom: Chordata – Reptiles, Aves, Mammals

Max. Marks : 120 **Time : 60 min.**

GENERAL INSTRUCTIONS

- The Daily Practice Problem Sheet contains 30 MCQ's. For each question only one option is correct. Darken the correct circle/ bubble in the Response Grid provided on each page.
- You have to evaluate your Response Grids yourself with the help of solution booklet.
- Each correct answer will get you 4 marks and 1 mark shall be deduced for each incorrect answer. No mark will be given/ deducted if no bubble is filled. Keep a timer in front of you and stop immediately at the end of 60 min.
- The sheet follows a particular syllabus. Do not attempt the sheet before you have completed your preparation for that syllabus. Refer syllabus sheet in the starting of the book for the syllabus of all the DPP sheets.
- After completing the sheet check your answers with the solution booklet and complete the Result Grid. Finally spend time to analyse your performance and revise the areas which emerge out as weak in your evaluation.

DIRECTIONS (Q.1-Q.30) : There are 30 multiple choice questions. Each question has 4 choices (a), (b), (c) and (d), out of which ONLY ONE choice is correct.

1. Only poisonous lizard of the world is
   (a) *Draco* (b) *Heloderma*
   (c) *Sphenodon* (d) *Varanus*

2. Members of class reptilia are
   (a) homeothermic and amniotic
   (b) homeothermic and anamniotic
   (c) poikilothermic and amniotic
   (d) poikilothermic and anamniotic

3. Which is a poisonous snake?
   (a) *Enhydrina*
   (b) *Typhlops*
   (c) *Python*
   (d) *Eryx*

4. Monocondylic skull, dry skin and procoelus vertebrae are present in
   (a) amphibia
   (b) reptilia
   (c) birds
   (d) mammals

RESPONSE GRID	1. ⓐⓑⓒⓓ	2. ⓐⓑⓒⓓ	3. ⓐⓑⓒⓓ	4. ⓐⓑⓒⓓ

Space for Rough Work

5. The most poisonous snake is
(a) *Krait* (b) tree snake
(c) *Python* (d) rat snake

6. Poisonous fangs of a snake are modified
(a) mandible (b) maxillary teeth
(c) canines (d) nasals

7. Large size scales fully extended from side to side on the belly are characteristics of
(a) krait and sea snake
(b) cobra and *Python*
(c) rat snake and cobra
(d) *Python* and krait

8. 3 or 4 chambered heart is present in which vertebrate group ?
(a) Fishes (b) Amphibia
(c) Reptilia (d) Aves

9. *Typhlops* is
(a) sea snake (b) grass snake
(c) glass snake (d) blind snake

10. Diapside skull is found in which of the following?
(a) *Natrix*, *Draco* and Turtle
(b) Crocodile, Turtle and *Seymouria*
(c) *Sphenodon*, Crocodile and *Viper*
(d) *Calotes*, Cobra and Varanosaurus

11. Which one of the following is a matching pair of an animal and its one of the characterstics?
(a) *Chamaelon* - Binocular vision
(b) *Heloderma* - Poison gland
(c) *Varanus* - Prehensile tail
(d) House lizard - 4 chambered heart

12. Which of the following feature is not common between *Newt & Hemidactylus*?
(a) Body is divisible into head, neck, trunk and tail.
(b) Head with pair of eyes and tympanic membrane.
(c) Trunk has 2 pairs of limb for locomotion.
(d) Heart is 3 chambered.

13. *Archaeopteryx* is called a connecting link, which carried the characteristics of
(a) reptile and bird.
(b) reptile and mammal.
(c) fish and amphibian.
(d) amphibian and reptile.

14. Which one of the following is the characteristic for birds?
(a) They are flying animals
(b) They are warm blooded
(c) They are bipedal and have feather
(d) They are quadruped and have scales

RESPONSE GRID					
	5. ⓐⓑⓒⓓ	6. ⓐⓑⓒⓓ	7. ⓐⓑⓒⓓ	8. ⓐⓑⓒⓓ	9. ⓐⓑⓒⓓ
	10. ⓐⓑⓒⓓ	11. ⓐⓑⓒⓓ	12. ⓐⓑⓒⓓ	13. ⓐⓑⓒⓓ	14. ⓐⓑⓒⓓ

Space for Rough Work

15. Monocondylic skull, warm blooded animals with air sacs are present in
(a) reptilia (b) birds
(c) amphibia (d) mammalia

16. Which one of the following is a flightless bird?
(a) *Passer* (b) *Corvus*
(c) *Aptenodytes* (d) *Pavo cristatus*

17. The wishbone of the bird is derived from
(a) skull (b) pectoral girdle
(c) pelvic girdle (d) hindlimb

18. Egg-laying mammals are grouped as
(a) eutheria (b) prototheria
(c) rodentia (d) metatheria

19. Eutherian mammals are
(a) oviparous
(b) viviparous
(c) ovoviviparous
(d) both oviparous and ovoviviparous

20. Why do mammals lack mucus glands in their skin?
(a) The skin is not slippery.
(b) This skin is tough.
(c) The epidermis has many layers of cells.
(d) The skin is not respiratory.

21. Milk glands are found in
(a) all vertebrates
(b) all mammals
(c) all placental mammals
(d) all prototherians

22. Which pair of the animal groups, has oxygenated and de-oxygenated blood in the heart separately?
(a) Amphibians and Reptiles
(b) Birds and Mammals
(c) Reptiles and Birds
(d) Reptiles and Mammals

23. Lingulates or odd-toed mammals (Horse, Ass and Zebra) belong to the order
(a) artiodactyla
(b) perissodactyla
(c) lagomorpha
(d) edentata

24 Which of the following statement is correct ?
(a) *Archaeopteryx* is connecting link between aves and mammals.
(b) *Duck-billed platypus* is connecting link between mammals and reptiles.
(c) *Sea horse* is connecting link between horse and fish.
(d) *Hydra* is connecting link between protozoa and metazoa.

RESPONSE GRID					
	15. ⓐⓑⓒⓓ	16. ⓐⓑⓒⓓ	17. ⓐⓑⓒⓓ	18. ⓐⓑⓒⓓ	19. ⓐⓑⓒⓓ
	20. ⓐⓑⓒⓓ	21. ⓐⓑⓒⓓ	22. ⓐⓑⓒⓓ	23. ⓐⓑⓒⓓ	24. ⓐⓑⓒⓓ

Space for Rough Work

25. "Sea lion" belongs to

(a) class reptilia (b) subclass prototheria

(c) superclass pisces (d) order carnivora

26. Which pair of characters are found without exception in all mammals?

(a) Hair and vivipartiy.

(b) Viviparity and mammary glands.

(c) Viviparity and internal fertilization.

(d) Mammary glands and internal fertilization.

27. Which of the following is not in mammals?

(a) Hepatic portal system

(b) Hypophyseal portal system

(c) Renal portal system

(d) Hepatic and hypophyseal portal system

28. The group 'amniota' includes

(a) birds and reptiles

(b) birds and mammals

(c) reptiles and mammals

(d) reptiles, birds and mammals

29. Bat can travel with

(a) eyes open

(b) eyes plugged and ears open

(c) ears plugged and eyes open

(d) ears closed and eyes plugged

30. Flippers of seal are

(a) fins (b) hind limb

(c) forelimb (d) Gills

RESPONSE GRID	
	25. ⓐⓑⓒⓓ 26. ⓐⓑⓒⓓ 27. ⓐⓑⓒⓓ 28. ⓐⓑⓒⓓ 29. ⓐⓑⓒⓓ
	30. ⓐⓑⓒⓓ

DAILY PRACTICE PROBLEM SHEET 12 - BIOLOGY			
Total Questions	30	Total Marks	120
Attempted		Correct	
Incorrect		Net Score	
Cut-off Score	40	Qualifying Score	60
Success Gap = Net Score – Qualifying Score			
Net Score = (Correct × 4) – (Incorrect × 1)			

Space for Rough Work

DPP - Daily Practice Problems

Name : Date :

Start Time : End Time :

BIOLOGY 13

SYLLABUS : Morphology of Flowering Plants Part I : Root, Stem, Inflorescence

Max. Marks : 120 **Time : 60 min.**

GENERAL INSTRUCTIONS

- The Daily Practice Problem Sheet contains 30 MCQ's. For each question only one option is correct. Darken the correct circle/ bubble in the Response Grid provided on each page.
- You have to evaluate your Response Grids yourself with the help of solution booklet.
- Each correct answer will get you 4 marks and 1 mark shall be deduced for each incorrect answer. No mark will be given/ deducted if no bubble is filled. Keep a timer in front of you and stop immediately at the end of 60 min.
- The sheet follows a particular syllabus. Do not attempt the sheet before you have completed your preparation for that syllabus. Refer syllabus sheet in the starting of the book for the syllabus of all the DPP sheets.
- After completing the sheet check your answers with the solution booklet and complete the Result Grid. Finally spend time to analyse your performance and revise the areas which emerge out as weak in your evaluation.

DIRECTIONS (Q.1-Q.30) : There are 30 multiple choice questions. Each question has 4 choices (a), (b), (c) and (d), out of which ONLY ONE choice is correct.

1. Roots developing from plant parts other than radicle are called
   (a) epiphyllous (b) epicaulous
   (c) adventitious (d) fibrous
2. Root cap takes part in
   (a) formation of new cells
   (b) absorption of water and minerals
   (c) protection of root meristem
   (d) storage of food
3. Assimilatory (Photosynthetic) roots is a characteristic of
   (a) *Trapa* and *Tinospora*
   (b) *Taeniophyllum* and *Podostemon*
   (c) both (a) and (b)
   (d) none of the above
4. Leguminous plants possess
   (a) napiform roots
   (b) nodulated roots
   (c) tuberous roots
   (d) fusiform roots

RESPONSE GRID 1. ⓐⓑⓒⓓ 2. ⓐⓑⓒⓓ 3. ⓐⓑⓒⓓ 4. ⓐⓑⓒⓓ

Space for Rough Work

5. Choose the correct statement about haustorial (parasitic) roots of *Cuscuta.*
   (a) These roots develop contact with xylem of host.
   (b) These develop contact with xylem and phloem of host.
   (c) These develop contact with phloem of host to get food.
   (d) These develop contact with pericycle and lateral roots. of host.
6. Roots help in clinging and climbing in
   (a) *Pothos* and *Tecoma*
   (b) *Hedera* (Ivy) and *Piper* (Betel)
   (c) black pepper
   (d) all of the above
7. Velamen takes part in
   (a) absorption of moisture from air.
   (b) absorption of water from soil.
   (c) exchange of gases.
   (d) transpiration.
8. Pneumatophores or breathing roots occur in
   (a) hydrophytes (b) epiphytes
   (c) xerophytes (d) mangrove plants
9. Which of the following pairs are correctly matched?
   (1) Assimilatory roots – Photosynthesis
   (2) Fasciculated roots – Food storage
   (3) Stilt root – Mechanical support
   (4) Sucking root – Absorption of moisture from the air
   Select the correct answer-
   (a) 2, 3 and 4 (b) 1, 2 and 4
   (c) 1, 3 and 4 (d) 1, 2 and 3
10. In *Amorphophallus* and *Colocasia* (Ariods), an extremely enlarged underground vertical stem meant for vegetative reproduction and storage is called
   (a) tuber (b) corm
   (c) bulb (d) rhizome
11. Buds typically occur at
   (a) leaf bases
   (b) leaf axils
   (c) tips of stems and roots
   (d) tips of branches and bases
12. Thorns, spines and prickles are
   (a) organs of defence.
   (b) organs of offence.
   (c) organs of clinging.
   (d) specialised for gaseous exchange.
13. Ginger is a stem which can be differentiated from root because it
   (a) grows parallel to ground.
   (b) stores food.
   (c) lacks chlorophyll.
   (d) has nodes and internodes.
14. Axillary buds develop
   (a) exogenously from epidermis.
   (b) exogenously from inner layer of cortex.
   (c) endogenously from inner layer of cortex.
   (d) endogenously from pericycle.

RESPONSE GRID					
	5. ⓐⓑⓒⓓ	6. ⓐⓑⓒⓓ	7. ⓐⓑⓒⓓ	8. ⓐⓑⓒⓓ	9. ⓐⓑⓒⓓ
	10. ⓐⓑⓒⓓ	11. ⓐⓑⓒⓓ	12. ⓐⓑⓒⓓ	13. ⓐⓑⓒⓓ	14. ⓐⓑⓒⓓ

Space for Rough Work

15. Potato and sweet potato
(a) have edible parts which are homologous organs.
(b) have edible parts which are analogous organs.
(c) have been introduced in India from the same place.
(d) are two species of the same genus.

16. Potato is (underground) stem because it
(a) possesses axillary buds (eyes)
(b) lacks chlorophyll
(c) does not bear roots
(d) contains reserve food

17. Green leaf-like one internode long stem branches are called
(a) phylloclades (b) phyllodes
(c) bulbils (d) cladodes

18. Phyllotaxis is
(a) mode of leaf arrangement of stem.
(b) types of roots.
(c) arrangement of sepals and petals in a flower.
(d) type of ovary.

19. Leaflets are called pinnules in
(a) twice pinnately compound leaf
(b) once pinnately compound leaf
(c) thrice pinnately compound leaf
(d) decompound leaf

20. Arrangement of floral leaves in a floral bud is called
(a) vernation (b) prefoliation
(c) aestivation (d) ptyxis

21. Phyllode is an adaptation to
(a) heterophylly environment
(b) halophytic environment
(c) mesophytic environment
(d) xerophytic environment

22. Isobilateral leaf is found in
(a) dicots
(b) monocots
(c) hydrophytes
(d) xerophytes

23. Parallel venation occurs in
(a) monocots (b) dicots
(c) all angiosperms (d) ferns

24. Thorn is a modified branch because
(a) it is hard, straight and pointed
(b) it is a part of the plant.
(c) it arises in the axil of a leaf.
(d) it is a defensive organ.

RESPONSE GRID					
	15. ⓐⓑⓒⓓ	16. ⓐⓑⓒⓓ	17. ⓐⓑⓒⓓ	18. ⓐⓑⓒⓓ	19. ⓐⓑⓒⓓ
	20. ⓐⓑⓒⓓ	21. ⓐⓑⓒⓓ	22. ⓐⓑⓒⓓ	23. ⓐⓑⓒⓓ	24. ⓐⓑⓒⓓ

Space for Rough Work

25. Inflorescence is
   (a) arrangement of flowers on peduncle
   (b) a system of branches bearing flowers
   (c) a branch bearing flowrers in definite manner
   (d) all of the above

26. The most important function of inflorescence is to help in
   (a) forming large number of fruits.
   (b) attracting insects for cross pollination.
   (c) dispersal of seeds.
   (d) release of pollen grains.

27. A racemose inflorescence with sessile acropetal flowers is called
   (a) spike (b) corymb
   (c) umbel (d) raceme

28. The special type of inflorescence found in *Ficus* where the female flower are at bottom and male flower near ostiole and enclosed within a cup shaped fleshy thalamus (receptacle) with ostiole is called
   (a) cyathium (b) verticillaster
   (c) spadix (d) hypanthodium

29. The edible part of cauliflower is
   (a) mesocarp (b) cotyledons
   (c) endosperms (d) inflorescence

30. The capitulum type of inflorescence is found in
   (a) *Marigold* (b) *Salvia*
   (c) *Euphorbia* (d) *Jasmine*

RESPONSE GRID	
	25. ⓐⓑⓒⓓ 26. ⓐⓑⓒⓓ 27. ⓐⓑⓒⓓ 28. ⓐⓑⓒⓓ 29. ⓐⓑⓒⓓ
	30. ⓐⓑⓒⓓ

DAILY PRACTICE PROBLEM SHEET 13 - BIOLOGY			
Total Questions	30	Total Marks	120
Attempted		Correct	
Incorrect		Net Score	
Cut-off Score	44	Qualifying Score	72
Success Gap = Net Score – Qualifying Score			
Net Score = (Correct × 4) – (Incorrect × 1)			

Space for Rough Work

DPP - Daily Practice Problems

Name :   Date :

Start Time :   End Time :

BIOLOGY 14

SYLLABUS : Morphology of Flowering Plants Part II : Flower, Seed, Fruits and their dispersal, Taxonomy of Angiospermic plants

Max. Marks : 120   **Time : 60 min.**

GENERAL INSTRUCTIONS

- The Daily Practice Problem Sheet contains 30 MCQ's. For each question only one option is correct. Darken the correct circle/ bubble in the Response Grid provided on each page.
- You have to evaluate your Response Grids yourself with the help of solution booklet.
- Each correct answer will get you 4 marks and 1 mark shall be deduced for each incorrect answer. No mark will be given/ deducted if no bubble is filled. Keep a timer in front of you and stop immediately at the end of 60 min.
- The sheet follows a particular syllabus. Do not attempt the sheet before you have completed your preparation for that syllabus. Refer syllabus sheet in the starting of the book for the syllabus of all the DPP sheets.
- After completing the sheet check your answers with the solution booklet and complete the Result Grid. Finally spend time to analyse your performance and revise the areas which emerge out as weak in your evaluation.

DIRECTIONS (Q.1-Q.30) : There are 30 multiple choice questions. Each question has 4 choices (a), (b), (c) and (d), out of which ONLY ONE choice is correct.

1. Floral formula represents
   (a) position of flower
   (b) symmetry of a flower
   (c) functions of a flower
   (d) diagrammatic notation of floral characters
2. In monadelphous condition, stamens have
   (a) filaments of all united in one group but anthers are free.
   (b) filaments united in groups but all anthers are free.
   (c) anthers are fused but filaments are free.
   (d) both anthers and filaments are fused.
3. Axis developing between androecium and gynoecium is
   (a) anthophore (b) androphore
   (c) gynophore (d) gynandrophore
4. Flower of *Hibiscus* is
   (a) actinomorphic and epigynous.
   (b) actinomorphic and hypogynous.
   (c) zygomorphic and hypogynous.
   (d) zygomorphic and epigynous.

RESPONSE GRID	1. ⓐⓑⓒⓓ	2. ⓐⓑⓒⓓ	3. ⓐⓑⓒⓓ	4. ⓐⓑⓒⓓ

Space for Rough Work

5. A flower is zygomorphic when
(a) any transverse section divides it into two equal halves.
(b) only one transverse section divides it into two equal halves.
(c) every vertical section passing through its centre divides it into two equal havles.
(d) only one vertical section passing through its centre divides it into two equal halves.

6. Flowers and stamens of compositae are
(a) hypogynous and inferior
(b) epigynous and superior
(c) hypogynous and superior
(d) epigynous and inferior

7. When two stamens are short and two are long, they are
(a) tetradynamous
(b) didynamous
(c) long and short stamens
(d) varied stamens

8. When both sexes are absent from a flower or are non-functional, the flower is said to be
(a) neutral (b) incomplete
(c) unisexual (d) intersexual

9. When two of the sepals or petals are outer, two are inner and one is partly outer partly inner, this condition is known as
(a) imbricate aestivation (b) quincuntial aestivation
(c) twisted aestivation (d) valvate aestivation

10. Corolla in China rose are
(a) 5, gamopetalous, twisted.
(b) 5, gamopetalous valvate.
(c) 5, polypetalous valvate.
(d) 5, polypetalous contorted.

11. After fertilization, the seed coats of seed develop from
(a) integuments (b) embryo sac
(c) chalaza (d) ovule

12. The plant whose seeds are known to have longest viability period is
(a) *Nelumbo nucifera* (lotus)
(b) *Triticum vulgare* (wheat)
(c) *Zizyphus jujuba* (bar)
(d) *Carica papaya* (papaya)

13. Viviparity means
(a) fruits are not formed.
(b) germination of seed on mother plant.
(c) formation of fruits directly by embryo.
(d) production of fruitless plant.

14. Aggregate fruit develops from
(a) multicarpellary, apocarpous ovary.
(b) multicarpellary ovary.
(c) multicarpellary, syncarpous ovary.
(d) monocarpellary ovary.

RESPONSE GRID				
5. ⓐⓑⓒⓓ	6. ⓐⓑⓒⓓ	7. ⓐⓑⓒⓓ	8. ⓐⓑⓒⓓ	9. ⓐⓑⓒⓓ
10. ⓐⓑⓒⓓ	11. ⓐⓑⓒⓓ	12. ⓐⓑⓒⓓ	13. ⓐⓑⓒⓓ	14. ⓐⓑⓒⓓ

Space for Rough Work

15. In caryopsis type of fruit
(a) seed is absent.
(b) three layers of pericarp are distinct.
(c) seed coat and pericarp are fused.
(d) autochory occurs.

16. Which one of the following is a dry indehiscent fruit?
(a) Caryopsis (b) Pod
(c) Follicle (d) Lomentum

17. Essential requirement for seed germination are
(a) H_2O and O_2
(b) O_2 and light
(c) H_2O and high temperature
(d) Scarification and vernalisation

18. Parachute mechanism of seed dispersal occurs in
(a) *Xanthium* (b) *Calotropis*
(c) mango (d) apple

19. The correct floral formula of Liliaceae is
(a) $\%$ ⚥ $P_{3+3} A_6 \underline{G_{(3)}}$
(b) Br $\%$ ⚥ $P_{3+3} A_{3+3} \underline{G_{(3)}}$
(c) $\oplus$ ⚥ $P_{3+3} A_{3+3} \underline{G_{(3)}}$
(d) $\oplus$ ⚥ $P_{3+3} A_{3+3} \underline{G_{(6)}}$

20. Which of the two families are dominate in having maximum useful plants ?
(a) Fabaceae and Poaceae
(b) Liliaceae and Solanaceae
(c) Malvaceae and Brassicaceae
(d) Liliaceae and Poaceae

21. Which statement is wrong for compositae?
(a) 5-lobed stamens (b) Syngenesious stamens
(c) Basal ovule (d) Ligulate ray florets

22. Axile placentation occurs in
(a) asteraceae and fabaceae.
(b) brassicaceae and solanaceae.
(c) solanaceae and liliaceae.
(d) brassicaceae and liliaceae.

23. Flower of fabaceae is
(a) complete, zygomorphic, pentamerous.
(b) complete, actinomorphic, trimerous.
(c) incomplete, zygomorphic, trimerous.
(d) incomplete, actinomorphic, pentamerous.

24. Most important character of *Brassica campesteris* is
(a) False septum (b) Parietal placentation
(c) Ebracteate (d) Imbricate aestivation

RESPONSE GRID					
	15. ⓐⓑⓒⓓ	16. ⓐⓑⓒⓓ	17. ⓐⓑⓒⓓ	18. ⓐⓑⓒⓓ	19. ⓐⓑⓒⓓ
	20. ⓐⓑⓒⓓ	21. ⓐⓑⓒⓓ	22. ⓐⓑⓒⓓ	23. ⓐⓑⓒⓓ	24. ⓐⓑⓒⓓ

Space for Rough Work

25. Bicarpellary, syncarpous, unilocular ovary with basal placentation occurs in
(a) liliaceae (b) solanaceae
(c) asteraceae (d) fabaceae

26. Which of the following is correct with reference to flowers of family Solanaceae?
(a) Pentamerous, actinomorphic, unisexual, hypogynous.
(b) Pentamerous, zygomorphic, bisexual, epigynous.
(c) Pentamerous, bisexual, actinomorphic, hypogynous.
(d) Trimerous, actinomorphic, bisexual, hypogynous.

27. Cruciferae differ from malvaceae in the presence of
(a) multicarpellary, unilocular ovary and siliqua fruit.
(b) multicarpellary, multilocular ovary and capsule fruit.
(c) monocarpellary, multilocular ovary with capsule fruit.
(d) multicarpellary, unilocular ovary and cypsella fruit.

28. The family containing mustard, and its main characters are:
(a) Brassicaceae-Tetramerous flowers, six stamens, bicarpellary gynoecium, siliqua type fruit.
(b) Brassicaceae-Pentamerous flowers, many stamens, pentacarpellary gynoecium, capsule type fruit.
(c) Solanaceae-Pentamerous flowers, five stamens, bicarpellary gynoecium, berry type fruit.
(d) Poaceae-Trimerous flowers, three stamens, monocarpellary gynoecium, caryopsis type of fruit.

29. Tetradynamous condition is characteristics of
(a) *Liliaceae/Allium/Asphodelus*
(b) Cruciferae/Mustard/*Iberis*
(c) Malvaceae/*Althea/Hibiscus*
(d) Solanaceae/*Nicotiana/Petunia*

30. Choose the correct description depicted by floral diagram.

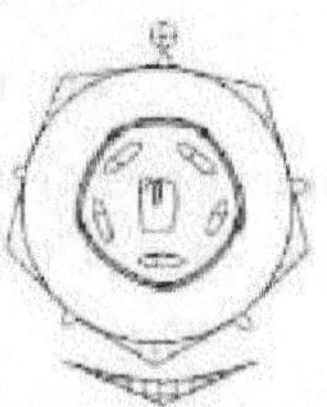

(a) United valvate sepals, free twisted petals, free stamens, unilocular ovary with marginal placenta.
(b) United valvate sepals, free imbricate petals, free stamens, unilocular ovary with axile placenta.
(c) United valvate sepals, free imbricate petals, epipetalous stamens, unilocular ovary with marginal placenta.
(d) United valvate sepals, free imbricate petals, free stamens, unilocular ovary with marginal placentation.

RESPONSE GRID	25. ⓐⓑⓒⓓ	26. ⓐⓑⓒⓓ	27. ⓐⓑⓒⓓ	28. ⓐⓑⓒⓓ	29. ⓐⓑⓒⓓ
	30. ⓐⓑⓒⓓ				

DAILY PRACTICE PROBLEM SHEET 14 - BIOLOGY			
Total Questions	30	Total Marks	120
Attempted		Correct	
Incorrect		Net Score	
Cut-off Score	44	Qualifying Score	64
Success Gap = Net Score – Qualifying Score			
Net Score = (Correct × 4) – (Incorrect × 1)			

Space for Rough Work

DPP - Daily Practice Problems

Name : 　　　　　　　　　　　　　　　　　　　Date :

Start Time : 　　　　　　　　　　　　　　　　End Time :

BIOLOGY 15

SYLLABUS : Anatomy of flowering Plants – Plant Tissues

Max. Marks : 120 　　　　　　　　　　　　　　　　**Time : 60 min.**

GENERAL INSTRUCTIONS

- The Daily Practice Problem Sheet contains 30 MCQ's. For each question only one option is correct. Darken the correct circle/ bubble in the Response Grid provided on each page.
- You have to evaluate your Response Grids yourself with the help of solution booklet.
- Each correct answer will get you 4 marks and 1 mark shall be deducted for each incorrect answer. No mark will be given/ deducted if no bubble is filled. Keep a timer in front of you and stop immediately at the end of 60 min.
- The sheet follows a particular syllabus. Do not attempt the sheet before you have completed your preparation for that syllabus. Refer syllabus sheet in the starting of the book for the syllabus of all the DPP sheets.
- After completing the sheet check your answers with the solution booklet and complete the Result Grid. Finally spend time to analyse your performance and revise the areas which emerge out as weak in your evaluation.

DIRECTIONS (Q.1-Q.30) : There are 30 multiple choice questions. Each question has 4 choices (a), (b), (c) and (d), out of which ONLY ONE choice is correct.

1. All xylem elements when mature are dead except
   (a) tracheids 　(b) vessels
   (c) xylem parenchyma 　(d) xylem fibres
2. Dermatogen, periblem and plerome are
   (a) permanent tissues 　(b) meristematic tissues
   (c) intercalary tissues 　(d) secondary tissues
3. A mature sieve tube differs from vessel in
   (a) being nearly dead.
   (b) lacking cytoplasm.
   (c) lacking a functional nucleus.
   (d) absence of lignified walls.
4. Mechanical tissue consisting of living cells is
   (a) sclerenchyma
   (b) collenchyma
   (c) chlorenchyma
   (d) parenchyma

RESPONSE GRID	1. ⓐⓑⓒⓓ	2. ⓐⓑⓒⓓ	3. ⓐⓑⓒⓓ	4. ⓐⓑⓒⓓ

Space for Rough Work

5. Collenchyma tissue is characterised by
   (a) elongated cells with thickening at the corners.
   (b) isodiametric cells with deposits of cellulose and pectin at the corners.
   (c) elongated cells with deposits of cellulose and pectin all over the wall.
   (d) isodiametric cells with thickening all over the cell wall.
6. Vessels differ from tracheids
   (a) in being derived from single cell.
   (b) in having vertical rows of cells with cross walls dissolved.
   (c) in being living.
   (d) in helping in the conduction of water.
7. The casparian strips of root endoderm is contain a mixture of
   (a) cellulose and cutin (b) cellulose and lignin
   (c) lignin and suberin (d) cellulose and suberin
8. Root cap in dicots is formed from
   (a) protoderm (b) ground meristem
   (c) calyptrogen (d) procambium
9. Meristematic cells are
   (a) thin-walled, isodiametric, nucleate and less protoplasmic.
   (b) thin-walled, isodiametric, non-nucleate and densely protoplasmic.
   (c) thick-walled, isodiametric, non-nucleate and densely protoplasmic.
   (d) thick-walled, isodiametric, nucleate and less protoplasmic.
10. The chief function of sieve tubes is
   (a) to translocate the organic materials manufactured in the leaves.
   (b) to conduct minerals.
   (c) to transport water from root to leaves.
   (d) to help the plant in forming wood.
11. Which meristem helps in increasing girth?
   (a) Lateral meristem (b) Intercalary meristem
   (c) Primary meristem (d) Apical meristem
12. Which of the following are simple tissues?
   (a) Parenchyma, xylem and phloem.
   (b) Parenchyma, collenchyma and sclerenchyma.
   (c) Parenchyma, xylem and collenchyma.
   (d) Parenchyma, xylem and sclerenchyma.
13. Which of the following tissues is present in the leaves of *Pinus* to conduct water and food?
   (a) Xylem
   (b) Phloem
   (c) Transfusion tissue
   (d) Conducting tissue
14. Histogen theory states that epidermis is derived from the
   (a) periblem
   (b) cambium
   (c) cortex
   (d) dermatogen

RESPONSE GRID	5. ⓐⓑⓒⓓ	6. ⓐⓑⓒⓓ	7. ⓐⓑⓒⓓ	8. ⓐⓑⓒⓓ	9. ⓐⓑⓒⓓ
	10. ⓐⓑⓒⓓ	11. ⓐⓑⓒⓓ	12. ⓐⓑⓒⓓ	13. ⓐⓑⓒⓓ	14. ⓐⓑⓒⓓ

Space for Rough Work

15. Lignin is the main constituent of
(a) woody tissues (b) growing tissues
(c) phloem (d) cortex

16. Tissue is
(a) first level of plant organisation.
(b) second level of plant organisation.
(c) third level of plant organisation.
(d) last level of plant organisation.

17. Angiosperms have
(a) tracheids only (b) both tracheid and vessels
(c) vessels (d) none of these

18. Meristematic tissues include
(a) leaf tips, cork cambium and vascular cambium.
(b) stem and root apices, cork cambium and mature fruits.
(c) stem and root apices, vascular cambium and cork cambium.
(d) mature fruits and leaf tips.

19. Vascular cambium of the root is an example of
(a) apical meristem
(b) intercalary meristem
(c) secondary meristem
(d) root apical meristem

20. A cap like region of slowly dividing or inactive cells in the middle of highly meristematic cells (between meristem and root cap) is
(a) somatic zone (b) vegetative zone
(c) quiescent centre (d) corpus centre

21. Collenchyma tissue is characterized by
(a) elongated cells with thickening at the corners.
(b) isodiametric cells with thickening all over the walls.
(c) isodiametric cells with deposits of cellulose and pectin at the corners.
(d) elongated cells with deposits of cellulose and pectin all over the walls.

22. Vessels are absent in
(a) teak wood (b) shisham wood
(c) *Pinus* wood (d) sal wood

23. Dermal tissue is derived from
(a) periblem (b) dermatogen
(c) procambium (d) pleurome

24. Lateral meristem is responsible for
(a) growth in length
(b) growth in parenchyma
(c) growth in thickness
(d) growth in cortex

RESPONSE GRID	15. ⓐⓑⓒⓓ	16. ⓐⓑⓒⓓ	17. ⓐⓑⓒⓓ	18. ⓐⓑⓒⓓ	19. ⓐⓑⓒⓓ
	20. ⓐⓑⓒⓓ	21. ⓐⓑⓒⓓ	22. ⓐⓑⓒⓓ	23. ⓐⓑⓒⓓ	24. ⓐⓑⓒⓓ

Space for Rough Work

25. Healing of wound in plants takes place by the activity of
    (a) ground tissue (b) callus deposition
    (c) secondary meristem (d) permanent tissue
26. The plant which reproduces by apical meristem is
    (a) gymnosperms (b) pteridophytes
    (c) angiosperms (d) algae
27. The function of a vessel is
    (a) conduction of food.
    (b) conduction of water and minerals.
    (c) conduction of hormones.
    (d) all of the above.
28. Porous wood contains mainly
    (a) fibres (b) vessels
    (c) tracheids (d) solid secretions
29. According to Histogen theory, the plerome gives rise to
    (a) epidermis (b) hypodermis
    (c) vascular bundles (d) external hair
30. At maturity which of the following is enucleate?
    (a) Sieve cell (b) Companion cell
    (c) Palisade cell (d) Cortical cell

RESPONSE GRID	25. ⓐⓑⓒⓓ	26. ⓐⓑⓒⓓ	27. ⓐⓑⓒⓓ	28. ⓐⓑⓒⓓ	29. ⓐⓑⓒⓓ
	30. ⓐⓑⓒⓓ				

DAILY PRACTICE PROBLEM SHEET 15 - BIOLOGY			
Total Questions	30	Total Marks	120
Attempted		Correct	
Incorrect		Net Score	
Cut-off Score	40	Qualifying Score	68
Success Gap = Net Score – Qualifying Score			
Net Score = (Correct × 4) – (Incorrect × 1)			

Space for Rough Work

DPP - Daily Practice Problems

Name : 　　　　　　　　Date :

Start Time : 　　　　　　　　End Time :

BIOLOGY 16

SYLLABUS : Anatomy of Flowering Plants Part II : Internal Structure of Root, Stem, Leaf and Secondary growth

Max. Marks : 120 　　　　　　　　**Time : 60 min.**

GENERAL INSTRUCTIONS

- The Daily Practice Problem Sheet contains 30 MCQ's. For each question only one option is correct. Darken the correct circle/ bubble in the Response Grid provided on each page.
- You have to evaluate your Response Grids yourself with the help of solution booklet.
- Each correct answer will get you 4 marks and 1 mark shall be deduced for each incorrect answer. No mark will be given/ deducted if no bubble is filled. Keep a timer in front of you and stop immediately at the end of 60 min.
- The sheet follows a particular syllabus. Do not attempt the sheet before you have completed your preparation for that syllabus. Refer syllabus sheet in the starting of the book for the syllabus of all the DPP sheets.
- After completing the sheet check your answers with the solution booklet and complete the Result Grid. Finally spend time to analyse your performance and revise the areas which emerge out as weak in your evaluation.

DIRECTIONS (Q.1-Q.30) : There are 30 multiple choice questions. Each question has 4 choices (a), (b), (c) and (d), out of which ONLY ONE choice is correct.

1. A stele with a central core of xylem surrounded by phloem is called
   (a) protostele (b) siphonostele
   (c) solenostele (d) dictyostele
2. Vascular bundles in which phloem is found on both sides of xylem are called
   (a) collateral (b) bicollateral
   (c) radial (d) amphicribral
3. A root hair is formed by
   (a) epidermal cell
   (b) endodermal cell
   (c) cortical cell
   (d) pericycle cell
4. In root, pericycle gives rise to
   (a) branch root and cork cambium
   (b) cortex and pith
   (c) epidermis and vascular bundles
   (d) xylem and pholem

RESPONSE GRID	1. ⓐⓑⓒⓓ	2. ⓐⓑⓒⓓ	3. ⓐⓑⓒⓓ	4. ⓐⓑⓒⓓ

Space for Rough Work

5. Vascular bundles having xylem and phloem sit at the same radius is termed as
   (a) concentric (b) radial
   (c) collateral (d) amphicribral
6. Amphiphloic (bicollateral) condition of stele means that
   (a) phloem is surrounded by xylem.
   (b) phloem is on both sides of xylem.
   (c) phloem is internal to xylem.
   (d) pholem is external to xylem.
7. The innermost layer of cortex which shows casparian thickenings in its cells is called as
   (a) epidermis (b) endodermis
   (c) pericycle (d) exodermis
8. A bicollateral vascular bundle has which of the following arrangement of tissues?
   (a) Outer phloem - outer xylem - middle cambium - inner xylem - inner phloem.
   (b) Outer cambium - outer phloem - middle xylem - inner phloem - inner cambium.
   (c) Outer phloem - outer cambium - middle xylem - inner cambium - inner phloem.
   (d) Outer xylem - outer cambium - middle phloem - inner cambium - inner xylem.
9. Ectophloic siphonostele is found in
   (a) *Osmunda* and *Equisetum*
   (b) *Marsilea* and *Botrychium*
   (c) *Adiantum* and *Cucurbitaceae*
   (d) *Dicksonia* and *Miadenhair fern*
10. In dicot root
   (a) vascular bundles are scattered and with cambium.
   (b) vascular bundles are arranged in a ring and have cambium.
   (c) xylem and phloem radially arranged.
   (d) xylem is always endarch.
11. Monocot root differs from dicot root in having
   (a) open vascular bundles.
   (b) scattered vascular bundles.
   (c) well developed pith.
   (d) radially arranged vascular bundles.
12. Which of the following is not characteristic feature of the anatomy of dicotyledonous root?
   (a) Radial vascular bundles (b) Secondary growth
   (c) Pith little or absent (d) Vascular bundles 15-20
13. Cortex in the monocot root constitutes
   (a) parenchymatous cells
   (b) collenchymatous cells
   (c) sclerenchymatous cells
   (d) companion cells
14. In leaves, the protoxylem elements
   (a) face towards abaxial surface.
   (b) face towards adaxial surface.
   (c) are surrounded by metaxylem elements.
   (d) are scattered in vascular bundle.

RESPONSE GRID					
	5. ⓐⓑⓒⓓ	6. ⓐⓑⓒⓓ	7. ⓐⓑⓒⓓ	8. ⓐⓑⓒⓓ	9. ⓐⓑⓒⓓ
	10. ⓐⓑⓒⓓ	11. ⓐⓑⓒⓓ	12. ⓐⓑⓒⓓ	13. ⓐⓑⓒⓓ	14. ⓐⓑⓒⓓ

Space for Rough Work

15. T.S. of stem of *Cucurbita* can be identified from the T.S. of sunflower stem by the presence of
(a) bicollateral vascular bundles.
(b) conjoint vascular bundles.
(c) scattered vascular bundles.
(d) cambium in the vascular bundles.

16. As the secondary growth takes place (proceeds) in a tree, thickness of
(a) heart wood increases
(b) sap-wood increases
(c) both increase
(d) both remain the same

17. In monocot roots, which types of vascular bundles are found?
(a) Collateral, conjoint and closed.
(b) Radial V.B. with exarch xylem.
(c) Bicollateral, conjoint and closed.
(d) Radial V.B. with endarch xylem.

18. Conjoint, collateral and closed vascular bundle is found in
(a) monocot stem (b) monocot root
(c) dicot stem (d) dicot root

19. Which of the following is correct sequence of layers in typical monocot root (from outer surface to inside)?
(a) Pericycle, cortex, endodermis, epiblema
(b) Epiblema, endodermis, cortex, pericycle
(c) Epiblema, cortex, endodermis, pericycle
(d) Epiblema, pericycle, cortex, endodermis

20. Largest number of chloroplast is found in
(a) palisade tissue
(b) spongy tissue
(c) transfusion tissue
(d) bundle sheath cells

21. Other names of secondary cortex, cork cambium and cork are
(a) phellem, phelloderm and phellogen
(b) phellogen, phellem and phelloderm
(c) phelloderm, phellogen and phellem
(d) phellogen, phelloderm and phellem

22. If four radial vascular bundles are present, then the structure will be
(a) monocot stem
(b) monocot root
(c) dicot stem
(d) dicot root

23. Secondary growth is absent in
(a) dicot stem (b) gymnosperms
(c) monocot stem (d) dicot root

24. Fascicular cambium found in dicot stem is a
(a) secondary meristem
(b) primary meristem
(c) intercalary meristem
(d) apical meristem

RESPONSE GRID					
	15. ⓐⓑⓒⓓ	16. ⓐⓑⓒⓓ	17. ⓐⓑⓒⓓ	18. ⓐⓑⓒⓓ	19. ⓐⓑⓒⓓ
	20. ⓐⓑⓒⓓ	21. ⓐⓑⓒⓓ	22. ⓐⓑⓒⓓ	23. ⓐⓑⓒⓓ	24. ⓐⓑⓒⓓ

Space for Rough Work

25. Tissue present in an annual ring is
(a) secondary xylem and phloem.
(b) primary xylem and phloem.
(c) secondary xylem only.
(d) primary phloem and secondary xylem.

26. The axillary buds arise
(a) endogenously from the pericycle.
(b) exogenously from the tissues of the main growing point.
(c) endogenously from the cambial tissues.
(d) exogenously from the innermost cortex.

27. Vascular cambium is a meristematic layer that cuts off
(a) primary xylem and primary phloem.
(b) xylem vessels and xylem tracheids.
(c) primary xylem and secondary xylem.
(d) secondary xylem and secondary phloem.

28. In woody dicotyledons, the arrangement of vessels is either diffuse porous or ring porous. Based on these data, which one of the following statements is correct?
(a) Ring porous vessels are specialised and are used for conducting more water for a shorter period only, when tyloses occur early in the vessels.
(b) Although diffuse porous vessels are not specialized as ring porous vessels, they conduct more water at all periods and through new xylem vessels added gradually during development.
(c) Diffuse porous vessels carry more water and also faster because of a greater number of small vessels having greater capillary force.
(d) Ring porous vessels conduct more water as they are formed early during development, when the need for water is great.

29. For a successful graft, the adhesion between stock and scion is a must. Which one of the following is the earlest event towards a good graft?
(a) Production of plasmodesmata in the cells at the interface of stock and scion.
(b) Coordinated differentiation of vascular tissue between the stock and scion.
(c) Regeneration of cortex and epidermis over the union of stock and scion.
(d) Production of callus tissue between the cells of stock and scion.

30. Considering the root apex of monocotyledons, there are three groups of initials. These are
(a) the terminal ones form the dermatogen and the cap, the median forms the periblem and the innermost ones form the plerome.
(b) the terminal ones form the cap, the median ones form the dermatogen and the innermost ones form the periblem and plerome.
(c) the terminal ones form the cap, the median ones form the dermatogen and periblem and the innermost ones form the plerome.
(d) the terminal ones form the cap and the dermatogen, the median ones are non-functional and the innermost ones form the periblem and plerome.

RESPONSE GRID	
	25. ⓐⓑⓒⓓ 26. ⓐⓑⓒⓓ 27. ⓐⓑⓒⓓ 28. ⓐⓑⓒⓓ 29. ⓐⓑⓒⓓ 30. ⓐⓑⓒⓓ

DAILY PRACTICE PROBLEM SHEET 16 - BIOLOGY			
Total Questions	30	Total Marks	120
Attempted		Correct	
Incorrect		Net Score	
Cut-off Score	40	Qualifying Score	68
Success Gap = Net Score – Qualifying Score			
Net Score = (Correct × 4) – (Incorrect × 1)			

Space for Rough Work

DPP - Daily Practice Problems

Name : Date :

Start Time : End Time :

BIOLOGY 17

SYLLABUS : Cockroach

Max. Marks : 120 **Time : 60 min.**

GENERAL INSTRUCTIONS

- The Daily Practice Problem Sheet contains 30 MCQ's. For each question only one option is correct. Darken the correct circle/ bubble in the Response Grid provided on each page.
- You have to evaluate your Response Grids yourself with the help of solution booklet.
- Each correct answer will get you 4 marks and 1 mark shall be deduced for each incorrect answer. No mark will be given/ deducted if no bubble is filled. Keep a timer in front of you and stop immediately at the end of 60 min.
- The sheet follows a particular syllabus. Do not attempt the sheet before you have completed your preparation for that syllabus. Refer syllabus sheet in the starting of the book for the syllabus of all the DPP sheets.
- After completing the sheet check your answers with the solution booklet and complete the Result Grid. Finally spend time to analyse your performance and revise the areas which emerge out as weak in your evaluation.

DIRECTIONS (Q.1-Q.30) : There are 30 multiple choice questions. Each question has 4 choices (a), (b), (c) and (d), out of which ONLY ONE choice is correct.

1. The principal function of blood vascular system in cockroach is –
   (a) Distribution of absorb nutrients
   (b) Distribution of O_2
   (c) Transportation of enzymes
   (d) Transportation of heat
2. Function of stomodael valve in the gut of cockroach is to prevent the entry of food from –
   (a) Gizzard into midgut (b) Preoral cavity
   (c) Hindgut to midgut (d) Midgut into gizzard
3. The total no.of ganglia present on the nerve cord of cockroach is –
   (a) 6
   (b) 9
   (c) 10
   (d) 12
4. A - *Periplaneta americana* is nocturnal, omnivorous, house hold pest.
   R - It is because it act as scavenger.
   (a) A is true but R is false.
   (b) A is false but R is true.
   (c) A and R is true and R is the right explanation of A.
   (d) A and R is true but R is not correct explanation of A.

RESPONSE GRID	1. ⓐⓑⓒⓓ	2. ⓐⓑⓒⓓ	3. ⓐⓑⓒⓓ	4. ⓐⓑⓒⓓ

Space for Rough Work

5. Alary muscles in cockroach occur in
(a) Heart wall and help in blood circulation
(b) Dorsal septum and connect the septum with heart and tergite
(c) Wall of gizzard and help in its contraction
(d) Intestinal wall and help in digestion

6. Two common Indian cockroaches are
(a) *Periplaneta american* and *Blatta indica*
(b) *Periplaneta orientalis* and *Blatta americana*
(c) *Periplaneta americana* and *Blatta orientalis*
(d) *Periplaneta indica* and *Blatta orientalis*

7. Number 16 in cockroach is concerned with
(a) Body segments
(b) Malpighian tubule
(c) Tracheae
(d) Optjeaca

8. Which part of cockroach has both exoskeleton and endoskeleton?
(a) Head
(b) Thorax
(c) Both of these
(d) None of these

9. Male and female cockroaches can be distinguished externally through
(a) Anal styles in male
(b) Anal cerci in female
(c) Anal style and antennae in female
(d) both (b) and (c)

10. The component units of compound eye of cockroach are called
(a) Eyelets
(b) Fenestra
(c) Ommatidia
(d) Ocelli

11. Metamorphosis in the life history of *Periplaneta* is ?
(a) Absent
(b) Incomplete
(c) Complete
(d) Pupa

12. Young cockroach is called –
(a) Maggot
(b) Ephyra
(c) Nymph
(d) Pupa

13. *Periplaneta americana* differs from *Blatta orientalis* in having –
(a) no wings
(b) only first pair of wings
(c) well developed wings
(d) only second pair of wings

14. Spiracles found in cockroach are
(a) 2 pairs in thorax and 10 pairs in abdomen
(b) 2 pairs in thorax and 6 pairs in abdomen
(c) 2 pairs in thorax and 8 pairs in abdomen
(d) 2 pairs in thorax and 4 pairs in abdomen

RESPONSE GRID					
	5. ⓐⓑⓒⓓ	6. 	7. ⓐⓑⓒⓓ	8. ⓐⓑⓒⓓ	9. ⓐⓑⓒⓓ
	10. ⓐⓑⓒⓓ	11. ⓐⓑⓒⓓ	12. ⓐⓑⓒⓓ	13. ⓐⓑⓒⓓ	14. ⓐⓑⓒⓓ

Space for Rough Work

15. The process of conversion of a small cockroach into an adult cockroach is called as
(a) Moulting (b) Metamorphosis
(c) Ecdysis (d) Transformation

16. Cockroach belongs to class
(a) Hexapoda (b) Apoda
(c) Metapoda (d) Cephalapoda

17. The image formed in the eyes of cockroach is
(a) Apposition (b) Superposition
(c) Both (a) and (b) (d) None of these

18. Tubular heart of cockroach has how many chambers ?
(a) 10 (b) 13
(c) 12 (d) 11

19. Mouth parts of cockroach are of
(a) Piercing and sucking (b) Sucking and siphoning
(c) Cutting and biting type (d) Sucking and rasping

20. In Cockroach, the number of ganglia are
(a) Two pairs thoracic and four pairs abdominal
(b) Three pairs thoracic and six pairs abdominal
(c) Three pairs thoracic and five pairs abdominal
(d) Two pairs thoracic and six pairs abdominal

21. The peculiar pungent smell of cockroach is produced by the secretions of ?
(a) Pheromones
(b) Flame cells
(c) Abdominal glands
(d) Cervical glands

22. Cockroach blood does not contain respiratory pigment. It means
(a) It does not respire
(b) Cockroach respires anaerobically
(c) Oxygen passes to all the tissues through diffusion
(d) Oxygen reaches tissue through tracheoles

23. Antennae of cockroach have
(a) Gustatory receptors
(b) Auditory receptors
(c) Tactile receptors
(d) Tactile and olfactory receptors

24. Organ of mastication in cockroach is
(a) Labrum (b) Labium
(c) Mandibles (d) Maxilla

RESPONSE GRID					
	15. ⓐⓑⓒⓓ	16. ⓐⓑⓒⓓ	17. ⓐⓑⓒⓓ	18. ⓐⓑⓒⓓ	19. ⓐⓑⓒⓓ
	20. ⓐⓑⓒⓓ	21. ⓐⓑⓒⓓ	22. ⓐⓑⓒⓓ	23. ⓐⓑⓒⓓ	24. ⓐⓑⓒⓓ

Space for Rough Work

25. Which one does not occur in cockroach leg ?
(a) Tibia (b) Femur
(c) Fibula (d) Coxa

26. Which structure is absent in male cockroach ?
(a) Labium (b) Phallomeres
(c) Spermatheca (d) None of these

27. Gizzard of cockraoch is a part of?
(a) Respiratory system
(b) Digestive system
(c) Immune system
(d) Circulatory system

28. Vision in cockroach is ?
(a) Monocular (b) Binocular
(c) Ultrasonic (d) Mosaic

29. The adhesive pads (soft-pads) in legs of cockroach are
(a) Galea (b) Lacinea
(c) Glossa (d) Plantulae

30. Open circulatory system is not of physiological hindrance in cockroach because
(a) Heart is simple but chambered.
(b) Blood is colourless.
(c) Circulatory and respiratory systems are not connected.
(d) Excretion occurs through malpighian tubules.

RESPONSE GRID	
	25. ⓐⓑⓒⓓ 26. ⓐⓑⓒⓓ 27. ⓐⓑⓒⓓ 28. ⓐⓑⓒⓓ 29. ⓐⓑⓒⓓ
	30. ⓐⓑⓒⓓ

DAILY PRACTICE PROBLEM SHEET 17 - BIOLOGY			
Total Questions	30	Total Marks	120
Attempted		Correct	
Incorrect		Net Score	
Cut-off Score	32	Qualifying Score	52
Success Gap = Net Score – Qualifying Score			
Net Score = (Correct × 4) – (Incorrect × 1)			

Space for Rough Work

DPP - Daily Practice Problems

Name : Date :

Start Time : End Time :

BIOLOGY 18

SYLLABUS : Structural Organisation in Animals: Animal Tissue

Max. Marks : 120 Time : 60 min.

GENERAL INSTRUCTIONS

- The Daily Practice Problem Sheet contains 30 MCQ's. For each question only one option is correct. Darken the correct circle/ bubble in the Response Grid provided on each page.
- You have to evaluate your Response Grids yourself with the help of solution booklet.
- Each correct answer will get you 4 marks and 1 mark shall be deduced for each incorrect answer. No mark will be given/ deducted if no bubble is filled. Keep a timer in front of you and stop immediately at the end of 60 min.
- The sheet follows a particular syllabus. Do not attempt the sheet before you have completed your preparation for that syllabus. Refer syllabus sheet in the starting of the book for the syllabus of all the DPP sheets.
- After completing the sheet check your answers with the solution booklet and complete the Result Grid. Finally spend time to analyse your performance and revise the areas which emerge out as weak in your evaluation.

DIRECTIONS (Q.1-Q.30) : There are 30 multiple choice questions. Each question has 4 choices (a), (b), (c) and (d), out of which ONLY ONE choice is correct.

1. Urethra, vagina and oesophagus have a common inner lining of
   (a) squamous epithelium
   (b) ciliated epithelium
   (c) columnar epithelium
   (d) stratified squamous epithelium
2. Pavement epithelium is another name for
   (a) cuboidal epithelium
   (b) ciliated epithelium
   (c) simple squamous epithelium
   (d) stratified epithelium
3. Stratified squamous epithelium is found in
   (a) trachea
   (b) epidermis
   (c) mouth cavity (buccal)
   (d) lining of blood vessels
4. Which of the following cells of connective tissue secrete antibodies?
   (a) Mast cells
   (b) Reticular cells
   (c) Adipose cells
   (d) Plasma cells

RESPONSE GRID 1. ⓐⓑⓒⓓ 2. ⓐⓑⓒⓓ 3. ⓐⓑⓒⓓ 4. ⓐⓑⓒⓓ

Space for Rough Work

5. Ligaments and tendons are
(a) connective tissue
(b) muscular tissue
(c) fibrous connective tissue
(d) skeletal tissue

6. Which of the following is not correct?
(a) The brown adipose tissue cell has a single large droplet surrounded by a small amount of cytoplasm, whereas the white adipose tissue cell has many small droplets of fat suspended in a larger amount of cytoplasm.
(b) Brown fat cells contain many mitochondria, while white fat cells have comparatively few.
(c) Brown fat has a larger capacity for generating heat than white fat.
(d) Brown fat is mainly found in new born mammals.

7. Reticular connective tissue is found in
(a) liver (b) spleen
(c) kidneys (d) skin

8. Connective tissue is
(a) ectodermal in origin with intercellular spaces.
(b) mesodermal in origin without intercellular spaces.
(c) ectodermal in origin without intercelluar spaces.
(d) mesodermal in origin with intercellular spaces.

9. The connective tissue that connects the skin to the underlying structures is
(a) areolar tissue (b) serous membrane
(c) reticular tissue (d) dense connective tissue

10. Bone marrow is made up of
(a) muscular fibre and fatty tissue.
(b) fatty tissue and areolar tissue.
(c) fatty tissue and cartilage.
(d) fatty tissue, areolar tissue and blood vessel.

11. In mammals, Haversian canals are connected with each other by transverse canals, which are called
(a) semicircular canals
(b) Volkman's canals
(c) inguinal canals
(d) bidder's canals

12. Intervertebral disc is made up of
(a) elastic cartilage
(b) fibrous cartilage
(c) calcified cartilage
(d) hyaline cartilage

13. The connective tissue of the vertebrate body is built up from fibres of the protein collagen, embedded in a polysaccharide matrix to form
(a) cartilage (b) blood vessel
(c) heart (d) lung

14. Humerous and muscles are connected with
(a) ligament
(b) tendons
(c) both of these
(d) none of these

RESPONSE GRID					
	5. ⓐⓑⓒⓓ	6. ⓐⓑⓒⓓ	7. ⓐⓑⓒⓓ	8. ⓐⓑⓒⓓ	9. ⓐⓑⓒⓓ
	10. ⓐⓑⓒⓓ	11. ⓐⓑⓒⓓ	12. ⓐⓑⓒⓓ	13. ⓐⓑⓒⓓ	14. ⓐⓑⓒⓓ

Space for Rough Work

15. Bone-forming cells are known as
   (a) chondroclasts (b) osteoblasts
   (c) chondroblasts (d) osteoclasts
16. The fibrous tissue which connects the two bone is
   (a) connective tissue (b) tendon
   (c) ligament (d) adipose tissue
17. Erythrocytes in camel are
   (a) oval and non-nucleated.
   (b) oval and nucleated.
   (c) circular, biconcave and non-nucleated.
   (d) circular, biconvex and nucleated.
18. The lymph resembles composition of blood plasma, but contains
   (a) less amount of protein, same number of leucocytes and a few erythrocytes.
   (b) more protein, same number of leucocytes but mainly lymphocytes and a few erythrocytes.
   (c) less amount of protein, large number of leucocytes chiefly granulocytes and a few erythrocytes.
   (d) less amount of protein, large number of leucocytes chiefly lymphocyte and a few granulocytes.
19. The chief difference between RBCs of human and frog is that
   (a) only human RBCs have haemoglobin.
   (b) human RBCs have more nuclei.
   (c) human RBCs are without nucleus.
   (d) frog RBCs are without nucleus.
20. Which is correct?
   (a) Blood has WBC and lymph has RBC.
   (b) Blood has WBC, RBC and lymph has nothing.
   (c) Blood has RBC, WBC and lymph has WBC.
   (d) Lymph has WBC, RBC and blood has RBC.
21. Which of the following substances, if introduced into the blood stream, cause coagulation of blood at the site of its introduction ?
   (a) Fibrinogen
   (b) Prothrombin
   (c) Heparin
   (d) Thromboplastin
22. The process of formation of blood corpuscles is called
   (a) haemopoiesis
   (b) haemolysis
   (c) haemozoin
   (d) none of these
23. G–6–P dehydrogenase deficiency is associated with haemolysis of
   (a) lymphocytes (b) RBCs
   (c) platelets (d) leucocytes
24. Which of the following makes heart wall more thick
   (a) Pericardium (b) Epicardium
   (c) Myocardium (d) Endocardium

RESPONSE GRID					
	15. ⓐⓑⓒⓓ	16. ⓐⓑⓒⓓ	17. ⓐⓑⓒⓓ	18. ⓐⓑⓒⓓ	19. ⓐⓑⓒⓓ
	20. ⓐⓑⓒⓓ	21. ⓐⓑⓒⓓ	22. ⓐⓑⓒⓓ	23. ⓐⓑⓒⓓ	24. ⓐⓑⓒⓓ

Space for Rough Work

25. Smooth muscle fibres are
    (a) cylindrical, unbranched, striated, multinucleate and voluntary.
    (b) spindle-shaped, unbranched, non-striated, uninucleate and involuntary.
    (c) cylindrical, unbranched, non-striated, multinucleate and involuntary.
    (d) spindle-shaped, unbranched, striated, uninucleate and voluntary.
26. At 0°C and below it
    (a) there is increase in muscle contraction.
    (b) there is decrease in muscle contraction.
    (c) there is loss of irritability in a muscle.
    (d) coagulation of muscle proteins take place.
27. The ready source of energy in living cells is
    (a) glucose (b) ATP
    (c) glycogen (d) ADP
28. In striated muscle contraction
    (a) H – band is lengthened.
    (b) H – band is obliterated.
    (c) A – band decrease in length.
    (d) Z – line moves away from A – band.
29. Shivering with cold in winter is caused by
    (a) voluntary action of striated muscles.
    (b) voluntary action of unstriated muscles.
    (c) involuntary action of striated muscles.
    (d) involuntary action of unstriated muscles.
30. Nissl's granules are characteristically found in
    (a) nephron (b) neuron
    (c) cyton (d) dendrite

RESPONSE GRID					
	25. ⓐⓑⓒⓓ	26. ⓐⓑⓒⓓ	27. ⓐⓑⓒⓓ	28. ⓐⓑⓒⓓ	29. ⓐⓑⓒⓓ
	30. ⓐⓑⓒⓓ				

DAILY PRACTICE PROBLEM SHEET 18 - BIOLOGY			
Total Questions	30	Total Marks	120
Attempted		Correct	
Incorrect		Net Score	
Cut-off Score	36	Qualifying Score	64
Success Gap = Net Score – Qualifying Score			
Net Score = (Correct × 4) – (Incorrect × 1)			

Space for Rough Work

DPP - Daily Practice Problems

Name : Date :

Start Time : End Time :

BIOLOGY 19

SYLLABUS : Cell-The Unit of life Part I : Cell wall, Plasma membrane, Protoplasm, Cytoplasm

Max. Marks : 120 **Time : 60 min.**

GENERAL INSTRUCTIONS

- The Daily Practice Problem Sheet contains 30 MCQ's. For each question only one option is correct. Darken the correct circle/ bubble in the Response Grid provided on each page.
- You have to evaluate your Response Grids yourself with the help of solution booklet.
- Each correct answer will get you 4 marks and 1 mark shall be deduced for each incorrect answer. No mark will be given/ deducted if no bubble is filled. Keep a timer in front of you and stop immediately at the end of 60 min.
- The sheet follows a particular syllabus. Do not attempt the sheet before you have completed your preparation for that syllabus. Refer syllabus sheet in the starting of the book for the syllabus of all the DPP sheets.
- After completing the sheet check your answers with the solution booklet and complete the Result Grid. Finally spend time to analyse your performance and revise the areas which emerge out as weak in your evaluation.

DIRECTIONS (Q.1-Q.30) : There are 30 multiple choice questions. Each question has 4 choices (a), (b), (c) and (d), out of which ONLY ONE choice is correct.

1. Who proposed the "Cell theory"?
   (a) Schleiden and Schwann (b) Waston and Crick
   (c) Mendel and Morgan (d) Robert Hooke
2. Difference between the prokaryotic and eukaryotic cells in having
   (a) cell wall (b) nuclear membrane
   (c) ribosome (d) none of these
3. Which of the following exists in two states – sol & gel ?
   (a) Protoplasm (b) Cytoplasm
   (c) Ribosome (d) Chloroplast
4. Who proposed the theory that "cells arise only from the pre-existing cells"
   (a) Mohl
   (b) Virchow
   (c) Haeckel
   (d) Brown

RESPONSE GRID	1. ⓐⓑⓒⓓ	2. ⓐⓑⓒⓓ	3. ⓐⓑⓒⓓ	4. ⓐⓑⓒⓓ

Space for Rough Work

5. Which one of the following is prokaryote ?

(a) *Agaricus* (b) *Salmonella*

(c) Green algae (d) Bacteriophage

6. The size of most of the cell is best expressed in

(a) Å (b) millimeters (*mm*)

(b) nanometers (*nm*) (d) micrometers (μ m)

7. The Cell theory states that

(a) all cells have nuclei.

(b) all cells are totipotent.

(c) cells reproduce by mitosis.

(d) cells are the basic structural units of living cell.

8. Cystolith is made up of

(a) calcium oxalate (b) calcium carbonate

(c) calcium hydroxide (d) calcium oxide

9. The inherent capacity of a cell to regenerate a new whole organism is called

(a) ontogeny (b) totipotency

(c) phycogeny (d) differentiation

10. If living cells similar to those found on earth, were found on another planet where there was no oxygen then which cell organelle would most probably be absent?

(a) Cell membrane

(b) Ribosomes

(c) Mitochondria

(d) Chromosomes

11. Which one of the following pairs is not correctly matched?

(a) Cristae	The "Shelves" Formed by the folding of the inner membrane of the mitochondrion.
(b) Plasmodesmata	The membrane surrounding the vacuole in plants.
(c) Grana	Membrane bound discs in chloroplasts that contain chlorophylls and carotenoids
(d) Middle lamella	Layer between adjacent cell wall in plants derived from cell plate

12. Which one of the following is the largest unicellular organism?

(a) *Planaria* (b) *Volvox*

(c) Blue green algae (d) *Acetabularia*

13. Middle lamella is made up of

(a) cellulose

(b) suberin

(c) calcium and magnesium pectate

(d) lignin

14. A mature plant cell has

(a) cell wall and protoplasm.

(b) protoplasm and vacuole.

(c) vacuole and cell wall.

(d) protoplasm, cell wall and vacuole.

RESPONSE GRID					
	5. ⓐⓑⓒⓓ	6. ⓐⓑⓒⓓ	7. ⓐⓑⓒⓓ	8. ⓐⓑⓒⓓ	9. ⓐⓑⓒⓓ
	10. ⓐⓑⓒⓓ	11. ⓐⓑⓒⓓ	12. ⓐⓑⓒⓓ	13. ⓐⓑⓒⓓ	14. ⓐⓑⓒⓓ

Space for Rough Work

15. Middle lamella is present
(a) inside the primary wall.
(b) inside the secondary wall.
(c) outside the primary wall.
(d) in between secondary and tertiary walls.

16. Plasma membrane is made of
(a) a protein layer between two lipid layers.
(b) a lipid layer between two protein layers.
(c) a protein, a lipid and a cellulose layer.
(d) bimolecular lipid layer surrounded by protein layers.

17. Which of the following layer is present nearest at plasma membrane in plant cell?
(a) Secondary wall (b) Middle lamella
(c) Primary wall (d) Tonoplast

18. The main difference between pinocytosis and phagocytosis is
(a) taking in of fluid substances and engulfing of food material and foreign bodies.
(b) taking in of small and large particles respectively.
(c) taking in of small amount of liquid and large volume of liquids.
(d) none of the above

19. According to mosaic model, plasma membrane is made up of
(a) cellulose and hemicellulose.
(b) phospholipid and integrate protein.
(c) phospolipid, extrinsic and intrinsic protein.
(d) phospholipid and hemicellulose.

20. Which is the latest model that is proposed to explain the structure of plasma membrane?
(a) Fluid mosaic model
(b) Molecule model
(c) Unit membrane model
(d) None of these

21. Biological membranes are composed of
(a) 40% proteins and 60% lipids.
(b) 50 % proteins and 50% lipids.
(c) 70% proteins and 30% lipids.
(d) 60% proteins and 40% lipids.

22. Single membrane bound organelles are
(a) centrioles
(b) ribosomes
(c) nucleolus
(d) vacuole

23. Cell organelles are embedded in
(a) nucleolus
(b) mitochondria
(c) protoplasm
(d) cytoplasmic membrane

24. Protein synthesis in an animal cell, takes place
(a) only in the cytoplasm.
(b) in the cytoplasm as well as in mitochondria.
(c) in the nucleolus as well as in the cytoplasm.
(d) only on ribose attached to nucleolus.

RESPONSE GRID					
	15. ⓐⓑⓒⓓ	16. ⓐⓑⓒⓓ	17. ⓐⓑⓒⓓ	18. ⓐⓑⓒⓓ	19. ⓐⓑⓒⓓ
	20. ⓐⓑⓒⓓ	21. ⓐⓑⓒⓓ	22. ⓐⓑⓒⓓ	23. ⓐⓑⓒⓓ	24. ⓐⓑⓒⓓ

Space for Rough Work

25. A protoplast is
    (a) a cell organelle (b) a cell with a wall
    (c) a single wall-less cell (d) cell without nucleus

26. The substance which makes up about 80% of cytoplasm and has unique structure, is
    (a) roteins (b) fats
    (c) minerals (d) water

27. Singer and Nicholson model of plasma membrane differs from Robert's model in the
    (a) number of lipid layers
    (b) arrangement of lipids layers.
    (c) arrangement of proteins.
    (d) absence of proteins.

28. Desmosome is modification of
    (a) plasma membrane (b) nucleolus
    (c) golgi bodies (d) ER-Nucleus complex

29. Contribution of protoplasm to the total weight of the body is
    (a) 10% (b) 45 %
    (c) 7% (d) 95 %

30. Which one of the following is a non-protoplasmic cell inclusion
    (a) Ribosomes (b) Mitochondria
    (c) Lysosomes (d) Cystoliths

RESPONSE GRID	25. ⓐⓑⓒⓓ	26. ⓐⓑⓒⓓ	27. ⓐⓑⓒⓓ	28. ⓐⓑⓒⓓ	29. ⓐⓑⓒⓓ
	30. ⓐⓑⓒⓓ				

DAILY PRACTICE PROBLEM SHEET 19 - BIOLOGY

Total Questions	30	Total Marks	120
Attempted		Correct	
Incorrect		Net Score	
Cut-off Score	40	Qualifying Score	68
Success Gap = Net Score – Qualifying Score			
Net Score = (Correct × 4) – (Incorrect × 1)			

Space for Rough Work

DPP - Daily Practice Problems

Name : Date :

Start Time : End Time :

BIOLOGY 20

SYLLABUS : Cell-The Unit of life Part II : Mitochondria, Plastids, E.R, Golgi body, Sphaerosomes, "Vacuoles, Centriole, Microtubules, Nucleus, Chromosomes

Max. Marks : 120 **Time : 60 min.**

GENERAL INSTRUCTIONS

- The Daily Practice Problem Sheet contains 30 MCQ's. For each question only one option is correct. Darken the correct circle/ bubble in the Response Grid provided on each page.
- You have to evaluate your Response Grids yourself with the help of solution booklet.
- Each correct answer will get you 4 marks and 1 mark shall be deduced for each incorrect answer. No mark will be given/ deducted if no bubble is filled. Keep a timer in front of you and stop immediately at the end of 60 min.
- The sheet follows a particular syllabus. Do not attempt the sheet before you have completed your preparation for that syllabus. Refer syllabus sheet in the starting of the book for the syllabus of all the DPP sheets.
- After completing the sheet check your answers with the solution booklet and complete the Result Grid. Finally spend time to analyse your performance and revise the areas which emerge out as weak in your evaluation.

DIRECTIONS (Q.1-Q.30) : There are 30 multiple choice questions. Each question has 4 choices (a), (b), (c) and (d), out of which ONLY ONE choice is correct.

1. Identify the correct match between types of chromosomes and their description.

	Chromosomes		Position of centromere
A.	Metacentric	1.	At the tip
B.	Sub metacentric	2.	Almost near tip
C.	Acrocentric	3.	At the middle
D.	Telocentric	4.	Slightly away from the middle

Codes
(a) A-1, B-3, C-2, D-4 (b) A-4, B-3, C-2, D-1
(c) A-1, B-2, C-3, D-4 (d) A-4, B-3, C-1, D-2

2. The number of mitochondria increases in cells of
(a) dormant seeds (b) germinating seeds
(c) dry seeds (d) dead seeds

3. In prokaryotes, the mitochondria are absent. Even then Krebs cycle takes place. What is the site of Krebs cyle in bacteria?
(a) Ribosomes (b) Nucleoid
(c) Cytoplasm (d) Mesosomes

4. A tetrad consists of
(a) four non-homologous chromatids.
(b) four non-homologous chromosomes.
(c) two sets of homologous chromosomes, each with two chromatids.
(d) four homologous pairs of chromosomes.

RESPONSE GRID	1. ⓐⓑⓒⓓ	2. ⓐⓑⓒⓓ	3. ⓐⓑⓒⓓ	4. ⓐⓑⓒⓓ

Space for Rough Work

5. Centre of phosphorylation is

(a) peroxisome (b) oxysome

(c) ribosome (d) mitochondria

6. Which of the following pair is correct?

(a) DNA synthesis — Ribosomes

(b) Protein synthesis — Smooth E.R.

(c) Aerobic respiration — Cristae

(d) Suicidal sacs — Dictyosomes

7. The presence of DNA in mitochondria and chloroplast supports the hypothesis that

(a) mitochondria and chloroplast both originated as independent free living organisms.

(b) glycolysis occurs in mitochondria and chloroplast both.

(c) ATP is produced in mitochondria and chloroplast both.

(d) mitochondria and chloroplast undergo meiosis and mitosis independent of nucleus.

8. The bright colours of ripe fruits are due to

(a) leucoplast (b) chloroplast

(c) amyloplast (d) chromoplast

9. Plant cell is differ animal cell is

(a) the presence of cell wall and absence of chlorophyll in plant cell.

(b) the presence of cell wall and chlorophyll in plant cell.

(c) the absence of cell wall and presence of chloroplast in animal cell.

(d) the absence of cell wall and presence of chlorophyll in plant cell.

10. Anthocyanins are pigments which are

(a) water soluble and located in chromoplasts.

(b) water insoluble and located in chromoplasts.

(c) water insoluble and located in cell sap.

(d) water soluble and located in cell sap.

11. The main difference between chlorophyll 'a' and 'b' is

(a) chlorophyll 'a' is linear chain compound and 'b' is branched chain compound.

(b) chlorophyll 'a' has no Mg^{+2} ion in centre of molecule.

(c) in chlorophyll 'a' there is CH_3 group whereas in 'b' it is –CHO group.

(d) all of the above

12. If we separate the cell organelles of a living cell, then which part should be alive

(a) endoplasmic reticulum (b) chloroplast

(c) cell wall (d) ribosome

13. Many cells function properly and divide mitotically even though they do not have

(a) plasma membrane (b) cytoskeleton

(c) mitochondria (d) plastid

14. The main organelle involved in modification and routing of newly synthesized proteins to their destinations is

(a) chloroplast (b) mitochondria

(c) lysosome (d) endoplasmic reticulum

RESPONSE GRID					
	5. ⓐⓑⓒⓓ	6. ⓐⓑⓒⓓ 	7. ⓐⓑⓒⓓ	8. ⓐⓑⓒⓓ	9. ⓐⓑⓒⓓ
	10. ⓐⓑⓒⓓ	11. ⓐⓑⓒⓓ	12. ⓐⓑⓒⓓ	13. ⓐⓑⓒⓓ	14. ⓐⓑⓒⓓ

Space for Rough Work

15. Which one of the following pairs is correctly matched?

(a) Microsomes	–	Participate in the process of photosynthesis.
(b) Lysosomes	–	Involved in synthesizing amino acids.
(c) E.R	–	Plays role in the formation of a new nuclear membrane during cell division.
(d) Centrosomes	–	Provide enzymes required in the digestive process.

16. Which of the following is responsible for mechanical support, enzyme transport and protein synthesis?
   (a) Dictyosomes (b) Cell membrane
   (c) Mitochondria (d) Endoplasmic reticulum

17. Dictyosomes are
   (a) class of ribosomes.
   (b) place of flagellar organelles.
   (c) respiratory particles.
   (d) golgi bodies (of plant cells).

18. Lysosomes are known as suicidal bags because of
   (a) catalytic enzymes
   (b) hydrolytic enzymes
   (c) parasitic on nucleus
   (d) proteolytic enzymes

19. Which cell organelles is having single layered unit membrane?
   (a) Centrosome (b) Lysosome
   (c) Mesosome (d) Nucleus

20. The larger sub-unit in 80 S ribosome is
   (a) 50 S (b) 60 S
   (c) 40 S (d) 0 S

21. Which of the following cell organelle lacks DNA and bounding membrane?
   (a) Ribosome (b) Plastid
   (c) Nucleolus (d) Plasmid

22. Centromere is required for
   (a) movement of chromosomes towards poles.
   (b) cytoplasmic cleavage.
   (c) crossing over.
   (d) transcription.

23. If all the ribosomes of a cell are destroyed
   (a) respiration will not takes place.
   (b) photosynthesis will not occur.
   (c) fat will not be stored.
   (d) protein will not be formed.

24. Eukaryotic 80 S ribosome breaks into
   (a) 40 S and 40 S
   (b) 60 S and 40 S
   (c) 60 S and 50 S
   (d) 50 S and 30 S

RESPONSE GRID					
	15. ⓐⓑⓒⓓ	16. ⓐⓑⓒⓓ	17. ⓐⓑⓒⓓ	18. ⓐⓑⓒⓓ	19. ⓐⓑⓒⓓ
	20. ⓐⓑⓒⓓ	21. ⓐⓑⓒⓓ	22. ⓐⓑⓒⓓ	23. ⓐⓑⓒⓓ	24. ⓐⓑⓒⓓ

Space for Rough Work

25. The site of protein synthesis in plants is the
    (a) chloroplast
    (b) ribosomes
    (c) pyrenoids
    (d) mitochondria
26. Tonoplast is a
    (a) covering layer of golgi complex.
    (b) covering layer of vacuoles.
    (c) covering layer of microbodies.
    (d) non-living cytoplasmic content.
27. The fluid part of cell (called cell sap) is the
    (a) non-living content of cell.
    (b) living content of a cell.
    (c) non-living content of the vacuole of cell.
    (d) living content of the vacoule cell.
28. Function of centriole is
    (a) formation of spindle fibres.
    (b) formation of nucleolus.
    (c) initiation of cell division.
    (d) formation of cell plate.
29. The function of centrosome is
    (a) inhibition of cell division.
    (b) initiation cell division.
    (c) to increase protein synthesis.
    (d) none of the above.
30. The nucleus is separated from surrounding cytoplasm by a nuclear membrane, which is
    (a) single layered with pores.
    (b) single layered without pores.
    (c) double layered with pores.
    (d) double layered without pores.

RESPONSE GRID	
	25. ⓐⓑⓒⓓ 26. ⓐⓑⓒⓓ 27. ⓐⓑⓒⓓ 28. ⓐⓑⓒⓓ 29. ⓐⓑⓒⓓ 30. ⓐⓑⓒⓓ

DAILY PRACTICE PROBLEM SHEET 20 - BIOLOGY			
Total Questions	30	Total Marks	120
Attempted		Correct	
Incorrect		Net Score	
Cut-off Score	40	Qualifying Score	64
Success Gap = Net Score – Qualifying Score			
Net Score = (Correct × 4) – (Incorrect × 1)			

Space for Rough Work

DPP - Daily Practice Problems

Name : Date :

Start Time : End Time :

BIOLOGY 21

SYLLABUS : Biomolecules

Max. Marks : 120 **Time : 60 min.**

GENERAL INSTRUCTIONS

- The Daily Practice Problem Sheet contains 30 MCQ's. For each question only one option is correct. Darken the correct circle/ bubble in the Response Grid provided on each page.
- You have to evaluate your Response Grids yourself with the help of solution booklet.
- Each correct answer will get you 4 marks and 1 mark shall be deduced for each incorrect answer. No mark will be given/ deducted if no bubble is filled. Keep a timer in front of you and stop immediately at the end of 60 min.
- The sheet follows a particular syllabus. Do not attempt the sheet before you have completed your preparation for that syllabus. Refer syllabus sheet in the starting of the book for the syllabus of all the DPP sheets.
- After completing the sheet check your answers with the solution booklet and complete the Result Grid. Finally spend time to analyse your performance and revise the areas which emerge out as weak in your evaluation.

DIRECTIONS (Q.1-Q.30) : There are 30 multiple choice questions. Each question has 4 choices (a), (b), (c) and (d), out of which ONLY ONE choice is correct.

1. Which one of the following is the sweetest sugar?
   (a) Fructose (b) Glucose
   (c) Galactose (d) Sucrose

2. Which of the following is the characteristic of plants?
   (a) Glucose and cellulose
   (b) Pyruvic acid and glucose
   (c) Cellulose and starch
   (d) Starch and pyruvic acid

3. Sucrose, a common table sugar, is composed of
   (a) glucose + fructose
   (b) glucose + galactose
   (c) fructose + galactose
   (d) none of these

4. A complex polysaccharide produced from sucrose by the bacterium *Leuconostic mesenteroides* is
   (a) chitin
   (b) starch
   (c) cellulose
   (d) dextran

RESPONSE GRID	1. ⓐⓑⓒⓓ	2. ⓐⓑⓒⓓ	3. ⓐⓑⓒⓓ	4. ⓐⓑⓒⓓ

Space for Rough Work

5. Which of the following chemical characteristics is not common to all living beings?
   (a) Similar triplet code for amino acids.
   (b) Energy is stored by high phosphate bonds.
   (c) Type of protein present in the body.
   (d) Ribosomes are the sites of protein synthesis.

6. Proteins consists of
   (a) carbon, hydrogen, chlorine, sulphur
   (b) carbon, hydrogen, oxygen, nitrogen
   (c) carbon, manganese, phosphorus, nitrogen
   (d) carbon, iodine, oxygen and inorganic phosphate

7. High content of lysine is present in
   (a) wheat (b) apple
   (c) maize (d) banana

8. Maltose consists of which one of the following?
   (a) β-glucose and β-galactose
   (b) α-glucose and α-fructose
   (c) α-sucrose and β-glucose
   (d) Glucose and glucose

9. Which of the following amino acids is not optically active?
   (a) Glycine
   (b) Valine
   (c) Leucine
   (d) Isoleucine

10. A ribose (but not deoxyribose) nucleotide is
   (a) cytosine — pentose sugar — phosphate
   (b) guanine — pentose sugar — phosphate
   (c) thymine — pentose sugar — phosphate
   (d) uracil — pentose sugar — phosphate

11. ATP is
   (a) adenosine D-ribose three phosphate.
   (b) adenosine L-ribose three phosphate.
   (c) adenine D-ribose three phosphate.
   (d) adenine L-ribose three phosphate.

12. DNA strands are antiparallel because of the presence of
   (a) H-bonds
   (b) peptide bonds
   (c) disulfied bonds
   (d) phosphate-diester bonds

13. The base pairs of DNA are correctly shown as
   (a) $A \equiv T$ and $C = G$
   (b) $A = T$ and $C = G$
   (c) $A = T$ and $C \equiv G$
   (d) $A \equiv$ and $C = G$

14. Ultraviolet light absorbed by nucleic acid is
   (a) 26 nm (b) 75 nm
   (c) 260 nm (d) 1500 nm

RESPONSE GRID					
	5. ⓐⓑⓒⓓ	6. ⓐⓑⓒⓓ	7. ⓐⓑⓒⓓ	8. ⓐⓑⓒⓓ	9. ⓐⓑⓒⓓ
	10. ⓐⓑⓒⓓ	11. ⓐⓑⓒⓓ	12. ⓐⓑⓒⓓ	13. ⓐⓑⓒⓓ	14. ⓐⓑⓒⓓ

Space for Rough Work

15. Nucleoside is made up of
    (a) Sugar only (b) Phosphate only
    (c) Sugar and phosphate (d) Sugar and base
16. RNA is absent in
    (a) plasmalemma (b) cytoplasm
    (c) chromosomes (d) ribosomes
17. Who got the Nobel prize working on enzymes in the year 1978?
    (a) W. Arber and D. Nathans
    (b) Nass and Nass
    (c) R. Misra
    (d) H.G. Khorana
18. Which one of the following statements is true for the enzymes?
    (a) All enzymes are proteins.
    (b) All proteins are enzymes.
    (c) All enzymes are not proteins.
    (d) All enzymes are vitamins.
19. Enzymes are different from inorganic catalysts by
    (a) not being used up in reactions.
    (b) being proteinaceous in nature.
    (c) having high diffusion rate.
    (d) working at high temperature.
20. The enzymes, which combines with non-protein part to form a functional enzyme known as
    (a) co-enzyme
    (b) holoenzyme
    (c) apoenzyme
    (d) prosthetic group
21. Coenzyme can function
    (a) in association with a vitamin.
    (b) in association with an apoenzyme.
    (c) independently of the apoenzyme.
    (d) in association with a protein.
22. In which one of the following enzymes, is copper necessarily associated as an activator?
    (a) Lactic dehydrogenase
    (b) Tyrosinase
    (c) Carbonic anhydrase
    (d) Trytophanase
23. To explain the mechanism of enzymatic action, who proposed *"Lock and key hypothesis"* ?
    (a) Fischer (b) Jacob
    (c) Koshland (d) Sumner
24. Jacob and Monad named those enzymes allosteric whose activity is regulated by
    (a) substrate (b) end product
    (c) coenzymes (d) prosthetic groups

RESPONSE GRID					
	15. ⓐⓑⓒⓓ	16. ⓐⓑⓒⓓ	17. ⓐⓑⓒⓓ	18. ⓐⓑⓒⓓ	19. ⓐⓑⓒⓓ
	20. ⓐⓑⓒⓓ	21. ⓐⓑⓒⓓ	22. ⓐⓑⓒⓓ	23. ⓐⓑⓒⓓ	24. ⓐⓑⓒⓓ

Space for Rough Work

25. Competitive inhibitors (like malonate) alter
(a) velocity of reaction
(b) value of *Km* (Michelis constant)
(c) active site
(d) structure of enzyme

26. Basically how many types of enzymes have been recognised by International Union of Biochemistry?
(a) 4
(b) 5
(c) 6
(d) 8

27. Which of the following has highest lactose content?
(a) Human milk
(b) Cow milk
(c) Buffalo milk
(d) Goat milk

28. The given graph shows the effect of substrate concentration on the rate of reaction of the enzyme green gram phosphatase.

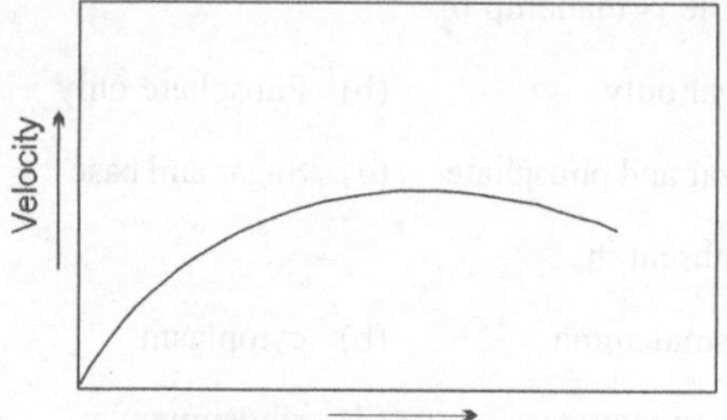

What does the graph indicate ?
(a) The rate of enzyme reaction is directly proportional to the substrate concentration.
(b) Presence of an enzyme inhibitor in the reaction mixture .
(c) Formation of an enzyme-substrate complex.
(d) At higher substrate concentration the pH increases.

29. Which one belong to hydrolase group?
(a) Amylase (b) Transaminase
(c) Citrate synthetase (d) Enolase

30. Enzymes which are slightly different in molecular structure but can perform identical activity are called
(a) isoenzymes (b) holoenzymes
(c) apoenzymes (d) coenzymes

RESPONSE GRID	
	25. ⓐⓑⓒⓓ 26. ⓐⓑⓒⓓ 27. ⓐⓑⓒⓓ 28. ⓐⓑⓒⓓ 29. ⓐⓑⓒⓓ
	30. ⓐⓑⓒⓓ

DAILY PRACTICE PROBLEM SHEET 21 - BIOLOGY			
Total Questions	30	Total Marks	120
Attempted		Correct	
Incorrect		Net Score	
Cut-off Score	44	Qualifying Score	68
Success Gap = Net Score – Qualifying Score			
Net Score = (Correct × 4) – (Incorrect × 1)			

Space for Rough Work

DPP - Daily Practice Problems

Name :   Date :

Start Time :   End Time :

BIOLOGY 22

SYLLABUS : Cell cycle and cell division

Max. Marks : 120   **Time : 60 min.**

GENERAL INSTRUCTIONS

- The Daily Practice Problem Sheet contains 30 MCQ's. For each question only one option is correct. Darken the correct circle/ bubble in the Response Grid provided on each page.
- You have to evaluate your Response Grids yourself with the help of solution booklet.
- Each correct answer will get you 4 marks and 1 mark shall be deduced for each incorrect answer. No mark will be given/ deducted if no bubble is filled. Keep a timer in front of you and stop immediately at the end of 60 min.
- The sheet follows a particular syllabus. Do not attempt the sheet before you have completed your preparation for that syllabus. Refer syllabus sheet in the starting of the book for the syllabus of all the DPP sheets.
- After completing the sheet check your answers with the solution booklet and complete the Result Grid. Finally spend time to analyse your performance and revise the areas which emerge out as weak in your evaluation.

DIRECTIONS (Q.1-Q.30) : There are 30 multiple choice questions. Each question has 4 choices (a), (b), (c) and (d), out of which ONLY ONE choice is correct.

1. Coiling of chromatids in mitotic and meiotic division is
   (a) paranemic in both.
   (b) plectonemic in both.
   (c) paranemic in mitosis and plectonemic in meiosis.
   (d) plectonemic in mitosis and paranemic in meiosis.
2. As there occurs more and more condensation of chromatin during cell division, there occurs
   (a) increase in heterochromatin.
   (b) increase in euchromatin .
   (c) differentiation of heterochromatin and euchromatin decreases.
   (d) differentiation of heterochromatin and euchromatin is clearly shown.
3. Condensation of chromosomes occurs in
   (a) prophase I
   (b) prophase II
   (c) anaphase
   (d) metaphase
4. The replication of nuclear DNA occurs in
   (a) G_1 phase   (b) G_2 phase
   (c) S phase   (d) M phase

RESPONSE GRID	1. ⓐⓑⓒⓓ	2. ⓐⓑⓒⓓ	3. ⓐⓑⓒⓓ	4. ⓐⓑⓒⓓ

Space for Rough Work

5. The number of DNA in chromosome at G_2 stage of cell cycle is
(a) one (b) two
(c) four (d) eight

6. Synapsis is pairing of
(a) any two chromosomes.
(b) non-homologous chromosome.
(c) acentric chromosomes.
(d) homologous chromosomes.

7. Synaptonemal complex is formed during
(a) meiosis
(b) amitosis
(c) mitosis
(d) cytokinesis

8. Bivalents in meiosis are
(a) tetrad
(b) pairs of non-homologous chromosomes
(c) pairs of several chromatids
(d) pairs of homozygous chromosomes

9. In pachytene stage of meiosis, the chromosomes appear
(a) single stranded
(b) double stranded
(c) three stranded
(d) four stranded

10. Recombination of genes occur at
(a) prophase in mitosis
(b) prophase I in meiosis
(c) prophase II in meiosis
(d) metaphase II in meiosis

11. If there were 4 chromosomes present during prophase, how many chromosomes are there in each cell at the end of anaphase II ?
(a) 16 (b) 4
(c) 2 (d) 8

12. How many mitotic divisions must occur in a cell of root tip to form 256 cells ?
(a) 256 (b) 8
(c) 128 (d) 64

13. In meiosis, the centromere divides during
(a) prophase-I
(b) metaphase-I
(c) anaphase-I
(d) anaphase-II

14. The no. of chromosome groups at the equatorial plate in metaphase-I of meiosis in plant with 2n = 50 shall be
(a) 50 (b) 25
(c) 30 (d) 100

Response Grid	5. ⓐⓑⓒⓓ	6. ⓐⓑⓒⓓ	7. ⓐⓑⓒⓓ	8. ⓐⓑⓒⓓ	9. ⓐⓑⓒⓓ
	10. ⓐⓑⓒⓓ	11. ⓐⓑⓒⓓ	12. ⓐⓑⓒⓓ	13. ⓐⓑⓒⓓ	14. ⓐⓑⓒⓓ

Space for Rough Work

15. Which one of the following pairs is correctly matched?

(a) Anaphase I — Homologous chromosomes are separated

(b) Metaphase I — Pairing of maternal and paternal homologous chromosomes takes place

(c) Interphase — A nuclear envelope encloses each haploid set of chromosomes

(d) Prophase I — Non-homologous chromosomes are separated

16. Chromosome number is halved in meiosis during

(a) metaphase-I (b) anaphase-I

(c) Metaphase-II (d) telophase-I

17. Chromosome start separating at which stage of mitosis?

(a) Early metaphase

(b) Late metaphase

(c) Early anaphase

(d) Early telophase

18. DNA replication occurs during

(a) prophase (b) metaphase

(c) anaphase (d) interphase

19. If a cell has a chromosome number after first meiosis equal to 48. The chromosome number in the daughter cell after the completion of meiosis will be

(a) 48 (b) 24

(c) 12 (d) 36

20. The number of chromatids in a chromosome at anaphase is

(a) 2 in mitosis and 1 in meiosis

(b) 1 in mitosis and 2 in meiosis

(c) 2 each in mitosis and meiosis

(d) 2 in mitosis and 4 in meiosis

21. The process by which the chromosome are separated in the sex cells and their number reduced from the diploid to haploid condition is known as

(a) division (b) mitosis

(c) conjugation (d) meiosis

22. Prophase of reduction division is divided into number of stages. The correct chronological sequence is

(a) leptotene–pachytene–zygotene–diplotene–diakinesis

(b) leptotene–diplotene–pachytene–zygotene–diakinesis

(c) leptotene–zygotene–diplotene–pachytene–diakinesis

(d) leptotene–zygotene–pachytene–diplotene–diakinesis

23. Regarding the sequence of cell cycle, which one is correct

(a) G_1, G_2, S and M (b) S, G_1, G_2, and M

(c) G_1, S, G_2, and M (d) G_2, S, G_1 and M

24. G_2 phase of mitosis takes

(a) 50% time of cell cycle.

(b) 25 to 33 % time of cell cycle.

(c) 12 to 16 % time of cell cycle.

(d) 4% time of cell cycle.

RESPONSE GRID					
	15. ⓐⓑⓒⓓ	16. ⓐⓑⓒⓓ	17. ⓐⓑⓒⓓ	18. ⓐⓑⓒⓓ	19. ⓐⓑⓒⓓ
	20. ⓐⓑⓒⓓ	21. ⓐⓑⓒⓓ	22. ⓐⓑⓒⓓ	23. ⓐⓑⓒⓓ	24. ⓐⓑⓒⓓ

Space for Rough Work

25. Meiosis is significant because
    (a) it produces identical cells.
    (b) it restores the original number of chromosomes (after fertilization).
    (c) there is doubling of DNA contents in the cell.
    (d) it occurs only in vegetative cells.

26. DNA replication takes place in

    or

    DNA molecules of each chromosome become double in
    (a) G_1 phase (b) G_2 phase
    (c) S phase (d) mitotic phase

27. Phragmoplast is precursor of
    (a) cell plate
    (b) chloroplast
    (c) chromoplast
    (d) colourless plastid

28. The study of chromosomes at meiotic diplotene shows that
    (a) the intimately paired chromosome repel each other and begin to separate.
    (b) the pairing of homologous chromosome which had initiated in the earlier stage is completed.
    (c) the homologous chromosomes remain united by chiasmata.
    (d) none of the above.

29. Which is not true for anaphase?
    (a) Golgi body and ER are reformed.
    (b) Chromosomes move to opposite poles.
    (c) Spindle poles move farther apart.
    (d) Centromeres split and chromatids separate.

30. DNA and histone proteins are synthesized during which of the following phase of cell cycle?
    (a) S phase (b) G_2 phase
    (c) G_1 phase (d) None of these

RESPONSE GRID	
	25. ⓐⓑⓒⓓ 26. ⓐⓑⓒⓓ 27. ⓐⓑⓒⓓ 28. ⓐⓑⓒⓓ 29. ⓐⓑⓒⓓ
	30. ⓐⓑⓒⓓ

DAILY PRACTICE PROBLEM SHEET 22 - BIOLOGY			
Total Questions	30	Total Marks	120
Attempted		Correct	
Incorrect		Net Score	
Cut-off Score	40	Qualifying Score	68
Success Gap = Net Score – Qualifying Score			
Net Score = (Correct × 4) – (Incorrect × 1)			

Space for Rough Work

DPP - Daily Practice Problems

Name : 　　　　　　　　　　 Date :

Start Time : 　　　　　　　　　　 End Time :

BIOLOGY 23

SYLLABUS : Transport in Plants Part I : Membrane osmosis, Diffusion, Imbibition, Plasmolysis, Wilting, DPD, Water potential, Absorption of water, Ascent of sap

Max. Marks : 120 　　　　　　　　　　 **Time : 60 min.**

GENERAL INSTRUCTIONS

- The Daily Practice Problem Sheet contains 30 MCQ's. For each question only one option is correct. Darken the correct circle/ bubble in the Response Grid provided on each page.
- You have to evaluate your Response Grids yourself with the help of solution booklet.
- Each correct answer will get you 4 marks and 1 mark shall be deduced for each incorrect answer. No mark will be given/ deducted if no bubble is filled. Keep a timer in front of you and stop immediately at the end of 60 min.
- The sheet follows a particular syllabus. Do not attempt the sheet before you have completed your preparation for that syllabus. Refer syllabus sheet in the starting of the book for the syllabus of all the DPP sheets.
- After completing the sheet check your answers with the solution booklet and complete the Result Grid. Finally spend time to analyse your performance and revise the areas which emerge out as weak in your evaluation.

DIRECTIONS (Q.1-Q.30) : There are 30 multiple choice questions. Each question has 4 choices (a), (b), (c) and (d), out of which ONLY ONE choice is correct.

1. The cell wall is permeable and not a semipermeable structure can be best deduced from the passage of water and mineral salts from
   (a) soil into periplasmic space of root hairs .
   (b) root hairs to cortical cells.
   (c) cortical cells to pericycle.
   (d) pericycle to trachea.
2. Selective permeability identifies the process of transmission through semipermeable membrane is called
   (a) diffusion 　 (b) osmosis
   (c) plasmolysis 　 (d) imbibition
3. A cell placed in strong solution will shrink because
   (a) cytoplasm will decompose.
   (b) mineral salt will break the cell wall.
   (c) salt water enter the cell.
   (d) water comes out by exosmosis.
4. Gaseous exchange in land plants takes place through stomatal opening, while in submerged hydrophytes it takes place by
   (a) lenticels
   (b) stomata
   (c) hydathodes
   (d) general surface of the cells by diffusion

RESPONSE GRID	1. ⓐⓑⓒⓓ	2. ⓐⓑⓒⓓ	3. ⓐⓑⓒⓓ	4. ⓐⓑⓒⓓ

Space for Rough Work

5. Movement of particles of matter in the cell due to its own kinetic energy is called

(a) imbibition (b) diffusion

(c) gummosis (d) none of these

6. During osmosis, water moves through a membrane

	FROM	TO
(a)	Low water potential	High water potential
(b)	High solute concen	Low solute concen.
(c)	High osmotic potential	Low osmotic potential
(d)	A hypotonic solution (less solute)	A hypertonic solution (high solute)

7. Osmosis cannot be demonstrated by a potato osmoscope using a solution of NaCl instead of sugar; because the potato tissue is

(a) permeable to na ion.

(b) permeable to cl ion.

(c) permeable to salt solution.

(d) impermeable to salt solution.

8. When pieces of potato tissue were placed in a solution, after some time a loss in weight was recorded. The possible reason could be

(a) death of tissue (b) exit of solutes

(c) exosmosis (d) respiration of tissue

9. A Professor kept some moist seeds in an airtight jar and started lecturing. At the end of the experiment an explosion occurred in the jar. What did the professor want to explain?

(a) osmosis (b) diffusion

(c) anaerobic respiration (d) imbibition

10. What would happen, if a thin slice of sugar beet is kept in NaCl ?

(a) It should lose water from the cell.

(b) It should become turgid.

(c) It does not absorb water from the soil solution.

(d) It should absorb water from the soil solution.

11. What shall be the sequence of events during wilting of aplant ?

(a) Exosmosis, deplasmolysis, wilting

(b) Endosmosis, plasmolysis, wilting

(c) Exosmosis, plasmolysis, wilting

(d) Endosmosis, deplasmolysis, wilting

12. Wilting occurs

(a) when rate of transpiration is higher than absorption.

(b) when rate of absorption is higher than transpiration.

(c) when excess root pressure.

(d) due to high relative humidity in air.

13. If a cell A with OP = 5, and TP = 4 is surrounded by other cell with OP = 3, and TP = 1, what will be the direction of water movement?

(a) From cell A to other cells.

(b) From other cells to cell A.

(c) Water will not move.

(d) Water will move up.

14. When the mesophyll cells lose water, their DPD

(a) increases (b) decreases

(c) remain constant (d) none of these

RESPONSE GRID					
	5. ⓐⓑⓒⓓ	6. 	7. ⓐⓑⓒⓓ	8. ⓐⓑⓒⓓ	9. ⓐⓑⓒⓓ
	10. ⓐⓑⓒⓓ	11. ⓐⓑⓒⓓ	12. ⓐⓑⓒⓓ	13. ⓐⓑⓒⓓ	14. ⓐⓑⓒⓓ

Space for Rough Work

15. In a fully turgid cells, the values of DPD, OP and TP will show which of the given tendency ?
(a) DPD = 10 atm, OP = 15 atm, TP = 5 atm.
(b) DPD = 5 atm, OP = 12 atm, TP = 7 atm.
(c) DPD = 2 atm, OP = 7 atm, TP = 5 atm.
(d) DPD = 0atm, OP = 15 atm, TP = 15 atm.

16. You are given three cells, a root hair, a cell of the inner cortical layer and a cell of the mesophyll.
Arrange them in the ascending order of DPD.
(a) Root hair <Cortical cell < Mesophyll
(b) Cortical cell < Mesophyll < Root hair
(c) Mesophyll < Root hair < Cortical cell
(d) Root hair < Mesophyll < Cortical cell

17. What will be the direction of movement of water, when a solution A having water potential of 9 bars and another solution B of 4 bars is separated by a semipermeable membrane ?
(a) B to A (b) A to B
(c) Both directions (d) None of these

18. Most widely accepted explanation for the ascent of sap in tree is
(a) capillarity
(b) role of atmospheric pressure.
(c) pulsating action of living cells.
(d) transpiration cohesion theory of Dixon.

19. Which of the following contributes most to transport of water from the ground to the leaves of a tall tree?

or

The factor or process which best explains the rise of water from roots to the top of tall tree is
(a) break down of ATP.
(b) root pressure.
(c) capillary rise of water in xylem.
(d) cohesion of water and transpiration pull.

20. Which of the following is the path of water from soil upto secondary xylem ?
(a) soil $\rightarrow$ root hair cell wall $\rightarrow$ cortex $\rightarrow$ endodermis $\rightarrow$ pericycle $\rightarrow$ protoxylem $\rightarrow$ metaxylem.
(b) metaxylem $\rightarrow$ protoxylem $\rightarrow$ cortex $\rightarrow$ soil $\rightarrow$ root hair.
(c) cortex $\rightarrow$ root hair $\rightarrow$ endodermis $\rightarrow$ pericycle $\rightarrow$ protoxylem $\rightarrow$ metaxylem.
(d) pericycle $\rightarrow$ soil $\rightarrow$ root hair $\rightarrow$ cortex $\rightarrow$ endodermis $\rightarrow$ protoxylem $\rightarrow$ metaxylem.

21. In soil, water available for plants is
(a) capillary water
(b) hygroscopic water
(c) gravitational water
(d) chemically bound water

22. Which of the following statements is incorrect?
(a) Water and salts are taken in simultaneously by root hairs.
(b) Plants absorb one thing at a time either water or inorganic salt.
(c) Plants absorb excess quantity of water.
(d) All of the above

23. The transport of water and salts takes place through
(a) phloem
(b) xylem
(c) sieve tubes
(d) sclerenchyma

24. Most water absorption in plants takes place through
(a) root cap
(b) root apex
(c) root hair zone
(d) meristematic zone

RESPONSE GRID					
	15. ⓐⓑⓒⓓ	16. ⓐⓑⓒⓓ	17. ⓐⓑⓒⓓ	18. ⓐⓑⓒⓓ	19. ⓐⓑⓒⓓ
	20. ⓐⓑⓒⓓ	21. ⓐⓑⓒⓓ	22. ⓐⓑⓒⓓ	23. ⓐⓑⓒⓓ	24. ⓐⓑⓒⓓ

Space for Rough Work

25. Water reaches the top of a plant due to
(a) root pressure
(b) capillarity
(c) transpiration
(d) diffusion

26. Maximum water loss takes place through
(a) cuticle (b) stomata
(c) hydathode (d) lenticel

27. Absorption of water by root is increased by
(a) increase in transpiration.
(b) increase in the rate of photosynthesis.
(c) decrease in transpiration.
(d) decrease in salt uptake.

28. Many transplanted seedling may not survive because
(a) they do not like the new soil.
(b) they do not get required mineral salts.
(c) most of the root hairs are lost during transplantation.
(d) the leaves get damaged.

29. Suitable temperature for active absorption of water by root is
(a) 40–45°C
(b) 10–15°C
(c) 20–35°C
(d) can take at any temperature

30. Water– logged condition will quickly occur in which type of soil?
(a) Sand (b) Clay
(c) Gravel (d) Loam

RESPONSE GRID	25. ⓐⓑⓒⓓ	26. ⓐⓑⓒⓓ	27. ⓐⓑⓒⓓ	28. ⓐⓑⓒⓓ	29. ⓐⓑⓒⓓ
	30. ⓐⓑⓒⓓ				

DAILY PRACTICE PROBLEM SHEET 23 - BIOLOGY			
Total Questions	30	Total Marks	120
Attempted		Correct	
Incorrect		Net Score	
Cut-off Score	32	Qualifying Score	56
Success Gap = Net Score – Qualifying Score			
Net Score = (Correct × 4) – (Incorrect × 1)			

Space for Rough Work

DPP - Daily Practice Problems

Name : Date :

Start Time : End Time :

BIOLOGY 24

SYLLABUS : Transport in Plants Part II : Transpiration, Stomata, Factor affecting transpiration, Mechanism of opening and closure of stomata, Guttation

Max. Marks : 120 **Time : 60 min.**

GENERAL INSTRUCTIONS

- The Daily Practice Problem Sheet contains 30 MCQ's. For each question only one option is correct. Darken the correct circle/ bubble in the Response Grid provided on each page.
- You have to evaluate your Response Grids yourself with the help of solution booklet.
- Each correct answer will get you 4 marks and 1 mark shall be deduced for each incorrect answer. No mark will be given/ deducted if no bubble is filled. Keep a timer in front of you and stop immediately at the end of 60 min.
- The sheet follows a particular syllabus. Do not attempt the sheet before you have completed your preparation for that syllabus. Refer syllabus sheet in the starting of the book for the syllabus of all the DPP sheets.
- After completing the sheet check your answers with the solution booklet and complete the Result Grid. Finally spend time to analyse your performance and revise the areas which emerge out as weak in your evaluation.

DIRECTIONS (Q.1-Q.30) : There are 30 multiple choice questions. Each question has 4 choices (a), (b), (c) and (d), out of which ONLY ONE choice is correct.

1. The chief role of transpiration in plants is to cause
   (a) loss of surplus water (b) cooling of the plant
   (c) rapid ascent of sap (d) rapid rise of minerals
2. The loss of water in the form of vapour from aerial plant parts is known as
   (a) osmosis (b) respiration
   (c) photosynthesis (d) transpiration
3. In both transpiration and evaporation, water is lost in the form of vapour yet they differ, because
   (a) both transpiration and evaporation are similar but the rate of water loss differs.
   (b) frequency of water loss is different in both of them.
   (c) transpiration is a physical process and evaporation is a physiological process.
   (d) transpiration is a physiological process and evaporation is a physical process.
4. Stomatal index (I) is equal to
   (a) $\frac{S}{E-S}$ (b) $\frac{S}{E+S}$
   (c) $\frac{E}{E+S}$ (d) $\frac{E+S}{E}$

RESPONSE GRID	1. ⓐⓑⓒⓓ	2. ⓐⓑⓒⓓ	3. ⓐⓑⓒⓓ	4. ⓐⓑⓒⓓ

Space for Rough Work

5. Of the processes which occur in leaves, the one which may lower their temperature (cooling effect) is
(a) respiration (b) photosynthesis
(c) transpiration (d) hydrolysis

6. Generally stomata found on the lower surface of leaf. Thus, approximate percentage of transpiration through abaxial surface is
(a) 47 % (b) 57 %
(c) 67 % (d) 97 %

7. Transpiration efficiency is the ratio of
(a) absorption to transpiration.
(b) rate of ascent of sap to transpiration.
(c) dry matter produced/kg of transpired water.
(d) transpiration to area of the leaf.

8. Transpiration ratio is the ratio of moles of H_2O transpired/ moles of CO_2 fixed. This ratio is measure of
(a) the efficiency of guard cells on stomatal movement.
(b) effectiveness of stomata is maximizing in photosynthesis while minimizing in water loss.
(c) distinguishing a xerophyte from glycophyte.
(d) stomatal pore size of the leaves.

9. Which one of the following will reduce the rate of transpiration ?
(a) Increase in wind velocity.
(b) Rise in temperature.
(c) Increase in water uptake by plants.
(d) Decrease in light intensity.

10. The condition under which transpiration would be most rapid is
(a) high humidity.
(b) excess of water in soil.
(c) low humidity, high temperature, guard cells are turgid (open) and moist soil.
(d) low velocity of wind.

11. Which of the following factors is the most important in the regulation of transpiration?
(a) Humidity (b) Light
(c) Temperature (d) Wind

12. Under what conditions the rate of transpiration increases?
(a) increase of humidity.
(b) increase of atmospheric pressure.
(c) decrease of temperature.
(d) decrease of humidity.

13. Stomata in angiosperms open and close due to
(a) their genetic constitution.
(b) effect of hormones.
(c) changes of turgor pressure in guard cells.
(d) pressure of gases inside the leaves.

14. The metal ion involved in the stomatal regulation is **or** Stomata will open, if there is accumulation of which of the following element in the guard cells?
(a) Iron (b) Magnesium
(c) Zinc (d) Potassium

Response Grid					
	5. ⓐⓑⓒⓓ	6. ⓐⓑⓒⓓ 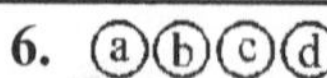	7. ⓐⓑⓒⓓ	8. ⓐⓑⓒⓓ	9. ⓐⓑⓒⓓ
	10. ⓐⓑⓒⓓ	11. ⓐⓑⓒⓓ	12. ⓐⓑⓒⓓ	13. ⓐⓑⓒⓓ	14. ⓐⓑⓒⓓ

Space for Rough Work

15. Stomata open during day time because the guard cells
(a) photosynthesize and produce osmotically active sugars or organic acids.
(b) are thin-walled.
(c) are bean shaped.
(d) help in gaseous exchange.

16. Which one of the following describes a possible reason for the opening and closing of the guard cells ?
(a) Stomata open in daylight because photosynthesis occurs in the guard cells producing sugar resulting in higher osmotic pressure.
(b) Stomata open in daylight because guard cells have chloroplasts and the epidermal cells do not, and so a differential osmotic pressure arises.
(c) Stomata open in daylight because starch is changed to osmotically active substances which increase the turgidity of the guard cells.
(d) Stomata close in darkness because the starch produced in daylight by photosynthesis, it is changed into sugars and translocated from the guard cells.

17. Guard cell movement furnishes an example of
(a) turgor movement
(b) tactic movement
(c) nastic Movement
(d) cyclosis movement

18. Opening and closing of stomata is controlled by
(a) K^+ (b) Mg^{2+}
(c) Ca^{2+} (d) N

19. Guttation is caused due to
(a) imbibition (b) osmosis
(c) positive root pressure (d) transpiration

20. A specialized multicellular structure in leaves which excretes water droplets is called as
(a) stomata (b) hydathodes
(c) lenticels (d) wounds

21. Guttation may be defined as a process in which plants can
(a) synthesize carbohydrates.
(b) excrete the salt.
(c) get rid of excess water in liquid form.
(d) compensate the loss of water.

22. Which of the following is used to determine the rate of transpiration in plants?
(a) Porometer (b) Potometer
(c) Auxanometer (d) Tensiometer

23. Principle of working of all potometer is
(a) absorption occurs by stem.
(b) absorption is a physical process.
(c) absorption is proportional to transpiration.
(d) transpiration depends on light.

24. Potometer and Clinostat are used to study
(a) photosynthesis and respiration
(b) transpiration and growth
(c) phototropism and geotropism
(d) transpiration and geotropism

RESPONSE GRID					
	15. ⓐⓑⓒⓓ	16. ⓐⓑⓒⓓ	17. ⓐⓑⓒⓓ	18. ⓐⓑⓒⓓ	19. ⓐⓑⓒⓓ
	20. ⓐⓑⓒⓓ	21. ⓐⓑⓒⓓ	22. ⓐⓑⓒⓓ	23. ⓐⓑⓒⓓ	24. ⓐⓑⓒⓓ

Space for Rough Work

25. According to one vital force theory, ascent of sap is due to active pulsation of innermost layer of cortex. This theory was given by
(a) J C Bose (b) Dixon
(c) Strasburger (d) Sacks

26. Match the theories given in column I with the names of scientist listed in column II. Choose the answer which gives the correct combination of the alphabets.

Column - I (Names of theories)	Column - II (Names of scientists)
A. Relay pump theory	p Stocking
B. Transpiration cohesion theory	q Sir J. C. Bose
C. Mass flow	r Godlewski
D. Pulsation theory	s Dixon and Jolly
	t Ernst Munch

(a) A = r; B = s; C = t; D = q
(b) A = s; B = r; C = p; D = q
(c) A = r; B = q; C = t; D = q
(d) A = q; B = p; C = t; D = r

27. Bidirectional translocation of minerals takes place in
(a) xylem (b) phloem
(c) parenchyma (d) cambium

28. Which theory explained the translocation of long interval ?
(a) protoplasmic streaming
(b) transcellular streaming
(c) cyclosis
(d) pressure flow

29. Starch is insoluble in water yet it is accumulated in large quantities in potato because
(a) it is synthesized in potato tuber itself.
(b) it is translocated from the leaves to the potato tuber in the form of sugar.
(c) soil micro-organism deposit starch in tuber.
(d) it is useful consumption.

30. In which form, the food is translocated from endosperm to plumule of a fatty seed, like castor?
(a) Fatty acids (b) Fat
(c) Glucose (d) Sucrose

RESPONSE GRID	
	25. ⓐⓑⓒⓓ 26. ⓐⓑⓒⓓ 27. ⓐⓑⓒⓓ 28. ⓐⓑⓒⓓ 29. ⓐⓑⓒⓓ 30. ⓐⓑⓒⓓ

DAILY PRACTICE PROBLEM SHEET 24 - BIOLOGY

Total Questions	30	Total Marks	120
Attempted		Correct	
Incorrect		Net Score	
Cut-off Score	44	Qualifying Score	72
Success Gap = Net Score – Qualifying Score			
Net Score = (Correct × 4) – (Incorrect × 1)			

Space for Rough Work

DPP - Daily Practice Problems

Name :

Date :

Start Time :

End Time :

BIOLOGY 25

SYLLABUS : Mineral Nutrition Part I : Micro and Macro Nutrients

Max. Marks : 120 **Time : 60 min.**

GENERAL INSTRUCTIONS

- The Daily Practice Problem Sheet contains 30 MCQ's. For each question only one option is correct. Darken the correct circle/ bubble in the Response Grid provided on each page.
- You have to evaluate your Response Grids yourself with the help of solution booklet.
- Each correct answer will get you 4 marks and 1 mark shall be deduced for each incorrect answer. No mark will be given/ deducted if no bubble is filled. Keep a timer in front of you and stop immediately at the end of 60 min.
- The sheet follows a particular syllabus. Do not attempt the sheet before you have completed your preparation for that syllabus. Refer syllabus sheet in the starting of the book for the syllabus of all the DPP sheets.
- After completing the sheet check your answers with the solution booklet and complete the Result Grid. Finally spend time to analyse your performance and revise the areas which emerge out as weak in your evaluation.

DIRECTIONS (Q.1-Q.30) : There are 30 multiple choice questions. Each question has 4 choices (a), (b), (c) and (d), out of which ONLY ONE choice is correct.

1. Which of the following elements is not required by plants for their normal healthy growth?

   (a) Calcium (b) Magnesium
   (c) Lead (d) Iron

2. Non-essential element in plant is

   (a) barium (b) magnesium
   (c) potassium (d) calcium

3. Which of the following, one element has not been proved to be essential in plants?

   (a) Potassium
   (b) Sodium
   (c) Zinc
   (d) Iron

4. An essential element is one which

   (a) improve plant growth.
   (b) is present in plant ash.
   (c) is indispensable for growth and is irreplaceable.
   (d) available in soil.

RESPONSE GRID	1. ⓐⓑⓒⓓ	2. ⓐⓑⓒⓓ	3. ⓐⓑⓒⓓ	4. ⓐⓑⓒⓓ

Space for Rough Work

5. Hydroponics are
   (a) growing of aquatic plants.
   (b) growing of floating aquatic plants.
   (c) growing of plants in sand.
   (d) growing of plants in aqueous balanced nutrient.

6. Which of the following minerals are required for ATP formation by plants?
   (a) N, P (b) N, Cu
   (c) N, Ca (d) K

7. Which one is not an essential element in plants ?
   (a) Iron
   (b) Boron
   (c) Sulphur
   (d) Cadmium

8. Which group of element is not essential for a normal plant?
   (a) K, Ca, Mg,
   (b) Fe, Zn, Mn, B
   (c) Pb, Ni, I, Na
   (d) Mg, Fe, Mo

9. Which of the following element is not essential for plants?
   (a) Iron (b) Zinc
   (c) Potassium (d) Iodine

10. Which of the following is used as green manure ?
    (a) *Azolla*
    (b) *Azadirachta indica*
    (c) *Crotolaria juncea*
    (d) *Hevea brassiliansis*

11. Which of the following ions of heavy metals participate in process of photosythesis in higher plants?
    (a) *Pb, Fe, Ni, Co*
    (b) *Mg, Zn, Cu, Hg*
    (c) *Mg, Mn, Co, Fe*
    (d) *Mg, Cu, Mn, Fe*

12. Which element forms part of structure of chlorophyll molecule?
    (a) *Fe* (b) *Mg*
    (c) *K* (d) *Mn*

13. Death of stem and root tips occur due to deficiency of
    (a) phosphorus (b) calcium
    (c) nitrogen (d) carbon

14. Which of the following is essential mineral element and is not a constituent of any enzyme but stimulate the activity of many enzymes?
    (a) *Zn* (b) *Mg*
    (c) *Mn* (d) *K*

RESPONSE GRID					
	5. ⓐⓑⓒⓓ	6. ⓐⓑⓒⓓ 	7. ⓐⓑⓒⓓ	8. ⓐⓑⓒⓓ	9. ⓐⓑⓒⓓ
	10. ⓐⓑⓒⓓ	11. ⓐⓑⓒⓓ	12. ⓐⓑⓒⓓ	13. ⓐⓑⓒⓓ	14. ⓐⓑⓒⓓ

Space for Rough Work

15. In plants a common symptom caused by deficiency of *P, K, Ca* and *Mg* is the
(a) bending of leaf tip.
(b) formation of anthocyanin.
(c) poor development of vasculature.
(d) appearance of dead necrotic areas.

16. The possible resource of phosphorus ions and nitrogen ions in soil generally get depleted because they are usually found as
(a) positively charged ions.
(b) negatively charged ions.
(c) a disproportionate mixture of negatively charged ions.
(d) particles carrying no charge.

17. In guard cells, presence of potassium is essential for
(a) maintaining osmotic pressure.
(b) in controlling cell division.
(c) in enzymatic reactions.
(d) all of the above.

18. Yellowing of tea leaf takes palce by the deficiency of
(a) chlorine (b) potassium
(c) oxygen (d) sulphur

19. Which of the following is not a macro-nutrient?
(a) *Mn* (b) *Ca*
(c) *Mg* (d) *P*

20. Essential macroelements are
(a) absorbed from soil.
(b) manufactured during photosynthesis.
(c) produced by enzymes.
(d) produced by growth hormones.

21. *Mg* and *Fe* are required by the plants for the
(a) energy transfer concerned with photosynthesis.
(b) synthesis of chlorophyll pigments in leaves.
(c) mechanism of stomatal opening and closing.
(d) translocation of carbohydrates from the leaf to the stem.

22. Which of the following element is used up in phosphorylation?
(a) Calcium and sulphur
(b) Chlorine and manganese
(c) Iron and phosphorous
(d) Magnesium and phosphate

23. The most abundant element present in the plants is
(a) manganese (b) iron
(c) carbon (d) nitrogen

24. A trace element is an element which
(a) is a radioactive and can be traced by Geiger counter.
(b) is required in very minute amounts.
(c) draws other element out of protoplasm.
(d) was one of the first to be discovered in protoplasm.

25. Deficiency of molybdenum cause

RESPONSE GRID					
	15. ⓐⓑⓒⓓ	16. ⓐⓑⓒⓓ	17. ⓐⓑⓒⓓ	18. ⓐⓑⓒⓓ	19. ⓐⓑⓒⓓ
	20. ⓐⓑⓒⓓ	21. ⓐⓑⓒⓓ	22. ⓐⓑⓒⓓ	23. ⓐⓑⓒⓓ	24. ⓐⓑⓒⓓ

Space for Rough Work

(a) poor development of vasculature.

(b) bending of leaf tip.

(c) yellowing of leaves.

(d) mottling and necrosis of leaves.

26. The cauliflower become brown due to deficiency of

(a) sodium (b) calcium

(c) boron (d) nitrogen

27. The deficiency of molybdenum induces

(a) citrus die back disease

(b) pea rossete disease

(c) cauliflower whip tail disease

(d) white bud of maize

28. Which of the following is not a micro or trace element?

(a) Boron (b) Molybdenum

(c) Manganese (d) Calcium

29. Function of zinc is

(a) synthesis of chlorophyll.

(b) biosynthesis of 3 IAA.

(c) closing of stomata.

(d) oxidation of carbohydrate.

30. Cytochrome oxidase has

(a) *Mo*

(b) *Fe*

(c) *Zn*

(d) *B*

RESPONSE GRID	25. ⓐⓑⓒⓓ	26. ⓐⓑⓒⓓ	27. ⓐⓑⓒⓓ	28. ⓐⓑⓒⓓ	29. ⓐⓑⓒⓓ
	30. ⓐⓑⓒⓓ				

DAILY PRACTICE PROBLEM SHEET 25 - BIOLOGY			
Total Questions	30	Total Marks	120
Attempted		Correct	
Incorrect		Net Score	
Cut-off Score	40	Qualifying Score	64
Success Gap = Net Score – Qualifying Score			
Net Score = (Correct × 4) – (Incorrect × 1)			

Space for Rough Work

DPP - Daily Practice Problems

Name :   Date :

Start Time :   End Time :

BIOLOGY 26

SYLLABUS : Mineral Nutrition Part II : Mineral absorption, Nitrogen nutrition, Special mode of nutrition

Max. Marks : 120   **Time : 60 min.**

GENERAL INSTRUCTIONS

- The Daily Practice Problem Sheet contains 30 MCQ's. For each question only one option is correct. Darken the correct circle/ bubble in the Response Grid provided on each page.
- You have to evaluate your Response Grids yourself with the help of solution booklet.
- Each correct answer will get you 4 marks and 1 mark shall be deduced for each incorrect answer. No mark will be given/ deducted if no bubble is filled. Keep a timer in front of you and stop immediately at the end of 60 min.
- The sheet follows a particular syllabus. Do not attempt the sheet before you have completed your preparation for that syllabus. Refer syllabus sheet in the starting of the book for the syllabus of all the DPP sheets.
- After completing the sheet check your answers with the solution booklet and complete the Result Grid. Finally spend time to analyse your performance and revise the areas which emerge out as weak in your evaluation.

DIRECTIONS (Q.1-Q.30) : There are 30 multiple choice questions. Each question has 4 choices (a), (b), (c) and (d), out of which ONLY ONE choice is correct.

1. Conduction of inorganic materials in plants occur mainly through or Minerals absorbed by roots move at the leaf through
   (a) xylem (b) phloem
   (c) sieve tube (d) none of these
2. Active transport from outside to inside of molecules across a membrane requires
   (a) cyclic AMP (b) acetyl chlorine
   (c) ATP (d) phloroglucinol
3. Ion uptake is called active because
   (a) ions are active
   (b) energy is expended
   (c) ions move freely
   (d) none of these
4. Which theory suggest that the CO_2 produced in respiration plays an important role in mineral absorption?
   (a) Contact exchange theory
   (b) Carbonic acid exchange theory
   (c) Active absorption theory
   (d) None of the above

RESPONSE GRID	1. ⓐⓑⓒⓓ	2. ⓐⓑⓒⓓ	3. ⓐⓑⓒⓓ	4. ⓐⓑⓒⓓ

Space for Rough Work

5. Entry of mineral ions in plant root cells by diffusion is
   (a) passive absorption
   (b) active absorption
   (c) osmosis
   (d) none of these

6. By which method ions are absorbed by plants?
   (a) By difference in DPD.
   (b) By difference in water potential.
   (c) By carriers and pumps.
   (d) By molecular diffusion.

7. Solutes are absorbed by a plant cell through
   (a) osmosis
   (b) diffusion
   (c) active absorption
   (d) passive absorption

8. Carrier proteins are involved in
   (a) active transport of ions.
   (b) passive transport of ions.
   (c) water transport.
   (d) water evaporation.

9. The transfer of minerals from top soil to sub soil through soil water is called
   (a) transpiration (b) conduction
   (c) percolation (d) leaching

10. Most of the plants obtain or absorb nitrogen from soil in the form of
    (a) free nitrogen gas
    (b) nitric acid
    (c) nitrite
    (d) nitrates

11. Which of the following essential element is necessary in plants for protein synthesis ?
    (a) Calcium
    (b) Phosphorus
    (c) Magnesium
    (d) Nitrogen

12. Which one of the following can fix atmospheric nitrogen directly?
    (a) Pea (b) *Brassica*
    (c) Castor (d) *Petunia*

13. Knot like bodies known as 'nodules' found in the roots of groundnut plant are produced by
    (a) *Azospirillum*
    (b) *Azotobacter*
    (c) *Pseudomonas*
    (d) *Rhizobium*

14. Symbiotic micro-organism is
    (a) *Clostridium* (b) *Azotobacter*
    (c) *Rhizobium* (d) *Chromatium*

RESPONSE GRID					
	5. ⓐⓑⓒⓓ	6. ⓐⓑⓒⓓ 	7. ⓐⓑⓒⓓ	8. ⓐⓑⓒⓓ	9. ⓐⓑⓒⓓ
	10. ⓐⓑⓒⓓ	11. ⓐⓑⓒⓓ	12. ⓐⓑⓒⓓ	13. ⓐⓑⓒⓓ	14. ⓐⓑⓒⓓ

Space for Rough Work

15. The converson of ammonia into nitrites and nitrates is called
(a) ammonification
(b) nitrification
(c) denitrification
(d) all of these

16. Which of the following pigments is essential for nitrogen fixation by leguminous plants?
(a) Anthocyanin
(b) Phycocyanin
(c) Phycoerythrin
(d) Leghaemoglobin

17. Plants which are unable to manufacture their food wholly or partially are
(a) autophytes (b) heterophytes
(c) halophytes (d) holophytes

18. Epiphytes are the plants which are dependent on other plants
(a) only for water (b) for water and food
(c) only for food (d) only for shelter (support)

19. Myrmecophily is a beneficial association between a flowering plant and
(a) ants (b) *mycoplasma*
(c) bacteria (d) viruses

20. Insectivorous plants usually grow in soils which are deficient in
(a) nitrogen (b) water
(c) organic matter (d) *Ca/Mg*

21. *Balanophora/Orobanche is a*
(a) total root parasite
(b) partial root parasite
(c) partial stem parasite
(d) total stem parasite

22. *Santalum album is*
(a) partial root parasite
(b) partial stem parasite
(c) total stem parasite
(d) total root parasite

23. *Loranthus is a*
(a) total stem parasite
(b) partial stem parasite
(c) total root parasite
(d) partial root parasite

24. A plant growing on another plant without drawing any nourishment is called
(a) ectoparasite (b) epiphyte
(c) symbiont (d) saprophyte

RESPONSE GRID					
	15. ⓐⓑⓒⓓ	16. ⓐⓑⓒⓓ	17. ⓐⓑⓒⓓ	18. ⓐⓑⓒⓓ	19. ⓐⓑⓒⓓ
	20. ⓐⓑⓒⓓ	21. ⓐⓑⓒⓓ	22. ⓐⓑⓒⓓ	23. ⓐⓑⓒⓓ	24. ⓐⓑⓒⓓ

Space for Rough Work

25. Examples of ammonifying bacteria are
    (a) *Nitrosomonas*
    (b) *Nitroscoccus*
    (c) *Pseudomonas*
    (d) *Bacillus vulgaris*
26. A plant living symbiotically inside another plant is
    (a) saprophyte
    (b) endophyte
    (c) semiparasite
    (d) parasite
27. A pair of insectivorous plants is
    (a) *Drosera* and *Rafflesia*
    (b) *Nepenthes* and *Bladderwort*
    (c) *Dionaea* and *Viscum*
    (d) Venus fly trap and *Rafflesia*
28. Insectivorous plants catch and digest insects for
    (a) obtaining nitrogen.
    (b) protecting their leaves.
    (c) protecting their fruits.
    (d) being heterotrophs of consumer level.
29. Match the column

I	II
(A) Magnesium	(i) Found in middle lamella.
(B) Sulphur	(ii) A structural component of chlorophyll.
(C) Calcium	(iii) Required for enzyme activity.
(D) Iodine	(iv) Found in some amino acids.
(E) Manganese	(v) A component of sugars.
	(vi) Not important for plants.

	A	B	C	D	E
(a)	(ii)	(iv)	(i)	(vi)	(iii)
(b)	(ii)	(iv)	(v)	(vi)	(iii)
(c)	(ii)	(vi)	(i)	(iv)	(iii)
(d)	(iii)	(iv)	(i)	(vi)	(ii)

30. According to the well known theory of transport of solutes across a cell membrane, what happens when sugar is passed through it ?
    (a) Na^+ flows in the direction of the sugar.
    (b) Na^+ flows independent of sugar molecules.
    (c) Na^+ flows against the sugar molecules.
    (d) Na^+ ions do not flow at all.

RESPONSE GRID	25. ⓐⓑⓒⓓ	26. ⓐⓑⓒⓓ	27. ⓐⓑⓒⓓ	28. ⓐⓑⓒⓓ	29. ⓐⓑⓒⓓ
	30. ⓐⓑⓒⓓ				

DAILY PRACTICE PROBLEM SHEET 26 - BIOLOGY			
Total Questions	30	Total Marks	120
Attempted		Correct	
Incorrect		Net Score	
Cut-off Score	40	Qualifying Score	60
Success Gap = Net Score – Qualifying Score			
Net Score = (Correct × 4) – (Incorrect × 1)			

Space for Rough Work

DPP - Daily Practice Problems

Name : Date :

Start Time : End Time :

BIOLOGY 27

SYLLABUS : Photosynthesis in Higher Plants Part I : Photosynthesis experiments, photosynthetic apparatus, light reactions

Max. Marks : 120 **Time : 60 min.**

GENERAL INSTRUCTIONS

- The Daily Practice Problem Sheet contains 30 MCQ's. For each question only one option is correct. Darken the correct circle/ bubble in the Response Grid provided on each page.
- You have to evaluate your Response Grids yourself with the help of solution booklet.
- Each correct answer will get you 4 marks and 1 mark shall be deduced for each incorrect answer. No mark will be given/ deducted if no bubble is filled. Keep a timer in front of you and stop immediately at the end of 60 min.
- The sheet follows a particular syllabus. Do not attempt the sheet before you have completed your preparation for that syllabus. Refer syllabus sheet in the starting of the book for the syllabus of all the DPP sheets.
- After completing the sheet check your answers with the solution booklet and complete the Result Grid. Finally spend time to analyse your performance and revise the areas which emerge out as weak in your evaluation.

DIRECTIONS (Q.1-Q.30) : There are 30 multiple choice questions. Each question has 4 choices (a), (b), (c) and (d), out of which ONLY ONE choice is correct.

1. The law of limiting factor for photosynthesis was enunciated by
   (a) Blackman (b) Hill
   (c) Ruben (d) Kalmen
2. Discovery of Emerson effect has clearly shown the existence of
   (a) two distinct photochemical reactions or processes.
   (b) light and dark reactions in photosynthesis.
   (c) photophosphorylation.
   (d) photorespiration.
3. "The amount of CO_2 absorbed and O_2 released during photosynthesis are in equal volumes" was proved by
   (a) Engleman
   (b) Robert Mayer
   (c) Priestley
   (d) Bousingault
4. The experimental material that has largely been responsible for making rapid advances in research on photosynthesis is
   (a) *Chlamydomonas*
   (b) Spinach leaf
   (c) *Chlorella*
   (d) *Hydrilla*

RESPONSE GRID	1. ⓐⓑⓒⓓ	2. ⓐⓑⓒⓓ	3. ⓐⓑⓒⓓ	4. ⓐⓑⓒⓓ

Space for Rough Work

5. The first experiment on photosynthesis in flashing light were carried out by
(a) F. F. Blackman
(b) Robert Meyer
(c) Melvin Calvin
(d) Robert Hill

6. Moll's experiment shows
(a) unequal transpiration from two surfaces of leaf.
(b) relation between transpiration and absorption.
(c) CO_2 is required for photosynthesis.
(d) chlorophyll is essential for photosynthesis.

7. Blackman demonstrated that increasing illumination increased the photosynthetic rate upto a point when CO_2 becomes limiting. If light was not limiting, temperature becomes limiting. Emerson found that maximum CO_2 fixation could be achieved with brief flashes of light. Mark the correct statement in the following.
(a) Only one factor can be limited in photosynthesis.
(b) Photosynthesis consists of a light and dark reaction.
(c) The trapping of light by chloroplast is temperature dependent.
(d) The trapping of light by chloroplast can occur only if CO_2 is present.

8. The core metal of chlorophyll is
(a) Fe (b) Mg
(c) Ni (d) Cu

9. In presence of light, green plants convert carbon dioxide and water into food stuffs. Which of the following is likely to be the first substance that a green plant makes in photosynthesis ?
(a) A simple sugar (b) Starch
(c) Cellulose (d) Fat

10. Grana refers to
(a) stacks of thylakoids in plastids of higher plants.
(b) a constant in quantum equation.
(c) glycolysis of glucose.
(d) byproduct of photosynthesis.

11. Photosynthesis is a
(a) exothermic process
(b) exergonic process
(c) anabolic process
(d) catabolic process

12. Quantasomes contain
(a) 200 chlorophyll molecules.
(b) 230 chlorophyll molecules.
(c) 250 chlorophyll molecules.
(d) 300 chlorophyll molecules.

13. For the process of photosynthesis all except one of the following items are essential. Point out the exception.
(a) Water, minerals
(b) Light, chlorophyll
(c) CO_2, optimum temperature
(d) Oxygen, sucrose

14. Algae float in the water during day time and sink during night time because
(a) they become buoyant due to consumption of food materials in respiration.
(b) they loose weight at night.
(c) they come up to enjoy sunshine.
(d) they become buoyant in light due to accumulation of oxygen bubbles released in photosynthesis.

RESPONSE GRID					
	5. ⓐⓑⓒⓓ	6. ⓐⓑⓒⓓ 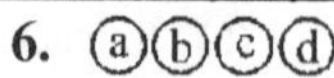	7. ⓐⓑⓒⓓ	8. ⓐⓑⓒⓓ	9. ⓐⓑⓒⓓ
	10. ⓐⓑⓒⓓ	11. ⓐⓑⓒⓓ	12. ⓐⓑⓒⓓ	13. ⓐⓑⓒⓓ	14. ⓐⓑⓒⓓ

Space for Rough Work

15. Photosynthesis is a process in which
    (a) CO_2 is reduced to carbohydrate.
    (b) NADH is reduced to NAD.
    (c) ATP is generated.
    (d) oxidative photophosphorylation occurs.
16. How many molecules of water are needed by a green plant to produce one molecule of hexose/reduce 6 molecules of CO_2
    (a) 6 (b) 12
    (c) 24 (d) One only
17. The full expansion of NADP is
    (a) nicotinamide adenine diphosphate
    (b) nicotinamide adenosine diphosphate
    (c) nicotinamide adenine dinucleotide phosphate
    (d) nicotinamide adenosine dinucleotide phosphate
18. Emerson effect explain the phenomenon of
    (a) transpiration
    (b) absorption of water by roots
    (c) photosynthesis
    (d) respiration
19. Which of the following is initiating step in photosynthesis?
    (a) ATP formation
    (b) Glucose formation
    (c) Photolysis of water
    (d) Activation of chlorophyll by sunlight
20. Photosynthetic pigments in chloroplast are embedded in membrane of
    (a) thylakoids
    (b) photoglobin
    (c) matrix
    (d) envelope of chloroplast
21. Which of the following pigment is yellow in colour?
    (a) Chlorophyll 'a'
    (b) Chlorophyll 'b'
    (c) Carotene
    (d) Xanthophyll
22. Which of the following wavelength occur in red part of the spectrum?
    (a) 470 nm
    (b) 390 nm
    (c) 680 nm
    (d) 830 nm
23. $NADPH_2$ is generated through
    (a) glycolysis (b) photosystem-I
    (c) photosystem-II (d) anaerobic respiration
24. In cyclic photophosphorylation which one of the following is formed?
    (a) *ATP*
    (b) *NADP* and *ATP*
    (c) $NADH_2$ and O_2
    (d) $NADPH_2$, *ATP* and O_2

RESPONSE GRID					
	15. ⓐⓑⓒⓓ	16. ⓐⓑⓒⓓ	17. ⓐⓑⓒⓓ	18. ⓐⓑⓒⓓ	19. ⓐⓑⓒⓓ
	20. ⓐⓑⓒⓓ	21. ⓐⓑⓒⓓ	22. ⓐⓑⓒⓓ	23. ⓐⓑⓒⓓ	24. ⓐⓑⓒⓓ

Space for Rough Work

25. Where does the primary photochemical reaction occur in chloropast or Where does the light reactions of photosynthesis take place or Light reaction takes place in
(a) stroma
(b) endoplasmic reticulum
(c) quantasome or thylakoids (grana)
(d) inner membrane of chloroplast

26. Photophosphorylation is a process in which
(a) light energy is converted into chemical energy in the form of *ATP*
(b) *NADP* is formed.
(c) chemical energy is used to produce ATP.
(d) CO_2 is reduced to carbohydrate.

27. During photochemical reactions of photosynthesis
(a) liberation of oxygen takes place.
(b) formation of *ATP* and $NADPH_2$ takes place.
(c) liberation of O_2 and formation of *ATP* and $NADPH_2$ takes place.
(d) assimilation of CO_2 takes place.

28. The main difference between the molecules of chlorophyll *a* and *b* is
(a) chlorophyll *a* has an aldehyde group while chlorophyll *b* has a methyl group.
(b) chlorophyll *a* has a methyl group while chlorophyll *b* has an aldehyde group.
(c) the phytol tail is absent in chlorophyll *a* and is present in chlorophyll *b*.
(d) the porphyrin group in chlorophyll *b* has no binding site but a binding site is present in chlorophyll *a*.

29. Which fractions of the visible spectrum of solar radiations are primarily absorbed by carotenoids of the higher plants?
(a) Violet and blue (b) Blue and green
(c) Green and red (d) Red and violet

30. Which of the following occurs first in the light reaction?
(a) Oxidation of chlorophyll.
(b) Photolysis of water.
(c) Evolution of *ATP* and O_2.
(d) Production of O_2 and $NADPH_2$.

RESPONSE GRID	25. ⓐⓑⓒⓓ	26. ⓐⓑⓒⓓ	27. ⓐⓑⓒⓓ	28. ⓐⓑⓒⓓ	29. ⓐⓑⓒⓓ
	30. ⓐⓑⓒⓓ				

DAILY PRACTICE PROBLEM SHEET 27 - BIOLOGY			
Total Questions	30	Total Marks	120
Attempted		Correct	
Incorrect		Net Score	
Cut-off Score	40	Qualifying Score	64
Success Gap = Net Score – Qualifying Score			
Net Score = (Correct × 4) – (Incorrect × 1)			

Space for Rough Work

DPP - Daily Practice Problems

Name : Date :

Start Time : End Time :

BIOLOGY 28

SYLLABUS : Photosynthesis in Higher Plants Part II : Dark reaction, C_4/ CAM, Photorespiration, Factors affecting Photosynthesis

Max. Marks : 120 **Time : 60 min.**

GENERAL INSTRUCTIONS

- The Daily Practice Problem Sheet contains 30 MCQ's. For each question only one option is correct. Darken the correct circle/ bubble in the Response Grid provided on each page.
- You have to evaluate your Response Grids yourself with the help of solution booklet.
- Each correct answer will get you 4 marks and 1 mark shall be deduced for each incorrect answer. No mark will be given/ deducted if no bubble is filled. Keep a timer in front of you and stop immediately at the end of 60 min.
- The sheet follows a particular syllabus. Do not attempt the sheet before you have completed your preparation for that syllabus. Refer syllabus sheet in the starting of the book for the syllabus of all the DPP sheets.
- After completing the sheet check your answers with the solution booklet and complete the Result Grid. Finally spend time to analyse your performance and revise the areas which emerge out as weak in your evaluation.

DIRECTIONS (Q.1-Q.30) : There are 30 multiple choice questions. Each question has 4 choices (a), (b), (c) and (d), out of which ONLY ONE choice is correct.

1. Which of the enzyme is involved in the conversion of 3PGA to 3PGAL?
   (a) Triose phosphate dehydrogenase
   (b) Carboxydismutase
   (c) Phosphatase
   (d) Aldolase
2. First stable product of Calvin cycle has
   (a) 2 carbon atoms (b) 3 carbon atoms
   (c) 4 carbon atoms (d) 6 carbon atoms
3. Which one of the following is the product of dark reaction of photosynthesis?
   (a) CO_2
   (b) *ATP*
   (c) Pyruvic acid
   (d) Phosphoglyceraldehyde
4. In C_4 plants, CO_2 combine with PEP in presence of
   (a) PEP carboxylase
   (b) RuBP carboxylase
   (c) RuBP oxygenase
   (d) hydrogenase

RESPONSE GRID	1. ⓐⓑⓒⓓ	2. ⓐⓑⓒⓓ	3. ⓐⓑⓒⓓ	4. ⓐⓑⓒⓓ

Space for Rough Work

5. Dark reaction of photosynthesis is called so because
(a) it can also occur in dark.
(b) it does not require light energy.
(c) cannot occur during day time.
(d) it occurs more rapidly in night.

6. During photosynthesis when PGA is changed into Phosphoglyceraldehyde, which of the following reaction occur?
(a) Oxidation (b) Reduction
(c) Electrolysis (d) Hydrolysis

7. Ribulose diphosphate carboxylase enzyme catalyses the carboxylation reaction between
(a) Oxaloacetic acid and acetyl CoA.
(b) CO_2 and ribulose 1, 5 diphosphate.
(c) ribulose diphosphate and phosphoglyceraldehyde.
(d) PGA and dihydroxy acetone phosphate.

8. During dark reaction of photosynthesis
(a) water split.
(b) CO_2 is reduced to organic compounds.
(c) chlorophyll is activated.
(d) 6 carbon sugar is broken down into 3 carbon sugar.

9. In C_3 plants, the first stable product of photosynthesis during dark reaction is
(a) 3–phosphoglyceric acid
(b) phosphoglyceraldehyde
(c) malic acid
(d) oxaloacetic acid

10. How many Calvin cycle form one hexose molecule?
(a) 2 (b) 6
(c) 4 (d) 8

11. CO_2 acceptor in C_3 plants is
(a) xylulose–5–phosphate.
(b) 3–phosphoglyceric acid.
(c) ribulose 1, 5 diphosphate.
(d) phosphoenol pyruvic acid.

12. The first step in dark reaction of photosynthesis is
(a) formation of ATP.
(b) Ionization of water.
(c) attachment of CO_2 to a pentose sugar.
(d) excitement of electron of chlorophyll by a photon of light.

13. Which of the following is required for conversion of 3PGAL and dihydroxy acetone phosphate to fructose 1, 6 diphosphate?
(a) Transketolase
(b) Aldolase
(c) Phosphatase
(d) Hexokinase

14. The family in which many plants are C_4 type is
(a) malvaceae
(b) solanaceae
(c) crucifereae
(d) graminae

RESPONSE GRID					
	5. ⓐⓑⓒⓓ	6. ⓐⓑⓒⓓ	7. ⓐⓑⓒⓓ	8. ⓐⓑⓒⓓ	9. ⓐⓑⓒⓓ
	10. ⓐⓑⓒⓓ	11. ⓐⓑⓒⓓ	12. ⓐⓑⓒⓓ	13. ⓐⓑⓒⓓ	14. ⓐⓑⓒⓓ

Space for Rough Work

15. Which of the following enzyme is responsible for the conversion of oxaloacetic acid into malic acid?
 (a) PEP oxidase (b) PEP reductase
 (c) PEP dismutase (d) Malate dehydrogenase
16. In succulent xerophytes, there is accumulation of malic acid in night, this path of CO_2 metabolism is called
 (a) beta carboxylation.
 (b) Hatch and slack cycle.
 (c) Crassulacean acid metabolism.
 (d) Calvin cycle.
17. In case of C_4 pathway
 (a) CO_2 combines with PGA.
 (b) CO_2 combines with PEP.
 (c) CO_2 combines with RuDP.
 (d) CO_2 combines with RMP.
18. In C_4 plants, Calvin cycle
 (a) occurs in stroma of bundle sheath chloroplast.
 (b) occurs in mesophyll chloroplast.
 (c) occurs in grana of bundle sheath chloroplast.
 (d) does not occur as CO_2 is fixed mainly by PEP and no CO_2 is left for Calvin cycle.
19. Which of the following is CO_2 acceptor in C_4 plants?
 (a) Phosphoenol pyruvate (PEP)
 (b) Ribulose 1, 5 diphosphate (RuDP)
 (c) Oxaloacetic acid (OAA)
 (d) Phosphoglyceric acid (PGA)
20. C_4 plants are adapted to
 (a) hot and dry climate (b) temperate climate
 (c) cold and dry climate (d) hot and humid climate
21. Which one of the following is C_4 plant?
 (a) *Spirogyra* (b) *Pinus*
 (c) *Sorghum* (d) *Funaria*
22. In Hatch and Slack pathway
 (a) have chloroplast of same type.
 (b) have in Kranz anatomy where mesophyll have small chloroplast whereas bundle sheath have agranal chloroplast.
 (c) have kranz anatomy where mesophyll and bundle sheath both have small chloroplast.
 (d) have kranz anatomy where mesophyll cell are diffused.
23. Photorespiration takes place in
 (a) chloroplast, mitochondria
 (b) mitochondria, peroxisome
 (c) chloroplasts, peroxisome, mitochondria
 (d) chloroplasts, cytoplasm, mitochondria
24. Which one is false about Kranz anatomy?
 (a) Bundle sheath have large chloroplast and less developed grana.
 (b) Mesophyll cells have large chloroplast and more developed grana.
 (c) It is found in *Atriplex*, sugarcane, maize.
 (d) Plant having it have better photosynthesizing power than C_3 plants.

RESPONSE GRID					
	15. ⓐⓑⓒⓓ	16. ⓐⓑⓒⓓ	17. ⓐⓑⓒⓓ	18. ⓐⓑⓒⓓ	19. ⓐⓑⓒⓓ
	20. ⓐⓑⓒⓓ	21. ⓐⓑⓒⓓ	22. ⓐⓑⓒⓓ	23. ⓐⓑⓒⓓ	24. ⓐⓑⓒⓓ

Space for Rough Work

25. Which of the following bacteria grow on isopropyl alcohol and convert it into acetone?
(a) Fermentative bacteria
(b) Chemosynthetic bacteria
(c) Photosynthetic purple non-sulphur bacteria
(d) Nitrifying bacteria

26. Compensation point is
(a) where there is neither photosynthesis nor respiration.
(b) when rate of photosynthesis is equal to the rate of respiration.
(c) when entire food synthesized into photosynthesis remain utilized.
(d) when there is enough water just to meet the requirements of plant.

27. Which of the following wavelength of light is absorbed maximum for photosynthesis?
(a) Red light (b) Blue light
(c) Green light (d) Yellow light

28. Which of the following conditions are favourable for cyclic photophosphorylation?
(a) Anaerobic condition
(b) Aerobic and optimum light
(c) Aerobic and low light intensity
(d) Anaerobic and low light intensity

29. Blackman's law of limiting factor is applied to
(a) growth (b) respiration
(c) transpiration (d) photosynthesis

30. Which of the following inhibits O_2 release in light phase?
(a) PMA (b) Zeatin
(c) DCMU (d) None of these

RESPONSE GRID					
	25. ⓐⓑⓒⓓ	26. ⓐⓑⓒⓓ	27. ⓐⓑⓒⓓ	28. ⓐⓑⓒⓓ	29. ⓐⓑⓒⓓ
	30. ⓐⓑⓒⓓ				

DAILY PRACTICE PROBLEM SHEET 28 - BIOLOGY			
Total Questions	30	Total Marks	120
Attempted		Correct	
Incorrect		Net Score	
Cut-off Score	36	Qualifying Score	60
Success Gap = Net Score – Qualifying Score			
Net Score = (Correct × 4) – (Incorrect × 1)			

Space for Rough Work

DPP - Daily Practice Problems

Name :   **Date :**

Start Time :   **End Time :**

BIOLOGY 29

SYLLABUS : Respiration in Plants

Max. Marks : 120   **Time : 60 min.**

GENERAL INSTRUCTIONS

- The Daily Practice Problem Sheet contains 30 MCQ's. For each question only one option is correct. Darken the correct circle/ bubble in the Response Grid provided on each page.
- You have to evaluate your Response Grids yourself with the help of solution booklet.
- Each correct answer will get you 4 marks and 1 mark shall be deduced for each incorrect answer. No mark will be given/ deducted if no bubble is filled. Keep a timer in front of you and stop immediately at the end of 60 min.
- The sheet follows a particular syllabus. Do not attempt the sheet before you have completed your preparation for that syllabus. Refer syllabus sheet in the starting of the book for the syllabus of all the DPP sheets.
- After completing the sheet check your answers with the solution booklet and complete the Result Grid. Finally spend time to analyse your performance and revise the areas which emerge out as weak in your evaluation.

DIRECTIONS (Q.1-Q.30) : There are 30 multiple choice questions. Each question has 4 choices (a), (b), (c) and (d), out of which ONLY ONE choice is correct.

1. During anaerobic conditions, the rate of glycolysis increases is called as
   (a) compensation point (b) extinction point
   (c) Warburg effect (d) Pasteur effect

2. The incomplete breakdown of sugars in anaerobic respiration results in the formation of
   (a) fructose and water (b) glucose and CO_2
   (c) alcohol and CO_2 (d) water and CO_2

3. When a molecule of pyruvic acid is subjected to anaerobic oxidation and forms lactic acid, there is
   (a) loss of 3 ATP molecules.
   (b) loss of 6 ATP molecules.
   (c) gain of 2 ATP molecules.
   (d) gain of 4 ATP molecules.

4. Conversion of pyruvic acid into ethyl alcohol is facilitated by the enzymes
   (a) carboxylase
   (b) dehydrogenase
   (c) decarboxylase and dehydrogenase
   (d) phosphatase

RESPONSE GRID	1. ⓐⓑⓒⓓ	2. ⓐⓑⓒⓓ	3. ⓐⓑⓒⓓ	4. ⓐⓑⓒⓓ

Space for Rough Work

5. During anaerobic respiration in yeast
(a) Water and CO_2 are end products.
(b) CO_2, C_2H_5OH and energy are end products.
(c) H_2S, $C_6H_{12}O_6$ and energy are the end products.
(d) H_2O, CO_2 and energy are the only end products.

6. Anaerobic respiration takes place in the
(a) mitochondria (b) cytoplasm
(c) lysosomes (d) ER

7. In anaerobic respiration in plants
(a) oxygen is taken in.
(b) oxygen is given out.
(c) carbon dioxide is given out.
(d) carbon dioxide is taken in.

8. Which one of the following reactions represents anaerobic respiration?
(a) $6CO_2 + 6H_2 \rightarrow C_6H_{12}O_6 + 3O_2$
(b) $C_6H_{12}O_6 + 6O_2 \rightarrow CO_2 + H_2O + \text{energy}$
(c) $C_6H_{12}O_6 \rightarrow 2C_2H_5OH + 2CO_2 + \text{energy}$
(d) $6CO_2 + 6H_2O \rightarrow C_6H_{12}O_6 + 6H_2O + 3O_2$

9. In both aerobic and anaerobic respiration which same product is formed?
(a) Lactic acid
(b) Pyruvic acid
(c) Citric acid
(d) Organic acid

10. Pneumatophores are responsible for a kind of respiration mainly present in
(a) hydrophytes (b) mangroove plants
(c) epiphytes (d) xerophytes

11. If a starved plant is provided with glucose, the rate of respiration would
(a) decrease (b) increase
(c) become constant (d) first rise and then fall

12. Oxidative phosphorylation is related to the formation of
(a) ATP by chloroplast *via* photosynthesis.
(b) ATP by chondriosome *via* photolysis.
(c) ATP by cytoplasm *via* glycolysis.
(d) ATP by chondriosome *via* respiration.

13. The energy yield as a result of total oxidation of one glucose molecule during cellular respiration is to convert
(a) 38 molecules of ADP into 38 molecules of ATP.
(b) 30 molecules of ADP into 30 molecules of ATP.
(c) 36 molecules of ADP into 36 molecules of ATP.
(d) 32 molecules of ADP into 32 molecules of ADP.

14. Aerobic respiration is more advantageous to a large organism than anaerobic respiration, because aerobic respiration
(a) does not require sunlight.
(b) produces oxygen as a waste product.
(c) does not require molecular oxygen and hydrogen.
(d) releases more energy from an equal amount of nutrients.

RESPONSE GRID					
	5. ⓐⓑⓒⓓ	6. ⓐⓑⓒⓓ	7. ⓐⓑⓒⓓ	8. ⓐⓑⓒⓓ	9. ⓐⓑⓒⓓ
	10. ⓐⓑⓒⓓ	11. ⓐⓑⓒⓓ	12. ⓐⓑⓒⓓ	13. ⓐⓑⓒⓓ	14. ⓐⓑⓒⓓ

Space for Rough Work

15. In oxidative phosphorylation, oxidation and phosphorylation take place simultaneously and form
(a) NADP (b) DPN
(c) Pyruvic acid (d) ATP

16. The aerobic respiration yields
(a) $8NADH_2, 2FADH_2, 2ATP$
(b) $10NADH_2, 2FADH_2, 38ATP$
(c) $12NADH_2, 30ATP, H_2O$
(d) $10NADH_2, 2FADH_2, 2GTP, 2ATP$

17. Which of the following forms the connecting link between glycolysis and Kreb's cycle?
(a) Glucose (b) Ethyl alcohol
(c) Lactic acid (d) Pyruvic acid

18. Ganong's respiroscope is used to demonstrate ....
(a) production of carbon dioxide during aerobic respiration.
(b) production of heat during aerobic respiration.
(c) evolution of oxygen during photosynthesis.
(d) evolution of carbon dioxides during fermentation.

19. ATP stands for
(a) adenosine triphosphate
(b) adenine diphosphate
(c) adenosine tetraphosphate
(d) adenine triphosphate

20. Which is common in respiration and photosynthesis?
(a) Oxidation
(b) Phytochrome
(c) Cytochrome
(d) Photosystem

21. Conversion of fructose-6-phosphate to fructose 1, 6 diphosphate is catalyzed by
(a) phosphofructose kinase
(b) aldolase
(c) hexokinase
(d) none of these

22. Besides the net gain of 2 ATP molecules in glycolysis which other molecules are simultaneously formed ?
(a) $FADH_2$ (b) $NADPH_2$
(c) $NADH_2$ (d) $FAMH_2$

23. In glycolysis, the end product is
(a) protein which is converted to glucose.
(b) glucose which is converted into fructose.
(c) starch which is converted into glucose.
(d) glucose which is converted into pyruvic acid.

24. Glycolysis occurs in
(a) cytoplasm (b) aerobic respiration
(c) anaerobic respiration (d) all of these

RESPONSE GRID					
	15. ⓐⓑⓒⓓ	16. ⓐⓑⓒⓓ	17. ⓐⓑⓒⓓ	18. ⓐⓑⓒⓓ	19. ⓐⓑⓒⓓ
	20. ⓐⓑⓒⓓ	21. ⓐⓑⓒⓓ	22. ⓐⓑⓒⓓ	23. ⓐⓑⓒⓓ	24. ⓐⓑⓒⓓ

Space for Rough Work

25. Which one of the following is the first step of glycolysis?
(a) Breakdown of glucose.
(b) Phosphorylation of glucose.
(c) Conversion of glucose into fructose.
(d) Dehydrogenation of glucose.

26. Which is not true for glycolysis?
(a) End product is CO_2, H_2O
(b) Substrate level phosphorylation
(c) Production of ATP
(d) Expenditure of ATP

27. Between which of the following stages GTP is formed by substrate level phosphorylation ?
(a) Succinate to fumarate
(b) Ketoglutarate to succinate
(c) Oxalosuccinate to glutarate
(d) Fumarate to malate

28. Kreb's cycle starts with the formation of six carbon compound by reaction between
(a) malic acid and acetyl CoA.
(b) succinic acid and pyruvic acid.
(c) fumaric acid and pyruvic acid.
(d) acetyl CoA and oxaloacetic acid.

29. The order of occurrence of the cytochromes in the F_1 particles is
(a) cyt b, cyt c, cyt a-cyt a_3
(b) cyt c, cyt b, cyt a-cyta_3
(c) cyt a, cyt b, cyt c-cyt a_3
(d) cyt a_3-cyt a, cyt c, cyt b

30. When an unripe banana is sealed in a polythene bag, it remains green for many days. But if an apple is also sealed in the same bag, the banana ripens and turns yellow within a few days. The reason is that apple
(a) removes O_2 released by the banana and thus promotes ripening.
(b) produces CO_2 which promotes ripening.
(c) removes CO_2 which inhibits ripening.
(d) releases ethylene which promotes ripening.

RESPONSE GRID	25. ⓐⓑⓒⓓ	26. ⓐⓑⓒⓓ	27. ⓐⓑⓒⓓ	28. ⓐⓑⓒⓓ	29. ⓐⓑⓒⓓ
	30. ⓐⓑⓒⓓ				

DAILY PRACTICE PROBLEM SHEET 29 - BIOLOGY			
Total Questions	30	Total Marks	120
Attempted		Correct	
Incorrect		Net Score	
Cut-off Score	32	Qualifying Score	56
Success Gap = Net Score – Qualifying Score			
Net Score = (Correct × 4) – (Incorrect × 1)			

Space for Rough Work

DPP - Daily Practice Problems

Name : Date :

Start Time : End Time :

BIOLOGY 30

SYLLABUS : Plant Growth and Development

Max. Marks : 120 **Time : 60 min.**

GENERAL INSTRUCTIONS

- The Daily Practice Problem Sheet contains 30 MCQ's. For each question only one option is correct. Darken the correct circle/ bubble in the Response Grid provided on each page.
- You have to evaluate your Response Grids yourself with the help of solution booklet.
- Each correct answer will get you 4 marks and 1 mark shall be deduced for each incorrect answer. No mark will be given/ deducted if no bubble is filled. Keep a timer in front of you and stop immediately at the end of 60 min.
- The sheet follows a particular syllabus. Do not attempt the sheet before you have completed your preparation for that syllabus. Refer syllabus sheet in the starting of the book for the syllabus of all the DPP sheets.
- After completing the sheet check your answers with the solution booklet and complete the Result Grid. Finally spend time to analyse your performance and revise the areas which emerge out as weak in your evaluation.

DIRECTIONS (Q.1-Q.30) : There are 30 multiple choice questions. Each question has 4 choices (a), (b), (c) and (d), out of which ONLY ONE choice is correct.

1. Where would you look for active cell division in plant?
   (a) In the pith cells.
   (b) In the cells of cortex.
   (c) In the internodal region.
   (d) At the tip of root and shoot.
2. Etiolation in plants is caused when they
   (a) are grown in dark.
   (b) have mineral deficiency.
   (c) are grown in intense light.
   (d) are grown in blue light.
3. Legume seeds exhibit dormancy because of
   (a) undeveloped embryos.
   (b) hard seed coat.
   (c) absence of cytokinins.
   (d) absence of gibberellic acid.
4. The correct sequence of cellular growth stages is
   (a) division $\rightarrow$ differentiation $\rightarrow$ elongation.
   (b) division $\rightarrow$ elongation $\rightarrow$ differentiation.
   (c) differentiation $\rightarrow$ division $\rightarrow$ elongation.
   (d) elongation $\rightarrow$ differentiation $\rightarrow$ division.

RESPONSE GRID 1. ⓐⓑⓒⓓ 2. ⓐⓑⓒⓓ 3. ⓐⓑⓒⓓ 4. ⓐⓑⓒⓓ

Space for Rough Work

5. To remove seed dormancy by mechanical removing of seed coat is called
(a) stratification (b) scarification
(c) vernalization (d) photoperiodism

6. Why the newly harvested potato tubers do not germinate even when placed in favourable conditions?
(a) Due to dormancy.
(b) Due to lack of water absorption.
(c) Due to difficulty of light penetration.
(d) Due to lack of photosynthetic apparatus.

7. Leaf fall occurs on abscission layer is formed when the content of
(a) auxin increases.
(b) auxin decreases.
(c) abscisic acid decreases.
(d) gibberellic acid decreases.

8. Which of the following movements is not related to change in auxin levels?
(a) Nyctinastic leaf movement.
(b) Movement of roots towards soil.
(c) Movement of sunflower tracking the direction of the sun.
(d) Movement of shoot towards light.

9. Which of the following is not naturally occurring plant hormone?
(a) 2, 4-D (b) GA_2
(c) Gibberellin (d) IAA

10. Bioassay for auxin is
(a) *Avena* curvature test
(b) green leaf test
(c) dwarf maize test
(d) cell division test

11. Flowering in pineapple is promoted by
(a) NAA
(b) GA_3
(c) ethylene
(d) cytokinin

12. Differentiation of shoot is controlled by
(a) high gibberellin : cytokinin ratio.
(b) high cytokinin: auxin ratio.
(c) high auxin : cytokinin ratio.
(d) high gibberellin : auxin ratio.

13. Specific property attributed to gibberellin is
(a) shortening of genetically tall plants.
(b) elongation of genetically dwarf plant.
(c) promotion of rooting.
(d) yellowing of young leaves.

14. Bioassay for gibberellin is
(a) green leaf test/richmond–Lang effect.
(b) cell division/bud induction test.
(c) dwarf maize/barley aleurone test.
(d) *Avena* curvature test.

Response Grid	5. ⓐⓑⓒⓓ	6. ⓐⓑⓒⓓ	7. ⓐⓑⓒⓓ	8. ⓐⓑⓒⓓ	9. ⓐⓑⓒⓓ
	10. ⓐⓑⓒⓓ	11. ⓐⓑⓒⓓ	12. ⓐⓑⓒⓓ	13. ⓐⓑⓒⓓ	14. ⓐⓑⓒⓓ

Space for Rough Work

15. The activity of α-amylase in the endosperm of barley germinating seed is induced by
    (a) ethylene
    (b) cytokinin
    (c) IAA
    (d) gibberellin
16. An excised leaf does not turn yellow if it is induced to root. This is attributed to synthesis in root of **or** Leaf ageing is retarded by
    (a) ethylene (b) cytokinins
    (c) gibberellins (d) auxins
17. Senescence is inhibited by
    (a) ethylene
    (b) gibberellic acid
    (c) cytokinin
    (d) abscissic acid
18. Which one of the following responses of plants to growth hormones is true for ethylene?
    (a) Increase in cell elongation.
    (b) Decrease in the formation of female flowers.
    (c) Increase in ripening of fruits.
    (d) Decrease in abscission of flowers.
19. Which combination of gases is suitable for fruit ripening?
    (a) 80% C_2H_4 and 20% CO_2
    (b) 80% CO_2 and 20% CH_2
    (c) 80% CH_4 and 20% CO_2
    (d) 80% CO_2 and 20% O_2
20. Presence of which of the following in seed is associated with dormancy?
    (a) Starch
    (b) Abscisic acid
    (c) Ethylene
    (d) Gibberellic acid
21. ABA is involved in
    (a) dormancy of seeds
    (b) root elongation
    (c) shoot elongation
    (d) increased cell division
22. Abscisic acid (ABA) is a stress related hormone and it also
    (a) aids in fruit development.
    (b) accelerates fruit ripening.
    (c) stimulates mitosis.
    (d) mediated stomatal closure.
23. Pigment phytochrome is involved in
    (a) phototropism (b) photorespiration
    (c) photoperiodism (d) geotropism
24. The pigment involved in red-far red light interconversion is
    (a) cytochrome
    (b) xanthophyll
    (c) lycopene
    (d) phytochrome

RESPONSE GRID					
	15. ⓐⓑⓒⓓ	16. ⓐⓑⓒⓓ	17. ⓐⓑⓒⓓ	18. ⓐⓑⓒⓓ	19. ⓐⓑⓒⓓ
	20. ⓐⓑⓒⓓ	21. ⓐⓑⓒⓓ	22. ⓐⓑⓒⓓ	23. ⓐⓑⓒⓓ	24. ⓐⓑⓒⓓ

Space for Rough Work

25. When the dark period of short day plants is interrupted by brief exposure of light, then the plant
   (a) will not flower at all.
   (b) flower immediately.
   (c) give more flowers.
   (d) turn into a long day plant.
26. A hypothetical chemical involved in the flowering of plants is **or** Chemical agent which has important role in the flowering is
   (a) gibberellin (b) kinetin
   (c) indole acetic acid (d) florigen
27. In SDP flowering does not occur
   (a) when intermediate light is given red $\rightarrow$ far red $\rightarrow$ red.
   (b) when interrupted by a flash of far light.
   (c) when interrupted by a flash red $\rightarrow$ far light.
   (d) all of the above.
28. A potted plant placed near the window bends outwardly due to
   (a) greater oxygen availability to the tip.
   (b) more auxin content on the shaded side.
   (c) greater light availability to tip.
   (d) availability of necessary warmth to the tip.
29. On touching' the leaves of *Mimosa pudica* droop down because of
   (a) seismonasty (b) hydrotropism
   (c) chemonasty (d) thigmotoropism
30. Example of positive geotropism is
   (a) closing of flowers.
   (b) upward growth of stem.
   (c) downward growth of root.
   (d) lateral growth of root.

RESPONSE GRID	
	25. ⓐⓑⓒⓓ 26. ⓐⓑⓒⓓ 27. ⓐⓑⓒⓓ 28. ⓐⓑⓒⓓ 29. ⓐⓑⓒⓓ
	30. ⓐⓑⓒⓓ

DAILY PRACTICE PROBLEM SHEET 30 - BIOLOGY			
Total Questions	30	Total Marks	120
Attempted		Correct	
Incorrect		Net Score	
Cut-off Score	36	Qualifying Score	64
Success Gap = Net Score – Qualifying Score			
Net Score = (Correct × 4) – (Incorrect × 1)			

Space for Rough Work

DPP - Daily Practice Problems

Name :   Date :

Start Time :   End Time :

BIOLOGY 31

SYLLABUS : Digestion and Absorption – PART I: Digestive system/Digestive glands, Physiology of digestion

Max. Marks : 120   **Time : 60 min.**

GENERAL INSTRUCTIONS

- The Daily Practice Problem Sheet contains 30 MCQ's. For each question only one option is correct. Darken the correct circle/ bubble in the Response Grid provided on each page.
- You have to evaluate your Response Grids yourself with the help of solution booklet.
- Each correct answer will get you 4 marks and 1 mark shall be deduced for each incorrect answer. No mark will be given/ deducted if no bubble is filled. Keep a timer in front of you and stop immediately at the end of 60 min.
- The sheet follows a particular syllabus. Do not attempt the sheet before you have completed your preparation for that syllabus. Refer syllabus sheet in the starting of the book for the syllabus of all the DPP sheets.
- After completing the sheet check your answers with the solution booklet and complete the Result Grid. Finally spend time to analyse your performance and revise the areas which emerge out as weak in your evaluation.

DIRECTIONS (Q.1-Q.30) : There are 30 multiple choice questions. Each question has 4 choices (a), (b), (c) and (d), out of which ONLY ONE choice is correct.

1. The length of the alimentary canal is more in herbivorous animals than the carnivorous because
   (a) herbivorous diet contains more fat to digest.
   (b) herbivorous diet contains more proteins to digest.
   (c) herbivorous diet contains more carbohydrates particularly cellulose which takes more time to digest.
   (d) herbivorous diet contains more vitamins to digest.
2. Dental formula of human being is
   (a) I_2, C_2, P_1, M_3 (b) I_2, C_1, P_2, M_3
   (c) I_3, C_1, P_2, M_2 (d) I_2, C_2, P_3, M_1
3. The mucosal lining of the oesophagus of mammals is made up of
   (a) simple columnar epithelium.
   (b) squamous epithelium.
   (c) stratified cuboidal epithelium.
   (d) stratified columnar epithelium.
4. The epithelial cells lining the stomach of vertebrates is protected from damage by HCl because
   (a) HCl is too dilute.
   (b) the epithelial cells are resistant to the action of HCl.
   (c) HCl is neutralised by alkaline gastric juice.
   (d) the epithelial cells are covered with a mucous secretion.

RESPONSE GRID	1. ⓐⓑⓒⓓ	2. ⓐⓑⓒⓓ	3. ⓐⓑⓒⓓ	4. ⓐⓑⓒⓓ

Space for Rough Work

5. Brunner's gland are found in which of the following layers?
(a) Submucosa of stomach
(b) Mucosa of ileum
(c) Submucosa of duodenum
(d) Mucosa of oesophagus

6. The vermiform appendix is situated at the junction of
(a) large intestine and small intestine.
(b) small intestine and duodenum.
(c) anus and rectum .
(d) stomach and duodenum.

7. Intestinal villi are more numerous and larger in posterior part of small intestine than in anterior part because
(a) digestion is faster in posterior part.
(b) blood supply is poor in posterior part.
(c) there is more digested food in posterior part.
(d) blood supply is rich in posterior part.

8. Lacteals are central lymph vessels which are found in
(a) liver (b) pancreas
(c) villi (d) spleen

9. In rabbit, the digestion of cellulose takes place in
(a) colon (b) lleum
(c) caecum (d) rectum

10. Parotid salivary gland are present
(a) below the tongue
(b) below the ear
(c) below the eye orbit
(d) in the angle between two jaws

11. Saccus entericus is the name given to
(a) junction between ileum and large intestine.
(b) intestinal juice.
(c) swelling in the gut.
(d) appendix.

12. Saliva is secreted from
(a) submaxillary and sublingual glands
(b) gastric gland
(c) pancreas
(d) gall bladder

13. Which of the following shows correct pairing for salivary glands in man?
(a) Parotid → Wharton's duct; submaxillary → duct of Stensen; sublingual → duct of Rivinus.
(b) Parotid → duct of Rivinus; submaxillary → duct of Stensen; sublingual → Wharton's duct.
(c) Parotid → duct of Rivinus; submaxillary → Wharton's duct; sublingual → duct of Stensen.
(d) Parotid → duct of Stensen; submaxillary → Wharton's duct; sublingual → duct of Rivinus.

14. Gastric juice contains
(a) pepsin, rennin, lipase
(b) pepsin, amylase, rennin
(c) pepsin, amylase, trypsin
(d) lipase, rennin, trypsin

RESPONSE GRID					
	5. ⓐⓑⓒⓓ	6. 	7. ⓐⓑⓒⓓ	8. ⓐⓑⓒⓓ	9. ⓐⓑⓒⓓ
	10. ⓐⓑⓒⓓ	11. ⓐⓑⓒⓓ	12. ⓐⓑⓒⓓ	13. ⓐⓑⓒⓓ	14. ⓐⓑⓒⓓ

Space for Rough Work

15. Which one of the following does not produce any digestive enzyme?
(a) Intestinal mucosa (b) Gastric mucosa
(c) Pancreas (d) Liver

16. Spiral valves are found in
(a) walls of cystic ducts (spleen).
(b) walls of cystic duct (gall bladder).
(c) hepatic duct (liver).
(d) common bile duct.

17. Match the names of glands listed under column-I with the location given under column-II and choose the answer which gives correct combination from the given options.

	Column-I (glands)		Column-II (location)
A	Crypts of Lieberkuhn	p	Loop of duodenum
B	Pancreas	q	Stomach
C	Adrenal gland	r	Intestine
D	Gastric gland	s	Kidney

(a) $A = r, B = p, C = q, D = s$
(b) $A = r, B = p, C = s, D = q$
(c) $A = q, B = s, C = r, D = p$
(d) $A = p, B = r, C = s, D = q$

18. Which one of the following statements about glycogen is correct?
(a) It is a disaccharide stored in liver which can react with ammonia to form proteins.
(b) It is synthesized in the liver and takes part in the formation of bile and lipase, besides being a source of energy.
(c) It is a polysaccharide which is synthesized and stored in liver cells.
(d) It is synthesized in blood and stored in liver and muscles to provide glucose in times of need.

19. Which of the following shows the correct sequence of events during digestion of protein ?
(a) Proteins → peptones → acid metaproteins and peptides.
(b) Proteins → proteoses and peptones → peptides → amino acids.
(c) Proteins → acid metaproteins → proteoses → amino acids → peptides.
(d) Proteins → primary proteins → peptides → amino acids.

20. Digestion of protein is necessary because
(a) proteins are not absorbed as such.
(b) proteins are large molecules.
(c) proteins have complex structure.
(d) proteins are made up of amino acids.

21. Which of the following is not present in pancreatic juice?
(a) Trypsinogen (b) Chymotrypsin
(c) Pepsin (d) Lipase

22. The function of erepsin in the process of digestion is to convert
(a) fats into amino acids.
(b) proteins into amino acids.
(c) polysaccharides into disaccharides.
(d) disaccharides into monosaccharides.

23. Bile salts are poured into the alimentary canal where they are necessary for the absorption of
(a) Na^+ and Ca^{++}
(b) fat soluble vitamins.
(c) amino acids and monosaccharides.
(d) all the nutrients contained in chyme.

24. Essentially the word 'digestion' means
(a) burning of food (b) oxidation of food
(c) hydrolysis of food (d) breakdown of food

RESPONSE GRID					
	15. ⓐⓑⓒⓓ	16. ⓐⓑⓒⓓ	17. ⓐⓑⓒⓓ	18. ⓐⓑⓒⓓ	19. ⓐⓑⓒⓓ
	20. ⓐⓑⓒⓓ	21. ⓐⓑⓒⓓ	22. ⓐⓑⓒⓓ	23. ⓐⓑⓒⓓ	24. ⓐⓑⓒⓓ

Space for Rough Work

25. Rennin acts on milk proteins and converts
    (a) caseingoen into casein
    (b) casein into paracasein
    (c) caseinogen into paracasein
    (d) paracasein into caseinogen
26. Just as hydrochloric acid is to pepsinogen, so is
    (a) enterokinase to trypsinogen
    (b) haemoglobin to oxygen
    (c) bile juice to fat
    (d) glucagon to glycogen
27. Duodenum has characteristic Brunner's glands which secrete two hormones called
    (a) prolactin, parathormone
    (b) estradoil, progesterone
    (c) kinase, estrogen
    (d) secretin, cholecystokinin
28. Bile salt
    (a) acts as emulsifying agent.
    (b) helps in absorption of fatty acids, cholesterol etc.
    (c) stimulates the bile production in liver.
    (d) all of the above
29. Salivary amylase is stimulated by
    (a) K^+ (b) N^+
    (c) HCO_3^- (d) Cl^-
30. The amount of bile released in proportion to the amount of
    (a) fat in meal
    (b) protein in meal
    (c) carbohydrate in meal
    (d) all of the above

RESPONSE GRID	25. ⓐⓑⓒⓓ	26. ⓐⓑⓒⓓ	27. ⓐⓑⓒⓓ	28. ⓐⓑⓒⓓ	29. ⓐⓑⓒⓓ
	30. ⓐⓑⓒⓓ				

DAILY PRACTICE PROBLEM SHEET 31 - BIOLOGY			
Total Questions	30	Total Marks	120
Attempted		Correct	
Incorrect		Net Score	
Cut-off Score	32	Qualifying Score	52
Success Gap = Net Score – Qualifying Score			
Net Score = (Correct × 4) – (Incorrect × 1)			

Space for Rough Work

DPP - Daily Practice Problems

Name : | Date :

Start Time : | End Time :

BIOLOGY 32

SYLLABUS : Digestion and Absorption - PART II : Gastrointestinal hormones/ Digestive enzymes/ Nutrition & Nutritional requirement

Max. Marks : 120 | **Time : 60 min.**

GENERAL INSTRUCTIONS

- The Daily Practice Problem Sheet contains 30 MCQ's. For each question only one option is correct. Darken the correct circle/ bubble in the Response Grid provided on each page.
- You have to evaluate your Response Grids yourself with the help of solution booklet.
- Each correct answer will get you 4 marks and 1 mark shall be deduced for each incorrect answer. No mark will be given/ deducted if no bubble is filled. Keep a timer in front of you and stop immediately at the end of 60 min.
- The sheet follows a particular syllabus. Do not attempt the sheet before you have completed your preparation for that syllabus. Refer syllabus sheet in the starting of the book for the syllabus of all the DPP sheets.
- After completing the sheet check your answers with the solution booklet and complete the Result Grid. Finally spend time to analyse your performance and revise the areas which emerge out as weak in your evaluation.

DIRECTIONS (Q.1-Q.30) : There are 30 multiple choice questions. Each question has 4 choices (a), (b), (c) and (d), out of which ONLY ONE choice is correct.

1. Which of the following statement is correct?
   (a) Though secretin is an enzyme, it is not involved in digestion.
   (b) Secretin is an enzyme and so it helps in digestion.
   (c) Secretin is a hormone but it plays a role in digestion.
   (d) Secretin is a hormone and hence it does not play any role in digestion.
2. The enzyme added to the starch solution may have been
   (a) maltase (b) invertase
   (c) lipase (d) amylase
3. Which group contains biocatalysts?
   (a) Peptidase, amylase, rennin
   (b) Myosin, oxytocin, adrenaline
   (c) Rhodopsin, pepsin, steapsin
   (d) Glucose, amino acids, fatty acids
4. Which of these is not an enzyme of digestive system?
   (a) Enterokinase
   (b) Amylase
   (c) Trypsin
   (d) Enterogastrone

RESPONSE GRID 1. ⓐⓑⓒⓓ 2. ⓐⓑⓒⓓ 3. ⓐⓑⓒⓓ 4. ⓐⓑⓒⓓ

Space for Rough Work

5. Which one of the following four secretions is correctly matched with its source, target and nature of action?

	Secretion	Source	Target	Action
(a)	Gastrin	Stomach lining	Oxyntic cells	Production of HCl
(b)	Inhibin	Sertoli cells	Hypothalamus	Inhibition of secretion of gonadotropin releasing hormone
(c)	Enterokinase	Duodenum	Gall bladder	Release of bile juice
(d)	Atrial Natriuretic Factor (ANF)	Sinu atrial node (SAN) M-cells of Atria	Juxta-glomerular apparatus (JGA)	Inhibition of release of renin

6. Cholecystokinin is a secretion of
   (a) stomach which stimulates pancreas to release the pancreatic juice.
   (b) liver synthesized from cholesterol and controls secondary sexual characters.
   (c) duodenum and makes the gall bladder to contract and release bile.
   (d) goblet cells of ileum and stimulates the secretion of succus entericus.
7. Trypsin converts
   (a) fats into fatty acids.
   (b) starch and glycogen into maltose.
   (c) proteins into peptones.
   (d) sucrose into glucose and fructose.
8. Enzymes, vitamins and hormones can be classified into a single category of biological chemicals, because all of them
   (a) are proteins.
   (b) enhance the oxidative metabolism.
   (c) aid the regulating mechanism.
   (d) are synthesized within the body of an organism.
9. What is common among amylase, rennin and trypsin?
   (a) These all are proteins.
   (b) These all are proteolytic enzymes.
   (c) These are product in stomach.
   (d) These act at a pH lower than 7.
10. The contraction of gall bladder is due to
   (a) Gastrin (b) Secretin
   (c) Cholecystokinin (d) Enterogastrone
11. Which of the following digest proteins into peptides?
   (a) Erepsin (b) Rennin
   (c) Pepsin (d) Lipase
12. Glucose, galactose and fructose all have the same molecular size and composition and their absorption through the mucosal cells takes place
   (a) at the same rate
   (b) glucose is absorbed most rapidly
   (c) fructose is absorbed most rapidly
   (d) galactose is absorbed most rapidly
13. What is common between amino acids, fatty acids and glycerol?
   (a) These are all rich source of calories.
   (b) These are the builders of protoplasm.
   (c) These are the end products of digestion of two categories of food constituents.
   (d) These can be stored in the form of fat.
14. During prolonged fasting
   (a) the first to be used up are carbohydrates, next fat is withdrawn and proteins are metabolised at the last.
   (b) the first to be used up are the fats, next carbohydrates are withdrawn from stored glycogen in the liver and muscles and proteins are withdrawn at the last.
   (c) first lipids are used up, then proteins and finally carbohydrate.
   (d) none of the above.

RESPONSE GRID				
5. ⓐⓑⓒⓓ	6. ⓐⓑⓒⓓ	7. ⓐⓑⓒⓓ	8. ⓐⓑⓒⓓ	9. ⓐⓑⓒⓓ
10. ⓐⓑⓒⓓ	11. ⓐⓑⓒⓓ	12. ⓐⓑⓒⓓ	13. ⓐⓑⓒⓓ	14. ⓐⓑⓒⓓ

Space for Rough Work

15. Lactose is composed of
(a) glucose + fructose
(b) glucose + glucose
(c) glucose + galactose
(d) fructose + galactose

16. Digestion of protein takes place in
(a) duodenum and stomach
(b) stomach and oesophagus
(c) small and large intestine
(d) intestine and rectum

17. Which of the following vitamins is water soluble as well as an anti-oxidant?
(a) Vitamin B_1
(b) Vitamin A
(c) Vitamin D
(d) Vitamin C

18. Which vitamin gets destroyed by heating?
(a) Vitamin A
(b) Vitamin E
(c) Vitamin C
(d) Vitamin K

19. Synthesis of vitamin A in the body takes place in
(a) blood
(b) pancreas
(c) spleen
(d) liver

20. In human body cobalt helps in
(a) erythropoiesis and in the activities of certain enzymes.
(b) maintaining the teeth healthy.
(c) the prevention from body deformities.
(d) enzyme activation only.

21. A person suffers from beri-beri, rickets and scurvy if he is not taking adequate amounts of
(a) Vitamin B_{12}, A and C
(b) Vitamin B_1, D and C
(c) Vitamin A, B and E
(d) Vitamin B_6, and K

22. Vitamins are
(a) inorganic substances that cannot be synthesized by animal.
(b) inorganic substances that can be synthesized by animal.
(c) organic substances that cannot be synthesized by animal.
(d) organic substances that can be synthesized by animal.

23. A doctor advises a person to have more meat, butter, milk, eggs in his diet. The person is
(a) suffering from scurvy.
(b) cannot see in dark.
(c) suffering from rickets.
(d) suffering from kwashirokar disease.

24. Which vitamin is destroyed by ultraviolet rays?
(a) A (b) D
(c) E (d) K

RESPONSE GRID					
	15. ⓐⓑⓒⓓ	16. ⓐⓑⓒⓓ	17. ⓐⓑⓒⓓ	18. ⓐⓑⓒⓓ	19. ⓐⓑⓒⓓ
	20. ⓐⓑⓒⓓ	21. ⓐⓑⓒⓓ	22. ⓐⓑⓒⓓ	23. ⓐⓑⓒⓓ	24. ⓐⓑⓒⓓ

Space for Rough Work

25. Which of the following is the matching pair of the deficient nutrient and the resulting disease in which the patient develops swollen lips, thick pigmented skin of hands and legs and irritability?
(a) Iodine-goitre (b) Niacin-pellagra
(c) Thiamine-beri-beri (d) Protein-kwashiorkar

26. Continuous bleeding from an injured part of body is due to deficiency of
(a) vitamin A (b) vitamin B
(c) vitamin K (d) vitamin E

27. Balanced diet should have approximately
(a) 1/5 protein, 3/5 fat and 1/5 carbohydrate.
(b) 3/5 protein, 1/5 fat and 1/5 carbohydrate.
(c) 1/5 protein, 1/5 fat and 3/5 carbohydrate.
(d) 1/2 protein, 1/4 fat and 3/5 carbohydrate.

28. A patient of diabetes mellitus excretes glucose in urine even when he is kept in a carbohydrate free diet. It is because
(a) fats are catabolised to form glucose.
(b) amino acids are catabolised in liver.
(c) amino acids are discharged in blood stream from liver.
(d) glycogen from muscles is released in blood stream.

29. Which of the following statements are true regarding beri-beri?
1. A crippling disease prevalent among the native population of sub-Saharan Africa.
2. A deficiency disease caused by lack of thiamine (vitamin B_1).
3. A nutritional disorder in infants and young children when the diet is persistently deficient in essential protein.
4. Occurs in those countries where the staple diet is polished rice.
5. The symptoms are pain from neuritis, paralysis, muscle wasting, progressive oedema, mental deterioration and finally heart failure.
(a) 2, 4 and 5 (b) 1, 2 and 4
(c) 1, 3 and 5 (d) 2, 3 and 5

30. The richest sources of vitamin B_{12} are
(a) rice and hen's egg.
(b) carrot and chicken's breast.
(c) goat's liver and *Spirulina.*
(d) chocolate and green gram.

RESPONSE GRID	25.ⓐⓑⓒⓓ 26.ⓐⓑⓒⓓ 27.ⓐⓑⓒⓓ 28.ⓐⓑⓒⓓ 29.ⓐⓑⓒⓓ 30.ⓐⓑⓒⓓ

DAILY PRACTICE PROBLEM SHEET 32 - BIOLOGY			
Total Questions	30	Total Marks	120
Attempted		Correct	
Incorrect		Net Score	
Cut-off Score	32	Qualifying Score	56
Success Gap = Net Score – Qualifying Score			
Net Score = (Correct × 4) – (Incorrect × 1)			

DPP - Daily Practice Problems

Name : ______ Date : ______

Start Time : ______ End Time : ______

BIOLOGY 33

SYLLABUS : Breathing and Exchange of gases

Max. Marks : 120 **Time : 60 min.**

GENERAL INSTRUCTIONS

- The Daily Practice Problem Sheet contains 30 MCQ's. For each question only one option is correct. Darken the correct circle/ bubble in the Response Grid provided on each page.
- You have to evaluate your Response Grids yourself with the help of solution booklet.
- Each correct answer will get you 4 marks and 1 mark shall be deduced for each incorrect answer. No mark will be given/ deducted if no bubble is filled. Keep a timer in front of you and stop immediately at the end of 60 min.
- The sheet follows a particular syllabus. Do not attempt the sheet before you have completed your preparation for that syllabus. Refer syllabus sheet in the starting of the book for the syllabus of all the DPP sheets.
- After completing the sheet check your answers with the solution booklet and complete the Result Grid. Finally spend time to analyse your performance and revise the areas which emerge out as weak in your evaluation.

DIRECTIONS (Q.1-Q.30) : There are 30 multiple choice questions. Each question has 4 choices (a), (b), (c) and (d), out of which ONLY ONE choice is correct.

1. In man and mammals, air passes from outside into the lungs through
   (a) nasal cavity, larynx, pharynx, trachea, bronchi, alveoli.
   (b) nasal cavity, larynx, pharynx, trachea, bronchioles, alveoli.
   (c) nasal cavity, pharynx, larynx, trachea, bronchioles, bronchi, alveoli.
   (d) nasal cavity, pharynx, larynx, trachea, bronchi, bronchioles, alveoli.
2. During forced expiration, actively contracting muscles include the
   (a) diaphragm
   (b) external intercostals
   (c) abdominal muscles and external intercostals
   (d) diaphragm and intestinal muscle
3. Carbon dioxide is transported *via* blood to lungs mostly
   (a) as carbaminohaemoglobin and as carbonic acid.
   (b) in the form of carbonic acid only.
   (c) in combination with haemoglobin only.
   (d) dissolved in blood plasma.
4. With respect to respiratory apparatus, pick out the incorrect pair/pairs.
   (a) Insects – Tracheoles
   (b) Amphibians – Skin and lungs
   (c) Starfish – Dermal bronchial
   (d) Crustacean – Lungs

RESPONSE GRID	1. ⓐⓑⓒⓓ	2. ⓐⓑⓒⓓ	3. ⓐⓑⓒⓓ	4. ⓐⓑⓒⓓ

Space for Rough Work

5. In mammals, the tracheal cartilaginous rings are
   (a) complete rings (b) incomplete rings
   (c) incomplete dorsally (d) incomplete laterally

6. Similarity between the trachea of cockroach and rabbit is that
   (a) both are paired and branched.
   (b) ciliated epithelium is present in both.
   (c) walls of both can not be deformed.
   (d) in both head originates from pharynx.

7. What is true about haemoglobin ?
   (a) It is a dipeptide and present in red blood corpuscles in bloodworm.
   (b) It is present in the dissolved state in blood plasma in earthworm.
   (c) It is a dipeptiede in mammals and localised in red blood corpuscles.
   (d) It is present in dissolved state in blood plasma in scorpions.

8. The vital capacity of the lung signifies the volume of air
   (a) breathed in during normal inspiration.
   (b) breathed out with forcible expiration.
   (c) breathed in with forcible inspiration.
   (d) with deep inspiration and forcible expiration.

9. In breathing, entering air becomes
   (a) warm and filtered.
   (b) cold and then warm.
   (c) warm, filtered and moistened.
   (d) cold and moistened.

10. The exchange of gases in the alveoli of the lungs takes place by
   (a) osmosis (b) simple diffusion
   (c) passive transport (d) active transport

11. Division of mammalian lungs into a very large number of tiny alveoli around alveolar ducts opening into bronchioles, is
   (a) an inefficient system of ventilation of alveoli though with very little residual air.
   (b) an inefficient system of ventilating the alveoli resulting in very high percentage of residual air in the lungs.
   (c) a very efficient system of ventilating the alveoli with no residual air.
   (d) an efficient system of ventilation of alveoli with little or no residual air.

12. Dissociation curve is connected with
   (a) oxyhaemoglobin (b) oxygen
   (c) carbon dioxide (d) none of the above

13. A small space called ________ lies in between the two lungs.
   (a) mediastinum
   (b) bronchioles
   (c) alveoli
   (d) cardiac notch

14. The site of gaseous exchange in lungs is
   (a) liver (b) alveoli
   (c) trachea (d) none of these

RESPONSE GRID					
	5. ⓐⓑⓒⓓ	6. ⓐⓑⓒⓓ 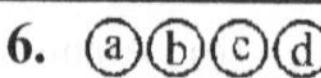	7. ⓐⓑⓒⓓ	8. ⓐⓑⓒⓓ	9. ⓐⓑⓒⓓ
	10. ⓐⓑⓒⓓ	11. ⓐⓑⓒⓓ	12. ⓐⓑⓒⓓ	13. ⓐⓑⓒⓓ	14. ⓐⓑⓒⓓ

Space for Rough Work

15. Oxyhaemoglobin is an unstable compound because
    (a) haemoglobin is a complex pigmented protein.
    (b) there is a physical bonding between oxygen and haemoglobin.
    (c) there is a chemical bonding between oxygen and haemoglobin.
    (d) one molecule of haemoglobin combines with four molecules of oxygen.
16. Respiration can be defined as
    (a) a catabolic process by which animal cells utilize carbon dioxide, produce oxygen and convert the released energy to ATP.
    (b) a catabolic process by which animals cells utilize oxygen, produce carbon dioxide and convert the released energy to ATP.
    (c) an anabolic process by which animal cells utilize oxygen and carbon dioxide to form ATP.
    (d) an anabolic process by which animal cells utilize oxygen, produce carbon dioxide and convert the released energy to ATP.
17. If O_2 concentration in tissue was almost as high as at the respiratory surface
    (a) oxyhaemoglobin would dissociate to supply to the tissue.
    (b) haemoglobin would combine with more O_2 at respiratory surface.
    (c) oxyhaemoglobin would not dissociate to supply O_2 to the tissue.
    (d) CO_2 will interfere the O_2 transport.
18. In lungs there is definite exchange of ions between RBC and plasma. Removal of CO_2 from blood involves
    (a) influx of Cl^- ions into RBC.
    (b) influx of HCO_3^- ions into RBC.
    (c) efflux of Cl^- ions into RBC.
    (d) efflux of HCO_3^- ions into RBC.
19. Which of the following statement correctly defines "Bohr effects"?
    (a) Rise in P_{CO_2} with a decrease in CO_2 concentration.
    (b) Rise in P_{CO_2} with an increase in CO_2 concentration.
    (c) Rise in P_{CO_2} with an increase in CO_2 and decrease in P_{O_2}.
    (d) Rise in P_{CO_2} with a decrease in pH (concentration at which 50% haemoglobin of blood is saturated with oxygen).
20. In lungs in order to facilitate the removal of CO_2 from the blood, there is an exchange of ions between RBC and the plasma which involves the movement of
    (a) chloride ions from RBC to plasma.
    (b) Cl^- ions from plasma to RBC.
    (c) bicarbonate ions from RBC to plasma.
    (d) Cl^- and bicarbonate ions as in (a) and (b).
21. Which is true for CO_2 concentration?
    (a) More in alveolar air than in expired air.
    (b) More in expired air than in alveolar air.
    (c) More in inspired air than in expired air.
    (d) More in inspired air than in alveolar air.
22. Respiration mechanism is controlled by
    (a) central nervous system.
    (b) autonomic nervous system.
    (c) sympathetic nervous system.
    (d) parasympathetic nervous system.
23. The respiratory centre in brain which controls inspiration and expiration, is situated in
    (a) medulla oblongata (b) cerebellum
    (c) hypothalamus (d) pericardium
24. During strenous exercise, which of the following change occurs?
    (a) Glucose is converted into glycogen.
    (b) Glucose is converted into pyruvic acid.
    (c) Starch is converted into glucose.
    (d) Pyruvic acid is converted into lactic acid.

RESPONSE GRID					
	15. ⓐⓑⓒⓓ	16. ⓐⓑⓒⓓ	17. ⓐⓑⓒⓓ	18. ⓐⓑⓒⓓ	19. ⓐⓑⓒⓓ
	20. ⓐⓑⓒⓓ	21. ⓐⓑⓒⓓ	22. ⓐⓑⓒⓓ	23. ⓐⓑⓒⓓ	24. ⓐⓑⓒⓓ

Space for Rough Work

25. Which of the following conditions is responsible for increase in ventilation rate of lungs?
(a) Increase of CO_2 content in inhaled air.
(b) Increase of CO_2 content in exhaled air.
(c) Decrease of O_2 content in inhaled air.
(d) Decrease of O_2 content in exhaled air.

26. The form of energy used in respiration is
(a) chemical energy
(b) electrical energy
(c) mechanical energy
(d) radiant energy

27. Foetal haemoglobin has a greater affinity for ______ than adult haemoglobin.
(a) CO_2
(b) CO
(c) H_2O
(d) O_2

28. Blood analysis of a patient reveals an unusually high quantity of carboxyhaemoglobin content. Which of the following is responsible for this high quantity?
(a) Carbon dioxide (b) Carbon monoxide
(c) Carbon disulphide (d) Chloroform

29. Ravi, who lived at sea level, had around 5 million RBC per cubic millimeter of his blood. Later when he lived at an altitude of 18,000 ft, showed around 8 million RBC per cubic millimeter of blood. This is an adaptation because
(a) at high altitude he ate more nutritive food.
(b) he had pollution free air to balance breathe.
(c) at high altitude O_2 level is less hence more RBCs were required to absorb enough oxygen.
(d) at high altitude there is more UV radiation which enhances RBCs production.

30. Ascent of high mountains may cause altitude sickness in men. Prime cause of this is
(a) excess of CO_2 in blood.
(b) decreased efficiency of haemoglobin.
(c) decreased partial pressure of oxygen.
(d) decreased proportion of oxygen in air.

RESPONSE GRID	25. ⓐⓑⓒⓓ	26. ⓐⓑⓒⓓ	27. ⓐⓑⓒⓓ	28. ⓐⓑⓒⓓ	29. ⓐⓑⓒⓓ
	30. ⓐⓑⓒⓓ				

DAILY PRACTICE PROBLEM SHEET 33 - BIOLOGY

Total Questions	30	Total Marks	120
Attempted		Correct	
Incorrect		Net Score	
Cut-off Score	40	Qualifying Score	68
Success Gap = Net Score – Qualifying Score			
Net Score = (Correct × 4) – (Incorrect × 1)			

Space for Rough Work

DPP - Daily Practice Problems

Name :

Date :

Start Time :

End Time :

BIOLOGY 34

SYLLABUS : Body fluids and circulation - PART I : Structure & function of heart / Blood vessel/ Blood pressure, ECG

Max. Marks : 120 **Time : 60 min.**

GENERAL INSTRUCTIONS

- The Daily Practice Problem Sheet contains 30 MCQ's. For each question only one option is correct. Darken the correct circle/ bubble in the Response Grid provided on each page.
- You have to evaluate your Response Grids yourself with the help of solution booklet.
- Each correct answer will get you 4 marks and 1 mark shall be deduced for each incorrect answer. No mark will be given/ deducted if no bubble is filled. Keep a timer in front of you and stop immediately at the end of 60 min.
- The sheet follows a particular syllabus. Do not attempt the sheet before you have completed your preparation for that syllabus. Refer syllabus sheet in the starting of the book for the syllabus of all the DPP sheets.
- After completing the sheet check your answers with the solution booklet and complete the Result Grid. Finally spend time to analyse your performance and revise the areas which emerge out as weak in your evaluation.

DIRECTIONS (Q.1-Q.30) : There are 30 multiple choice questions. Each question has 4 choices (a), (b), (c) and (d), out of which ONLY ONE choice is correct.

1. Systemic heart refers to
   (a) the two ventricles together in humans.
   (b) the heart that contracts under stimulation from nervous system.
   (c) left auricle and left ventricle in higher vertebrates.
   (d) entire heart in lower vertebrates.
2. The first heart sound is
   (a) 'lubb' sound at the end of systole.
   (b) 'dub' sound at the end of systole.
   (c) 'lubb' sound at the beginning of systole.
   (d) 'dub' sound at the beginning of systole.
3. In the heart of mammal, the bicuspid valve is situated between
   (a) left auricle and left ventricle.
   (b) post caval and right caval.
   (c) right auricle and left auricle.
   (d) right ventricle and pulmonary aorta.
4. Systolic pressure is higher than diastolic pressure because
   (a) arteries are contracting during systole.
   (b) blood is pumped with increased pressure in the arteries by the heart during systole but not during diastole.
   (c) arteries resist during systole only.
   (d) volume of blood is higher in systole than that of diastole in the heart.

RESPONSE GRID	1. ⓐⓑⓒⓓ	2. ⓐⓑⓒⓓ	3. ⓐⓑⓒⓓ	4. ⓐⓑⓒⓓ

Space for Rough Work

5. The murmuring sound in heart takes place due to
(a) the defective and leaking valve.
(b) the thrombosis in coronary artery.
(c) defect in SA node.
(d) the arterial pulse.

6. The beating of heart of man is heared on the left side because
(a) the left ventricle is towards the left side.
(b) both the ventricles are towards the left side.
(c) entire heart is on the left side.
(d) the aorta is on the left side.

7. The atrio-ventricular valves of the heart is prevented from turning inside out by tough strands of connective tissue. This connective tissue is called as
(a) tendinous cords
(b) tricuspid valve
(c) pocket valve
(d) bicuspid valve

8. The typical lubb–dup sounds heard in the heart beat of a healthy person are due to
(a) closing of the tricuspid and bicuspid valve.
(b) blood flow through the aorta.
(c) closing of the bicuspid-tricuspid and semilunar valves.
(d) closing of the semilunar valves.

9. Cardiac output signifies
(a) the amount of blood entering the heart per unit time.
(b) the amount of blood entering the lung per unit time.
(c) the amount of blood leaving the heart per unit time.
(d) the amount of blood leaving the lung per unit time.

10. Heart beats are accelerated by
(a) cranial nerves and acetylcholine.
(b) sympathetic nerves and acetylcholine.
(c) cranial nerves and adrenaline.
(d) sympathetic nerves and epinephrine.

11. Which one is the correct route through which pulse making impulse travels in the heart?
(a) SA node $\rightarrow$ Purkinje fibres $\rightarrow$ Bundle of His $\rightarrow$ AV node $\rightarrow$ Heart muscles.
(b) AV node $\rightarrow$ SA node $\rightarrow$ Purkinje fibres $\rightarrow$ Bundle of His $\rightarrow$ Heart muscles.
(c) AV node $\rightarrow$ Bundle of His $\rightarrow$ SA node $\rightarrow$ Purkinje fibres $\rightarrow$ Heart muscles.
(d) SA node $\rightarrow$ AV node $\rightarrow$ Bundle of His $\rightarrow$ Purkinje fibres $\rightarrow$ Heart muscles.

12. The pace–setter in the heart is called
(a) purkinje fibres
(b) sino–atrial node (SAN)
(c) papillary muscle
(d) atrio–ventricular node (AVN)

13. Valves are necessary in veins but not in arteries because
(a) blood flows with greater force in veins.
(b) blood in veins flows without jerk.
(c) blood from heart may not be pushed back into vein.
(d) pressure in veins is low, which can flow the blood.

14. Heart beat originates from
(a) pacemaker (b) cardiac muscles
(c) left atrium (d) right ventricle

15. Bundle of His is a network of

RESPONSE GRID					
	5. ⓐⓑⓒⓓ	6. ⓐⓑⓒⓓ	7. ⓐⓑⓒⓓ	8. ⓐⓑⓒⓓ	9. ⓐⓑⓒⓓ
	10. ⓐⓑⓒⓓ	11. ⓐⓑⓒⓓ	12. ⓐⓑⓒⓓ	13. ⓐⓑⓒⓓ	14. ⓐⓑⓒⓓ

Space for Rough Work

(a) nerve fibres found throughout the heart.

(b) muscle fibres distributed throughout the heart walls.

(c) muscle fibres found only in the ventricle wall.

(d) nerve fibres distributed in ventricles.

16. What is correct about sinus venosus?

(a) It is situated on dorsal surface of rabbit's heart.

(b) It is situated ventrally in frog's heart.

(c) It sends blood to dorsal aorta.

(d) It opens into right auricle.

17. In rabbit oxygenated blood flows from

(a) left auricle to left ventricle during auricular systole.

(b) right auricle to right ventricle during ventricular systole.

(c) right ventricle to aorta during ventricular systole.

(d) pulmonary vein to left auricle during auricular systole.

18. The blood vascular system of mammals is known as double vascular system because

(a) a group of veins carry oxygenated and other group conduct deoxygenated blood.

(b) oxygenated blood runs from heart to different organs by one set of veins while deoxygenated blood runs from heart to lung by another set.

(c) the two different systems never meet.

(d) all of the above.

19. Open vascular system is found in

(a) man

(b) fish

(c) prawn

(d) snake

20. Which of the following is the correct statement about the circulatory system of cockroach?

(a) It is closed type of circulatory system.

(b) It is a complicated type of circulatory system.

(c) It takes place without the participation of tissue.

(d) It has 13 chambered heart and in each segment one pair of ostia are present.

21. Mammals are said to have a "double circulatory system". This means that

(a) the blood vessels are paired.

(b) there are two types of blood vessels attached to every organ; an artery and a vein.

(c) there are two system; one from the heart to the lungs and back to the heart and other to and from rest of the body.

(d) the blood circulates twice as quickly.

22. The heart of a crocodile consists of

(a) a single auricle and two ventricles.

(b) two auricles and a single ventricle.

(c) two auricles and two ventricles.

(d) a single auricle and a single ventricle.

23. Which one of the following is a matching pair?

(a) Lubb – Sharp closure of AV valves at the beginning of ventricular systole.

(b) Dupp – sudden opening of semilunar valves at the beginning of ventricular diastole.

(c) Pulsation of the radial artery – Valves in the blood vessels

(d) Initiation of the heart beat – Purkinje fibres

24. Purkinje's fibres of the vertebrate heart are modified

(a) parasympathetic nerves

(b) sympathetic nerves

(c) motor nerves

(d) muscle cells

25. Which of the following has thickest wall?

(a) Right atrium (b) Left atrium

RESPONSE GRID	15. ⓐⓑⓒⓓ	16. ⓐⓑⓒⓓ 	17. ⓐⓑⓒⓓ	18. ⓐⓑⓒⓓ	19. ⓐⓑⓒⓓ
	20. ⓐⓑⓒⓓ	21. ⓐⓑⓒⓓ	22. ⓐⓑⓒⓓ	23. ⓐⓑⓒⓓ	24. ⓐⓑⓒⓓ

Space for Rough Work

(c) Right ventricle (d) Left ventricle

26. Cardiac output is determined by
(a) heart rate (b) stroke volume
(c) blood flow (d) both (a) and (b)

27. 'Heart of Heart' is
(a) SA node
(b) AV node
(c) Bundle of His
(d) Purkinje fibres

28. Human heart is enclosed in a double walled sac called
(a) peritome (b) pericardium
(c) pericardial sinus (d) perineural sinus

29. When the right ventricle contracts the blood goes into
(a) aorta (b) brain
(c) pulmonary artery (d) none of these

30. The opening of upper chamber of the heart into the lower chamber is known as
(a) sinu-auricular aperture
(b) auriculo - ventricular aperture
(c) atrioventricular septum
(d) atrioventricular node.

RESPONSE GRID	
	25. ⓐⓑⓒⓓ 26. ⓐⓑⓒⓓ 27. ⓐⓑⓒⓓ 28. ⓐⓑⓒⓓ 29. ⓐⓑⓒⓓ 30. ⓐⓑⓒⓓ

DAILY PRACTICE PROBLEM SHEET 34 - BIOLOGY			
Total Questions	30	Total Marks	120
Attempted		Correct	
Incorrect		Net Score	
Cut-off Score	32	Qualifying Score	52
Success Gap = Net Score – Qualifying Score			
Net Score = (Correct × 4) – (Incorrect × 1)			

Space for Rough Work

DPP - Daily Practice Problems

Name :   Date :

Start Time :   End Time :

BIOLOGY 35

SYLLABUS : Body fluids and Circulation - PART II : Blood/ Lymphatic System/ Portal System

Max. Marks : 120   **Time : 60 min.**

GENERAL INSTRUCTIONS

- The Daily Practice Problem Sheet contains 30 MCQ's. For each question only one option is correct. Darken the correct circle/ bubble in the Response Grid provided on each page.
- You have to evaluate your Response Grids yourself with the help of solution booklet.
- Each correct answer will get you 4 marks and 1 mark shall be deduced for each incorrect answer. No mark will be given/ deducted if no bubble is filled. Keep a timer in front of you and stop immediately at the end of 60 min.
- The sheet follows a particular syllabus. Do not attempt the sheet before you have completed your preparation for that syllabus. Refer syllabus sheet in the starting of the book for the syllabus of all the DPP sheets.
- After completing the sheet check your answers with the solution booklet and complete the Result Grid. Finally spend time to analyse your performance and revise the areas which emerge out as weak in your evaluation.

DIRECTIONS (Q.1-Q.30) : There are 30 multiple choice questions. Each question has 4 choices (a), (b), (c) and (d), out of which ONLY ONE choice is correct.

1. Serotonin in the blood
   (a) relaxes blood vessels.
   (b) prevents clotting of blood.
   (c) helps in clotting of blood.
   (d) constricts blood vessels.
2. Carotid artery carries
   (a) impure blood from brain.
   (b) oxygenated blood to anterior region of body or to brain.
   (c) impure blood to kidney.
   (d) oxygenated blood to heart.
3. Which of the following blood vessels carry blood from lungs to heart?
   (a) Pulmonary artery
   (b) Pulmonary vein
   (c) Azygous vein
   (d) Coronary artery
4. ECG records
   (a) electric current of the body.
   (b) potential differences that accompany cardiac cycle.
   (c) pulse rate.
   (d) quantity of blood pumped per minute.

RESPONSE GRID	1. ⓐⓑⓒⓓ	2. ⓐⓑⓒⓓ	3. ⓐⓑⓒⓓ	4. ⓐⓑⓒⓓ

Space for Rough Work

5. QRST is related with
   (a) ventricular contraction or depolarization.
   (b) auricular contraction.
   (c) auricular relaxation.
   (d) cardiac cycle.
6. Normal diastolic pressure in young man is about
   (a) 20 mm Hg (b) 80 mm Hg
   (c) 100 mm Hg (d) 130 mm Hg
7. During systole of ventricle
   (a) blood enters the heart.
   (b) blood leaves the heart.
   (c) blood leaves the ventricle.
   (d) blood enters lungs.
8. During diastole
   (a) blood enters lungs.
   (b) blood leaves the ventricle.
   (c) blood leaves the heart.
   (d) blood enters the heart.
9. If pH of blood is artificially changed to 8, what is most likely to happen?
   (a) The tissues will not get oxygen as oxyhaemoglobin will not dissociate into oxygen and haemoglobin.
   (b) The carbon dioxide will not be released from carbonic acid and carbonates.
   (c) The carbonic anhydrase will be completely inhibited.
   (d) It will generate carbon monoxide and the animal will die of carbon monoxide poisoning.
10. Haemoglobin is a
    (a) copper containing pigmentX
    (b) iron containing pigment
    (c) magnesium containing pigment
    (d) calcium containing pigment
11. A yellow substance oozing out from wound has
    (a) lymph + RBC + WBC
    (b) lymph + RBC + dead bacteria
    (c) lymph + WBC + dead bacteria
    (d) lymph + dead leucocytes
12. In haemoglobin iron is attached with globin protein by
    (a) hydrogen bond
    (b) ionic bond
    (c) covalent bond
    (d) coordinate bond
13. Haemoglobin contains
    (a) 70% globin + 30% haematin
    (b) 80% globin + 20% haematin
    (c) 95% globin + 05% haematin
    (d) 90% globin + 10% haematin
14. The lymphocytes protect from
    (a) pathogens
    (b) lymph
    (c) leucocytes
    (d) toxins

RESPONSE GRID					
	5. ⓐⓑⓒⓓ	6. ⓐⓑⓒⓓ	7. ⓐⓑⓒⓓ	8. ⓐⓑⓒⓓ	9. ⓐⓑⓒⓓ
	10. ⓐⓑⓒⓓ	11. ⓐⓑⓒⓓ	12. ⓐⓑⓒⓓ	13. ⓐⓑⓒⓓ	14. ⓐⓑⓒⓓ

Space for Rough Work

15. The principle function of the lymph node in the man is
    (a) destruction of old RBC.
    (b) destruction of old WBC.
    (c) collection and destruction of pathogens present in blood.
    (d) production of WBC.
16. Antigen –A and antibody –b are present in which blood group?
    (a) B (b) A
    (c) AB (d) O
17. Lymph is colourless because
    (a) WBC are absent.
    (b) WBC are present.
    (c) haemoglobin is absent.
    (d) RBC are present.
18. Which vertebrate organ receives only oxygenated blood?
    (a) Gill (b) Lung
    (c) Spleen (d) Liver
19. Spleen is
    (a) lymphoid organ
    (b) reproductive organ
    (c) celluloid organ
    (d) none of these
20. All arteries carry oxygenated blood except
    (a) systemic (b) hepatic
    (c) pulmonary (d) cardiac
21. A portal system is a system in which
    (a) a vein starts from an organ and ends up in heart.
    (b) an artery breaks up in an organ and restarts by the union of its capillaries.
    (c) the blood from the gut is brought into the kidney before it is poured into posterior vena cava.
    (d) a vein breaks up in an organ into capillaries and restarts by their union as a new vein in the same organ.
22. When a vein instead of carrying the blood directly into the heart, first carries it to some intermediate organ like liver, is known as
    (a) pulmonary artery
    (b) renal portal vein
    (c) hepatic portal vein
    (d) pulmonary vein
23. Hepatic portal system starts from
    (a) digestive system to liver.
    (b) kidney to liver.
    (c) liver to heart.
    (d) liver to kidney.
24. Blood circulation that starts in capillaries and ends in capillaries is called
    (a) portal circulation
    (b) hepatic circulation
    (c) cardiac circulation
    (d) none of these

RESPONSE GRID					
	15. ⓐⓑⓒⓓ	16. ⓐⓑⓒⓓ	17. ⓐⓑⓒⓓ	18. ⓐⓑⓒⓓ	19. ⓐⓑⓒⓓ
	20. ⓐⓑⓒⓓ	21. ⓐⓑⓒⓓ	22. ⓐⓑⓒⓓ	23. ⓐⓑⓒⓓ	24. ⓐⓑⓒⓓ

Space for Rough Work

25. Which of the following carries glucose from digestive tract to liver?
(a) Hepatic artery (b) Hepatic portal vein
(c) Pulmonary vein (d) None of these

26. The renal portal system is made of
(a) femoral, renel portal veins
(b) sciatic, renal portal veins
(c) renal portal veins
(d) femoral, sciatic, renal portal veins

27. Which of the following is the largest blood vessel in the body?
(a) Carotid artery (b) Dorsal aorta
(c) Phrenic artery (d) Coronary artery

28. Heart beats are controlled by a nodal tissue which is composed of
(a) Purkinje fibres (b) myonemes
(c) collagen fibres (d) telodendrites

29. Stimulation of the vagus nerve will make the heart beat
(a) faster (b) 70 times/minute
(c) slower (d) normal

30. Mammals have biconcave RBC. The physiological use for it is
(a) to decrease the surface area of respiration.
(b) to increase the surface area of respiration.
(c) to be packed like coins.
(d) none of the above.

RESPONSE GRID	25. ⓐⓑⓒⓓ 26. ⓐⓑⓒⓓ 27. ⓐⓑⓒⓓ 28. ⓐⓑⓒⓓ 29. ⓐⓑⓒⓓ 30. ⓐⓑⓒⓓ

DAILY PRACTICE PROBLEM SHEET 35 - BIOLOGY			
Total Questions	30	Total Marks	120
Attempted		Correct	
Incorrect		Net Score	
Cut-off Score	36	Qualifying Score	52
Success Gap = Net Score – Qualifying Score			
Net Score = (Correct × 4) – (Incorrect × 1)			

Space for Rough Work

DPP - Daily Practice Problems

Name : 　　　　Date :

Start Time : 　　　　End Time :

BIOLOGY 36

SYLLABUS : Excretory products and their elimination

Max. Marks : 120 　　　　**Time : 60 min.**

GENERAL INSTRUCTIONS

- The Daily Practice Problem Sheet contains 30 MCQ's. For each question only one option is correct. Darken the correct circle/ bubble in the Response Grid provided on each page.
- You have to evaluate your Response Grids yourself with the help of solution booklet.
- Each correct answer will get you 4 marks and 1 mark shall be deduced for each incorrect answer. No mark will be given/ deducted if no bubble is filled. Keep a timer in front of you and stop immediately at the end of 60 min.
- The sheet follows a particular syllabus. Do not attempt the sheet before you have completed your preparation for that syllabus. Refer syllabus sheet in the starting of the book for the syllabus of all the DPP sheets.
- After completing the sheet check your answers with the solution booklet and complete the Result Grid. Finally spend time to analyse your performance and revise the areas which emerge out as weak in your evaluation.

DIRECTIONS (Q.1-Q.30) : There are 30 multiple choice questions. Each question has 4 choices (a), (b), (c) and (d), out of which ONLY ONE choice is correct.

1. Waste products of adenine and guanine metabolism are excreted by man as
   (a) ammonia
   (b) urea
   (c) uric acid
   (d) allantois
2. In man, the urea is mainly produced in
   (a) liver 　　(b) kidney
   (c) gall bladder 　　(d) spleen
3. Aquatic reptiles are
   (a) ammonotelic
   (b) ureotelic
   (c) uricotelic
   (d) none of these
4. The major nitrogenous waste (s) in the respective animals is/are
   (a) urea and uric acid in cow, frog and lion.
   (b) uric acid in parrot, cockroach and garden lizard.
   (c) ammonia in elephant, sea horse and turtle.
   (d) urea in scorpio, earthworm and shark.

RESPONSE GRID	1. ⓐⓑⓒⓓ	2. ⓐⓑⓒⓓ	3. ⓐⓑⓒⓓ	4. ⓐⓑⓒⓓ

Space for Rough Work

5. A person is undergoing prolonged fasting. His urine will be found to contain abnormal quantities of
(a) fats (b) aminoacids
(c) glucose (d) ketones

6. In glomerules of rabbit
(a) afferent glomerular capillary is wider than efferent glomerular capillary.
(b) afferent glomerular capillary is narrower than efferent glomerular capillary.
(c) afferent glomerular arteriole is narrower than the efferent glomerular arteriole.
(d) afferent glomerular arteriole is wider than the efferent glomerular arteriole.

7. Correct order of excretory organs in cockroach, earthworm and rabbit respectively are
(a) skin, malpighian tubules, kidney.
(b) malpighian tubules, nephridia, kidney.
(c) nephridia, malpighian tubule, kidney.
(d) nephridia, kidney, green gland.

8. "Columns of Bertini" in the kidney of mammals are formed as the extension of
(a) medulla into cortex (b) cortex into medulla
(c) medulla into pelvis (d) pelvis into ureter

9. The kidneys resemble the contractile vacuoles of protozoans in
(a) expelling out excess of water.
(b) expelling out glucose.
(c) expelling out urea and uric acid.
(d) expelling out salts.

10. The Bowman's capsule is
(a) a part of uriniferous tubule and is the site of filtration of various blood constituents during the formation of urine.
(b) a part of uriniferous tubule and is the site of reabsorption of water and glucose.
(c) present in the liver and is the site of secretion of the bile juice.
(d) associated with liver and is the site of the storage of the bile juice.

11. What is the characteristic of metanephric kidney?
(a) Hypotonic urine production
(b) Excess secretion of uric acid
(c) Loop of Henle
(d) Hormone production

12. Match the excretory organs listed under colum I with the animals given under column II. Choose the answer which gives the correct combination of alphabets of the two columns.

Column I (Excretory organs)		Column II (Animals)	
A.	Nephridia	p	*Hydra*
B.	Malpighian tubules	q	Leech
C.	Protonephridia	r	Shark
D.	Kidneys	s	Round worms
		t	Cockroach

(a) A = q; B = t; C = s; D = r
(b) A = s; B = q; C = p; D = t
(c) A = t; B = q; C = s; D = r
(d) A = q; B = s; C = t; D = p

13. When a person is suffering from poor renal reabsorption then which of the following will not help in the maintenance of blood volume?
(a) Decreased glomerular filtration
(b) Increased ADH secretion
(c) Decreased arterial pressure in kidney
(d) Increased arterial pressure in kidney

14. Filtration into the kidney tubule is accomplished by means of
(a) active transport by renal tubule.
(b) hydrostatic blood pressure in glomerulus.
(c) an osmotic potential gradient.
(d) secretion by renal tubule.

RESPONSE GRID					
	5. ⓐⓑⓒⓓ	6. ⓐⓑⓒⓓ	7. ⓐⓑⓒⓓ	8. ⓐⓑⓒⓓ	9. ⓐⓑⓒⓓ
	10. ⓐⓑⓒⓓ	11. ⓐⓑⓒⓓ	12. ⓐⓑⓒⓓ	13. ⓐⓑⓒⓓ	14. ⓐⓑⓒⓓ

Space for Rough Work

15. In distal convoluted tubule of the nephrons
(a) Na ions reabsorption requires energy.
(b) secretion of K ions does not require energy.
(c) water reabsorption requires energy.
(d) ammonia is secreted.

16. Which one of the following pair of waste substances is removed from blood in ornithine cycle?
(a) CO_2 and urea (b) Ammonia and urea
(c) CO_2 and ammonia (d) Urea and sodium salt

17. The glomerular filtrate contains
(a) blood minus cells and proteins.
(b) blood minus cells.
(c) blood minus proteins.
(d) plasma minus cells and proteins.

18. Reabsorption of useful substances back into the blood from the filtrate in a nephron occurs in
(a) proximal convoluted tubule
(b) loop of Henle
(c) distal convoluted tubule
(d) collecting duct

19. Ornithine is converted into citrulline by an enzyme
(a) glutamic dehydrogenase
(b) aspartic glutamic transaminase
(c) carbamyl phosphate synthetase
(d) ornithine carbamyl transferase

20. In which of the following ways does the blood leaving the glomerulus of a mammalian kidney tubule differ from the blood entering the glomerulus?
(a) It has a lower concentration of plasma proteins.
(b) It contains fewer corpuscles per unit volume.
(c) It has a lower concentration of crystalloids.
(d) It has a higher concentration of crystalloids.

21. In excessive bleeding, profuse sweating or prolonged delay in fluid intake
(a) glomerular filtration increases.
(b) glomerular filtration slows down.
(c) glomerular filtration remains constant.
(d) none of the above.

22. Which of the following is correct?
(a) Afferent arteriole is narrower than efferent arteriole.
(b) Afferent venule is narrower than efferent venule.
(c) Efferent arteriole is narrower than afferent arteriole.
(d) Efferent venule is narrower than afferent venule.

23. In micturition
(a) urethra relaxes. (b) ureter contracts.
(c) ureter relaxes. (d) urethra contracts.

24. Choose the correct statement for biosynthesis of urea.
(a) Uric acid is starting material for biosynthesis of urea.
(b) Urea is synthesized inside lysosomes.
(c) Urea cycle enzyme are located inside mitochondria.
(d) Urea is synthesized in kidney.

RESPONSE GRID					
	15. ⓐⓑⓒⓓ	16. ⓐⓑⓒⓓ	17. ⓐⓑⓒⓓ	18. ⓐⓑⓒⓓ	19. ⓐⓑⓒⓓ
	20. ⓐⓑⓒⓓ	21. ⓐⓑⓒⓓ	22. ⓐⓑⓒⓓ	23. ⓐⓑⓒⓓ	24. ⓐⓑⓒⓓ

Space for Rough Work

25. ADH controls water permeability of
(a) collecting tube.
(b) distal convolulated tubule and collecting duct.
(c) proximal convolulated tubule.
(d) Bowman's capsule.

26. Which one is component of ornithine cycle?
(a) Ornithine, citrulline and alanine.
(b) Ornithine, citrulline and arginine.
(c) Amino acid are not used.
(d) Ornithine, citrulline and fumaric acid.

27. Which of the following amino acids is present in ornithine cycle?
(a) Valine and cystine
(b) Arginine and citrulline
(c) Glycine and methionine
(d) None of the above

28. The main function of Henle's loop is
(a) conservation of water.
(b) filtration of blood.
(c) passage of urine.
(d) formation of urine.

29. The main function of pyramids of kidney is to
(a) contain collecting tubules of kidney.
(b) direct the urine to flow in ureter.
(c) support the openings of collecting canals.
(d) store fats and protein.

30. Due to insufficient filtration in the Bowman's capsule, all are likely to happen except
(a) accumulation of fluid in the body.
(b) increase in blood pressure.
(c) increase in blood urea level.
(d) loss of glucose through urine.

RESPONSE GRID	25. ⓐⓑⓒⓓ 26. ⓐⓑⓒⓓ 27. ⓐⓑⓒⓓ 28. ⓐⓑⓒⓓ 29. ⓐⓑⓒⓓ 30. ⓐⓑⓒⓓ

DAILY PRACTICE PROBLEM SHEET 36 - BIOLOGY			
Total Questions	30	Total Marks	120
Attempted		Correct	
Incorrect		Net Score	
Cut-off Score	44	Qualifying Score	68
Success Gap = Net Score – Qualifying Score			
Net Score = (Correct × 4) – (Incorrect × 1)			

Space for Rough Work

DPP - Daily Practice Problems

Name : Date :

Start Time : End Time :

BIOLOGY 37

SYLLABUS : Locomotion and movement

Max. Marks : 120 **Time : 60 min.**

GENERAL INSTRUCTIONS

- The Daily Practice Problem Sheet contains 30 MCQ's. For each question only one option is correct. Darken the correct circle/ bubble in the Response Grid provided on each page.
- You have to evaluate your Response Grids yourself with the help of solution booklet.
- Each correct answer will get you 4 marks and 1 mark shall be deduced for each incorrect answer. No mark will be given/ deducted if no bubble is filled. Keep a timer in front of you and stop immediately at the end of 60 min.
- The sheet follows a particular syllabus. Do not attempt the sheet before you have completed your preparation for that syllabus. Refer syllabus sheet in the starting of the book for the syllabus of all the DPP sheets.
- After completing the sheet check your answers with the solution booklet and complete the Result Grid. Finally spend time to analyse your performance and revise the areas which emerge out as weak in your evaluation.

DIRECTIONS (Q.1-Q.30) : There are 30 multiple choice questions. Each question has 4 choices (a), (b), (c) and (d), out of which ONLY ONE choice is correct.

1. The total number of vertebrae in man is
   (a) 30 (b) 31
   (c) 32 (d) 33
2. The first vertebra, the atlas by its articulation with axis vertebra facilitates a
   (a) noding movement
   (b) sideway movement
   (c) rotatory movement
   (d) backward movement
3. The opening at the base of the skull for the spinal cord is called
   (a) foramen Magnum
   (b) foramen of Monro
   (c) obturator foramen
   (d) foramen of Magendie
4. What is correct about human body?
   (a) There are 5 vertebra in the neck.
   (b) Brain box is made up of 4 bones.
   (c) There are 15 pairs of ribs.
   (d) There are 12 thoracic vertebra.

RESPONSE GRID	1. ⓐⓑⓒⓓ	2. ⓐⓑⓒⓓ	3. ⓐⓑⓒⓓ	4. ⓐⓑⓒⓓ

Space for Rough Work

5. In mammals,the zygomatic arch is formed by
   (a) maxilla, premaxilla and squamosal.
   (b) periotic, jugal and palatine.
   (c) maxilla, squamosal and jugal.
   (d) jugal, maxilla and periotic.
6. The last two pairs of ribs are named floating ribs because
   (a) their sternal parts are attached to the sternum directly.
   (b) their sternal parts are attached on the 7th pair of ribs.
   (c) their sternal parts remain free and do not even reach the sternum.
   (d) they float in the body cavity.
7. Vertebral formula for human beings is
   (a) $C_5T_{12}L_7S_5C_{3-5} = 33 - 55$
   (b) $C_7T_{12}L_5S_5C_{3-5} = 33 - 35$
   (c) $C_5T_{10}L_5S_5C_{3-5} = 33$
   (d) $C_7T_{10}L_5S_5C_{3-5} = 33$
8. How many ribs are present in human beings?
   (a) 6 pairs
   (b) 9 pairs
   (c) 12 pairs
   (d) 15 pairs
9. The lower jaw of a mammal is made up
   (a) dentary
   (b) maxilla
   (c) premaxilla
   (d) post maxilla
10. The cup-shaped structure of pelvic girdle, the acetabulum in man is formed by
    (a) ilium, ischium and pubis.
    (b) ilium, ischium and cotyloid.
    (c) ilium and ischium.
    (d) ilium and cotyloid.
11. The articulation of vertebrae in vertebral column occurs when
    (a) post-zygapophyses of a vertebra in front fit beneath the pre-zygapophyses of the another vertebra behind.
    (b) post-zygapophyses of a vertebra in front fit over the pre-zygapophyses of another vertebra behind.
    (c) pre-zygapophyses of a vertebra fit over the post-zygapophyses of the vertebra in front.
    (d) pre and post-zygapophyses of vertebra simply touch one another.
12. Phalangeal formula of hand of man is
    (a) 1, 2, 2, 2, 2, (b) 2, 1, 1, 1, 1
    (c) 2, 3, 3, 3, 3 (d) 2, 3, 3, 2, 2
13. Structure responsible for formation of sigmoid notch is
    (a) olecranon process of humerus.
    (b) olecranon process of femur.
    (c) olecranon process of radius ulna.
    (d) olecranon process of tibia fibula.
14. Haversian canals are found in
    (a) spinal cord (b) brain
    (c) long bones (d) sponge

RESPONSE GRID					
	5. ⓐⓑⓒⓓ	6. ⓐⓑⓒⓓ	7. ⓐⓑⓒⓓ	8. ⓐⓑⓒⓓ	9. ⓐⓑⓒⓓ
	10. ⓐⓑⓒⓓ	11. ⓐⓑⓒⓓ	12. ⓐⓑⓒⓓ	13. ⓐⓑⓒⓓ	14. ⓐⓑⓒⓓ

Space for Rough Work

15. Hinge joint is present between
(a) humerus and pectoral girdle.
(b) femur and acetabulum.
(c) humerus and radio-ulna.
(d) femur and pelvic girdle.

16. Sutural joints are found between
(a) parietals of skull.
(b) humerus and radio-ulna.
(c) glenoid cavity and pectoral girdle.
(d) thumb and metatarsal.

17. Which of the following pairs is correctly matched?
(a) Hinge joint – Between vertebrae
(b) Gliding joint – Between zygapophyses of the successive vertebrae
(c) Cartilaginous joint– Skull bones
(d) Fibrous joint – Between phalanges

18. The contraction of muscle of shortest duration is seen in
(a) heart (b) jaws
(c) intestine (d) eyelids

19. What is sprain?
(a) More pulling of tendon.
(b) Less pulling of tendon.
(c) More pulling of ligament.
(d) Less pulling of liagament.

20. Muscles are red because of the presence of
(a) myoglobin and mitochondria.
(b) haemoglobin and golgi bodies.
(c) globulin and mitochondria.
(d) protein and lysosome.

21. An example of unstriated muscles which are entirely involuntary are located
(a) in the diaphragm (b) in the eyelid
(c) at the pylorus (d) at the base of external ear

22. Electron microscopic studies of the sarcomeres have revealed that during muscle contraction
(a) the width of A–band remains constant.
(b) the width of the H–zone becomes smaller.
(c) the width of I–band increases.
(d) the diameter of the fibre increases.

23. Which one of the following events is thought to be most closely related to the sliding process between the two types of filaments which bring about contraction of the fibril?
(a) The liberation of acetyl choline at the motor end plates.
(b) Splitting of an ATP myosin complex.
(c) The influx of sodium ions through the sarcolemma.
(d) Binding of calcium ions.

24. Striated muscles contract by
(a) sliding of actin filament upon myosin filaments.
(b) sliding of myosin filaments upon actin filaments.
(c) pulling together of myosin filaments.
(d) pulling together of actin filaments.

RESPONSE GRID					
	15. ⓐⓑⓒⓓ	16. ⓐⓑⓒⓓ	17. ⓐⓑⓒⓓ	18. ⓐⓑⓒⓓ	19. ⓐⓑⓒⓓ
	20. ⓐⓑⓒⓓ	21. ⓐⓑⓒⓓ	22. ⓐⓑⓒⓓ	23. ⓐⓑⓒⓓ	24. ⓐⓑⓒⓓ

Space for Rough Work

25. A ...... muscle rotates the forearm to turn the palm downwards.
(a) pronator
(b) supinator
(c) adductor
(d) abductor

26. In muscular hypertrophy
(a) cells increase in size by synthesizing the length of a myofibril.
(b) cells multiply in number.
(c) cells decrease in size.
(d) old cells are damaged.

27. The zygomatic process in mammals arises from
(a) premaxilla (b) maxilla
(c) mandible (d) squamosal

28. Bone is distinguished from the cartilage by the presence of
(a) Collagen (b) Blood vessels
(c) Lymph vessels (d) Haversian canals

29. Which one of the following is a viral disease that weakens the muscles?
(a) Atrophy (b) Poliomyelitis
(c) Dystrophy (d) Duchene dystrophy

30. Smallest muscle in the human body is
(a) sartorius (b) spinal muscle
(c) stapes (d) stapedius

RESPONSE GRID	25. ⓐⓑⓒⓓ 26. ⓐⓑⓒⓓ 27. ⓐⓑⓒⓓ 28. ⓐⓑⓒⓓ 29. ⓐⓑⓒⓓ 30. ⓐⓑⓒⓓ

DAILY PRACTICE PROBLEM SHEET 37 - BIOLOGY			
Total Questions	30	Total Marks	120
Attempted		Correct	
Incorrect		Net Score	
Cut-off Score	36	Qualifying Score	60
Success Gap = Net Score – Qualifying Score			
Net Score = (Correct × 4) – (Incorrect × 1)			

Space for Rough Work

DPP - Daily Practice Problems

Name : 　　　　　　　　　　　　Date :

Start Time : 　　　　　　　　　End Time :

BIOLOGY 38

SYLLABUS : Neural control and coordination Part I : Parts of nervous system, Biochemical aspect of nervous physiology

Max. Marks : 120 　　　　　　　　　　**Time : 60 min.**

GENERAL INSTRUCTIONS

- The Daily Practice Problem Sheet contains 30 MCQ's. For each question only one option is correct. Darken the correct circle/ bubble in the Response Grid provided on each page.
- You have to evaluate your Response Grids yourself with the help of solution booklet.
- Each correct answer will get you 4 marks and 1 mark shall be deduced for each incorrect answer. No mark will be given/ deducted if no bubble is filled. Keep a timer in front of you and stop immediately at the end of 60 min.
- The sheet follows a particular syllabus. Do not attempt the sheet before you have completed your preparation for that syllabus. Refer syllabus sheet in the starting of the book for the syllabus of all the DPP sheets.
- After completing the sheet check your answers with the solution booklet and complete the Result Grid. Finally spend time to analyse your performance and revise the areas which emerge out as weak in your evaluation.

DIRECTIONS (Q.1-Q.30) : There are 30 multiple choice questions. Each question has 4 choices (a), (b), (c) and (d), out of which ONLY ONE choice is correct.

1. Which one of the following is responsible for the control of reflex action?
(a) Sensory nerves
(b) Motor nerves
(c) Sympathetic nervous system
(d) Central nervous system

2. The hind brain consists of
(a) Pons + cerebellum
(b) Hypothalamus + cerebellum
(c) Medulla oblongata + cerebellum
(d) Medulla oblongata + cerebellum + pons

3. Which of the following cranial nerves are involved in the movement of eye?
(a) Optic, occulomotor, abducens
(b) Occulomotor, abducens, trochlear
(c) Trochlear, abducens and optic
(d) Abducens, vagus, trochlear, occulomotor

4. The smallest cranial nerve in human being is
(a) trochlear 　　(b) opthalamic
(c) abducens 　　(d) vagus

RESPONSE GRID	1. ⓐⓑⓒⓓ	2. ⓐⓑⓒⓓ	3. ⓐⓑⓒⓓ	4. ⓐⓑⓒⓓ

Space for Rough Work

5. Foramen of Monro is an aperture between
(a) 2nd and 3rd ventricle
(b) diocoel and metacoel
(c) rhinocoel and diocoel
(d) 3rd and 4th ventricle

6. Olfactory lobe of rabbit is
(a) fused and solid
(b) fused and hollow
(c) free and solid
(d) free and hollow

7. Which one of the following occurs without the help of brain?
(a) Cranial reflex
(b) Spinal reflex
(c) Efferent reflex
(d) Afferent reflex

8. The nervous strip connecting both the cerebral hemispheres in the rabbit is
(a) corpus callosum
(b) corpus albicans
(c) corpus striatum
(d) corpus spongiosum

9. Space between the two adjoining neurons where the chemical transmitter is released known as
(a) synaptic vesicle
(b) synapse
(c) synaptic cleft
(d) terminal button

10. IV, V and IX cranial nerves are
(a) olfactory, spinal accessary and vagus.
(b) trigeminal, vagus and glossopharyngeal.
(c) occulomotor, trigeminal and hypoglossal.
(d) pathetic, trigeminal and glossopharyngeal.

11. The innermost meninx surrounding the central nervous system in frog and man respectively are
(a) piamater and piamater
(b) arachnoid and piamater
(c) piamater and duramater
(d) arachanoid and duramater

12. The spinal cord extends from the brain through
(a) foramen magnum (b) iter
(c) anterior commissure (d) foramen of monro

13. Bell experiment on dog for salivation is associated with
(a) simple reflex (b) conditioned reflex
(c) voluntary action (d) none of these

14. The trigeminal nerve (V cranial nerve) arises from brain in the region of
(a) cerebellum and divides into palatine, chorda tympani and hyomandibular.
(b) medulla and divides into palatine, chorda tympani and hyomandibular.
(c) cerebellum and divides into ophthalmic, maxillary and mandibular.
(d) medulla and divides into opthalamic, maxillary and mandibular.

RESPONSE GRID					
	5. ⓐⓑⓒⓓ	6. ⓐⓑⓒⓓ	7. 	8. ⓐⓑⓒⓓ	9. ⓐⓑⓒⓓ
	10. ⓐⓑⓒⓓ	11. ⓐⓑⓒⓓ	12. ⓐⓑⓒⓓ	13. ⓐⓑⓒⓓ	14. ⓐⓑⓒⓓ

Space for Rough Work

15. One common example of simple reflex is
 (a) tying your shoe laces while talking to another person and not looking at them.
 (b) watering of mouth at the sight of a favourite food.
 (c) climbing up a stairs in dark without stumbling.
 (d) closing of eyelids when an object suddenly approaches the eye.
16. Cerebral aqua is
 (a) a cavity in the ventricle of a brain.
 (b) a fluid filled in the posterior chamber of eye.
 (c) a fluid filled in the sacculus of ear.
 (d) an aperture in the auricle of heart.
17. Cranial nerves originating from medulla oblongata are
 (a) III, VII, IX (b) IX, X, XI, XII, III
 (c) I, VIII, IX, X (d) VII, IX, X, XI, XII
18. Increase in gastro-intestinal secretion and movement after ingestion of food is mainly brought about by
 (a) sympathetic nervous system.
 (b) parasympathetic nervous system.
 (c) central nervous system.
 (d) hormone secreted by thyroid.
19. Secretion of which of the following is under neurosecretory nerve axons ?
 (a) Pineal (b) Adrenal cortex
 (c) Anterior pituitary (d) Posterior pituitary
20. Each spinal nerve in a mammal arises from the spinal cord by two roots, a dorsal and a ventral. Of these the ventral root is composed of
 (a) somatic motor and visceral motor fibres.
 (b) somatic sensory and visceral motor fibres.
 (c) somatic motor and visceral sensory fibres.
 (d) somatic sensory and visceral sensory fibres.
21. In a man, abducens nerve is injured. Which one of the following functions will be affected?
 (a) Movement of eye ball
 (b) Swallowing
 (c) Movement of the tongue
 (d) Movement of the neck
22. Four healthy people in their twenties got involved in injuries resulting in damage and death of few cells of the following. Which of the cells are least likely to be replaced by new cells ?
 (a) Osteocytes
 (b) Malpighian layer of the skin
 (c) Liver cells
 (d) Neurons
23. Which one of the following statement in regard to nerve activity is true ?
 (a) The synaptic cleft does not prevent direct propagation of action potential from presynaptic neuron to post synaptic cell.
 (b) Information across the synaptic cleft is transmitted by means of a chemical neurotransmitter in small vesicle.
 (c) Combination of neurotransmitter with receptor site changes membrane potential without changing membrane potentiality.
 (d) In tetanus the excitatory impulse to muscles are inhibited leading to lock jaw.
24. When a neuron is stimulated to generate nerve impulse, the electrical potential on the inside of the nerve membrane changes
 (a) from negative to positive and remains positive.
 (b) from negative to positive and back to negative.
 (c) from positive to negative and remains negative.
 (d) from positive to negative and back to positive.

RESPONSE GRID					
	15. ⓐⓑⓒⓓ	16. ⓐⓑⓒⓓ	17. ⓐⓑⓒⓓ	18. ⓐⓑⓒⓓ	19. ⓐⓑⓒⓓ
	20. ⓐⓑⓒⓓ	21. ⓐⓑⓒⓓ	22. ⓐⓑⓒⓓ	23. ⓐⓑⓒⓓ	24. ⓐⓑⓒⓓ

Space for Rough Work

25. During the transmission of nerve impulse, which of the following takes place ?
    (a) flux of Na^+ inwards and K^+ outwards.
    (b) flux of K^+ inwards and Na^+ outwards.
    (c) flux of K^+ inwards and Na^+ inwards.
    (d) Flux of K^+ outwards and Na^+ outwards

26. When an impulse passes, the membrane is depolarized and the charge of the cells is
    (a) outside positive and inside negative.
    (b) inside positive and outside negative.
    (c) both sides have zero potential.
    (d) both sides are electronegative.

27. The potential difference between outside and inside of a nerve before excitation is known as
    (a) resting potential
    (b) action potential
    (c) spike potential
    (d) reaction potential

28. The action potential of a nerve cell is
    (a) 60 mV (b) 55 mV
    (c) 80 mV (d) 75 mV

29. Unidirectional transmission of a nerve impulse through nerve fibre is due to the fact that
    (a) nerve fibre is insulated by a medullary sheath.
    (b) sodium pump starts operating only at the cyton and then continues into the nerve fibre.
    (c) neurotransmitters are released by dendrites and not by axon endings.
    (d) neurotransmitters are released by the axon endings and not by dendrites.

30. Parkinson's disease (characterized by tremors and progressive rigidity of limb) is caused by degeneration of brain neurons that are involved in movement control and make use of neurotransmitter
    (a) acetylcholine (b) nor epinephrine
    (c) dopamine (d) GABA

RESPONSE GRID	
	25. ⓐⓑⓒⓓ 26. ⓐⓑⓒⓓ 27. ⓐⓑⓒⓓ 28. ⓐⓑⓒⓓ 29. ⓐⓑⓒⓓ
	30. ⓐⓑⓒⓓ

DAILY PRACTICE PROBLEM SHEET 38 - BIOLOGY			
Total Questions	30	Total Marks	120
Attempted		Correct	
Incorrect		Net Score	
Cut-off Score	32	Qualifying Score	56
Success Gap = Net Score – Qualifying Score			
Net Score = (Correct × 4) – (Incorrect × 1)			

Space for Rough Work

DPP - Daily Practice Problems

Name : | Date :

Start Time : | End Time :

BIOLOGY 39

SYLLABUS : Neural control and coordinations Part II : Sense organs – Eye and Ear

Max. Marks : 120 | **Time : 60 min.**

GENERAL INSTRUCTIONS

- The Daily Practice Problem Sheet contains 30 MCQ's. For each question only one option is correct. Darken the correct circle/ bubble in the Response Grid provided on each page.
- You have to evaluate your Response Grids yourself with the help of solution booklet.
- Each correct answer will get you 4 marks and 1 mark shall be deduced for each incorrect answer. No mark will be given/ deducted if no bubble is filled. Keep a timer in front of you and stop immediately at the end of 60 min.
- The sheet follows a particular syllabus. Do not attempt the sheet before you have completed your preparation for that syllabus. Refer syllabus sheet in the starting of the book for the syllabus of all the DPP sheets.
- After completing the sheet check your answers with the solution booklet and complete the Result Grid. Finally spend time to analyse your performance and revise the areas which emerge out as weak in your evaluation.

DIRECTIONS (Q.1-Q.30) : There are 30 multiple choice questions. Each question has 4 choices (a), (b), (c) and (d), out of which ONLY ONE choice is correct.

1. The pacinian corpuscle present in the skin is for ?
   (a) pain (b) pressure
   (c) movement (d) temperature

2. Jacobson's organ in man
   (a) functions as smelling organs.
   (b) functions as gustatory organ.
   (c) is a vestigial organ.
   (d) functions as pain sensory organ.

3. Meissner's corpuscles are located in
   (a) pancreas and secrete trypsinogen.
   (b) adrenal and secrete epinephrine.
   (c) spleen and destroy worn out erythrocytes.
   (d) skin and perceive gentle pressure.

4. The sweet and acidic tastes are better detected by
   (a) tip of the tongue.
   (b) base of the tongue.
   (c) middle of the tongue.
   (d) lateral sides of the tongue.

RESPONSE GRID 1. ⓐⓑⓒⓓ 2. ⓐⓑⓒⓓ 3. ⓐⓑⓒⓓ 4. ⓐⓑⓒⓓ

Space for Rough Work

5. The part of eye acting as diaphragm of a camera is
   (a) pupil
   (b) lens
   (c) iris
   (d) ciliary body

6. Sense of smell is perceived by
   (a) pituitary body
   (b) hypothalamus
   (c) olfactory lobe
   (c) cerebrum

7. Which part of the eye controls the amount of light entering in it?
   (a) Cornea
   (b) Ciliary body
   (c) Iris
   (d) Suspensory ligament

8. The friction between the eyelids and the cornea is avoided by the secretion of
   (a) lacrymal glands
   (b) conjunctiva and eyelids
   (c) hardarian glands
   (d) meibomian glands

9. Cavity of vitreous humour in the eyes are situated
   (a) behind the lens
   (b) infront of the lens
   (c) behind the retina
   (d) between the retina and sclerotic

10. In mammalian eye, the power of accommodation is controlled by changing the shape of the lens. It is governed by
    (a) cornea (b) pupil
    (c) iris (d) ciliary body

11. The binocular vision in mammals is due to the
    (a) eyes adapted for each vision.
    (b) decussation of optic nerve at optic chiasma but no change of fibres of left and right optic nerve.
    (c) exchange of left and right optic nerve fibres after dicussation at optic chiasma.
    (d) none of the above.

12. Differentiation of colour is the property of the cone, it is because of the
    (a) types of cones sensory of different colours only.
    (b) different types of cones alongwith the different type of colour sensory centres in the brain.
    (c) single type of cones but it has the properties to differentiate the colour.
    (d) the colour differentiation is entirely a property of the colour sensory centre in the brain.

13. Rhodopsin (visual purple) of eye will require?
    (a) guava (b) mango
    (c) carrot (d) wheat

14. The eyes bipolar cells are present in
    (a) Sclerotic (b) choroid
    (c) retina (d) yellow spot

RESPONSE GRID					
	5. ⓐⓑⓒⓓ	6. ⓐⓑⓒⓓ 	7. ⓐⓑⓒⓓ	8. ⓐⓑⓒⓓ	9. ⓐⓑⓒⓓ
	10. ⓐⓑⓒⓓ	11. ⓐⓑⓒⓓ	12. ⓐⓑⓒⓓ	13. ⓐⓑⓒⓓ	14. ⓐⓑⓒⓓ

Space for Rough Work

15. The vitreous chamber is perforated from front to behind by a narrow tube which is known as
(a) vitreous canal
(b) hyaloid canal
(c) vitreous aqueous canal
(d) posterior canal

16. In mammals, the colour is perceived by or which is responsible for colour detection
(a) Rod cells of retina
(b) Cone cells of retina
(c) Cornea lens complex of eye
(d) Rods and cones

17. Human eye lens is
(a) spherical and can be moved forward.
(b) biconvex and cannot be moved forward.
(c) spherical and cannot be moved forward.
(d) biconvex and can be moved forward.

18. The central opening of iris is called as
(a) pupil
(b) cornea
(c) lens
(d) fovea centralis

19. Chief functions of the rods and cones in the eye of a vertebrate are
(a) rods are important for vision in bright light and cones in dim light.
(b) rods are important for vision in dim light and cones in bright light.
(c) rods are involved in colour vision and cones in distinguishing intensities of light.
(d) rods and cones are both important for vision in dim light.

20. Transmission of light into nerve impulse is a
(a) mechanical process (b) chemical process
(c) biophysical process (d) biological process

21. The cornea and lens of the mammalian eyes are both
(a) richly supplied by nerves.
(b) richly supplied by blood vessels.
(c) transparent and they diverge the light rays to form an image on retina.
(d) transparent and they contribute in the formation of image on retina.

22. What is the alternate name for the yellow spot in retina?
(a) Canalis centralis (b) Corpus lutea
(c) Fovea centralis (d) Macula lutea

23. In hypermetropia, the image is formed
(a) before retina and is corrected by use of convex lens.
(b) behind retina and is corrected by use of convex lens.
(c) behind retina and is corrected by use of concave lens.
(d) before retina and is corrected by use of concave lens.

24. The black pigment in the eye which reduces the internal reflection is located in
(a) retina (b) iris
(c) sclerotic (d) cornea

RESPONSE GRID					
	15. ⓐⓑⓒⓓ	16. ⓐⓑⓒⓓ	17. ⓐⓑⓒⓓ	18. ⓐⓑⓒⓓ	19. ⓐⓑⓒⓓ
	20. ⓐⓑⓒⓓ	21. ⓐⓑⓒⓓ	22. ⓐⓑⓒⓓ	23. ⓐⓑⓒⓓ	24. ⓐⓑⓒⓓ

Space for Rough Work

25. The human eye is sensitive only to light having wave length ranging from
(a) 80 to 280 nanometres
(b) 380 to 760 nanometres
(c) 780 to 870 nanometres
(d) 880 to 980 nanometres

26. Scala vestibuli is connected with
(a) fenestra rotundus (b) fenestra ovalis
(c) scala tympani (d) scala media

27. The perception of sound by a mammal involves the stimulation of the mechano receptors located in the internal ear
(a) on the organ of corti.
(b) on the reissner's membrane.
(c) in the sacculus.
(d) in the semicircular canal.

28. Vibrations of fenestra ovalis are transmitted to
(a) perilymph of scala vestibuli.
(b) perilymph of scala tympani.
(c) endolymph of scala media.
(d) endolymph of scala vestibuli.

29. Malleus is present in the
(a) inner ear (b) outer ear
(c) middle ear (d) eye

30. Utriculus is the part of internal ear or membranous labyrinth which forms
(a) lower chamber and is concerned with maintenance of equilibrium.
(b) lower chamber and is concerned with transmission of sound waves.
(c) upper chamber and is concerned with maintenance of equilibrium.
(d) upper chamber and is concerned with perception.

RESPONSE GRID	25. ⓐⓑⓒⓓ	26. ⓐⓑⓒⓓ	27. ⓐⓑⓒⓓ	28. ⓐⓑⓒⓓ	29. ⓐⓑⓒⓓ
	30. ⓐⓑⓒⓓ				

DAILY PRACTICE PROBLEM SHEET 39 - BIOLOGY			
Total Questions	30	Total Marks	120
Attempted		Correct	
Incorrect		Net Score	
Cut-off Score	40	Qualifying Score	60
Success Gap = Net Score – Qualifying Score			
Net Score = (Correct × 4) – (Incorrect × 1)			

Space for Rough Work

DPP - Daily Practice Problems

Name : | Date :

Start Time : | End Time :

BIOLOGY 40

SYLLABUS : Chemical coordination and integration

Max. Marks : 120 **Time : 60 min.**

GENERAL INSTRUCTIONS

- The Daily Practice Problem Sheet contains 30 MCQ's. For each question only one option is correct. Darken the correct circle/ bubble in the Response Grid provided on each page.
- You have to evaluate your Response Grids yourself with the help of solution booklet.
- Each correct answer will get you 4 marks and 1 mark shall be deduced for each incorrect answer. No mark will be given/ deducted if no bubble is filled. Keep a timer in front of you and stop immediately at the end of 60 min.
- The sheet follows a particular syllabus. Do not attempt the sheet before you have completed your preparation for that syllabus. Refer syllabus sheet in the starting of the book for the syllabus of all the DPP sheets.
- After completing the sheet check your answers with the solution booklet and complete the Result Grid. Finally spend time to analyse your performance and revise the areas which emerge out as weak in your evaluation.

DIRECTIONS (Q.1-Q.30) : There are 30 multiple choice questions. Each question has 4 choices (a), (b), (c) and (d), out of which ONLY ONE choice is correct.

1. What is hormone?
   (a) Glandular secretion
   (b) Enzyme
   (c) Chemical messenger
   (d) Organic complex substance
2. The effect of different hormones on the body can be best said to bring about
   (a) stimulation of organs.
   (b) release of inherent capacities.
   (c) proper growth.
   (d) co-ordination of functions.
3. Rate of hormone synthesis and secretion depends upon
   (a) functional efficiency of the feedback system.
   (b) amount of excitation in target tissue.
   (c) degree of inhibition caused.
   (d) functional state of the tissue/organ.
4. Which of the following statement is not false?
   (a) Hormone produced in thyroid stimulates metabolism.
   (b) Hormone produced in ovary affects the uterine contraction.
   (c) Hormone produced in small intestine stimulates heart.
   (d) Hormone produced in adrenal cortex stimulates heart beat.

RESPONSE GRID	1. ⓐⓑⓒⓓ	2. ⓐⓑⓒⓓ	3. ⓐⓑⓒⓓ	4. ⓐⓑⓒⓓ

Space for Rough Work

5. According to the accepted concept of hormone action, if receptor molecules are removed from target organs
(a) the target organ will continue to respond to the hormone without any difference.
(b) the target organ will continue to respond to the hormone but will require higher concentration.
(c) the target organ will not respond to the hormone.
(d) the target organ will continue to respond to the hormone but in the opposite way.

6. The hormones of posterior pituitary are oxytocin and vasopressin; but latter is better known as
(a) antidiuretic hormone
(b) growth hormone
(c) corticotrophic hormone
(d) neurohypophyseal

7. Which hormone stops the release of FSH from the pituitary after fertilization?
(a) Placental hormone
(b) Fertilizin
(c) Estradiol
(d) Luteinizing hormone

8. The intermediate lobe of the pituitary gland produces a secretion which causes a dramatic darkening of the skin of many fishes, amphibians and reptiles. It is
(a) ACTH (b) FSH
(c) MSH (d) LH

9. Hormone released by posterior lobe of pituitary is concerned with
(a) metabolism of carbohydrates.
(b) stimulation of thyroid.
(c) secondary sexual characters.
(d) contraction of uterus.

10. Which is true for melanocyte stimulating hormone?
(a) It is same as melatonin.
(b) It is secreted by pars tuberalis of anterior pituitary.
(c) Its secretion is stimulated by hydrocortisone.
(d) It is secreted by pars intermedia of anterior pituitary.

11. Which of the following groups of endocrine glands are involved in carbohydrate metabolism?
(a) Pancreas, neurohypophysis and adrenal.
(b) Pancreas, pituitary and thyroid.
(c) Pancreas, pituitary and liver.
(d) Pancreas, adenohypophysis and adrenal.

12. Secretin is secreted from
(a) endocrine gland and acts on an endocrine gland.
(b) exocrine gland and acts on an exocrine gland.
(c) endocrine gland and acts on an exocrine gland.
(d) exocrine gland and acts on an endocrine gland.

13. 'Exophthalmic goitre' (Grave's disease) is caused due to
(a) Hypofunction of the thyroid.
(b) Hyperfunction of the thyroid.
(c) Hypofunction of the parathyroid.
(d) Hyperfunction of the parathyroid.

14. Why thyroxine is a hormone not an enzyme?
(a) It is secreted in small quantity.
(b) It is not a polypeptide.
(c) It has no special effect.
(d) It is directly poured into blood.

RESPONSE GRID					
	5. ⓐⓑⓒⓓ	6. ⓐⓑⓒⓓ	7. ⓐⓑⓒⓓ	8. ⓐⓑⓒⓓ	9. ⓐⓑⓒⓓ
	10. ⓐⓑⓒⓓ	11. ⓐⓑⓒⓓ	12. ⓐⓑⓒⓓ	13. ⓐⓑⓒⓓ	14. ⓐⓑⓒⓓ

Space for Rough Work

15. Hypoparathyroidism results in
(a) upset in metabolism.
(b) improper gonodial function.
(c) convulsions and tetany.
(d) nervousness and wasting.

16. Which of the following condition is responsible for human female to start developing male characteristics like beared, degeneration of uterus and ovaries, enlargement of clitoris etc.?
(a) Over production of estrogen and testosterone.
(b) Damage to posterior pituitary.
(c) Over production of adrenal androgens.
(d) Surgical removal of mammary gland.

17. Which of the following endocrine glands functions under nervous control?
(a) Cortex of adrenal glands.
(b) Medulla of adrenal glands.
(c) Anterior pituitary glands.
(d) Posterior pituitary gland.

18. Pineal body originates from
(a) dorsal part of diencephalon.
(b) ventral part of diencephalon.
(c) ventral part of cerebellum.
(d) dorsal part of cerebellum.

19. Which one of the following statement about sex hormones is correct?
(a) Testosterone is produced by Leydig cells under the influence of luteinizing hormone (LH).
(b) Progesterone is secreted by corpus luteum and soften ligaments during child birth.
(c) Estrogen is secreted by both Sertoli cells and corpus luteum.
(d) The progesterone produced by corpus luteum is biologically different from the one produced by placenta.

20. Testosterone, a hormone responsible for the development of secondary sexual characteristics in male is produced by the
(a) spermatogonia.
(b) seminiferous tubules.
(c) anterior lobe of the pituitary.
(d) cells that lie between seminiferous tubules.

21. Progesterone is
(a) an enzyme for digesting proteins.
(b) a hormone to initiate uterine contraction during child birth.
(c) an amino acid which may cause alkaptonuria.
(d) a hormone concerned with retention and growth of pregnancy.

22. Correct hormonal sequence in the increase of menstruation is
(a) estrogen, FSH and progesterone.
(b) estrogen, progesterone and FSH.
(c) FSH, progesterone and estrogen.
(d) FSH, estrogen and progesterone.

23. High increase in oxytocin level in a pregnant lady results in
(a) increased synthesis of milk.
(b) decrease in haemoglobin %.
(c) abortion.
(d) high blood pressure.

24. Which hormone(s) of the following endocrine glands lacks peptides, amines and sulphur?
(a) Hormone of anterior pituitary.
(b) Hormone of posterior pituitary and pancreas.
(c) Hormone of thyroid and adrenal gland.
(d) Hormone of testes and ovary.

RESPONSE GRID					
	15. ⓐⓑⓒⓓ	16. ⓐⓑⓒⓓ	17. ⓐⓑⓒⓓ	18. ⓐⓑⓒⓓ	19. ⓐⓑⓒⓓ
	20. ⓐⓑⓒⓓ	21. ⓐⓑⓒⓓ	22. ⓐⓑⓒⓓ	23. ⓐⓑⓒⓓ	24. ⓐⓑⓒⓓ

Space for Rough Work

25. Which one of the following pair is the matching pair of the part and the hormone it secretes?
(a) Thyroid - Epinephrine
(b) Alpha cells of pancreas - Glucagon
(c) Anterior pituitary - Adrenaline
(d) Stomach epithelium - Secretin

26. The hormone glucagon
(a) has the opposite effect as that of insulin.
(b) is produced in the beta cells of pancreas.
(c) converts glucose into glycogen.
(d) is used in the treatment of diabetes mellitus.

27. Blood sugar level can be decreased by
(a) insulin given from mouth.
(b) glucagon given through mouth.
(c) intraveinal injection of insulin.
(d) intraveinal injection of glucagon.

28. If the pancreating duct of a healthy dog is blocked an hour after it had its food, which one of the following functions of the pancreas will not be affected?
(a) Carbohydrate digestion
(b) Neutralization of chyme
(c) Break down of protein
(d) Maintenance of normal blood sugar level

29. The structure of insulin consists of two polypeptide chains *A* and *B*. Which of the following statements regarding the two polypeptide chains are true?
(a) Both the chains have equal number of amino acids.
(b) Chain *A* has 21 amino acids, while chain *B* has 30 amino acids.
(c) Chain *A* has 30 amino acids, while chain *B* has 21 amino acids.
(d) Chain *A* has 11 amino acids, while chain *B* has 40 amino acids.

30. Which one of the following hormone is anti-inflammatory?
(a) Secretin (b) Epinephrine
(c) Glucoprotein (d) Glucocorticoid

RESPONSE GRID	25. ⓐⓑⓒⓓ	26. ⓐⓑⓒⓓ	27. ⓐⓑⓒⓓ	28. ⓐⓑⓒⓓ	29. ⓐⓑⓒⓓ
	30. ⓐⓑⓒⓓ				

DAILY PRACTICE PROBLEM SHEET 40 - BIOLOGY

Total Questions	30	Total Marks	120
Attempted		Correct	
Incorrect		Net Score	
Cut-off Score	40	Qualifying Score	64
Success Gap = Net Score – Qualifying Score			
Net Score = (Correct × 4) – (Incorrect × 1)			

Space for Rough Work

DPP - Daily Practice Problems

Name :

Date :

Start Time :

End Time :

BIOLOGY 41

SYLLABUS : Reproduction in Plants – Asexual reproduction, Vegetative propagation, Parthenogenesis

Max. Marks : 120 **Time : 60 min.**

GENERAL INSTRUCTIONS

- The Daily Practice Problem Sheet contains 30 MCQ's. For each question only one option is correct. Darken the correct circle/ bubble in the Response Grid provided on each page.
- You have to evaluate your Response Grids yourself with the help of solution booklet.
- Each correct answer will get you 4 marks and 1 mark shall be deduced for each incorrect answer. No mark will be given/ deducted if no bubble is filled. Keep a timer in front of you and stop immediately at the end of 60 min.
- The sheet follows a particular syllabus. Do not attempt the sheet before you have completed your preparation for that syllabus. Refer syllabus sheet in the starting of the book for the syllabus of all the DPP sheets.
- After completing the sheet check your answers with the solution booklet and complete the Result Grid. Finally spend time to analyse your performance and revise the areas which emerge out as weak in your evaluation.

DIRECTIONS (Q.1-Q.30) : There are 30 multiple choice questions. Each question has 4 choices (a), (b), (c) and (d), out of which ONLY ONE choice is correct.

1. Apomixis in plant means development of a plant
   (a) from root cuttings
   (b) without fusion of gametes
   (c) from fusion of gametes
   (d) from stem cuttings
2. Micropropagation is
   (a) germination of seed with cotyledons above the soil.
   (b) a technique to obtain new plants by cultivating the cells or tissues in culture medium.
   (c) to manufacture hormones.
   (d) the mature stage of endosperm.
3. Juvenile and adult plants are
   (a) morphologically different
   (b) physiologically same
   (c) both (a) and (b)
   (d) none of the above
4. Which type of propagation is better for the plant?
   (a) By seeds
   (b) By tubers
   (c) By bulbs
   (d) By rhizome

RESPONSE GRID	1. ⓐⓑⓒⓓ	2. ⓐⓑⓒⓓ	3. ⓐⓑⓒⓓ	4. ⓐⓑⓒⓓ

Space for Rough Work

5. A clone is a group of individuals obtained through
(a) self pollination
(b) hybridization
(c) vegetative propagation
(d) cross pollination

6. Among the following which one is not a method of vegetative propagation ?
(a) Budding (b) Layering
(c) Sowing (d) Tissue culture

7. Reproducing new plants by cells instead of seeds is known as
(a) biofertilizer
(b) mutation
(c) tissue culture
(d) antibiotics

8. The development of a sporophyte without fertilization from the vegetative cells of the gametophyte is called
(a) zygospory (b) aplanospory
(c) apospory (d) apogamy

9. Genetically identical progeny is produced when an individual
(a) practices self-fertilization.
(b) produces identical gametes.
(c) practices reproduction.
(d) practices by breeding without meiosis.

10. Vegetative reproduction by layering is found in
(a) jasmine (b) mango
(c) rose (d) all of these

11. Development of embryo from the cells of the nucellus is called
(a) parthenocarpy
(b) apocarpy
(c) adventive embryony
(d) apospory

12. The development of gametophyte from the vegetative parts of the sporophyte without the intervention of spores is called
(a) parthenocarpy (b) parthenogenesis
(c) apogamy (d) apospory

13. The reason of formation of embryoid from pollen grain in a tissue culture medium is
(a) organogenesis
(b) double fertilization
(c) test tube culture
(d) cellular totipotency

14. Which of the following propagates through leaf-tip?
(a) *Marchantia*
(b) Moss
(c) Walking fern
(d) Sprout-leaf plant

RESPONSE GRID					
	5. ⓐⓑⓒⓓ	6. ⓐⓑⓒⓓ	7. ⓐⓑⓒⓓ	8. ⓐⓑⓒⓓ	9. ⓐⓑⓒⓓ
	10. ⓐⓑⓒⓓ	11. ⓐⓑⓒⓓ	12. ⓐⓑⓒⓓ	13. ⓐⓑⓒⓓ	14. ⓐⓑⓒⓓ

Space for Rough Work

15. Grafting of tissue or organ between individuals of different species is called
(a) autograft (b) isograft
(c) xenograft (d) allograft

16. One of the plants using 'Foliar adventitious buds' as method for vegetative propagation is
(a) banana (b) ginger
(c) *Bryophyllum* (d) *Calocasia*

17. In which pair both the plants can be vegetatively propagated by leaf pieces?
(a) *Bryophyllum* and *Kalanchoe*
(b) *Chrysanthemum* and *Agave*
(c) *Agave* and *Kalanchoe*
(d) *Asparagus* and *Bryophyllum*

18. In a type of apomixis, known as adventive embryony, embryos develop directly from the
(a) nucellus or integuments.
(b) synergids or antipodals in a embryo sac.
(c) accessary embryo sacs in the ovule.
(d) zygote.

19. Stem cuttings are commonly used for the propagation of
(a) banana (b) rose
(c) mango (d) cotton

20. Carrot micropropagate through
(a) embryo
(b) embryoids
(c) shoot culture
(d) callus

21. The part which is grafted on stalk of another tree is called
(a) graft (b) bulbil
(c) bud (d) scion

22. The formation of gametophyte from the sporophyte (without meiosis) is called
(a) apospory
(b) apogamy
(c) agametospermy
(d) none of these

23. New Banana plants develop from
(a) rhizome (b) sucker
(c) stolon (d) seed

24. Vegetative reproduction of *Agave* occurs through
(a) rhizome (b) stolon
(c) bulbils (d) sucker

RESPONSE GRID					
	15. ⓐⓑⓒⓓ	16. ⓐⓑⓒⓓ	17. ⓐⓑⓒⓓ	18. ⓐⓑⓒⓓ	19. ⓐⓑⓒⓓ
	20. ⓐⓑⓒⓓ	21. ⓐⓑⓒⓓ	22. ⓐⓑⓒⓓ	23. ⓐⓑⓒⓓ	24. ⓐⓑⓒⓓ

Space for Rough Work

25. A perennial plant differs from biennial in
   (a) having underground perennating structure.
   (b) having asexual reproductive structures.
   (c) being tree species.
   (d) not dying after seasonal production of flowers.

26. Which of the following is eye of potato?
   (a) Axillary bud
   (b) Accessory bud
   (c) Adventitious bud
   (d) Apical bud

27. Bananas are seedless because they
   (a) reproduce asexually.
   (b) are triploid.
   (c) are cross-pollinated.
   (d) are parthenocarpic.

28. The parthenogenesis refers to a fruit
   (a) with seeds after pollination.
   (b) without seed after pollination.
   (c) with viable seed after fertilization.
   (d) seedless without fertilization.

29. Seedless fruits in *Vitis* are formed due to
   (a) sterility
   (b) double fertilization
   (c) syngamy
   (d) parthenocarpy

30. Which plant will lose its economic value, if its fruits are produced by induced parthenocarpy?
   (a) Grape
   (b) Pomegranate
   (c) Orange
   (d) Banana

RESPONSE GRID	25.ⓐⓑⓒⓓ	26.ⓐⓑⓒⓓ	27.ⓐⓑⓒⓓ	28.ⓐⓑⓒⓓ	29.ⓐⓑⓒⓓ
	30.ⓐⓑⓒⓓ				

DAILY PRACTICE PROBLEM SHEET 41 - BIOLOGY			
Total Questions	30	Total Marks	120
Attempted		Correct	
Incorrect		Net Score	
Cut-off Score	44	Qualifying Score	72
Success Gap = Net Score – Qualifying Score			
Net Score = (Correct × 4) – (Incorrect × 1)			

Space for Rough Work

DPP - Daily Practice Problems

Name :

Date :

Start Time :

End Time :

BIOLOGY 42

SYLLABUS : Reproduction in Animals – Asexual reproduction, Parthenogenesis

Max. Marks : 120 **Time : 60 min.**

GENERAL INSTRUCTIONS

- The Daily Practice Problem Sheet contains 30 MCQ's. For each question only one option is correct. Darken the correct circle/ bubble in the Response Grid provided on each page.
- You have to evaluate your Response Grids yourself with the help of solution booklet.
- Each correct answer will get you 4 marks and 1 mark shall be deduced for each incorrect answer. No mark will be given/ deducted if no bubble is filled. Keep a timer in front of you and stop immediately at the end of 60 min.
- The sheet follows a particular syllabus. Do not attempt the sheet before you have completed your preparation for that syllabus. Refer syllabus sheet in the starting of the book for the syllabus of all the DPP sheets.
- After completing the sheet check your answers with the solution booklet and complete the Result Grid. Finally spend time to analyse your performance and revise the areas which emerge out as weak in your evaluation.

DIRECTIONS (Q.1-Q.30) : There are 30 multiple choice questions. Each question has 4 choices (a), (b), (c) and (d), out of which ONLY ONE choice is correct.

1. Which are exclusively viviparous?
   (a) Bony fishes (b) Cartilaginous fishes
   (c) Sharks (d) Whales
2. When both the fusing gametes come from the same parent, then it is called
   (a) exogamy (b) endogamy
   (c) isogamy (d) anisogamy
3. In all the methods of asexual reproduction
   (a) offsprings produced are genetically identical to the parents.
   (b) offsprings produced are genetically different from the parents.
   (c) offsprings produced may or may not be identical to the parents.
   (d) none of the above
4. A person which shows the secondary sexual characters of both male and female is called
   (a) intersex (b) hermaphrodite
   (c) bisexual (d) gynandromorph

RESPONSE GRID	1. ⓐⓑⓒⓓ	2. ⓐⓑⓒⓓ	3. ⓐⓑⓒⓓ	4. ⓐⓑⓒⓓ

Space for Rough Work

5. In sexual reproduction, offsprings resemble the parents
   (a) structurally but not functionally.
   (b) functionally but not structurally.
   (c) both structurally and functionally.
   (d) neither structurally nor functionally.
6. The polyestrous mammal is
   (a) man (b) rabbit
   (c) dog (d) horse
7. Viviparity is found in
   (a) frog
   (b) lizard
   (c) snake
   (d) rabbit
8. Common method of asexual reproduction is by
   (a) regeneration
   (b) budding
   (c) archeocytes
   (d) gemmulation
9. Asexual reproduction takes place in
   (a) higher animals
   (b) lower animals
   (c) plants
   (d) all of these
10. As a result of binary fission number of individuals produced by one fission is
   (a) two (b) three
   (c) four (d) five
11. An alternation of asexual and sexual generations where both the generations are diploid and the haploid stage is represented only by the gametes is called
   (a) alternation of generation
   (b) metagenesis
   (c) paedogenesis
   (d) parthenogenesis
12. Differentiation in morphology of the two sexes of the same species is called
   (a) hermaphrodite
   (b) heteromorphosis
   (c) sexual dimorphism
   (d) unisexual
13. Fertilization is internal in
   (a) toads
   (b) frogs
   (c) dog fish
   (d) cat fish
14. Which type of reproduction is found in *Hydra*?
   (a) Polyembryony
   (b) Sexual and asexual
   (c) Parthenogenesis
   (d) Encystment

RESPONSE GRID					
	5. ⓐⓑⓒⓓ	6. 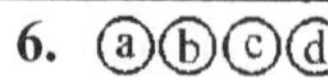	7. ⓐⓑⓒⓓ	8. ⓐⓑⓒⓓ	9. ⓐⓑⓒⓓ
	10. ⓐⓑⓒⓓ	11. ⓐⓑⓒⓓ	12. ⓐⓑⓒⓓ	13. ⓐⓑⓒⓓ	14. ⓐⓑⓒⓓ

Space for Rough Work

15. Gemmule formation in sponges is helpful in
(a) parthenogenesis (b) sexual reproduction
(c) only dissemination (d) asexual reproduction

16. Which of the following is the mode of reproduction in *Amoeba*?
(a) Binary fission only.
(b) Binary fission and multiple fission.
(c) Binary fission and conjugation.
(d) Multiple fission only.

17. Which of the following shows the sexual dimorphism?
(a) *Hydra* and *Ascaris*
(b) *Hydra* and *Oryctolagus*
(c) *Ascaris* and *Pheretima*
(d) *Ascaris* and *Oryctolagus*

18. Drones in a colony of honey bees originate by
(a) thelytoky
(b) arrhenotoky
(c) cyclic parthenogenesis
(d) diploid parthenogenesis

19. Plasmotomy occurs in
(a) *Opalina* (b) *Plasmodium*
(c) *Aurelia* (d) *Amoeba*

20. In earthworms, self fertilization does not occur due to
(a) hypogyny (b) protogyny
(c) protandry (d) epigyny

21. In some species parthenogenesis may alternate with sexual reproduction, this process is called
(a) complete parthenogenesis
(b) incomplete or cyclic parthenogenesis
(c) both (a) and (b)
(d) none of the above

22. Haploid parthenogenesis among insects is shown by order
(a) hymenoptera (b) homoptera
(c) coleoptera (d) all of these

23. Cessation of menstrual cycle in the human female is known as
(a) ovulation
(b) puberty
(c) menopause
(d) maturation

24. Arrhenotoky is related to
(a) parthenogenesis
(b) sexual reproduction
(c) both (a) and (b)
(d) none of these

RESPONSE GRID					
	15. ⓐⓑⓒⓓ	16. ⓐⓑⓒⓓ	17. ⓐⓑⓒⓓ	18. ⓐⓑⓒⓓ	19. ⓐⓑⓒⓓ
	20. ⓐⓑⓒⓓ	21. ⓐⓑⓒⓓ	22. ⓐⓑⓒⓓ	23. ⓐⓑⓒⓓ	24. ⓐⓑⓒⓓ

Space for Rough Work

25. Binary fission is found in
    (a) *Amoeba* (c) *Paramecium*
    (c) *Planaria* (d) All of these

Direction (Q. 26 to Q. 30): Read the Statement-1 and Statement-2 carefully to mark the correct option out of the options given below .

(a) If both the Statement-1 and the Statement-2 are true and the Statement-2 is a correct explanation of the Statement-1.
(b) If both the Statement-1 and Statement-2 are true but the Statement-2 is not a correct explanation of the Statement-1.
(c) If the Statement-1 is true but the Statement-2 is false.
(d) If both the Statement-1 and Statement-2 are false.
(e) If the Statement-1 is false but Statement-2 is true.

26. **Statement-1 :** Sexual reproduction is advantageous than asexual reproduction.
    Statement-2 : It is rapid mode of reproduction while asexual reproduction is slow mode of reproduction.
27. **Statement-1 :** *Amoeba* shows multiple fission during unfavourable conditions.
    Statement-2 : Chances of survival are less during unfavourable conditions.
28. **Statement-1 :** Asexual reproduction is also called blastogenesis.
    Statement-2 : In asexual reproduction, there is no formation and fusion of gametes.
29. **Statement-1 :** Individuals produced by asexual reproduction are genetically similar to parents.
    Statement-2 : Asexual reproduction involves only mitotic divisions.
30. **Statement-1 :** In honey bee, life history shows natural and complete parthenogenesis.
    Statement-2 : In honey bee, both drones and workers are parthenogenetically developed.

RESPONSE GRID	
	25. ⓐⓑⓒⓓ 26. ⓐⓑⓒⓓ 27. ⓐⓑⓒⓓ 28. ⓐⓑⓒⓓ 29. ⓐⓑⓒⓓ 30. ⓐⓑⓒⓓ

DAILY PRACTICE PROBLEM SHEET 42 - BIOLOGY			
Total Questions	30	Total Marks	120
Attempted		Correct	
Incorrect		Net Score	
Cut-off Score	36	Qualifying Score	60
Success Gap = Net Score – Qualifying Score			
Net Score = (Correct × 4) – (Incorrect × 1)			

Space for Rough Work

DPP - Daily Practice Problems

Name : Date :

Start Time : End Time :

BIOLOGY 43

SYLLABUS : Sexual reproduction in flowering plants – Pollination, Fertilization, Embryo, Endosperm

Max. Marks : 120 **Time : 60 min.**

GENERAL INSTRUCTIONS

- The Daily Practice Problem Sheet contains 30 MCQ's. For each question only one option is correct. Darken the correct circle/ bubble in the Response Grid provided on each page.
- You have to evaluate your Response Grids yourself with the help of solution booklet.
- Each correct answer will get you 4 marks and 1 mark shall be deduced for each incorrect answer. No mark will be given/ deducted if no bubble is filled. Keep a timer in front of you and stop immediately at the end of 60 min.
- The sheet follows a particular syllabus. Do not attempt the sheet before you have completed your preparation for that syllabus. Refer syllabus sheet in the starting of the book for the syllabus of all the DPP sheets.
- After completing the sheet check your answers with the solution booklet and complete the Result Grid. Finally spend time to analyse your performance and revise the areas which emerge out as weak in your evaluation.

DIRECTIONS (Q.1-Q.30) : There are 30 multiple choice questions. Each question has 4 choices (a), (b), (c) and (d), out of which ONLY ONE choice is correct.

1. Which one of the following character causes cross pollination?
   (a) Cleistogamy (b) Dichogamy
   (c) Homogamy (d) All of these
2. In which characters air pollinated flowers differ from insect pollinated ones?
   (a) Due to small perianth and sticky pollen.
   (b) Small coloured perianth and heavy pollen grains.
   (c) Coloured perianth and large pollen grains.
   (d) Without perianth and light pollen grains.
3. Pollen kit is present in
   (a) anemophilous flowers
   (b) entomophilous flowers
   (c) ornithophilous flowers
   (d) hydrophilious flowers
4. Pollination by slug and snails is called
   (a) ornithophilous
   (b) malacophilous
   (c) anemophilous
   (d) chiropterophilous

RESPONSE GRID	1. ⓐⓑⓒⓓ	2. ⓐⓑⓒⓓ	3. ⓐⓑⓒⓓ	4. ⓐⓑⓒⓓ

Space for Rough Work

5. Dichogamy which helps in cross pollination is a floral mechanism in which
   (a) pollen sac and stigma are at different heights.
   (b) anther and stigma mature at different times.
   (c) structure of pollen sac and stigma functions as hurdles.
   (d) pollen grain is unable to germinate on the stigma of the same flower.

6. Correct definition of pollination is
   (a) transfer of pollen grain from anther to stigma.
   (b) germination of pollen grain.
   (c) growth of pollen tube in ovule.
   (d) visits of insects in flower.

7. Self pollination means
   (a) occurrence of male and female sex organs in the same flower.
   (b) germination of pollens within the anther.
   (c) transference of pollens from anther to the stigma within the same flower.
   (d) transference of pollens from one flower to another on the same plant.

8. Part of the gynoecium which receives the pollen is called
   (a) ovary
   (b) ovule
   (c) style
   (d) stigma

9. Pollination characteristically occurs in
   (a) angiosperms and fungi.
   (b) angiosperms and gymnosperms.
   (c) pteridophytes and angiosperms.
   (d) bryophytes and angiosperms.

10. Syngamy means
    (a) fusion of similar spores.
    (b) fusion of dissimilar spores.
    (c) fusion of cytoplasm.
    (d) fusion of gametes.

11. Which one of the following is an example of cleistogamy?
    (a) Sunflower (b) *Vallisneria*
    (c) *Commelina* (d) *Calotropis*

12. Number of nuclei taking part in double fertilization is
    (a) 2 (b) 3
    (c) 4 (d) 5

13. The process of fusion between male nucleus and egg nucleus is called as
    (a) syngamy (b) triple fusion
    (c) double fertilization (d) conjugation

14. In angiosperm, triple fusion is necessary for the formation of
    (a) seed coat (b) fruit wall
    (c) embryo (d) endosperm

RESPONSE GRID					
	5. ⓐⓑⓒⓓ	6. 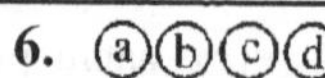 ⓐⓑⓒⓓ	7. ⓐⓑⓒⓓ	8. ⓐⓑⓒⓓ	9. ⓐⓑⓒⓓ
	10. ⓐⓑⓒⓓ	11. ⓐⓑⓒⓓ	12. ⓐⓑⓒⓓ	13. ⓐⓑⓒⓓ	14. ⓐⓑⓒⓓ

Space for Rough Work

15. Double fertilization involves
(a) fertilization of the egg by two male gametes.
(b) fertilization of two eggs in the same embryo sac by two sperms brought by one pollen tube.
(c) fertilization of the egg and the central cell by two sperms brought by different pollen tubes.
(d) fertilization of the egg and the central cell by two sperms brought by the same pollen tube.

16. In angiosperms, free nuclear divisions occurs during
(a) gamete formation
(b) embryo formation
(c) endosperm formation
(d) flower formation

17. Milky water of green coconut is
(a) liquid nucellus
(b) liquid of female gametophyte
(c) liquid endosperm
(d) liquid embryo

18. Dicot embryo consists of
(a) radicle and plumule.
(b) radicle, plumule, cotyledons and sometimes endosperm.
(c) radicle, plumule, cotyledons and tegmen.
(d) radicle, plumule, cotyledons, tegmen and testa.

19. The outermost layer of endosperm of maize grain is called
(a) epidermis (b) tunica
(c) aleurone (d) pericarp

20. In angiosperms endosperm is formed by
(a) free nuclear divisions of megaspore.
(b) division of fused polar nuclei.
(c) division of fused polar nuclei and male gamete.
(d) division of fused synergids and male gamete.

21. Endospermic nucleus is usually
(a) haploid (b) diploid
(c) triploid (d) tetraploid

22. Turn pipe mechanism is present in
(a) *Salvia* (b) *Ophrys*
(c) *Yucca* (d) *Vallisneria*

23. In agamospermy, the embryo sac is diploid because it is formed without meiosis. Such embryo sac may develop from
(a) megaspore mother cell
(b) micropore mother cell
(c) megaspores
(d) microspores

24. Presence of many embroys (Polyembryony) is a characteristic feature of
(a) *Citrus* (b) mango
(c) banana (d) none of these

RESPONSE GRID	15. ⓐⓑⓒⓓ	16. ⓐⓑⓒⓓ	17. ⓐⓑⓒⓓ	18. ⓐⓑⓒⓓ	19. ⓐⓑⓒⓓ
	20. ⓐⓑⓒⓓ	21. ⓐⓑⓒⓓ	22. ⓐⓑⓒⓓ	23. ⓐⓑⓒⓓ	24. ⓐⓑⓒⓓ

Space for Rough Work

25. Which of the following workers are related as ecologist, palaeobotanist and embryologist?
    (a) B. Sahni, R. Mishra, P. Maheshwari
    (b) R. Mishra, B. Sahni, P. Maheshwari
    (c) B. Sahni, P. Maheshwari, R. Mishra
    (d) P. Maheshwari, R. Mishra, B. Sahni
26. In a seed of maize, scutellum is considered as cotyledon because it
    (a) protects the embryo.
    (b) contains food for the embryo.
    (c) absorbs food materials and supplies them to the embryo.
    (d) converts itself into a monocot leaf.
27. In which one of the following would you expect to find glyoxysomes?
    (a) Endosperm of wheat
    (b) Endosperm of castor
    (c) Palisade cells in leaf
    (d) Root hairs
28. Even after killing the generative cell with a laser beam, the pollen grain of a flowering plant germinates and produces normal pollen tube because
    (a) laser beam stimulates pollen germination and pollen tube growth.
    (b) the laser beam does not damage the region from which pollen tube emerges.
    (c) the contents of killed generative cell permit germination and pollen tube growth.
    (d) the vegetative cell has not been damaged.
29. When the pollen grains are not transferred from anthers to the stigma in flower due to the barrier or fence, it is referrred as
    (a) heterostyly (b) herkogamy
    (c) dichogamy (d) cleistogamy
30. Despite high level of heterozygosity, the progeny derived from seed of a cross pollinated plant was found to be completely uniform. One reason for this may be the phenomenon of
    (a) parthenocarpy (b) apomixis
    (c) induced mutation (d) polyploidy

RESPONSE GRID	
	25. ⓐⓑⓒⓓ 26. ⓐⓑⓒⓓ 27. ⓐⓑⓒⓓ 28. ⓐⓑⓒⓓ 29. ⓐⓑⓒⓓ 30. ⓐⓑⓒⓓ

DAILY PRACTICE PROBLEM SHEET 43 - BIOLOGY			
Total Questions	30	Total Marks	120
Attempted		Correct	
Incorrect		Net Score	
Cut-off Score	44	Qualifying Score	72
Success Gap = Net Score – Qualifying Score			
Net Score = (Correct × 4) – (Incorrect × 1)			

Space for Rough Work

DPP - Daily Practice Problems

Name :   Date :

Start Time :   End Time :

BIOLOGY 44

SYLLABUS : Sexual reproduction in flowering plants – Pollination, Fertilization, Embryo, Endosperm

Max. Marks : 120   **Time : 60 min.**

GENERAL INSTRUCTIONS

- The Daily Practice Problem Sheet contains 30 MCQ's. For each question only one option is correct. Darken the correct circle/ bubble in the Response Grid provided on each page.
- You have to evaluate your Response Grids yourself with the help of solution booklet.
- Each correct answer will get you 4 marks and 1 mark shall be deduced for each incorrect answer. No mark will be given/ deducted if no bubble is filled. Keep a timer in front of you and stop immediately at the end of 60 min.
- The sheet follows a particular syllabus. Do not attempt the sheet before you have completed your preparation for that syllabus. Refer syllabus sheet in the starting of the book for the syllabus of all the DPP sheets.
- After completing the sheet check your answers with the solution booklet and complete the Result Grid. Finally spend time to analyse your performance and revise the areas which emerge out as weak in your evaluation.

DIRECTIONS (Q.1-Q.30) : There are 30 multiple choice questions. Each question has 4 choices (a), (b), (c) and (d), out of which ONLY ONE choice is correct.

1. Cryptorchidism is the condition in man when
   (a) there are two testis in each scrotum.
   (b) testis do not descend into the scrotum.
   (c) testis enlarge in the scrotum.
   (d) testis degenerate in the scrotum.
2. If the vas deferens of a man is surgically disconnected
   (a) sperms in the semen will be without nuclei.
   (b) semen will be without sperms.
   (c) spermatogenesis will not occur.
   (d) sperms in the semen will be non-motile.
3. Gubernaculum cordis is a contractile structure that
   (a) pulls down the testis during breeding season into the scrotal sac.
   (b) allows daily migration of the testis from the abdominal cavity into the scrotum.
   (c) facilitates ejaculation of spermatozoa from the testis.
   (d) keeps the testis in position.
4. The narrow middle part of the epididymis is called
   (a) caput epididymis
   (b) cauda epididymis
   (c) corpus epididymis
   (d) all of these

RESPONSE GRID	1. ⓐⓑⓒⓓ	2. ⓐⓑⓒⓓ	3. ⓐⓑⓒⓓ	4. ⓐⓑⓒⓓ

Space for Rough Work

5. Bidder's canal is found in
(a) testes of frog
(b) kidney of frog
(c) ovary of mammal
(d) kidney of mammal

6. The function of prostate gland is
(a) storage of semen.
(b) provide motility to sperms.
(c) formation of semen.
(d) release of hormones.

7. In female mammals, Bartholin's glands open into the
(a) vestibule and release a lubricating fluid in the vagina.
(b) uterus and release a lubricating fluid during the birth of young ones.
(c) urinary bladder and assist in release of urine.
(d) fallopian tubes and release a secretion which makes sperms motile.

8. If after ovulation, no pregnancy results, the corpus luteum
(a) is maintained by the presence of progesterone.
(b) degenerates in a short time.
(c) becomes active and secretes lot of FSH and LH.
(d) produces lot of oxytocin and relaxin.

9. Egg is liberated from ovary in
(a) oogonial stage
(b) mature ovum stage
(c) secondary oocyte stage
(d) primary oocyte stage

10. Bartholin's glands in rabbit are found in
(a) male and produce a viscous alkaline fluid which neutralizes acidity in the urethra.
(b) male and produce the clear liquid part of the spermatic fluid.
(c) female and produce the hormone estrogen which regulates secondary sexual characters.
(d) female and produce a clear fluid which lubricates the vestibule during copulation.

11. Ovulation in mammals is caused by
(a) FSH and TSH (b) FSH and LH
(c) FSH and LTH (d) LTH and LH

12. The structure formed after release of ova from graafian follicles and secretory in nature, is
(a) corpus callosum (b) corpus luteum
(c) corpus albicans (d) corpus stratum

13. Ovulation does not occur during pregnancy in the human female because
(a) The follicles are not influenced by the level of progesterone in the blood
(b) The corpus luteum and later the placenta produce large amounts of progesterone
(c) The corpus luteum generates a luteinising hormone that not produced for a long time.
(d) The embryo produces hormones which retard the production of maternal follicle stimulating hormones

14. The first stage of graafian follicular growth involves
(a) addition of layers of granulosa cells.
(b) moderate enlargement of the ovum.
(c) formation of theca interna.
(d) formation of theca externa.

Response Grid	5. ⓐⓑⓒⓓ	6. ⓐⓑⓒⓓ	7. ⓐⓑⓒⓓ	8. ⓐⓑⓒⓓ	9. ⓐⓑⓒⓓ
	10. ⓐⓑⓒⓓ	11. ⓐⓑⓒⓓ	12. ⓐⓑⓒⓓ	13. ⓐⓑⓒⓓ	14. ⓐⓑⓒⓓ

Space for Rough Work

15. Note the relationship between the first two words and suggest a suitable word for the fourth place.
Mesovarium : Ovary :: Mesometrium : ....
(a) Clitoris
(b) Uterus
(c) Testis
(d) Penis

16. Bartholin's glands are situated
(a) on either side of vas deferens in humans.
(b) on the sides of the head of frog.
(c) at the reduced tail end of birds.
(d) on either side of vagina in humans.

17. Ovulation in the human female normally takes place during the menstrual cycle
(a) at the beginning of the proliferative phase.
(b) at the end of the proliferative phase.
(c) at the mid secretory phase.
(d) just before the end of the secretory phase.

18. Which of the following statements is incorrect? Human spermatozoa normally
(a) at a concentration of 1 million/ml of ejaculate will fertilize the ovum.
(b) are released in large numbers after a few days of abstinence.
(c) contain enzyme in their head which helps in penetration of ovum.
(d) move with the help of their tails.

19. Sperm's acrosome has
(a) hyaluronic acid and proacrosine.
(b) hyaluronic acid and fertilizin.
(c) hyaluronidase and proacrosin.
(d) fertilizin and proacrosin.

20. Sperms formed from 4 primary spermatocytes are
(a) 4 (b) 1
(c) 16 (d) 32

21. How many ova and sperms would be produced from 100 secondary oocytes and 100 secondary spermatocytes during gametogenesis in human?
(a) 100 ova, 100 sperms (b) 100 ova, 200 sperms
(c) 50 ova, 100 sperms (d) 200 ova, 200 sperms

22. The period of preparation with reference to developmental phenomena in vertebrates means
(a) formation of gastrula.
(b) formation of germ layers.
(c) tissue differentiation.
(d) parents preparation and elaboration of gametes.

23. The sperm produces substances of enzymatic nature called sperm lysin. In mammals, it is called
(a) hyaluronidase (b) hyaluronic acid
(c) androgamone (d) cryanogamone

24. What is true in the process of fertilization ?
(a) Only one sperm reaches the egg and enters it.
(b) The entry of sperm activates the egg for completing meiosis.
(c) Two haploid nuclei fuse and immediately divide to produce two nuclei which are again haploid.
(d) Only the acrosome of the sperm enters the egg.

RESPONSE GRID					
	15. ⓐⓑⓒⓓ	16. ⓐⓑⓒⓓ	17. ⓐⓑⓒⓓ	18. ⓐⓑⓒⓓ	19. ⓐⓑⓒⓓ
	20. ⓐⓑⓒⓓ	21. ⓐⓑⓒⓓ	22. ⓐⓑⓒⓓ	23. ⓐⓑⓒⓓ	24. ⓐⓑⓒⓓ

Space for Rough Work

25. Amphimixis means
(a) fusion of male and female gametes.
(b) fusion of male and female pronuclei.
(c) movement of the sperm towards ovum.
(d) movement of the ovum towards sperm.

26. The disadvantage of parthenogenesis is
(a) establishment of polyploid generation.
(b) elimination of variety in population.
(c) means of reproduction.
(d) does not encourage the appearance of new and advantageous combinations of genes.

27. In sexually reproducing animals the union of male and female gamete forms a cell which is called
(a) ovarian cell (b) oocyte
(c) zygote (d) graafian follicle

28. Epiboly is the process of
(a) mass migration of cells from the animal hemisphere so that upper micromeres begin to migrate over the edge of the dorsal lip and roll inside and are tucked beneath the outer layer.
(b) over growth when the micromeres divide rapidly and begin to spread downwards over the megameres except at the yolk plug.
(c) rotation of gastrula within the vitelline membrane so that the animal pole becomes anterior.
(d) formation of a small slit like invagination occurring on the grey crescent.

29. Cleavage divisions differ from normal mitotic divisions in that
(a) there is no nuclear division during cleavage.
(b) there is no division of the cytoplasm during cleavage.
(c) there is no period of growth in between the divisions.
(d) the division of the cytoplasm follows nuclear divisions.

30. Which is the correct sequence in the development of man?
(a) Fertilization, zygote, cleavage, morula, blastula, gastrula.
(b) Zygote, morula, blastula, differentiation.
(c) Fertilization, cleavage, gastrula, morula, blastula.
(d) Cleavage, zygote, blastula, morula, gastrula.

RESPONSE GRID	25. ⓐⓑⓒⓓ	26. ⓐⓑⓒⓓ	27. ⓐⓑⓒⓓ	28. ⓐⓑⓒⓓ	29. ⓐⓑⓒⓓ
	30. ⓐⓑⓒⓓ				

DAILY PRACTICE PROBLEM SHEET 44 - BIOLOGY			
Total Questions	30	Total Marks	120
Attempted		Correct	
Incorrect		Net Score	
Cut-off Score	32	Qualifying Score	52
Success Gap = Net Score – Qualifying Score			
Net Score = (Correct × 4) – (Incorrect × 1)			

Space for Rough Work

DPP - Daily Practice Problems

Name : Date :

Start Time : End Time :

BIOLOGY 45

SYLLABUS : Reproductive Health

Max. Marks : 120 **Time : 60 min.**

GENERAL INSTRUCTIONS

- The Daily Practice Problem Sheet contains 30 MCQ's. For each question only one option is correct. Darken the correct circle/ bubble in the Response Grid provided on each page.
- You have to evaluate your Response Grids yourself with the help of solution booklet.
- Each correct answer will get you 4 marks and 1 mark shall be deduced for each incorrect answer. No mark will be given/ deducted if no bubble is filled. Keep a timer in front of you and stop immediately at the end of 60 min.
- The sheet follows a particular syllabus. Do not attempt the sheet before you have completed your preparation for that syllabus. Refer syllabus sheet in the starting of the book for the syllabus of all the DPP sheets.
- After completing the sheet check your answers with the solution booklet and complete the Result Grid. Finally spend time to analyse your performance and revise the areas which emerge out as weak in your evaluation.

DIRECTIONS (Q.1-Q.30) : There are 30 multiple choice questions. Each question has 4 choices (a), (b), (c) and (d), out of which ONLY ONE choice is correct.

1. Legally acceptable term of abortion is
   (a) MTP (b) MMTP
   (c) MTTP (d) none of these

2. The most important component of the oral contraceptive pills is
   (a) progesterone (b) growth hormone
   (c) thyroxin (d) luteinizing hormone

3. Amniocentesis is the withdraw of amniotic fluid in
   (a) menopause
   (b) lactation
   (c) gestation
   (d) pregnancy

4. What is the function of copper-T ?
   (a) Checks mutation
   (b) Stops fertilization
   (c) Stops zygote formation
   (d) Stops obliteration of blastocoel

RESPONSE GRID 1. ⓐⓑⓒⓓ 2. ⓐⓑⓒⓓ 3. ⓐⓑⓒⓓ 4. ⓐⓑⓒⓓ

Space for Rough Work

5. Action of vaginal diaphragm is
   (a) to prevent the ova to come in the uterus.
   (b) to prevent the sperm to come in contact with ova.
   (c) spermicidal
   (d) anti-implantational
6. Surgical removal or cutting and ligation of the ends of oviduct is known as
   (a) tubectomy
   (b) oviductomy
   (c) vasectomy
   (d) ovariectomy
7. Surgical removal of testes is known as
   (a) testectomy
   (b) gonadectomy
   (c) castration
   (d) none of these
8. Removal of a segment surgically and ligation of cut ends of vas deferens is known as
   (a) tubectomy (b) vasectomy
   (c) gonadectomy (d) castration
9. Test tube baby means a baby born when
   (a) it develops from a non-fertilized egg.
   (b) it developed in a test tube.
   (c) it is developed through tissue culture method.
   (d) the ovum is fertilized externally and thereafter implanted in the uterus.
10. The latest technique to produce the child is GIFT. The full form is
   (a) gametic internal fertilization and transfer.
   (b) gametic intra-fallopian transfer.
   (c) gametic internal fallopian transplant.
   (d) general internal fallopian transfer.
11. The mechanical measure of population control includes –
   (a) condom only (b) diaphragm only
   (c) IUD only (d) all of these
12. The chemical method of contraception includes
   (a) jellies only (b) creams only
   (c) foams only (d) all of these
13. Which of the following is the main reason in population explosion ?
   (a) Early marriages
   (b) Decreased food production
   (c) High literacy
   (d) Decline in death rate and increase in longevity
14. The success of birth control programmes in controlling population growth is dependent on
   (a) use of contraceptives
   (b) tubectomy
   (c) vasectomy
   (d) acceptability of the above by the people

RESPONSE GRID					
	5. ⓐⓑⓒⓓ	6. ⓐⓑⓒⓓ	7. ⓐⓑⓒⓓ	8. ⓐⓑⓒⓓ	9. ⓐⓑⓒⓓ
	10. ⓐⓑⓒⓓ	11. ⓐⓑⓒⓓ	12. ⓐⓑⓒⓓ	13. ⓐⓑⓒⓓ	14. ⓐⓑⓒⓓ

Space for Rough Work

15. Trade name of weekly oral contraceptive pill is
    (a) Mala
    (b) Saheli
    (c) Mala A
    (d) Mala D
16. Daily oral contraceptive pill is
    (a) Mala C
    (b) Mala N and Mala D
    (c) Mala A
    (d) Mala D
17. Test tube babies are produced by
    (a) Fertilizing the egg removed from the body of the female with the husband's sperm outside *in vitro* culture. The zygote is transferred back.
    (b) External fertilization and development *in vitro* culture till 32 cells stage before putting the embryo back into mother's uterus.
    (c) Complete development of a baby *in vitro.*
    (d) Development upto 32 cells stage and transplanting embryo in the uterus of a surrogate mother.
18. Which one of the following is tested by the technique of amniocentesis?
    (a) Biochemical abnormalities in the foetus.
    (b) Errors of metabolism in the foetus.
    (c) Chromosomal abnormalities in the foetus.
    (d) All of the above.
19. Foetal sex can be determined by examining cells from the amniotic fluid by looking for
    (a) barr bodies (b) autosomes
    (c) chiasmata (d) kinetochore
20. Which of the following is a method for birth control?
    (a) IUDs (b) GIFT
    (c) HTF (d) IVE-ET
21. In amniocentesis, the fluid is taken from
    (a) foetal blood
    (b) mother's blood
    (c) body fluid of mother
    (d) fluid surrounding foetus
22. Amniocentesis involves the analysis of
    (a) amnion
    (b) body fluid of amniotes
    (c) amino acids of protein
    (d) amniotic fluid
23. Surrogate mother is used for
    (a) induction of lactation.
    (b) artificially inseminated female.
    (c) future mother with transplanted embryo.
    (d) artificial insemination.
24. Progesterone pill helps in preventing pregnancy by not allowing
    (a) ova formation (b) fertilization
    (c) implantation (d) none of these

RESPONSE GRID					
	15. ⓐⓑⓒⓓ	16. ⓐⓑⓒⓓ	17. ⓐⓑⓒⓓ	18. ⓐⓑⓒⓓ	19. ⓐⓑⓒⓓ
	20. ⓐⓑⓒⓓ	21. ⓐⓑⓒⓓ	22. ⓐⓑⓒⓓ	23. ⓐⓑⓒⓓ	24. ⓐⓑⓒⓓ

Space for Rough Work

25. The method of sterilization is
(a) loop (b) IUD
(c) diaphragm (d) tubectomy
26. MTP means
(a) many transferable pregnancies.
(b) medically temporary pregnancy.
(c) medical termination of pregnancy.
(d) multiple temporary pregnancy.
27. Action of jelly and cream is
(a) spermicidal and immovalizing the sperms also.
(b) entangles the sperms.
(c) preventing the ova to be released.
(d) enables the sperms to reach towards ovum speedly.
28. Action of contraceptive is
(a) to prevent the ovulation only.
(b) prevention of ovulation and fertilization only.
(c) prevention of ovulation, fertilization and implantation only.
(d) to prevent the rapid passing of eggs in oviduct.

***Directions* (Q. 29 – Q. 30) :** Each questions contain STATEMENT-1 (Assertion) and STATEMENT-2 (Reason). Each question has 4 choices (1), (2), (3) and (4) out of which ONLY ONE is correct.

(a) Statement- 1 is True, Statement-2 is True, Statement-2 is a correct explanation for Statement -1
(b) Statement -1 is True, Statement -2 is True ; Statement-2 is NOT a correct explanation for Statement - 1
(c) Statement - 1 is True, Statement- 2 is False
(d) Statement -1 is False, Statement -2 is False

29. **Statement 1 :** Test tube baby has raised several legal problems.
Statement 2 : It involves *in vitro* fertilization followed by embryo transfer.
30. **Statement 1 :** Copper-T is an effective contraceptive device in human females.
Statement 2 : Copper-T prevents passage of sperms from vagina upwards into fallopian tubes.

RESPONSE GRID	25. ⓐⓑⓒⓓ	26. ⓐⓑⓒⓓ	27. ⓐⓑⓒⓓ	28. ⓐⓑⓒⓓ	29. ⓐⓑⓒⓓ
	30. ⓐⓑⓒⓓ				

DAILY PRACTICE PROBLEM SHEET 45 - BIOLOGY			
Total Questions	30	Total Marks	120
Attempted		Correct	
Incorrect		Net Score	
Cut-off Score	44	Qualifying Score	72
Success Gap = Net Score – Qualifying Score			
Net Score = (Correct × 4) – (Incorrect × 1)			

Space for Rough Work

DPP - Daily Practice Problems

Name : Date :

Start Time : End Time :

BIOLOGY 46

SYLLABUS : Principles of inheritance and variation-Part I : Mendelism, Interaction of gene, Cytoplasmic inheritance, Linkage and Crossing over

Max. Marks : 120 **Time : 60 min.**

GENERAL INSTRUCTIONS

- The Daily Practice Problem Sheet contains 30 MCQ's. For each question only one option is correct. Darken the correct circle/ bubble in the Response Grid provided on each page.
- You have to evaluate your Response Grids yourself with the help of solution booklet.
- Each correct answer will get you 4 marks and 1 mark shall be deduced for each incorrect answer. No mark will be given/ deducted if no bubble is filled. Keep a timer in front of you and stop immediately at the end of 60 min.
- The sheet follows a particular syllabus. Do not attempt the sheet before you have completed your preparation for that syllabus. Refer syllabus sheet in the starting of the book for the syllabus of all the DPP sheets.
- After completing the sheet check your answers with the solution booklet and complete the Result Grid. Finally spend time to analyse your performance and revise the areas which emerge out as weak in your evaluation.

DIRECTIONS (Q.1-Q.30) : There are 30 multiple choice questions. Each question has 4 choices (a), (b), (c) and (d), out of which ONLY ONE choice is correct.

1. The branch of botany dealing with heredity and variation is called ?
   (a) geobotany (b) sericulture
   (c) genetics (d) evolution
2. Mendel proposed which of the following term for hereditary units?
   (a) Factor (Determiner) (b) Genome
   (c) Genetic particle (d) None of these
3. In 1900 A.D. three biologists independently discovered Mendel's principles. They are
   (a) De Vries, Correns and Tschermak.
   (b) Sutton, Morgan and Bridges.
   (c) Avery, McLeod and McCarthy.
   (d) Bateson, Punnet and Bridgess.
4. Which of the following Mendel has selected for his experiment ?
   (a) Garden pea (b) Pigeon pea
   (c) Sweet pea (d) Moong

RESPONSE GRID	1. ⓐⓑⓒⓓ	2. ⓐⓑⓒⓓ	3. ⓐⓑⓒⓓ	4. ⓐⓑⓒⓓ

Space for Rough Work

5. If an offspring shows any one of the characters of either of parents, the
   (a) character is said to be hereditary.
   (b) offspring is hereditary.
   (c) the character is related with cytoplasm.
   (d) none of the above.

6. What is the correct sequence of the following events?
   1. Formation of the chromosome theory of heredity.
   2. Experiments which proved that DNA is the hereditary material.
   3. Mendel's laws of inheritance-discovery.

   Code:
   (a) 1, 3 and 2 (b) 1, 2 and 3
   (c) 3, 1 and 2 (d) 2, 1 and 3

7. What type of gametes will form by genotype RrYy?
   (a) RY, Ry, rY, ry (b) RY, Ry, ry, ry
   (c) Ry, Ry, Yy, ry (d) Rr, RR, Yy, YY

8. The term 'allelomorphic' implies
   (a) any two characters.
   (b) a pair of contrasting characters.
   (c) sex linked characters.
   (d) a pair of non-contrasting characters.

9. The alleles are
   (a) a pair of genes governing a specific character such as tallness or dwarfness.
   (b) multiple forms of genes.
   (c) genes governing eye characters.
   (d) genes present in allosomes.

10. Organisms phenotypically similar but genotypically different are said to be
   (a) heterozygous
   (b) monozygous
   (c) multizygous
   (d) homozygous

11. When heterozygous tall plants are self-pollinated than tall and dwarf plants are obtained, this is an explanation to
   (a) law of purity of gamete.
   (b) segregation of law.
   (c) division in spores.
   (d) independent assortment.

12. Change of male and female parents during crosses is known as
   (a) back cross
   (b) test cross
   (c) reciprocal cross
   (d) monohybrid cross

13. In F_2 generation, a phenotypic ratio of 1:1:1:1 exhibits
   (a) back cross
   (b) monohybrid test cross
   (c) lethality
   (d) dihybrid test cross

14. Which of the statement is correct?
   (a) Each back cross is a test corss.
   (b) Each test cross is a back cross.
   (c) Both tests have the same meaning.
   (d) Reappearence of similar characters in a test cross is called a back cross.

RESPONSE GRID					
	5. ⓐⓑⓒⓓ	6. ⓐⓑⓒⓓ	7. ⓐⓑⓒⓓ	8. ⓐⓑⓒⓓ	9. ⓐⓑⓒⓓ
	10. ⓐⓑⓒⓓ	11. ⓐⓑⓒⓓ	12. ⓐⓑⓒⓓ	13. ⓐⓑⓒⓓ	14. ⓐⓑⓒⓓ

Space for Rough Work

15. Mendel crossed a pure white-flowered recessive pea plant with a dominant pure red-flowered plant. The first generation of hydrids from the cross should show
   (a) 50% white-flowered and 50% red-flowered plants.
   (b) All red-flowered plants.
   (c) 75% red-flowered and 25% white-flowered plants.
   (d) All white-flowered plants.
16. If in a dihybrid cross Mendel had used two such characters which are linked, he would have faced difficulty in explaining the results on the basis of his
   (a) Law of segregation.
   (b) Law of multiple factor hypothesis.
   (c) Law of independent assortment.
   (d) Law of dominance.
17. A farmer crossed a walnut combed chicken with a single combed one and obtained all walnut combed chickens in F_1. The genotype of the parents was
   (a) $Rr\,Pp \times rr\,pp$ (b) $RR\,PP \times rr\,pp$
   (c) $RR\,pp \times rr\,pp$ (d) $RR\,Pp \times rr\,pp$
18. In sweet pea plants the presence of dominant C and P genes is essential for development of purple colour. The ratio of plants producing flowers of different colours in the progeny of the cross $Cc\,Pp \times Cc\,pp$ will be
   (a) 2 white and 6 purple coloured flowers.
   (b) 2 purple and 6 white coloured flowers.
   (c) 3 white and 5 purple coloured flowers.
   (d) 3 purple and 5 white coloured flowers.
19. From a single ear of corn, a farmer planted 200 kernels which produced 140 tall and 40 dwarf plants. The genotype of these offsprings are most likely
   (a) TT, Tt and tt (b) TT and tts
   (c) TT and Tt (d) Tt and tt
20. While experimenting with the factor of tallness in *Pisum sativum,* Mendel obtained the results 73.97% tall and 26.03% dwarf from one of the cross. This shows that the parents are
   (a) hybrid tall and pure dwarf plant.
   (b) hybrid tall and pure tall plant.
   (c) hybrid tall and hybrid tall plant.
   (d) pure tall and pure dwarf plant.
21. The genotypes of a plant variety were $TtHh$, $Tthh$, $ttHh$ and $tthh$, where T = tallness and H = hairy stem. Which one of the following crosses would produce progeny giving a phenotypic ratio approximately 1 : 1 : 1 : 1?
   (a) $TtHh \times TtHh$
   (b) $TtHh \times Tthh$
   (c) $TtHh \times ttHh$
   (d) $TtHh \times tthh$
22. A self-fertilizing trihybrid plant forms
   (a) 8 different gametes and 16 different zygotes.
   (b) 8 different gametes and 32 different zygotes.
   (c) 8 different gametes and 64 different zygotes.
   (d) 4 different gametes and 16 different zygotes.
23. Pure tall plants are crossed with pure dwarf plants. In the F_1 generation all plants were tall. These tall plants of F_1 generation were selfed and the ratio of tall to dwarf plants obtained was 3 : 1. This is called
   (a) Dominance (b) Inheritance
   (c) Co-dominance (d) Heredity
24. Mendel's law of segregation is applicable to
   (a) dihybrid cross only.
   (b) monohybrid cross only.
   (c) both dihybrid and monohybrid cross.
   (d) dihybrid but not monohybrid.

RESPONSE GRID					
	15. ⓐⓑⓒⓓ	16. ⓐⓑⓒⓓ	17. ⓐⓑⓒⓓ	18. ⓐⓑⓒⓓ	19. ⓐⓑⓒⓓ
	20. ⓐⓑⓒⓓ	21. ⓐⓑⓒⓓ	22. ⓐⓑⓒⓓ	23. ⓐⓑⓒⓓ	24. ⓐⓑⓒⓓ

Space for Rough Work

25. What will be the ratio in F_2 generation if red flowered variety of *Mirabilis jalapa* is crossed with white flowered variety?
(a) 1 : 1 : 1 : 1 (b) 1 : 2 : 1
(c) 2 : 1 (d) 3 : 1

26. Kappa particles indicate
(a) nuclear inheritance
(b) cytoplasmic inheritance
(c) mutation
(d) nucleo-cytoplasmic inheritance

27. Cytoplasmic male sterility in maize is manifestation of interaction between
(a) chloroplast and nuclear genes.
(b) mitochondrial and nuclear genes.
(c) chloroplast and mitochondrial genes.
(d) cytoplasmic factors and male sterile genes.

28. Grain colour in wheat is determined by three pairs of polygenes. Following the cross AABBCC (dark colour) × aabbcc (light colour), in F_2 generation what proportion of the progeny likely to resemble either parent?
(a) None (b) Less than 5 per cent
(c) One third (d) Half

29. The evidence for crossing over which occurs at four stranded stage and not at two stranded stage of the chromosomes, comes from
(a) 2 : 2 : 2 : 2 arrangement of ascospores in *Neurospora.*
(b) 4 : 4 arrangement of ascospores in *Neurospora.*
(c) Studies of meiosis in maize.
(d) Studies on linkage maps of chromosomes in *Drosophila.*

30. The percentage of crossing over will be more if
(a) linked genes are located far apart from each other
(b) linked genes are located close to each other
(c) genes are not linked
(d) genes are located in a different cell

RESPONSE GRID	25. ⓐⓑⓒⓓ	26. ⓐⓑⓒⓓ	27. ⓐⓑⓒⓓ	28. ⓐⓑⓒⓓ	29. ⓐⓑⓒⓓ
	30. ⓐⓑⓒⓓ				

DAILY PRACTICE PROBLEM SHEET 46 - BIOLOGY			
Total Questions	30	Total Marks	120
Attempted		Correct	
Incorrect		Net Score	
Cut-off Score	44	Qualifying Score	68
Success Gap = Net Score – Qualifying Score			
Net Score = (Correct × 4) – (Incorrect × 1)			

Space for Rough Work

DPP - Daily Practice Problems

Name : Date :

Start Time : End Time :

BIOLOGY 47

SYLLABUS : Principles of inheritance and variation-Part II : Genetic variation, Sex determination, Sex-linked disorder

Max. Marks : 120 **Time : 60 min.**

GENERAL INSTRUCTIONS

- The Daily Practice Problem Sheet contains 30 MCQ's. For each question only one option is correct. Darken the correct circle/ bubble in the Response Grid provided on each page.
- You have to evaluate your Response Grids yourself with the help of solution booklet.
- Each correct answer will get you 4 marks and 1 mark shall be deduced for each incorrect answer. No mark will be given/ deducted if no bubble is filled. Keep a timer in front of you and stop immediately at the end of 60 min.
- The sheet follows a particular syllabus. Do not attempt the sheet before you have completed your preparation for that syllabus. Refer syllabus sheet in the starting of the book for the syllabus of all the DPP sheets.
- After completing the sheet check your answers with the solution booklet and complete the Result Grid. Finally spend time to analyse your performance and revise the areas which emerge out as weak in your evaluation.

DIRECTIONS (Q.1-Q.30) : There are 30 multiple choice questions. Each question has 4 choices (a), (b), (c) and (d), out of which ONLY ONE choice is correct.

1. Number of sex chromosomes is normal in
   (a) super female
   (b) Turner's syndrome
   (c) Klinefelter's syndrome
   (d) Down's syndrome

2. If an albino man marries with a normal woman and 50 offsprings are albino and 50 are normal, the woman is
   (a) heterozygous normal (b) homozygous normal
   (c) heterozygous carrier (d) none of these

3. Mongolism syndrome is caused by
   (a) one extra chromosome.
   (b) one extra sex chromosome.
   (c) one extra chromosome in 21st pair.
   (d) one less sex chromosome.

4. Why some mutations which are harmful do not get eliminated from the gene pool?
   (a) They have future survival value.
   (b) They are recessive and carried by heterozygous individuals.
   (c) They are dominant and show up more frequently.
   (d) Genetic drift occurs because of a small population area.

RESPONSE GRID	1. ⓐⓑⓒⓓ	2. ⓐⓑⓒⓓ	3. ⓐⓑⓒⓓ	4. ⓐⓑⓒⓓ

Space for Rough Work

5. An abnormal male child is born to normal parents. This child shows a short neck, stubby hands, slanting eyes, permanently open mouth with projecting upper lip and short stature. This abnormality can be attributed to
   (a) an extra *Y* chromosome
   (b) an extra *X* chromosome
   (c) an extra chromosome in the 21st pair of autosomes
   (d) absence of *X* chromosome leading to a total of 45 chromosomes instead of the normal 46

6. The monosomic condition in human beings depicted as XO is referred to as
   (a) Criminal syndrome (b) Down's syndrome
   (c) Klinefelter's syndrome (d) Turner's syndrome

7. Gene mutation is caused by
   (a) reproduction.
   (b) linkage.
   (c) change in the sequence of nitrogenous base.
   (d) change in the sequence of genes in DNA.

8. Genotype of a Down's syndrome is
   (a) 45 + *XX* (b) 44 + *XY*
   (c) 44 + *XXY* (d) 22 + *XY*

9. The point mutations A to G, C to T, C to G and T to A in DNA are
   (a) transition, transition, transversion and transversion respectively.
   (b) transition, transversion, transition and transversion respectively.
   (c) transversion, transversion, transition and transition respectively.
   (d) all four are transition.

10. Albinism is
   (a) hereditary character
   (b) non-hereditary character
   (c) sex-linked character
   (d) acquired character

11. A man homozygous for brown colour marries a lady heterozygous for brown colour is (dominant). What will be the fate of their children?
   (a) All brown
   (b) Three brown
   (c) Two brown
   (d) None of these

12. What would be the number of chromosomes in the ovum (fertilized by a normal sperm) that resulted in the appearance of Klinefelter's syndrome in the offspring?
   (a) 23 (b) 22
   (c) 21 (d) 24

13. The most striking example of point mutation is found in a disease called
   (a) night blindness
   (b) thalassemia
   (c) Down's syndrome
   (d) sickle-cell anaemia

14. Edward syndrome is on account of ......
   (a) 45 chromosomes instead of 46.
   (b) presence of three chromosomes on 18th pair of autosome.
   (c) presence of three chromosomes on 21st pair of autosome.
   (d) presence of three pair of sex chromosomes.

RESPONSE GRID	5. ⓐⓑⓒⓓ	6. 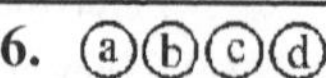	7. ⓐⓑⓒⓓ	8. ⓐⓑⓒⓓ	9. ⓐⓑⓒⓓ
	10. ⓐⓑⓒⓓ	11. ⓐⓑⓒⓓ	12. ⓐⓑⓒⓓ	13. ⓐⓑⓒⓓ	14. ⓐⓑⓒⓓ

Space for Rough Work

15. Phenylketonuria is a genetic disorder of
   (a) trisomic condition
   (b) monosomic condition
   (c) autosomal dominant gene
   (d) autosomal recessive gene

16. A mutation is most likely to have a selective advantage in evolution if
   (a) it affects dominant genes.
   (b) it affects recessive genes.
   (c) it affects whole chromosomes.
   (d) the environment remains stable.

17. Which of the following is not *X*-linked recessive ?
   (a) Haemophilia-A
   (b) Colour blindness
   (c) β-Thalassemia
   (d) G-6 PD deficiency

18. Polydactyly in man is due to
   (a) autosomal dominant gene.
   (b) autosomal recessive gene.
   (c) sex-linked dominant gene.
   (d) sex-linked recessive gene.

19. Given below is the representation of a kind of chromosomal mutation

   What kind of mutation is represented?

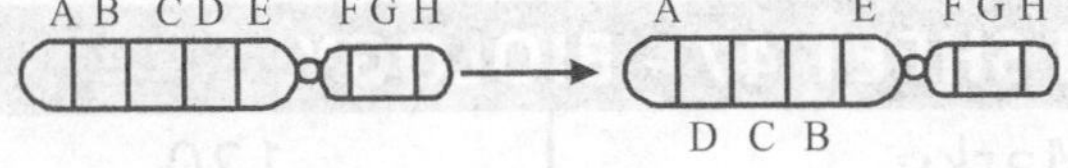

   (a) Deletion (b) Duplication
   (c) Inversion (d) Reciprocal translocation

20. In man, which of the following genotypes and phenotypes may be the correct result of aneuploidy in sex chromosomes
   (a) 22 pairs + XXY males
   (b) 22 pairs + XX females
   (c) 22 pairs + XXXY females
   (d) 22 pairs + Y females

21. Albinism is known to be due to an autosomal recessive mutation. The first child of a couple with normal skin pigmentation was an albino. What is the probability that their second child will also be an albino?
   (a) 100% (b) 25%
   (c) 50% (d) 75%

22. Barr bodies (seen in saliva test in Olympic games) are found in human and are associated with
   (a) male autosome
   (b) female autosome
   (c) female sex chromosome
   (d) male sex chromosome

23. Mule is an offspring of
   (a) male and female donkey
   (b) cow and ox
   (c) male ass and mare
   (d) male horse and female ass

24. Sex of a human child is determined by
   (a) size of the egg at the time of fertilization.
   (b) size of the sperm at the time of fertilization.
   (c) sex chromosome of father.
   (d) sex chromosome of mother.

RESPONSE GRID					
	15. ⓐⓑⓒⓓ	16. ⓐⓑⓒⓓ	17. ⓐⓑⓒⓓ	18. ⓐⓑⓒⓓ	19. ⓐⓑⓒⓓ
	20. ⓐⓑⓒⓓ	21. ⓐⓑⓒⓓ	22. ⓐⓑⓒⓓ	23. ⓐⓑⓒⓓ	24. ⓐⓑⓒⓓ

Space for Rough Work

25. Animal which remains male initially, then changes to female (*Tapeworm proglottides*) is called
(a) protandrous (b) apomixis
(c) profixation (d) none of these

26. The theory where ratio between the number of *X*- chromosomes and number of complete sets of autosomes will determine the sex is known as
(a) chromosome theory of sex determination.
(b) genic balance theory of sex determination.
(c) hormonal balance theory of sex determination.
(d) environmental sex determination theory.

27. Sex determination in *Drosophila melanogaster* is based on
(a) *X-Y* chromosome mechanism.
(b) genetic balance between the *X*-chromosome and autosomes.
(c) chromosome - environment interaction.
(d) pseudoalleles.

28. A colour blind man marries the daughter of a colour blind person. Then in their progeny
(a) none of their daughters are colour blind.
(b) all the sons are colour blind.
(c) all the daughters are colour blind.
(d) half of their sons are colour blind.

29. If mother is a carrier for colour blindness and father is normal, then in the offsprings this disease may be seen in
(a) all the sons.
(b) all the daughters.
(c) 50% sons and 50% daughters (carrier).
(d) all the sons and not in daughters.

30. A man known to be victim of haemophilia marries a normal woman whose father was known to be a bleeder. Then it is expected that
(a) all their children will be bleeders.
(b) half of their children will be bleeders.
(c) one fourth of their children will be bleeders.
(d) none of their children will be bleeders.

RESPONSE GRID	
	25. ⓐⓑⓒⓓ 26. ⓐⓑⓒⓓ 27. ⓐⓑⓒⓓ 28. ⓐⓑⓒⓓ 29. ⓐⓑⓒⓓ
	30. ⓐⓑⓒⓓ

DAILY PRACTICE PROBLEM SHEET 47 - BIOLOGY

Total Questions	30	Total Marks	120
Attempted		Correct	
Incorrect		Net Score	
Cut-off Score	32	Qualifying Score	52
Success Gap = Net Score – Qualifying Score			
Net Score = (Correct × 4) – (Incorrect × 1)			

Space for Rough Work

DPP - Daily Practice Problems

Name :

Date :

Start Time :

End Time :

BIOLOGY 48

SYLLABUS : Molecular basis of inheritance

Max. Marks : 120 **Time : 60 min.**

GENERAL INSTRUCTIONS

- The Daily Practice Problem Sheet contains 30 MCQ's. For each question only one option is correct. Darken the correct circle/ bubble in the Response Grid provided on each page.
- You have to evaluate your Response Grids yourself with the help of solution booklet.
- Each correct answer will get you 4 marks and 1 mark shall be deduced for each incorrect answer. No mark will be given/ deducted if no bubble is filled. Keep a timer in front of you and stop immediately at the end of 60 min.
- The sheet follows a particular syllabus. Do not attempt the sheet before you have completed your preparation for that syllabus. Refer syllabus sheet in the starting of the book for the syllabus of all the DPP sheets.
- After completing the sheet check your answers with the solution booklet and complete the Result Grid. Finally spend time to analyse your performance and revise the areas which emerge out as weak in your evaluation.

DIRECTIONS (Q.1-Q.30) : There are 30 multiple choice questions. Each question has 4 choices (a), (b), (c) and (d), out of which ONLY ONE choice is correct.

1. Which one of the following is a base analogue ?
   (a) Nitrous acid (b) Colchicine
   (c) 5 - bromouracil (d) Caffeine
2. At which site of a t-RNA molecule hydrogen bonds to a m-RNA molecule?
   (a) Codon
   (b) Anticodon
   (c) 5′ end of the t-RNA molecule
   (d) 3′ end of the t-RNA molecule
3. Feulgen reaction is a special test for
   (a) RNA
   (b) DNA
   (c) protein
   (d) carbohydrate
4. Semiconservative model of DNA replication was proposed by which workers in eukaryotes?
   (a) Taylor, Woods and Huges
   (b) Messelson and Stahl
   (c) Nirenberg and Khorana
   (d) Watson and Crick

RESPONSE GRID	1. ⓐⓑⓒⓓ	2. ⓐⓑⓒⓓ	3. ⓐⓑⓒⓓ	4. ⓐⓑⓒⓓ

Space for Rough Work

5. Process of DNA replication is effected by an enzyme called
(a) ligase (b) mutase
(c) ribonuclease (d) polymerase-I

6. Which form of RNA is most heterogeneous?
(a) tRNA (b) mRNA
(c) rRNA (d) hnRNA

7. Which one of the following is called polynucleotide joining enzyme?
(a) Polymerase-I (b) Polymerase-II
(c) Ligase (d) Ribonuclease

8. A DNA molecule in which both strands have radioactive thymidine is allowed to duplicate in an environment containing non-radioactive thymidine. What will be the correct number of DNA molecules that contain some radioactive thymidine after three duplications?
(a) There will be four such molecules.
(b) There will be eight such molecules.
(c) There will be only one such molecule.
(d) There will be two such molecules.

9. Which one of the following ratio is constant in DNAs of different species?
(a) $A+T/C+G$
(b) $A+G/T+C$
(c) $A+C/U+G$
(d) $A+U/C+G$

10. Chemically DNA differs from RNA in having
(a) thymine in place of uracil.
(b) single strand.
(c) ribose sugar instead of deoxyribose sugar.
(d) all of the above.

11. Which one of the following pair is correctly matched?
(a) Frederick Griffith Discovered the phenomenon of transformation.
(b) Linus Pauling Isolated the DNA for the first time.
(c) Francis Crick Proposed one gene one polypeptide hypothesis.
(d) George Beadle Proposed the concept of inborn errors.

12. Replication of DNA is in
(a) $3' \rightarrow 5'$ direction.
(b) $2' \rightarrow 5'$ direction.
(c) both $3' \rightarrow 5'$ and $5' \rightarrow 3'$ direction.
(d) none of the above

13. Adenine is 30% then what would be the % of guanine?
(a) 10% (b) 20%
(c) 30% (d) 40%

14. Okazaki fragments are joined in correct sequence by
(a) DNA polymerase (b) DNA ligase
(c) RNA polymerase (d) primase

RESPONSE GRID					
	5. ⓐⓑⓒⓓ	6. ⓐⓑⓒⓓ	7. ⓐⓑⓒⓓ	8. ⓐⓑⓒⓓ	9. ⓐⓑⓒⓓ
	10. ⓐⓑⓒⓓ	11. ⓐⓑⓒⓓ	12. ⓐⓑⓒⓓ	13. ⓐⓑⓒⓓ	14. ⓐⓑⓒⓓ

Space for Rough Work

15. If the base sequence in one polynucleotide series of DNA is G-C-A-T-G, what shall be the sequence in the replicated complementary strand?
(a) G-C-A-T-G (b) C-G-T-A-C
(c) A-T-G-C-G (d) G-C-A-T-C

16. Eukaryotes differ from prokaryotes in the mechanism of DNA replication due to
(a) different enzymes (instead of same enzyme) for synthesis of lagging and leading strands.
(b) discontinuous rather than semi-discontinuous replication.
(c) use of DNA primers rather than RNA primers.
(d) unidirectional rather than bidirectional replication.

17. Okazaki fragments are synthesized on
(a) leading strands of DNA only.
(b) lagging strands of DNA only.
(c) both leading and lagging strands of DNA.
(d) complementary DNA.

18. Restriction endonucleases are
(a) synthesized by bacteria.
(b) present in mammalian cells for degradation of DNA.
(c) used in genetic engineering.
(d) used for *in vitro* DNA synthesis.

19. Which is not correctly matched?
(a) Lipase - Hydrolysis of fats.
(b) Isomerases - Joining of similar substrate and management of substrate.
(c) Polymerase - Chain elongation.
(d) DNA ligase - Breaks DNA strand into two segments.

20. Which form of RNA has a structure resembling clover leaf ?
(a) m RNA (b) t RNA
(c) r RNA (d) hn RNA

21. The eukaryotic genome differs from the prokaryotic genome because
(a) repetitive sequences are present in eukaryotes.
(b) genes in the former case are organized into operons.
(c) the DNA is complexed with histones in prokaryotes.
(d) the DNA is circular and single stranded in prokaryotes.

22. Which one of the following mRNA can be translated completely?
(a) AUG UUC UCC UGG UAA UAU
(b) AUG UUC UCC UGA UGG UAU
(c) AUC ACG UAU UUC UGA CUC
(d) AUG UAU UUC UGC CUC UAG

23. In Operon concept, the regulator gene regulates chemical reactions in the cell by
(a) inactivating enzymes in the reaction.
(b) inhibiting transcription of mRNA.
(c) inhibiting migration of mRNA into cytoplasm.
(d) inhibiting the substrate in the reaction.

24. The regulatory genes are located
(a) along with the structural genes.
(b) in between operator and the structural genes.
(c) in the middle of structural genes.
(d) at the end of structural genes.

RESPONSE GRID					
	15. ⓐⓑⓒⓓ	16. ⓐⓑⓒⓓ	17. ⓐⓑⓒⓓ	18. ⓐⓑⓒⓓ	19. ⓐⓑⓒⓓ
	20. ⓐⓑⓒⓓ	21. ⓐⓑⓒⓓ	22. ⓐⓑⓒⓓ	23. ⓐⓑⓒⓓ	24. ⓐⓑⓒⓓ

Space for Rough Work

25. Operon is
 (a) a set of closely linked genes regulating a metabolic pathways in prokaryotes.
 (b) the sequence of three nitrogen bases determining a single amino acid.
 (c) the sequence of nitrogen bases in *m*RNA which codes for a single amino acid.
 (d) a gene responsible for switching on or off other genes
 (e) a segment of DNA specifying one polypeptide chain in protein synthesis.

26. In the lac operon, the structural genes are switched off when
 (a) repressor binds to operator.
 (b) repressor binds to promotor.
 (c) repressor binds to regulator.
 (d) repressor binds to inducer.

27. A polypeptide of five amino acids is synthesized during protein synthesis in eukaryotic cells. Which one of the following can be the correct polypeptide?
 (a) Glycine –valine – methionine – histidine – lysine
 (b) Methionine – glycine – histidine –lysine – valine
 (c) Valine – methionine – glycine – histidine –lysine
 (d) Histidine –lysine – methionine – valine – glycine

28. Which of the following is genetically inert and does not transcribe messenger RNA for protein synthesis ?
 (a) DNA of euchromatin
 (b) DNA of heterochromatin
 (c) DNA of euchromatin and heterochromatin
 (d) None of the above

29. After a mutation at a genetic locus the character of an organism changes due to the change in
 (a) protein synthesis pattern.
 (b) RNA transcription pattern.
 (c) protein structure.
 (d) DNA replication.

30. The central dogma of protein synthesis is
 (a) DNA Translation *m*RNA Transcription Protein
 (b) DNA Transcription RNA Translation Protein
 (c) RNA Replication DNA Transcription Protein
 (d) RNA Transcription RNA Translation Protein

RESPONSE GRID	
	25. ⓐⓑⓒⓓ 26. ⓐⓑⓒⓓ 27. ⓐⓑⓒⓓ 28. ⓐⓑⓒⓓ 29. ⓐⓑⓒⓓ
	30. ⓐⓑⓒⓓ

DAILY PRACTICE PROBLEM SHEET 48 - BIOLOGY			
Total Questions	30	Total Marks	120
Attempted		Correct	
Incorrect		Net Score	
Cut-off Score	40	Qualifying Score	60
Success Gap = Net Score – Qualifying Score			
Net Score = (Correct × 4) – (Incorrect × 1)			

Space for Rough Work

DPP - Daily Practice Problems

Name : Date :

Start Time : End Time :

BIOLOGY 49

SYLLABUS : Evolution

Max. Marks : 120 Time : 60 min.

GENERAL INSTRUCTIONS

- The Daily Practice Problem Sheet contains 30 MCQ's. For each question only one option is correct. Darken the correct circle/ bubble in the Response Grid provided on each page.
- You have to evaluate your Response Grids yourself with the help of solution booklet.
- Each correct answer will get you 4 marks and 1 mark shall be deduced for each incorrect answer. No mark will be given/ deducted if no bubble is filled. Keep a timer in front of you and stop immediately at the end of 60 min.
- The sheet follows a particular syllabus. Do not attempt the sheet before you have completed your preparation for that syllabus. Refer syllabus sheet in the starting of the book for the syllabus of all the DPP sheets.
- After completing the sheet check your answers with the solution booklet and complete the Result Grid. Finally spend time to analyse your performance and revise the areas which emerge out as weak in your evaluation.

DIRECTIONS (Q.1-Q.30) : There are 30 multiple choice questions. Each question has 4 choices (a), (b), (c) and (d), out of which ONLY ONE choice is correct.

1. An experiment to prove that organic compounds were the basis of life, was performed by

   (a) Oparin (b) Miller

   (c) Melvin (d) Fox

2. Which of the following was most likely to have been absent in free form in the primordial atmosphere at the time of origin of life?

   (a) O_2 (b) CH_4

   (c) H_2 (d) NH_3

3. In his classic experiment on the formation of amino acids, Stanley Miller passed an electric discharge in a mixture of

   (a) steam, CH_4, H_2 and NH_3

   (b) CH_4, CO_2, O_2, and H_2

   (c) CH_4, O_2, H_2 and steam

   (d) CH_4, H_2, N_2 and steam

4. Abiogenesis means

   (a) origin of life from non-living organisms.

   (b) origin of life from living organisms.

   (c) origin of viruses and microbes.

   (d) none of the above

RESPONSE GRID	1. ⓐⓑⓒⓓ	2. ⓐⓑⓒⓓ	3. ⓐⓑⓒⓓ	4. ⓐⓑⓒⓓ

Space for Rough Work

5. Which one of the following is the correct sequence of chemical substances produced during the origin of life on the earth?
   (a) Water, amino acid, nucleic acid and enzyme
   (b) Glucose, amino acid, nucleic acid and protein
   (c) Amino acid, ammonium phosphate and nucleic acid
   (d) Ammonia, amino acid, protein and nucleic acid
6. The idea that life originates from pre-existing life is referred as
   (a) biogenesis theory
   (b) special creation theory
   (c) abiogenesis theory
   (d) extra terrestrial theory
7. The genetic material in viruses is
   (a) only RNA
   (b) only DNA
   (c) RNA and DNA both
   (d) RNA or DNA *i.e.* one nucleic acid in a virus
8. Which of the following is false?
   (a) Most plant viruses are RNA viruses.
   (b) Most animal viruses are DNA viruses.
   (c) TMV has double stranded RNA molecule.
   (d) T_4 bacteriophage has double stranded DNA molecule.
9. Temin worked on which virus?
   (a) Herpes virus
   (b) Rhino virus
   (c) Retro virus
   (d) Dengu virus
10. Bacteriophage is made up of
   (a) carbon and nitrogen
   (b) DNA
   (c) nucleoprotein (Nucleic acid + protein)
   (d) protein only
11. Which of the following sequence is found in *Rous sarcoma virus* ?
   (a) DNA → RNA → Protein
   (b) RNA → RNA → Protein
   (c) RNA → DNA → RNA → Protein
   (d) DNA → DNA → Protein
12. A bacteriophage is
   (a) a virus attacking a bacterium.
   (b) a bacterium attacking a virus.
   (c) a stage in the life-cycle of bacterium.
   (d) a virus attacking another virus.
13. Tobacco Mosaic Virus (TMV) has
   (a) a single stranded RNA molecule.
   (b) a double stranded RNA molecule.
   (c) a single stranded DNA molecule.
   (d) a double stranded DNA molecule.
14. Haeckal's theory of recapitulation (Biogenetic law) means that
   (a) all organisms start as an egg.
   (b) life history of an animal reflects its evolutionary history.
   (c) progeny of an organism resembles its parents.
   (d) body parts once lost are regenerated.

RESPONSE GRID					
	5. ⓐⓑⓒⓓ	6. ⓐⓑⓒⓓ	7. ⓐⓑⓒⓓ	8. ⓐⓑⓒⓓ	9. ⓐⓑⓒⓓ
	10. ⓐⓑⓒⓓ	11. ⓐⓑⓒⓓ	12. ⓐⓑⓒⓓ	13. ⓐⓑⓒⓓ	14. ⓐⓑⓒⓓ

Space for Rough Work

15. Recapitulation theory (Biogenetic law) explains "*Ontogeny repeats Phylogeny*". It was given by
    (a) Lamarck (b) Haeckel
    (c) Darwin (d) Robert Hooke
16. Mule is a hybrid. Which of the following statement is correct?
    (a) Mule is not a species.
    (b) Mule is a new species.
    (c) Horse and ass are two populations.
    (d) Mule represents a common ancestor of horse and ass.
17. The pioneers in the field of 'organic evolution' are
    (a) Karl landsteiner, Hugo de Vries, Malthus
    (b) Darwin, Hugo de Vries, Lamarck, Huxley
    (c) Lamarck, Karl landsteiner, Malthus, DeVries
    (d) Darwin, Lamarck, Karl landsteiner, De Vries
18. Evolutionary convergence is characterized by
    (a) development of characteristics by random mating.
    (b) replacement of common characteristics in different group.
    (c) development of dissimilar characteristics in closely related groups.
    (d) development of a common set of characteristics in group of different ancestry.
19. Analogous organs are those which are
    (a) structurally similar.
    (b) structurally and functionally similar.
    (c) functionally similar.
    (d) normally non-functionable.
20. Which one of the following sets includes all homologous organs ?
    (a) Hind legs of pig, duck and kangaroo.
    (b) Wings of bats, butterfly and bird.
    (c) Sting of honey bee, scorpion and snake.
    (d) Tail of rat, peacock and cockroach
21. A living connective link which provides evidence for organic evolution is
    (a) *Archaeopteryx* between reptiles and mammals.
    (b) *Lung fish* between pisces and reptiles.
    (c) *Duck bill platypus* between reptiles and mammals.
    (d) *Sphenodon* between reptiles and birds.
22. Fossil *X* is older than fossil *Y*. Most evident answer is
    (a) Fossil *Y* has got some of the vestigial organs which are functional in *X*.
    (b) Fossil *Y* has got homologous and analogous organs of fossile *X*.
    (c) Fossil *X* is found in deeper sedimentation than *Y*.
    (d) Fossil *Y* was found in better state than that of *X*.
23. Convergent evolution of two species is associated with
    (a) analogous organs (b) recent common ancestor
    (c) homologous organs (d) different habitat
24. Which of the following are not analogous organs?
    (a) Fins of fishes and flippers of whales
    (b) Stings of honey bee and scorpion
    (c) Thorn of *Bougainvillia* and tendril of *Cucurbita*
    (d) Wings of insect and wing of pterodactyly

RESPONSE GRID					
	15. ⓐⓑⓒⓓ	16. ⓐⓑⓒⓓ	17. ⓐⓑⓒⓓ	18. ⓐⓑⓒⓓ	19. ⓐⓑⓒⓓ
	20. ⓐⓑⓒⓓ	21. ⓐⓑⓒⓓ	22. ⓐⓑⓒⓓ	23. ⓐⓑⓒⓓ	24. ⓐⓑⓒⓓ

Space for Rough Work

25. Hand of man, wing of bat and flipper of seal represents
    (a) vestigeal organs (b) analogous organs
    (c) evolutionary organs (d) homologous organs
26. Germplasm theory against Lamarck's principle, was given by
    (a) Weismann (b) Darwin
    (c) Lamarck (d) Hugo de Vries
27. Lamarck's theory of evolution is also called
    (a) theory of special creation
    (b) inheritance of acquired characters
    (c) survival of the fittest
    (d) none of the above
28. In forming the theory of evolution by natural selection, Darwin was greatly influenced by
    (a) mutations of Hugo de Vries
    (b) Malthus idea of population control
    (c) environmental factors
    (d) Lamarck acquired characters
29. Which fossil man has been known from Shivalik hills in India?
    (a) *Pithecanthropus*
    (b) *Ramapithecus*
    (c) *Sinanthropus*
    (d) *Zinjanthropus*
30. Which of the following is the correct order of evolutionary history of man?
    (a) *Peking man, Heidelberg man, Neanderthal, Cro-magnon.*
    (b) *Peking man, Homo sapiens, Cro-magnon, Neanderthal.*
    (c) *Peking man, Neanderthal, Homo sapiens, Heidelberg.*
    (d) *Peking man, Cro-magnon, Homo sapiens, Neanderthal.*

RESPONSE GRID	25. ⓐⓑⓒⓓ	26. ⓐⓑⓒⓓ	27. ⓐⓑⓒⓓ	28. ⓐⓑⓒⓓ	29. ⓐⓑⓒⓓ
	30. ⓐⓑⓒⓓ				

DAILY PRACTICE PROBLEM SHEET 49 - BIOLOGY			
Total Questions	30	Total Marks	120
Attempted		Correct	
Incorrect		Net Score	
Cut-off Score	36	Qualifying Score	56
Success Gap = Net Score – Qualifying Score			
Net Score = (Correct × 4) – (Incorrect × 1)			

Space for Rough Work

DPP - Daily Practice Problems

Name :

Date :

Start Time :

End Time :

BIOLOGY 50

SYLLABUS : Human health and disease

Max. Marks : 120 **Time : 60 min.**

GENERAL INSTRUCTIONS

- The Daily Practice Problem Sheet contains 30 MCQ's. For each question only one option is correct. Darken the correct circle/ bubble in the Response Grid provided on each page.
- You have to evaluate your Response Grids yourself with the help of solution booklet.
- Each correct answer will get you 4 marks and 1 mark shall be deduced for each incorrect answer. No mark will be given/ deducted if no bubble is filled. Keep a timer in front of you and stop immediately at the end of 60 min.
- The sheet follows a particular syllabus. Do not attempt the sheet before you have completed your preparation for that syllabus. Refer syllabus sheet in the starting of the book for the syllabus of all the DPP sheets.
- After completing the sheet check your answers with the solution booklet and complete the Result Grid. Finally spend time to analyse your performance and revise the areas which emerge out as weak in your evaluation.

DIRECTIONS (Q.1-Q.30) : There are 30 multiple choice questions. Each question has 4 choices (a), (b), (c) and (d), out of which ONLY ONE choice is correct.

1. Immune activities (defence mechanism) for different diseases is known as
   (a) immune action (b) immune reaction
   (c) first line of defence (d) immunity

2. The function of vaccine is the production and storage of
   (a) antigens (b) immune bodies
   (c) immune reactions (d) antibodies

3. 'Second line of defence' is known as
   (a) antibodies (b) WBC
   (c) liver (d) blood

4. T–cells are lymphocytes which produce the cellular immunity. These cells are developed from
   (a) thymus
   (b) liver
   (c) spleen
   (d) endothelium of blood vessels

RESPONSE GRID	1. ⓐⓑⓒⓓ	2. ⓐⓑⓒⓓ	3. ⓐⓑⓒⓓ	4. ⓐⓑⓒⓓ

Space for Rough Work

5. The binding of antibodies to the antigens to produce a large insoluble complex is known as
(a) antibody–antigen complex
(b) agglutination
(c) immunization
(d) suppressor cell reaction

6. Immune deficiency syndrome in human could develop as a consequence of
(a) AIDS virus infection
(b) defective liver
(c) defective thymus
(d) weak immune system

7. Antibody formation and immunity production is done by a protein called globulin present in the
(a) stroma of RBC
(b) haemoglobin of RBC
(c) plasma
(d) blood platelets

8. An antibody is a
(a) molecule that specifically inactivates an antigen.
(b) WBC which invades bacteria.
(c) secretion of mammalian RBC.
(d) component of blood.

9. AIDS virus has
(a) single stranded RNA.
(b) double stranded RNA.
(c) single stranded DNA.
(d) double stranded DNA.

10. Some people who have suffered from a disease may not be affected again during their life time; such immunity is called
(a) natural immunity (b) acquired immunity
(c) active immunity (d) passive immunity

11. Which one of the following pairs of disease can spread through blood transfusion?
(a) Cholera and hepatitis
(b) Hepatitis and AIDS
(c) Diabetes mellitus and malaria
(d) Hay fever and AIDS

12. Which cell of immune system cause pore formation at the surface of the plasma membrane?
(a) Helper T-cell
(b) Killer T-cell
(c) Suppressor T-cell
(d) B-cell

13. DPT vaccine is given for
(a) tetanus, polio, plague
(b) diphtheria, whooping cough, leprosy
(c) diphtheria, pneumonia, tetanus
(d) diphtheria, whooping cough, tetanus

14. Recognition and digestion by the phagocytes due to the coated surface of antigens by the antibodies is known as
(a) opsonization
(b) immunization
(c) T–cells immunization reaction
(d) B–cells immunization reaction

RESPONSE GRID					
	5. ⓐⓑⓒⓓ	6. ⓐⓑⓒⓓ	7. ⓐⓑⓒⓓ	8. ⓐⓑⓒⓓ	9. ⓐⓑⓒⓓ
	10. ⓐⓑⓒⓓ	11. ⓐⓑⓒⓓ	12. ⓐⓑⓒⓓ	13. ⓐⓑⓒⓓ	14. ⓐⓑⓒⓓ

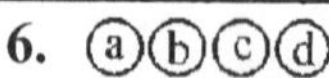

Space for Rough Work

15. Thalidomide is a non-barbiturate sedative drug which was to be given to pregnant women. Its use was withdrawn in 1961 because it resulted in phocomelia. This condition illustrates
(a) early abortion of foetus
(b) poor development of foetus
(c) malformation of foetus
(d) erythroblastosis foetalis

16. Severe Acute Respiratory Syndrome (SARS)
(a) is caused by a variant of *Pneumococcus*.
(b) is caused by a variant of the common cold virus (corona virus).
(c) is an acute form of asthma.
(d) affects non-vegetarians much faster than the vegetarians.

17. ELISA is used to detect viruses, where
(a) alkaline phosphatase is the key reagent.
(b) catalase is the key reagent.
(c) DNA-probes are required.
(d) southern blotting is done.

18. 'Polio' is caused by a
(a) bacteriophage.
(b) virus with single strand RNA.
(c) virus with single strand DNA.
(d) virus with double strand DNA.

19. 'Chicken pox' is caused by
(a) adeno virus (b) varicella virus
(c) *SV*–40 virus (d) bacteriophage T-2

20. Which of the disease is not transmitted by house flies ?
(a) Typhoid (b) Yellow fever
(c) Cholera (d) Dysentery

21. Common symptoms of measles are
(a) dew drop-like rashes on skin and high fever.
(b) erupting of small red spots and inflammation of mucous membrane of nose.
(c) lacerating ulcers.
(d) none of the above.

22. In human beings retrovirus is considered as a cause of cancer because
(a) in their genome oncogene is present.
(b) their hereditary material made up of single stranded RNA.
(c) they have a gene for reverse transcriptase.
(d) in their genome there may be cellular proto-oncogene.

23. If a person shows production of interferons in his body, the chances are that he has got an infection of
(a) typhoid (b) measles
(c) tetanus (d) malaria

24. Leprosy is diagnosed by which of the following set of symptoms?
(a) Fever, loss of pigmentation.
(b) Deformity of fingers, scales, ulcers, loss of pigmentation, wasting of body parts.
(c) Frequent watery stools and deformities in fingers and toes.
(d) White spots on the skin without any scales or ulcer.

RESPONSE GRID	15. ⓐⓑⓒⓓ	16. ⓐⓑⓒⓓ	17. ⓐⓑⓒⓓ	18. ⓐⓑⓒⓓ	19. ⓐⓑⓒⓓ
	20. ⓐⓑⓒⓓ	21. ⓐⓑⓒⓓ	22. ⓐⓑⓒⓓ	23. ⓐⓑⓒⓓ	24. ⓐⓑⓒⓓ

Space for Rough Work

25. Which one of the following causes plague?
(a) *Salmonella typhimurum*
(b) *Trichinella spiralis*
(c) *Yersinia pestis*
(d) *Leishmania donovani*

26. The main reason why antibiotics could not solve all the problems of bacteria mediated diseases is
(a) insensitivity of the individual following prolonged exposure to antibiotics.
(b) inactivation of antibiotics by bacterial enzymes.
(c) decreased efficiency of the immune system.
(d) the development of mutant strains resistant to antibiotics.

27. Match the following bacteria with the diseases and choose the correct option

	Column I		Column II
A.	*Treponema pallidum*	1.	Plague
B.	*Yersinia pestis*	2.	Anthrax
C.	*Bacillus anthracis*	3.	Syphilis
D.	*Vibrio*	4.	Cholera

(a) A-1, B-3, C-2, D-4 (b) A-3, B-1, C-2, D-4
(c) A-2, B-3, C-1, D-4 (d) A-4, B-3, C-1, D-2

28. Malarial parasites could be best obtained from a patient
(a) an hour before rise of temperature.
(b) when temperature rises with rigor.
(c) when temperature comes to normal.
(d) few hours after the temperature reaches to normal.

29. Which of the following set is of vector host?
(a) Sand fly, deer fly, tse–tse fly, house fly
(b) Frog, lizard, snake, rabbit
(c) *Leishmania*, sand fly, *Trypanosoma*, tse–tse fly
(d) Sand fly, frog, house fly

30. Chemicals, which can induce cancer are called
(a) mutagenic agents and produce benign tumour.
(b) carcinogens and produce non-malignant tumour.
(c) mutagenic agents and do not produce malignant tumour.
(d) carcinogens and produce malignant tumour.

RESPONSE GRID	
	25. ⓐⓑⓒⓓ 26. ⓐⓑⓒⓓ 27. ⓐⓑⓒⓓ 28. ⓐⓑⓒⓓ 29. ⓐⓑⓒⓓ
	30. ⓐⓑⓒⓓ

DAILY PRACTICE PROBLEM SHEET 50 - BIOLOGY			
Total Questions	30	Total Marks	120
Attempted		Correct	
Incorrect		Net Score	
Cut-off Score	36	Qualifying Score	56
Success Gap = Net Score – Qualifying Score			
Net Score = (Correct × 4) – (Incorrect × 1)			

Space for Rough Work

DPP - Daily Practice Problems

Name :

Date :

Start Time :

End Time :

BIOLOGY

51

SYLLABUS : Strategies for enhancement in food production

Max. Marks : 120 **Time : 60 min.**

GENERAL INSTRUCTIONS

- The Daily Practice Problem Sheet contains 30 MCQ's. For each question only one option is correct. Darken the correct circle/ bubble in the Response Grid provided on each page.
- You have to evaluate your Response Grids yourself with the help of solution booklet.
- Each correct answer will get you 4 marks and 1 mark shall be deduced for each incorrect answer. No mark will be given/ deducted if no bubble is filled. Keep a timer in front of you and stop immediately at the end of 60 min.
- The sheet follows a particular syllabus. Do not attempt the sheet before you have completed your preparation for that syllabus. Refer syllabus sheet in the starting of the book for the syllabus of all the DPP sheets.
- After completing the sheet check your answers with the solution booklet and complete the Result Grid. Finally spend time to analyse your performance and revise the areas which emerge out as weak in your evaluation.

DIRECTIONS (Q.1-Q.30) : There are 30 multiple choice questions. Each question has 4 choices (a), (b), (c) and (d), out of which ONLY ONE choice is correct.

1. The dwarf varieties of wheat brought from Mexico into India were
   (a) Sonara–64 and Sonalika
   (b) Sonara–64 and Lerma Rojo–64
   (c) Sharbati sonara and Pusa Lerma
   (d) Sonalika
2. *Gossypium hirsutum* is
   (a) Old world diploid (b) Old world tetraploid
   (c) New world diploid (d) New world tetraploid
3. Apart from high yield, other main objective of plant breeding is
   (a) improvement of quality.
   (b) development of resistance.
   (c) establishment of change in duration.
   (d) all of the above.
4. *Sonara*–64 and *Lerma rojo* are varieties of
   (a) wheat
   (b) rice
   (c) pea
   (d) maize

RESPONSE GRID	1. ⓐⓑⓒⓓ	2. ⓐⓑⓒⓓ	3. ⓐⓑⓒⓓ	4. ⓐⓑⓒⓓ

Space for Rough Work

5. IR–36 was developed through breeding
   (a) 6 rice varieties and *Oryza nivara*
   (b) 13 rice varieties and *Oryza nivara*
   (c) *Oryza indica* and *Oryza nivara*
   (d) *Oryza indica* and *Oryza sativa*
6. Plants can be disease resistant by
   (a) breeding with their wild relatives
   (b) colchicine treatment
   (c) hormone treatment
   (d) heat treatment
7. Aims of plant breeding are to produce
   (a) disease–free varieties
   (b) high yielding varieties
   (c) early maturing varieties
   (d) all of the above
8. Crosses between the plants of the same variety are called
   (a) interspecific
   (b) intervarietal
   (c) intravarietal
   (d) intergeneric
9. A plant cell has potential to develop into full plant. This property of the plant cell is called
   (a) tissue culture
   (b) totipotency
   (c) pleuripotency
   (d) gene cloning
10. The process of mating of individuals, which are more closely related than the average of the population to which they belong, is called
   (a) inbreeding
   (b) hybridization
   (c) heterosis
   (d) self breeding
11. Desired improved varieties of economically useful crops are raised by
   (a) migration
   (b) biofertilizer
   (c) hybridization
   (d) natural selection
12. Haploid from anther culture were first obtained in
   (a) *Brassica*
   (b) *Gossypium*
   (c) *Nicotiana*
   (d) *Datura*
13. Dwarf wheat was developed by
   (a) Gurudev S. Khush
   (b) M. S. Swaminathan
   (c) Norman Borlaug
   (d) none of these
14. Somatic hybridization of potato and tomato forms
   (a) *Triticale* (b) Pomato
   (c) *Secale* (d) *Altonia*

RESPONSE GRID					
	5. ⓐⓑⓒⓓ	6. ⓐⓑⓒⓓ 	7. ⓐⓑⓒⓓ	8. ⓐⓑⓒⓓ	9. ⓐⓑⓒⓓ
	10. ⓐⓑⓒⓓ	11. ⓐⓑⓒⓓ	12. ⓐⓑⓒⓓ	13. ⓐⓑⓒⓓ	14. ⓐⓑⓒⓓ

Space for Rough Work

15. What is a clone?
(a) A heterozygote produced by sexual means.
(b) A homozygote produced by asexual means.
(c) A heterozygote produced by asexual means.
(d) A homozygote produced by sexual means.

16. One of the popular method employed in the production of new varieties of plants is
(a) selection and vegetative propagation.
(b) chemical treatment and selection.
(c) exposure to radiation and selection.
(d) selection and hybridization.

17. In crop improvement programme haploids are of great importance because they
(a) are useful in studies on meiosis.
(b) require only about half the amount of chemical fertilizers as compare to diploids.
(c) give homozygous lines following diploidization.
(d) grow better under adverse conditions.

18. The *Triticum aestivum* wheat is
(a) haploid (7 chromosome)
(b) diploid (14 chromosome)
(c) tetraploid (30 chromosome)
(d) hexaploid (42 chromosome)

19. Heterosis means
(a) hybrid vigour
(b) hybrids are weak
(c) hybrids are weak as well as vigorous
(d) hybrids are neither weak nor vigorous

20. Which of the following condition is hybrid breakdown?
(a) Failure of hybrid adult to produce functional gametes.
(b) Failure of the fusion of ova and sperm plant breed of two species.
(c) Failure of hybrid zygote to develop into an offspring.
(d) None of the above.

21. In which crops is the method of mass selection applied?
(a) Cross-pollinated
(b) Self-pollinated
(c) Both self and cross-pollinated
(d) Potato and sugarcane

22. The reason for vegetatively reproducing crop plants to suit for maintaining hybrid vigour is that
(a) they can be easily propagated.
(b) they have a longer life span.
(c) they are more resistant to diseases.
(d) once a desired hybrid is produced, no chances of losing it.

23. Which of the following is the most important edible fish?
(a) Rohu
(b) Catla
(c) Wallago
(d) *Clarius*

24. *Triticale* is the hybrid between wheat and
(a) maize
(b) barley
(c) rye
(d) bean

RESPONSE GRID	15. ⓐⓑⓒⓓ	16. ⓐⓑⓒⓓ	17. ⓐⓑⓒⓓ	18. ⓐⓑⓒⓓ	19. ⓐⓑⓒⓓ
	20. ⓐⓑⓒⓓ	21. ⓐⓑⓒⓓ	22. ⓐⓑⓒⓓ	23. ⓐⓑⓒⓓ	24. ⓐⓑⓒⓓ

Space for Rough Work

25. What is meant by apiculture?
   (a) Culture of water animals.
   (b) Culture of lac insects.
   (c) Raising of honey bees.
   (d) Raising of prawn.
26. Varieties developed by pureline method are
   (a) homozygous and not uniform.
   (b) homozygous and uniform.
   (c) heterozygous and not uniform.
   (d) heterozygous and uniform.
27. In high yielding 'hybrid crop varieties', to exploit hybrid vigour, the farmers need to purchase fresh hybrid seed every year, because
   (a) they are not allowed to grow their own seed.
   (b) the hybrid vigour is lost due to inbreeding depression.
   (c) the government of India has accepted Dunkel's proposals.
   (d) it is cheaper to purchase fresh seed.
28. Process of hybridization is difficult because of
   (a) susceptibility to mutations
   (b) incompatibility and sterility
   (c) selection of suitable parents and hybrids
   (d) all of the above
29. As a general rule, inbreeding is possible between
   (a) any two members of a order
   (b) any two members of a family
   (c) any two members of a genus
   (d) any two members of a species
30. Axenic culture is best defined as
   (a) cell cultures
   (b) cell cultures free from micro-organisms
   (c) cell cultures of insectivorous plants
   (d) cell cultures free from other micro-organisms

RESPONSE GRID				
25. ⓐⓑⓒⓓ	26. ⓐⓑⓒⓓ	27. ⓐⓑⓒⓓ	28. ⓐⓑⓒⓓ	29. ⓐⓑⓒⓓ
30. ⓐⓑⓒⓓ				

DAILY PRACTICE PROBLEM SHEET 51 - BIOLOGY

Total Questions	30	Total Marks	120
Attempted		Correct	
Incorrect		Net Score	
Cut-off Score	32	Qualifying Score	52
Success Gap = Net Score – Qualifying Score			
Net Score = (Correct × 4) – (Incorrect × 1)			

Space for Rough Work

DPP - Daily Practice Problems

Name :   Date :

Start Time :   End Time :

BIOLOGY 52

SYLLABUS : Strategies for enhancement in food production

Max. Marks : 120   **Time : 60 min.**

GENERAL INSTRUCTIONS

- The Daily Practice Problem Sheet contains 30 MCQ's. For each question only one option is correct. Darken the correct circle/ bubble in the Response Grid provided on each page.
- You have to evaluate your Response Grids yourself with the help of solution booklet.
- Each correct answer will get you 4 marks and 1 mark shall be deduced for each incorrect answer. No mark will be given/ deducted if no bubble is filled. Keep a timer in front of you and stop immediately at the end of 60 min.
- The sheet follows a particular syllabus. Do not attempt the sheet before you have completed your preparation for that syllabus. Refer syllabus sheet in the starting of the book for the syllabus of all the DPP sheets.
- After completing the sheet check your answers with the solution booklet and complete the Result Grid. Finally spend time to analyse your performance and revise the areas which emerge out as weak in your evaluation.

DIRECTIONS (Q.1-Q.30) : There are 30 multiple choice questions. Each question has 4 choices (a), (b), (c) and (d), out of which ONLY ONE choice is correct.

1. Conversion of sugar into alcohol during fermentation is due to the direct action of
   (a) temperature
   (b) micro-organisms
   (c) zymase
   (d) concentration of sugar solution
2. The term antibiotic was coined by
   (a) Alexander Fleming (b) S. Waksman
   (c) Louis Pasteur (d) Edward Jenner
3. Rennin used in cheese industry is
   (a) antibiotics
   (b) enzyme
   (c) alkaloid
   (d) inhibitor
4. A compound which is produced by an organism and inhibits the growth of other organism is called
   (a) antibody
   (b) antigen
   (c) anti-allergic
   (d) antibiotic

RESPONSE GRID	1. ⓐⓑⓒⓓ	2. ⓐⓑⓒⓓ	3. ⓐⓑⓒⓓ	4. ⓐⓑⓒⓓ

Space for Rough Work

5. Genetically engineered bacteria have been used in commercial production of
   (a) thyroxine
   (b) testosterone
   (c) human insulin
   (d) melatonin

6. Which one of the micro-organism is used for production of citric acid in industries ?
   (a) *Lactobacillus bulgaricus*
   (b) *Penicillium citrinum*
   (c) *Aspergillus niger*
   (d) *Rhizopus nigricans*

7. Which one of the following is used in the manufacture of alcohol ?
   (a) Bacteria
   (b) Water molds
   (c) Yeasts
   (d) Slime molds

8. Propionic bacteria is an important source of
   (a) vitamin-C (b) cobalamin
   (c) sugars (d) proteins

9. Formation of vinegar from alcohol is caused by
   (a) *Bacillus subtilis*
   (b) *Clostridium*
   (c) *Acetobacter aceti*
   (d) *Azotobacter*

10. Biogas consists of
    (a) carbon monooxide, methane and hydrogen.
    (b) carbon dioxide, methane and hydrogen.
    (c) carbon monooxide, ethane and hydrogen.
    (d) carbon dioxide, ethane and hydrogen.

11. For biogas production besides dung, an extensive use of which weed is recommended in our country?
    (a) *Mangifera indica*
    (b) *Hydrilla*
    (c) *Eicchornia crassipes*
    (d) *Solanum*

12. Which of the following can be controlled by using biopesticides?
    (a) Insects
    (b) Diseases
    (c) Weeds
    (d) All of these

13. Biogas is produced by anaerobic breakdown of biomass of agricultural waste by methanogenic bacteria. It is a
    (a) one step process
    (b) two step process
    (c) three step process
    (d) multistep process

14. Beer is obtained from
    (a) molasses (b) grapes
    (c) barley (d) rye

RESPONSE GRID	5. ⓐⓑⓒⓓ	6. ⓐⓑⓒⓓ 	7. ⓐⓑⓒⓓ	8. ⓐⓑⓒⓓ	9. ⓐⓑⓒⓓ
	10. ⓐⓑⓒⓓ	11. ⓐⓑⓒⓓ	12. ⓐⓑⓒⓓ	13. ⓐⓑⓒⓓ	14. ⓐⓑⓒⓓ

Space for Rough Work

15. Which of the following microorganisms is used for preparation of swiss cheese?
    (a) *Propionibacterium*
    (b) *Geotrichum*
    (c) *Penicillium*
    (d) *Streptococcus*

16. Which country has successfully used alcohol as a motor fuel ?
    (a) India (b) U.S.A.
    (c) Brazil (d) Canada

17. Which is not a character of good fire wood ?
    (a) Highly combustible
    (b) Soft wood
    (c) Non resinous and non smoky
    (d) High caloric value

18. Which one is not a component of organic farming ?
    (a) Chemical fertilizer
    (b) Compost & farmyard manure
    (c) Crop rotation
    (d) Green manure

19. Vitamin B_{12} is formed during fermentation of
    (a) *Ashbya gossipii*
    (b) *Rhizopus stolonifer*
    (c) *Propionibacteria*
    (d) *Saccharomyces cerevisiae*

20. Cheese and yoghurt are products of
    (a) pasteurisation
    (b) distillation
    (c) dehydration
    (d) fermentation

21. Mycorrhiza is used in
    (a) synthesis of food.
    (b) increasing soil fertility.
    (c) providing resistance against different regulators.
    (d) obtaining nutrients from soil.

22. *Saccharomyces cerevisiae* is used in the formation of
    (a) ethanol
    (b) methanol
    (c) acetic acid
    (d) antibiotics

23. Indicator of water pollution is
    (a) *E. coli*
    (b) *Chlorella*
    (c) *Beggiatoa*
    (d) *Ulothrix*

24. Sewage purification is done by
    (a) microbes
    (b) fertilisers
    (c) antibiotics
    (d) antiseptics

RESPONSE GRID					
	15. 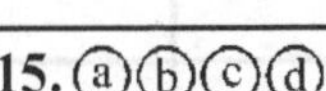	16. 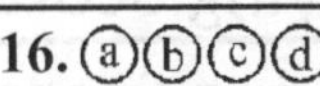	17. 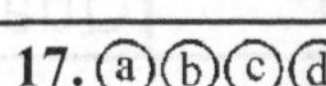	18. 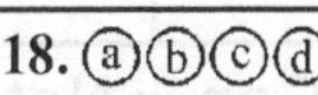	19. ⓐⓑⓒⓓ
	20. ⓐⓑⓒⓓ	21. ⓐⓑⓒⓓ	22. ⓐⓑⓒⓓ	23. ⓐⓑⓒⓓ	24. ⓐⓑⓒⓓ

Space for Rough Work

25. Which one of the following is helpful in paddy fields?
(a) *Rhizobium* (b) Ferrobacterium
(c) *Anabaena* (d) Archaebacteria

26. Which of the following is the pair of biofertilizers ?
(a) *Azolla* and BGA
(b) *Nostoc* and legume
(c) *Rhizobium* and grasses
(d) *Salmonella* & *E. coli*

27. Which of the following bacteria is utilized in gober gas plant ?
(a) Methanogens
(b) Nitrifying bacteria
(c) Ammonifying bacteria
(d) Denitrifying bacteria

28. *Escherichia coli* is used as an indicator organism to determine pollution of water with
(a) heavy metals
(b) faecal matter
(c) industrial effluents
(d) pollen of aquatic plants

29. A major component of gobar gas is
(a) butane (b) ammonia
(c) methane (d) ethane

30. A free living nitrogen-fixing cyanobacterium which can also form symbiotic association with the water fern *Azolla* is
(a) *Anabaena* (b) *Tolypothrix*
(c) *Chlorella* (d) *Nostoc*

RESPONSE GRID	25. ⓐⓑⓒⓓ 26. ⓐⓑⓒⓓ 27. ⓐⓑⓒⓓ 28. ⓐⓑⓒⓓ 29. ⓐⓑⓒⓓ 30. ⓐⓑⓒⓓ

DAILY PRACTICE PROBLEM SHEET 52 - BIOLOGY			
Total Questions	30	Total Marks	120
Attempted		Correct	
Incorrect		Net Score	
Cut-off Score	40	Qualifying Score	60
Success Gap = Net Score – Qualifying Score			
Net Score = (Correct × 4) – (Incorrect × 1)			

Space for Rough Work

DPP - Daily Practice Problems

Name : Date :

Start Time : End Time :

BIOLOGY 53

SYLLABUS : Strategies for enhancement in food production

Max. Marks : 120 **Time : 60 min.**

GENERAL INSTRUCTIONS

- The Daily Practice Problem Sheet contains 30 MCQ's. For each question only one option is correct. Darken the correct circle/ bubble in the Response Grid provided on each page.
- You have to evaluate your Response Grids yourself with the help of solution booklet.
- Each correct answer will get you 4 marks and 1 mark shall be deduced for each incorrect answer. No mark will be given/ deducted if no bubble is filled. Keep a timer in front of you and stop immediately at the end of 60 min.
- The sheet follows a particular syllabus. Do not attempt the sheet before you have completed your preparation for that syllabus. Refer syllabus sheet in the starting of the book for the syllabus of all the DPP sheets.
- After completing the sheet check your answers with the solution booklet and complete the Result Grid. Finally spend time to analyse your performance and revise the areas which emerge out as weak in your evaluation.

DIRECTIONS (Q.1-Q.30) : There are 30 multiple choice questions. Each question has 4 choices (a), (b), (c) and (d), out of which ONLY ONE choice is correct.

1. A genetic clone is
   (a) a plant produced by asexual means.
   (b) hybrid produced by sexual means.
   (c) homozygous plant produced by sexual means.
   (d) heterozygous plant produced by sexual means.
2. PCR (Polymerase Chain Reaction) method is useful for
   (a) amplification of DNA for forming billions of copies of itself.
   (b) monoclonal antibody production.
   (c) hybridoma production.
   (d) all of the above.
3. Collection of organisms as bacteria that all contain the same piece of recombinant DNA is called
   (a) gene cloning
   (b) cell cloning
   (c) organ cloning
   (d) all of these
4. *Agrobacterium tumefaciens* is used in genetics engineering for
   (a) DNA-mapping
   (b) DNA-modification
   (c) vector
   (d) DNA finger printing

RESPONSE GRID	1. ⓐⓑⓒⓓ	2. ⓐⓑⓒⓓ	3. ⓐⓑⓒⓓ	4. ⓐⓑⓒⓓ

Space for Rough Work

5. A genetically engineered bacteria used for clearing oil spills is
   (a) *Escherichia coli*
   (b) *Bacillus subtilis*
   (c) *Agrobacterium tumifaciens*
   (d) *Pseudomonas putida*

6. Genetic engineering is
   (a) study of extra nuclear gene.
   (b) manipulation of genes by artificial method.
   (c) manipulation of RNA.
   (d) manipulation of enzymes.

7. Which of the following enzymes cut the DNA molecule at specific nucleotide sequence?
   (a) Restriction endonuclease
   (b) DNA - ligase
   (c) RNA - polymerase
   (d) Exonuclease

8. 'Molecular scissor' used in genetic engineering is
   (a) DNA ligase
   (b) helicase
   (c) DNA polymerase
   (d) restriction endonuclease

9. Which of the following is the first cloned animal ?
   (a) Dolly sheep
   (b) Polly sheep
   (c) Molly sheep
   (d) Dog

10. Who among the following discovered the restriction endonuclease?
    (a) Haminton Othanel Smith
    (b) F. Jacob
    (c) Andre Lwoff
    (d) None of the above

11. Which one of the following pair is incorrect?
    (a) Plasmid - Small piece of extra-chromosomal DNA in bacteria.
    (b) Interferon - An enzyme that interferes with DNA replication.
    (c) Cosmid - A vector for carrying large DNA fragments into host cells.
    (d) Myeloma - Antibody producing tumor cells.

12. Restriction enzyme 'Eco RI' nicks at which one of the following sites on DNA?
    (a) 5' G $\downarrow$ AATTC 3' (b) 5' GG$\uparrow$CC 3'
    (c) 5'A$\downarrow$AGCTT 3' (d) none of these

13. Genetic engineering is possible because
    (a) we can cut DNA at specific sites by endonuclease like DNAase.
    (b) restriction endonuclease purified from bacteria is well understood.
    (c) the phenomenon of transduction in bacteria is well understood.
    (d) we can see DNA by electron microscope.

14. Which one of the following bacteria has found extensively used in genetic engineering work in plants?
    (a) *Clostridium septicum*
    (b) *Xanthomonas citri*
    (c) *Bacillus coagulens*
    (d) *Agrobacterium tumefaciens*

RESPONSE GRID					
	5. ⓐⓑⓒⓓ	6. 	7. ⓐⓑⓒⓓ	8. ⓐⓑⓒⓓ	9. ⓐⓑⓒⓓ
	10. ⓐⓑⓒⓓ	11. ⓐⓑⓒⓓ	12. ⓐⓑⓒⓓ	13. ⓐⓑⓒⓓ	14. ⓐⓑⓒⓓ

Space for Rough Work

15. Plasmids are autonomously replicating mini chromosomes, found in
    (a) Bacteriophage Lambda
    (b) *Leishmania donovani*
    (c) *Escherichia coli*
    (d) *Paramecium caudatum*
16. The bacterium generally used for genetic engineering is
    (a) *Agrobacterium* (b) *Bacillus*
    (c) *Pseudomonas* (d) *Clostridium*
17. Which one of the following techniques is employed in human genetic counselling ?
    (a) Serological technique
    (b) Polyploidy
    (c) Genetic engineering
    (d) Pedigree analysis
18. The process of joining together different DNA fragments is often referred to as
    (a) transcription (b) cloning
    (c) gene splicing (d) DNA amplification
19. A distinct mechanism that usually involves a short segment of DNA with remarkable capacity to move from one location in a chromosome to another, is called
    (a) DNA replication (b) DNA transposition
    (c) DNA hybridization (d) DNA recombination
20. The term genetic engineering is used for
    (a) blotting technique
    (b) RNA reaction technique
    (c) protein synthesis technique
    (d) recombinant DNA technique
21. DNA fingerprinting refers to
    (a) techniques used for molecular analysis of different specimens of DNA.
    (b) techniques used for identification of fingerprints of individuals.
    (c) molecular analysis of profiles of DNA samples.
    (d) analysis of DNA samples using imprinting devices.
22. Restriction endonucleases
    (a) are used for *in vitro* DNA synthesis.
    (b) are synthesized by bacteria as a part of defense mechanism.
    (c) are present in mammalian cells for degradation of DNA when the cells dies.
    (d) are used in genetic engineering for ligating two DNA molecules.
23. In bacteria, genes for antibiotic resistance are usually located in
    (a) plasmid
    (b) chromosomal DNA
    (c) mitochondria
    (d) cytoplasm
24. What is the first step in the Southern blot technique?
    (a) Denaturation of DNA on the gel for hybridization with specific probe.
    (b) Production of a group of genetically identical cells.
    (c) Digestion of DNA by restriction enzyme.
    (d) Isolation of DNA from a nucleated cell such as the one from the scene of crime.

RESPONSE GRID					
	15. ⓐⓑⓒⓓ	16. ⓐⓑⓒⓓ	17. ⓐⓑⓒⓓ	18. ⓐⓑⓒⓓ	19. ⓐⓑⓒⓓ
	20. ⓐⓑⓒⓓ	21. ⓐⓑⓒⓓ	22. ⓐⓑⓒⓓ	23. ⓐⓑⓒⓓ	24. ⓐⓑⓒⓓ

Space for Rough Work

25. cDNA probes are copied from the messenger RNA molecules with the help of
(a) restriction enzymes
(b) reverse transcriptase
(c) DNA polymerase
(d) adenosine deaminase

26. Southern blotting is done for
(a) DNA (b) RNA
(c) mRNA (d) protein

27. Which one of the following pairs of terms/names mean one and the same thing?
(a) Gene pool- Genome
(b) Codon - Gene
(c) Cistron -Triplet
(d) DNA fingerprinting-DNA profiling

28. The agent used to carry new genes into cells is called____.
(a) vector
(b) exonuclease
(c) protein
(d) gene gun

29. After 4 PCR cycles how many DNA molecules are formed from one DNA template molecule ?
(a) 4 (b) 32
(c) 16 (d) 8

30. Which of the following discoveries resulted in a Nobel Prize?
(a) X-rays induced sex-linked recessive lethal mutations.
(b) Cytoplasmic inheritance.
(c) Recombination of linked genes.
(d) Genetic engineering.

RESPONSE GRID	25. ⓐⓑⓒⓓ	26. ⓐⓑⓒⓓ	27. ⓐⓑⓒⓓ	28. ⓐⓑⓒⓓ	29. ⓐⓑⓒⓓ
	30. ⓐⓑⓒⓓ				

DAILY PRACTICE PROBLEM SHEET 53 - BIOLOGY			
Total Questions	30	Total Marks	120
Attempted		Correct	
Incorrect		Net Score	
Cut-off Score	36	Qualifying Score	60
Success Gap = Net Score – Qualifying Score			
Net Score = (Correct × 4) – (Incorrect × 1)			

Space for Rough Work

DPP - Daily Practice Problems

Name : Date :

Start Time : End Time :

BIOLOGY 54

SYLLABUS : Biotechnology and its applications in Animals

Max. Marks : 120 **Time : 60 min.**

GENERAL INSTRUCTIONS

- The Daily Practice Problem Sheet contains 30 MCQ's. For each question only one option is correct. Darken the correct circle/ bubble in the Response Grid provided on each page.
- You have to evaluate your Response Grids yourself with the help of solution booklet.
- Each correct answer will get you 4 marks and 1 mark shall be deduced for each incorrect answer. No mark will be given/ deducted if no bubble is filled. Keep a timer in front of you and stop immediately at the end of 60 min.
- The sheet follows a particular syllabus. Do not attempt the sheet before you have completed your preparation for that syllabus. Refer syllabus sheet in the starting of the book for the syllabus of all the DPP sheets.
- After completing the sheet check your answers with the solution booklet and complete the Result Grid. Finally spend time to analyse your performance and revise the areas which emerge out as weak in your evaluation.

DIRECTIONS (Q.1-Q.30) : There are 30 multiple choice questions. Each question has 4 choices (a), (b), (c) and (d), out of which ONLY ONE choice is correct.

1. The new strain of bacteria produced by biotechnology in alcohol industry is
   (a) *Escherichia coli*
   (b) *Saccharomyces cerevisiae*
   (c) *Bacillus subtilis*
   (d) *Pseudomonas putida*
2. The name of drug used in cancer treatment produced by biotechnology is
   (a) interferon
   (b) human growth hormone [hGH]
   (c) TSH
   (d) insulin
3. The pre-requisites for biotechnological production of antibiotics is
   (a) to search an antibiotic producing micro-organism.
   (b) to isolate the antibiotic gene.
   (c) to join antibiotic gene with *E.coli* plasmid.
   (d) all of the above.
4. Vaccine prepared by recombinant DNA technology is called
   (a) first generation vaccine.
   (b) second generation vaccine.
   (c) third generation vaccine.
   (d) synthesized vaccine.

RESPONSE GRID	1. ⓐⓑⓒⓓ	2. ⓐⓑⓒⓓ	3. ⓐⓑⓒⓓ	4. ⓐⓑⓒⓓ

Space for Rough Work

5. Which of the following vaccine is related with tissue culture?
   (a) Chicken pox
   (b) Hepatitis -B
   (c) Anti-rabies
   (d) Measles

6. Humulin is prepared by which type of transgenic micro-organism?
   (a) *Escherichia coli*
   (b) *Entamoeba coli*
   (c) *Pseudomonas*
   (d) *Clostridium*

7. Which of following is not related with bio weapons?
   (a) *Bacillus anthracis*
   (b) *Clostridium botulinum*
   (c) *E. coli*
   (d) *Yersinia pestis*

8. Transgenic animal has
   (a) foreign DNA in all of its cells.
   (b) foreign RNA in all of its cells.
   (c) foreign DNA in some of the cells.
   (d) both (b) and (c).

9. *E. coli* are used in production of
   (a) rifampicin
   (b) LH
   (c) ecdyson
   (d) interferon

10. The bacteria generally used for genetic engineering is
   (a) *Agrobacterium*
   (b) *Bacillus*
   (c) *Pseudomonas*
   (d) *Clostridium*

11. Maximum application of animal cell culture technology today is in the production of
   (a) insulin
   (b) interferons
   (c) vaccines
   (d) edible proteins

12. A genetically manipulated organism containing in its genome one or more inserted gene of another species is called
   (a) transposon
   (b) gene expression
   (c) transgenic organism
   (d) retroposons

13. Production of human insulin is from
   (a) super bug (b) gut bacterium
   (c) smallest bacterium (d) all of these

14. Find the incorrect match.
   (a) TPA–Blood clot preventer
   (b) Erythropoietin – Formation of RBCs
   (c) Reopro– Transplantation rejection
   (d) Interferon – Treatment of cancer

RESPONSE GRID					
	5. ⓐⓑⓒⓓ	6. ⓐⓑⓒⓓ 	7. ⓐⓑⓒⓓ	8. ⓐⓑⓒⓓ	9. ⓐⓑⓒⓓ
	10. ⓐⓑⓒⓓ	11. ⓐⓑⓒⓓ	12. ⓐⓑⓒⓓ	13. ⓐⓑⓒⓓ	14. ⓐⓑⓒⓓ

Space for Rough Work

15. Which one of the following is not used in the production of yoghurt?
(a) *Streptococcus lactis*
(b) *Streptococcus thermophilus*
(c) *Lactobacillus bulgaricus*
(d) *Acetobacter aceti*

16. Which one of following vaccines is not made from recombinant DNA technology?
(a) FMD (b) BCG
(c) Hepatitis (d) Anthrax

17. In tissue culture medium the embryoids formed from pollen grain is due to
(a) cellular totipotency (b) organogenesis
(c) double fertilization (d) test tube culture

18. Kohler and Milstein developed biotechnology for the production of
(a) myelomas
(b) steroid conversion
(c) monoclonal antibodies
(d) immobilised enzymes

19. Ethanol is produced from starch as a raw material by the activity of
(a) yeast
(b) bacteria
(c) virus
(d) none of these

20. Maximum utilization of the techniques of biotechnology has been made in the field of
(a) agriculture
(b) medicines
(c) industries
(d) biogas production

21. In which bacteria by the use of biotechnology the production of B_{12} vitamins has been increased to about 20,000 times?
(a) *Ashbya gossypi*
(b) *E. coli*
(c) *Pseudomonas denitrificans*
(d) *Propionibacterium shermanii*

22. Somatostatin or human growth hormone (hGH) for the first time was genetically isolated by
(a) Banting and Best
(b) Ross
(c) Pasteur
(d) Goeddel

23. First cloned animal is
(a) Dog (b) Molly
(c) Dolly sheep (d) Polly sheep

24. Genetically engineered human insulin is prepared by using
(a) *E. coli* (b) *Rhizopus*
(c) *Pseudomonas* (d) Yeast

RESPONSE GRID					
	15. ⓐⓑⓒⓓ	16. ⓐⓑⓒⓓ	17. ⓐⓑⓒⓓ	18. ⓐⓑⓒⓓ	19. ⓐⓑⓒⓓ
	20. ⓐⓑⓒⓓ	21. ⓐⓑⓒⓓ	22. ⓐⓑⓒⓓ	23. ⓐⓑⓒⓓ	24. ⓐⓑⓒⓓ

Space for Rough Work

25. ANDI is a cloned
(a) sheep (b) bull
(c) monkey (d) cat

26. The transfer of protein from electrophoretic gel to nitrocellulose membrane is known as
(a) transferase (b) northern blotting
(c) western blotting (d) southern blotting

27. An example of gene therapy is
(a) production of injectable Hepatitis B vaccine.
(b) production of vaccines in food crops like potatoes which can be eaten.
(c) introduction of gene for adenosine deaminase in persons suffering from Severe Combined Immuno deficiency.
(d) production of test tube babies by artificial insemination and implantation of fertilized eggs.

28. First hormone prepared by genetic engineering is
(a) oxytocin (b) somatotropin
(c) adrenaline (d) insulin

29. Whose biological activity can be used to convert milk into curd ?
(a) Fungi
(b) Algae
(c) *Lactobacillus*
(d) Virus

30. Hybridomas are the result of fusion of
(a) male reproductive cells.
(b) female reproductive cells.
(c) normal antibody producing cells with myeloma.
(d) abnormal antibody producing cells with myeloma.

RESPONSE GRID	25. ⓐⓑⓒⓓ	26. ⓐⓑⓒⓓ	27. ⓐⓑⓒⓓ	28. ⓐⓑⓒⓓ	29. ⓐⓑⓒⓓ
	30. ⓐⓑⓒⓓ				

DAILY PRACTICE PROBLEM SHEET 54 - BIOLOGY			
Total Questions	30	Total Marks	120
Attempted		Correct	
Incorrect		Net Score	
Cut-off Score	32	Qualifying Score	56
Success Gap = Net Score – Qualifying Score			
Net Score = (Correct × 4) – (Incorrect × 1)			

Space for Rough Work

DPP - Daily Practice Problems

Name : Date :

Start Time : End Time :

BIOLOGY 55

SYLLABUS : Biotechnology and its applications in plants

Max. Marks : 120 **Time : 60 min.**

GENERAL INSTRUCTIONS

- The Daily Practice Problem Sheet contains 30 MCQ's. For each question only one option is correct. Darken the correct circle/ bubble in the Response Grid provided on each page.
- You have to evaluate your Response Grids yourself with the help of solution booklet.
- Each correct answer will get you 4 marks and 1 mark shall be deduced for each incorrect answer. No mark will be given/ deducted if no bubble is filled. Keep a timer in front of you and stop immediately at the end of 60 min.
- The sheet follows a particular syllabus. Do not attempt the sheet before you have completed your preparation for that syllabus. Refer syllabus sheet in the starting of the book for the syllabus of all the DPP sheets.
- After completing the sheet check your answers with the solution booklet and complete the Result Grid. Finally spend time to analyse your performance and revise the areas which emerge out as weak in your evaluation.

DIRECTIONS (Q.1-Q.30) : There are 30 multiple choice questions. Each question has 4 choices (a), (b), (c) and (d), out of which ONLY ONE choice is correct.

1. Cry protein is obtained from
   (a) *Bacillus thuringiensis*
   (b) *Bacillus subtilis*
   (c) *Clostridium welchi*
   (d) *E. coli*
2. First transgenic plant is
   (a) potato (b) tomato
   (c) tobacco (d) maize
3. *Bacillus thuringiensis* (Bt) strains have been used for designing novel
   (a) bio-metallurgical techniques
   (b) bio-insecticidal plants
   (c) bio-mineralization processes
   (d) bio-fertilizers
4. Which one of the following is a correct statement ?
   (a) "Bt" in "Bt-cotton" indicates that it is a genetically modified organism produced through biotechnology.
   (b) Somatic hybridization involves fusion of two complete plant cells carrying desired genes.
   (c) The anticoagulant hirudin is being produced from transgenic *Brassica napus* seeds.
   (d) "Flavr Savr" variety of tomato has enhanced the production of ethylene which improves its taste.

RESPONSE GRID	1. ⓐⓑⓒⓓ	2. ⓐⓑⓒⓓ	3. ⓐⓑⓒⓓ	4. ⓐⓑⓒⓓ

Space for Rough Work

5. Use of transgenic plants as biological factories for the production of special chemicals is called
(a) molecular farming (b) molecular genetics
(c) molecular mapping (d) dry farming

6. Important objective of biotechnology in agriculture section is to
(a) produce pest resistant varieties of plants.
(b) increase the nitrogen content.
(c) decrease the seed number.
(d) increase the plant weight.

7. In tissue culture medium the embryoids formed from pollen grain is due to
(a) cellular totipotency (b) organogenesis
(c) double fertilization (d) test tube culture

8. Green banana of Bombay (Mumbai) is the result of
(a) mass selection (b) pure line selection
(c) clonal selection (d) natural selection

9. The main technique involved in agricultural biotechnology is called
(a) tissue culture (b) transformation
(c) plant breeding (d) DNA replication

10. Golden rice is a transgenic crop of the future with which of the following improved trait?
(a) High lysine (essential amino acid) content
(b) Insect resistance
(c) High protein content
(d) High vitamin-A content

11. Find the incorrect statement
(a) Gene therapy is a genetic engineering technique used to treat diseases at molecular level by replacing defective genes with normal genes.
(b) Calcitonin is a medically useful recombinant product in the treatment of infertility.
(c) Bt toxin is a biodegradable insecticide obtained from *Bacillus thuringiensis.*
(d) *Trichoderma* species is a biocontrol agent for fungal diseases of plants.

12. Which of the following is false for Bt transgenic plant ?
(a) Disease resistance
(b) Prepared by *Bacillus thuringiensis*
(c) It is recombinant type
(d) No such plant is known

13. Which of the following is a tumour inducing plasmid widely used in the production of transgenic plants ?
(a) *Escherichia coli*
(b) *Bacillus thuringiensis*
(c) *Staphylococcus aureus*
(d) *Agrobacterium tumefaciens*

14. Crown gall disease in plants is caused by
(a) Ti-plasmid
(b) Pi-plasmid
(c) bacteria
(d) virus

RESPONSE GRID					
	5. ⓐⓑⓒⓓ	6. 	7. ⓐⓑⓒⓓ	8. ⓐⓑⓒⓓ	9. ⓐⓑⓒⓓ
	10. ⓐⓑⓒⓓ	11. ⓐⓑⓒⓓ	12. ⓐⓑⓒⓓ	13. ⓐⓑⓒⓓ	14. ⓐⓑⓒⓓ

Space for Rough Work

15. In transgenics, expression of transgene in target tissue is determined by
(a) enhancer
(b) promotor
(c) transgene
(d) reporter

16. Which one of the following is regarded as a natural genetic engineer ?
(a) *Klebsiella frosteri*
(b) *Bacillus subtilis*
(c) *Agrobacterium tumefaciens*
(d) *Neurospora crassa*

17. Which of the following is important in genetic engineering?
(a) DNA polymerase
(b) Restriction endonuclease
(c) RNA polymerase
(d) Nuclease

18. Two bacteria found to be very useful in genetic engineering experiment are
(a) *Nitrosomonas* and *Klebsiella*
(b) *Escherichia* and *Agrobacterium*
(c) *Nitrobacter* and *Azotobacter*
(d) *Rhizobium* and *Diplococcus*

19. The latest trend in plant disease control is
(a) chemical control
(b) biological control
(c) use of fertilizers
(d) use of disease resistant varieties

20. Somaclonal variations are produced
(a) by mutagens.
(b) in tissue culture during differentiation.
(c) by gamma rays.
(d) by sexual reproduction.

21. Plants are genetically engineered with novel genes by
(a) protoplast fusion
(b) recombinant DNA technology
(c) embryo rescue technique
(d) recombination breeding

22. When two plants growing in different season and different geographical area, they can produce hybrid by
(a) pollen culture
(b) tissue culture
(c) somatic embryogenesis
(d) *In vitro* synthesis

23. The genetically engineered crop which has been recently introduced in India is
(a) herbicide tolerant maize
(b) Bt cotton
(c) slow ripening tomato
(d) golden rice

24. Cultivation of Bt Cotton has been much in the news. The prefix "Bt" means
(a) "Barium-treated" cotton seeds.
(b) "Bigger thread" variety of cotton with better tensile strength.
(c) Production by "biotechnology" using restriction enzymes and ligases.
(d) Carrying an endotoxin gene from *Bacillus thuringiensis.*

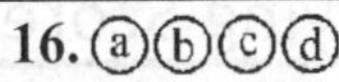

RESPONSE GRID					
	15. ⓐⓑⓒⓓ	16. ⓐⓑⓒⓓ	17. ⓐⓑⓒⓓ	18. ⓐⓑⓒⓓ	19. ⓐⓑⓒⓓ
	20. ⓐⓑⓒⓓ	21. ⓐⓑⓒⓓ	22. ⓐⓑⓒⓓ	23. ⓐⓑⓒⓓ	24. ⓐⓑⓒⓓ

Space for Rough Work

25. Abnormal gene is replaced by normal genes through
(a) gene therapy (b) medicines
(c) cloning (d) radiation

26. The technique of obtaining large number of plantlets by tissue culture method is called
(a) micropropagation
(b) macropropagation
(c) plantlet culture
(d) organ culture

27. A cybrid is a hybrid carrying
(a) cytoplasms of two different plants.
(b) genomes and cytoplasms of two different plants.
(c) cytoplasms of two different plants and genome of one plant.
(d) genomes of two different plants.

28. Protoplasts of two different species are fused in ?
(a) micropropagation
(b) somatic hybridization
(c) clonal propagation
(d) organography

29. Somatic hybrids are produced by
(a) protoplast fusion
(b) tissue culture
(c) pollen culture
(d) hybridoma process

30. Callus is
(a) undifferentiated mass of tissue.
(b) root formation in culture media.
(c) plant hormones.
(d) plant byproduct.

RESPONSE GRID					
	25. ⓐⓑⓒⓓ	26. ⓐⓑⓒⓓ	27. ⓐⓑⓒⓓ	28. ⓐⓑⓒⓓ	29. ⓐⓑⓒⓓ
	30. ⓐⓑⓒⓓ				

DAILY PRACTICE PROBLEM SHEET 55 - BIOLOGY			
Total Questions	30	Total Marks	120
Attempted		Correct	
Incorrect		Net Score	
Cut-off Score	36	Qualifying Score	56
Success Gap = Net Score – Qualifying Score			
Net Score = (Correct × 4) – (Incorrect × 1)			

Space for Rough Work

DPP - Daily Practice Problems

Name : Date :

Start Time : End Time :

BIOLOGY 56

SYLLABUS : Organisms and population

Max. Marks : 120 **Time : 60 min.**

GENERAL INSTRUCTIONS

- The Daily Practice Problem Sheet contains 30 MCQ's. For each question only one option is correct. Darken the correct circle/ bubble in the Response Grid provided on each page.
- You have to evaluate your Response Grids yourself with the help of solution booklet.
- Each correct answer will get you 4 marks and 1 mark shall be deduced for each incorrect answer. No mark will be given/ deducted if no bubble is filled. Keep a timer in front of you and stop immediately at the end of 60 min.
- The sheet follows a particular syllabus. Do not attempt the sheet before you have completed your preparation for that syllabus. Refer syllabus sheet in the starting of the book for the syllabus of all the DPP sheets.
- After completing the sheet check your answers with the solution booklet and complete the Result Grid. Finally spend time to analyse your performance and revise the areas which emerge out as weak in your evaluation.

DIRECTIONS (Q.1-Q.30) : There are 30 multiple choice questions. Each question has 4 choices (a), (b), (c) and (d), out of which ONLY ONE choice is correct.

1. Biotic potential refers to
   (a) increase of population under optimum conditions.
   (b) increase of population under given conditions.
   (c) increase of population under natural conditions.
   (d) increase of population under climatic conditions.
2. The term 'niche' of a species refers to
   (a) specific and habitual function.
   (b) specific place where an organism lives.
   (c) competitive power of an organism.
   (d) specific function of organism.
3. Group of two or more than two plant species is called as
   (a) plant community
   (b) animal ecosystem
   (c) plant ecosystem
   (d) ecological niche
4. Study of environment and animal relation is
   (a) ecosystem
   (b) phytosociology
   (c) biotic community
   (d) ecology

RESPONSE GRID	1. ⓐⓑⓒⓓ	2. ⓐⓑⓒⓓ	3. ⓐⓑⓒⓓ	4. ⓐⓑⓒⓓ

Space for Rough Work

5. What is a keystone species?
   (a) A rare species that has minimal impact on the biomass and on other species in the community.
   (b) A dominant species that constitutes a large proportion of the biomass and which affects many other species.
   (c) A species which makes up only a small proportion of the total biomass of a community, yet has a huge impact on the community's organization and survival.
   (d) A common species that has plenty of biomass, yet has a fairly low impact on the community's organization.

6. Identify the correctly matched pair.
   (a) Basal convention – Biodiversity conservation
   (b) Kyoto protocol – Climatic change
   (c) Montreal protocol – Global warming
   (d) Ramsar convention – Ground water pollution

7. Plants grow sparsely in arid regions of the world
   (a) only because of high temperature.
   (b) because no seeds fall on their soil.
   (c) because of several factors combining together.
   (d) because the soil is sandy.

8. If carbon dioxide is withdrawn from the biosphere, which organism would first experience negative effects?
   (a) Primary producers
   (b) Producers
   (c) Secondary consumers
   (d) Tertiary consumers

9. The sun loving plants are referred to as
   (a) halophytes
   (b) heliophytes
   (c) heterotrophs
   (d) sciophytes

10. An area of soil is thoroughly wetted and allowed to drain until capillary movement of water is stopped. The water contents of the soil will give an estimate of its
   (a) capillary water (b) storage water
   (c) field capacity (d) gravitational water

11. Salinity of the soil is often caused due to
   (a) accumulation of soluble minerals near or in the surface in arid region.
   (b) excessive addition of water.
   (c) rapid evaporation of surface water.
   (d) excessive leaching.

12. When one organism is benefitted without affecting the others it is called as
   (a) parasitism (b) commensalism
   (c) saprophytism (d) symbiosis

13. Mycorrhiza, a relationship between fungi and roots of higher plants is
   (a) parasitic relationship
   (b) saprophytic relationship
   (c) symbiotic relationship
   (d) epiphytic relationship

14. Which of the following statements is incorrect ?
   (a) Lichen, an association of fungus and algae is an example of mutualism.
   (b) Those epiphytes which use other plants for support only and not for water or food suppyly are examples of commensation.
   (c) Sea-anemone on hermit-crab is an example of protocooperation.
   (d) Mutualism, protocooperation, commensalism cannot be included under symbiosis.

RESPONSE GRID					
	5. ⓐⓑⓒⓓ	6. 	7. ⓐⓑⓒⓓ	8. ⓐⓑⓒⓓ	9. ⓐⓑⓒⓓ
	10. ⓐⓑⓒⓓ	11. ⓐⓑⓒⓓ	12. ⓐⓑⓒⓓ	13. ⓐⓑⓒⓓ	14. ⓐⓑⓒⓓ

Space for Rough Work

15. A bird enters the mouth of crocodile and feed on parasitic leeches. The bird gets food and crocodile get rid of blood sucking leeches. Both the partners can live independently. Such an association is called
(a) mutualism (b) amensalism
(c) commensalism (d) proto-cooperation

16. Which one of the following is a right matching pair of certain organims (s) and the kind of association?
(a) Shark and sucker fish-Amensalism
(b) Algae and fungi in lichens-Mutualism
(c) Orchids growing on trees-Parasitism
(d) *Cuscuta* (dodder) growing in other flowering plants-epiphytism.

17. Acclimatization is
(a) Pure-line breeding
(b) Pure-line selection
(c) Introduction
(d) Adaptation to new environment

18. If the high altitude birds become rare or extinct, the plants which may disappear along with them are
(a) pine (b) oak
(c) orchids (d) *Rhododendrons*

19. Which is not a part of atmosphere?
(a) Light
(b) Temperature
(c) Edaphic factor
(d) Precipitation

20. Species which occupy a similar ecological niche in similar communities occupying different biogeographical regions are
(a) ecotones
(b) biomes
(c) ecological equivalents
(d) ecads

21. The closely related morphologically similar sympatric populations, but reproductively isolated, are designated as
(a) clines (b) demes
(c) clones (d) sibling species

22. The abundance of a species population within its habitat is called
(a) niche density (b) regional density
(c) relative density (d) absolute density

23. Which of the following statements is correct for species?
(a) The members of a species occupy the same habitat.
(b) They are morphologically similar.
(c) They can interbreed among themselves.
(d) They cannot interbreed with members of the other species.

24. Which of the following is not true for a species?
(a) Members of a species can interbreed.
(b) Gene flow does not occur between the populations of a species.
(c) Each species is reproductively isolated from every other species.
(d) Variations occur among members of a species.

RESPONSE GRID					
	15. ⓐⓑⓒⓓ	16. 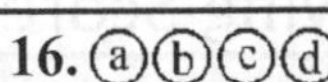	17. ⓐⓑⓒⓓ	18. ⓐⓑⓒⓓ	19. ⓐⓑⓒⓓ
	20. ⓐⓑⓒⓓ	21. ⓐⓑⓒⓓ	22. ⓐⓑⓒⓓ	23. ⓐⓑⓒⓓ	24. ⓐⓑⓒⓓ

Space for Rough Work

25. The side effects of heavy and constant grazing is that the
    (a) upper layer of the soil are loosened and are eroded.
    (b) lower layer are made more compact and root respiration is interfered with.
    (c) lower compact layer makes root penetration difficult.
    (d) all the above.
26. The change of the lighter coloured variety of peppered moth *Biston betularia* to its darker variety (carbonaria) is due to
    (a) mutation of single Mendelian gene for survival in the smoke-laden industrial environment.
    (b) deletion of a segment of gene due to industrial pollution.
    (c) industrial carbon deposited on the wings of moth resulting in darker variety.
    (d) translocation of a block of genes in chromosomes in response to heavy carbons.
27. In recent years, there has been an increasing incidence of floods in the plains of northern India because
    (a) there has been an increase in annual rainfall.
    (b) the rate of silting of dams has gone up.
    (c) there has been increased deforestation in the catchment areas.
    (d) increased areas of land is being self cultivated.
28. In the mountains, there is a lot of moisture and water and still the plants are xerophytes, because
    (a) water turns into snow on mountains.
    (b) water runs away due to the presence of slopes and can not be used by plant (physiological dryness).
    (c) rocks of mountains cannot absorb water.
    (d) none of them is incorrect.
29. Which of the following changes would be likely to make terrestrial life on this planet impossible?
    (a) A decrease in mean temperature by 10° *C*.
    (b) Change in atmosphere permitting all the solar radiation reaching the upper atmosphere to penetrate to the surface of the earth.
    (c) Change in the orbit of the earth from an ellipse to a circle.
    (d) Disappearance of the moon.
30. At which latitude, heat gain through insulation approximately equals heat loss through terrestrial radiation?
    (a) $22\frac{1}{2}°$ North and South
    (b) 40° North and South
    (c) $42\frac{1}{2}°$ North and South
    (d) 66° North and South

RESPONSE GRID	25. ⓐⓑⓒⓓ	26. ⓐⓑⓒⓓ	27. ⓐⓑⓒⓓ	28. ⓐⓑⓒⓓ	29. ⓐⓑⓒⓓ
	30. ⓐⓑⓒⓓ				

DAILY PRACTICE PROBLEM SHEET 56 - BIOLOGY			
Total Questions	30	Total Marks	120
Attempted		Correct	
Incorrect		Net Score	
Cut-off Score	40	Qualifying Score	60
Success Gap = Net Score – Qualifying Score			
Net Score = (Correct × 4) – (Incorrect × 1)			

Space for Rough Work

DPP - Daily Practice Problems

Name : **Date :**

Start Time : **End Time :**

BIOLOGY 57

SYLLABUS : Ecosystem – Structure and function

Max. Marks : 120 **Time : 60 min.**

GENERAL INSTRUCTIONS

- The Daily Practice Problem Sheet contains 30 MCQ's. For each question only one option is correct. Darken the correct circle/ bubble in the Response Grid provided on each page.
- You have to evaluate your Response Grids yourself with the help of solution booklet.
- Each correct answer will get you 4 marks and 1 mark shall be deduced for each incorrect answer. No mark will be given/ deducted if no bubble is filled. Keep a timer in front of you and stop immediately at the end of 60 min.
- The sheet follows a particular syllabus. Do not attempt the sheet before you have completed your preparation for that syllabus. Refer syllabus sheet in the starting of the book for the syllabus of all the DPP sheets.
- After completing the sheet check your answers with the solution booklet and complete the Result Grid. Finally spend time to analyse your performance and revise the areas which emerge out as weak in your evaluation.

DIRECTIONS (Q.1-Q.30) : There are 30 multiple choice questions. Each question has 4 choices (a), (b), (c) and (d), out of which ONLY ONE choice is correct.

1. The net primary productivity of an ecosystem is
   (a) Total weight of green plants – respiratory losses
   (b) Total weight of green plants + respiratory losses
   (c) Respiratory losses alone
   (d) None of the above
2. Which group of living organisms given below is a link in food chain between green plants and hawk?
   (a) Grasshopper, frog and snake
   (b) Grasshopper, rat and snake
   (c) Millipedes, centipedes and sparrow
   (d) Earthworm, hen and rat
3. Which one statement is correct regarding man made ecosystem?
   (a) They are highly efficient.
   (b) They are poor in diversity.
   (c) They are vulnerable to drought, floods and diseases.
   (d) All are correct
4. Dynamic aspect of an ecosystem is
   (a) producer and energy flow.
   (b) producer and mineral cycles.
   (c) consumer and mineral cycles.
   (d) energy flow and mineral cycles.

RESPONSE GRID	1. ⓐⓑⓒⓓ	2. ⓐⓑⓒⓓ	3. ⓐⓑⓒⓓ	4. ⓐⓑⓒⓓ

Space for Rough Work

5. In an ecosystem, there are more prey than predators. This relationship is called
   (a) food webs
   (b) predator–prey relationship
   (c) pyramid of number
   (d) succession

6. The order of organism in an aquatic food chain is
   (a) Bacteria → Seal → Diatom → Fish →→ Crustacea
   (b) Crustacea → Seal → Fish → Polar Bear → Diatom
   (c) Polar Bear → Diatom → Seal → Crustacea
   (d) Diatom → Crustacea → Fish → Seal → Bacteria

7. Under what condition, the upright pyramid of biomass becomes inverted?
   (a) Grassland ecosystem
   (b) Terrestrial ecosystem
   (c) Pond ecosystem
   (d) Tree ecosystem

8. The total rate of photosynthesis including the organic matter used up in respiration during definite period is known as
   (a) gross primary productivity.
   (b) net primary productivity.
   (c) secondary productivity.
   (d) none of the above.

9. An ecosystem is a complex interacting system of
   (a) individual
   (b) population
   (c) communities and their physical environment
   (d) communities and their soil conditions

10. In an ecosystem, the population of
   (a) primary producers are more than that of primary consumers.
   (b) secondary consumers are largest because they are powerful.
   (c) primary consumers out number primary producers.
   (d) primary consumers are least dependent upon primary producers.

11. The pyramid of number in a grassland ecosystem is
   (a) always erect
   (b) always inverted
   (c) either erect or inverted
   (d) irregular

12. Driving force of an ecosystem is
   (a) producer
   (b) carbohydrates in plants
   (c) biomass
   (d) solar energy

13. Biological equilibrium is an equilibrium among the
   (a) producers
   (b) producers and consumers
   (c) decomposers and producers
   (d) producers, consumers and decomposers

14. Which of the following must be preserved in an ecosystem, if the system is to be maintained?
   (a) Producers and carnivores
   (b) Producers and decomposers
   (c) Carnivores and decomposers
   (d) Herbivores and carnivores

RESPONSE GRID	5. ⓐⓑⓒⓓ	6. 	7. ⓐⓑⓒⓓ	8. ⓐⓑⓒⓓ	9. ⓐⓑⓒⓓ
	10. ⓐⓑⓒⓓ	11. ⓐⓑⓒⓓ	12. ⓐⓑⓒⓓ	13. ⓐⓑⓒⓓ	14. ⓐⓑⓒⓓ

Space for Rough Work

15. The trophic level of lion in a forest ecosystem is
(a) T_3 (b) T_4
(c) T_2 (d) T_1

16. Transfer of energy from one trophic level to other trophic level is according to the second law of thermodynamics. The efficiency of energy transfer from herbivorous to carnivorous is
(a) 25% (b) 50%
(c) 10% (d) 5%

17. If forest area is reduced to half, which one of the following will be a long term effect?
(a) The natives (tribals) of that area will die on account of hunger.
(b) Cattles of that area will die due to scarcity of fodder.
(c) The diversity in germplasm will affect the crop breeding.
(d) It will be converted into large desert.

18. The correct path of energy flow in an ecosystem is
(a) Producers $\rightarrow$ Carnivores $\rightarrow$ Herbivores $\rightarrow$ Decomposers
(b) Producers $\rightarrow$ Herbivores $\rightarrow$ Carnivores $\rightarrow$ Decomposers
(c) Herbivores $\rightarrow$ Carnivores $\rightarrow$ Producers $\rightarrow$ Decomposers
(d) Herbivores $\rightarrow$ Producers $\rightarrow$ Carnivores $\rightarrow$ Decomposers

19. Tip of ecological pyramid is occupied by
(a) herbivores
(b) carnivores
(c) producers
(d) decomposers

20. During food chain, the maximum energy is stored in
(a) producers
(b) decomposers
(c) herbivores
(d) carnivores

21. When peacock eats snakes which eat insects thriving on green plants, the peacock is
(a) a primary consumer
(b) a primary decomposer
(c) final decomposer
(d) the apex of food pyramid

22. In a food web, each successive trophic level has
(a) increased total energy
(b) less total energy content
(c) more total energy content
(d) non estimated energy content

23. In forest ecosystem, fungi is grouped as
(a) producer
(b) consumer
(c) secondary consumer
(d) decomposer

24. If decomposers are removed what will happen to ecosystem?
(a) Energy cycle is stopped.
(b) Mineral cycle is stopped.
(c) Consumers cannot absorb solar energy.
(d) Rate of decomposition of mineral increases.

RESPONSE GRID					
	15. ⓐⓑⓒⓓ	16. ⓐⓑⓒⓓ	17. ⓐⓑⓒⓓ	18. ⓐⓑⓒⓓ	19. ⓐⓑⓒⓓ
	20. ⓐⓑⓒⓓ	21. ⓐⓑⓒⓓ	22. ⓐⓑⓒⓓ	23. ⓐⓑⓒⓓ	24. ⓐⓑⓒⓓ

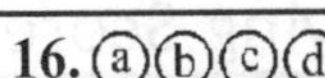

Space for Rough Work

25. Inverted pyramid is found in
(a) biomass pyramid of aquatic system.
(b) energy pyramid of grassland.
(c) biomass pyramid of grassland.
(d) pyramid of number of aquatic system.

26. An ecosystem which can be easily damaged but can recover after some time if damaging effect stops will be having
(a) low stability and low resilience.
(b) high stability and high resilience.
(c) low stability and high resilience.
(d) high stability and low resilience.

27. Area where two ecosystems overlap each other is called
(a) niche (b) ecotype
(c) edge line (d) ecotone

28. Energy flow and energy transformations in living systems strictly conform to the
(a) Law of limiting factors
(b) Liebig's law of minimum
(c) Laws of thermodynamics
(d) Shelford's law of tolerance

29. Given below is one of the types of ecological pyramids. This type represents

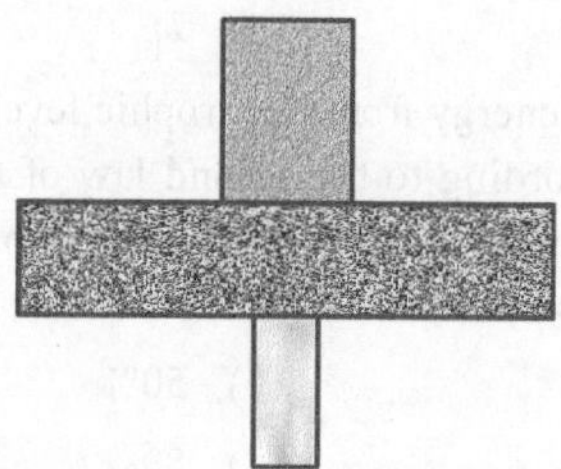

(a) pyramid of numbers in grassland.
(b) pyramid of biomass in a fallow land.
(c) pyramid of biomass in a lake.
(d) energy pyramid in a spring.

30. Mr. X is eating curd/yoghurt. For this food intake in a food chain he should be considered as occupying
(a) first trophic level (b) second trophic level
(c) third trophic level (d) fourth trophic level

RESPONSE GRID	
	25. ⓐⓑⓒⓓ 26. ⓐⓑⓒⓓ 27. ⓐⓑⓒⓓ 28. ⓐⓑⓒⓓ 29. ⓐⓑⓒⓓ
	30. ⓐⓑⓒⓓ

DAILY PRACTICE PROBLEM SHEET 57 - BIOLOGY			
Total Questions	30	Total Marks	120
Attempted		Correct	
Incorrect		Net Score	
Cut-off Score	40	Qualifying Score	64
Success Gap = Net Score – Qualifying Score			
Net Score = (Correct × 4) – (Incorrect × 1)			

Space for Rough Work

DPP - Daily Practice Problems

Name :

Date :

Start Time :

End Time :

BIOLOGY 58

SYLLABUS : Ecosystem – Ecological succession, Biome, Biosphere, Biogeochemical Cycle

Max. Marks : 120 **Time : 60 min.**

GENERAL INSTRUCTIONS

- The Daily Practice Problem Sheet contains 30 MCQ's. For each question only one option is correct. Darken the correct circle/ bubble in the Response Grid provided on each page.
- You have to evaluate your Response Grids yourself with the help of solution booklet.
- Each correct answer will get you 4 marks and 1 mark shall be deduced for each incorrect answer. No mark will be given/ deducted if no bubble is filled. Keep a timer in front of you and stop immediately at the end of 60 min.
- The sheet follows a particular syllabus. Do not attempt the sheet before you have completed your preparation for that syllabus. Refer syllabus sheet in the starting of the book for the syllabus of all the DPP sheets.
- After completing the sheet check your answers with the solution booklet and complete the Result Grid. Finally spend time to analyse your performance and revise the areas which emerge out as weak in your evaluation.

DIRECTIONS (Q.1-Q.30) : There are 30 multiple choice questions. Each question has 4 choices (a), (b), (c) and (d), out of which ONLY ONE choice is correct.

1. In succession the seral forms are
   (a) large sized (b) long lived
   (c) small sized (d) slow growing
2. Primary succession is development of communities on
   (a) newly exposed habitat.
   (b) cleared forest area.
   (c) freshly harvested crop field.
   (d) pond filled after a dry season.
3. Ecological succession on sand is called
   (a) psammosere
   (b) xerosere
   (c) halosere
   (d) hydrosere
4. "Biomes" describe
   (a) desert vegetation.
   (b) major ecological grouping of plants.
   (c) major ecological groupings of animals.
   (d) relation between plants, animals and climate.

RESPONSE GRID	1. ⓐⓑⓒⓓ	2. ⓐⓑⓒⓓ	3. ⓐⓑⓒⓓ	4. ⓐⓑⓒⓓ

Space for Rough Work

5. Which one is the main source of minerals essential for man and other organisms ?
   (a) Lithosphere
   (b) Ionosphere
   (c) Mesosphere
   (d) Stratosphere

6. Succulent xerophytes are likely to be found in
   (a) tropical rain forest
   (b) deciduous forest
   (c) desert
   (d) tundra

7. MAB stands for
   (a) man and biosphere.
   (b) man, antibiotics and bacteria.
   (c) man and biotic community.
   (d) Mayer, Anderson and Bishby.

8. The sphere of living matter together with water, air and soil on the surface of earth is
   (a) lithosphere
   (b) biosphere
   (c) hydrosphere
   (d) atmosphere

9. Which one is not dangerous for life and atmosphere?
   (a) Biopollutants
   (b) Ozone layer
   (c) Nuclear blast
   (d) Deforestation

10. The salinity of the oceans is
   (a) increasing
   (b) decreasing
   (c) constant
   (d) decreasing now but was increasing in past

11. Deciduous forests have
   (a) variety of grasses
   (b) broad-leaved trees
   (c) narrow-leaved trees
   (d) variety of crocodiles

12. The conversion of nitrate to nitrous oxide and nitrogen gas is termed as
   (a) nitrification
   (b) denitrification
   (c) nitrogen-fixation
   (d) none of these

13. Which one of the following atoms often limit the primary productivity of an ecosystem?
   (a) Carbon (b) Nitrogen
   (c) Phosphorus (d) Sulphur

14. The phosphorus cycle differs from those of carbon and nitrogen as well as those of sulphur, oxygen and hydrogen in that it lacks
   (a) Sater
   (b) dust particles
   (c) gaseous phase
   (d) all of these

RESPONSE GRID					
	5. ⓐⓑⓒⓓ	6. 	7. ⓐⓑⓒⓓ	8. ⓐⓑⓒⓓ	9. ⓐⓑⓒⓓ
	10. ⓐⓑⓒⓓ	11. ⓐⓑⓒⓓ	12. ⓐⓑⓒⓓ	13. ⓐⓑⓒⓓ	14. ⓐⓑⓒⓓ

Space for Rough Work

15. Biogeochemical cycles are of
(a) two types (b) three types
(c) four types (d) five types

16. Sedimentary cycle having a small gaseous component also is found in
(a) 'P' cycle (b) 'S' cycle
(c) 'N' cycle (d) 'C' cycle

17. Burning of fossil fuels affects
(a) nitrogen cycle
(b) carbon cycle
(c) phosphorus cycle
(d) water cycle

18. The process of evapo-transpiration and precipitation is known as
(a) carbon cycle
(b) hydrological cycle
(c) nitrogen cycle
(d) all of these

19. The agent/s which make/s free atmoshperic nitrogen available to plants is/are
(a) lightning
(b) free living nitrogen fixing bacteria
(c) symbiotic nitrogen fixing bacteria
(d) all of the above

20. Phosphorus cycle is
(a) gaseous cycle
(b) perfect cycle
(c) sedimentary cycle
(d) partly gaseous and partly sedimentary

21. Phosphorus cycles is in the form of
(a) HPO^{3-} (b) P^2 (gas)
(c) PO_4^{3-} (d) $Al_2(PO_4)_3$

22. A biogeochemical cycle without an atmospheric component is
(a) carbon cycle
(b) phosphorus cycle
(c) nitrogen cycle
(d) sulphur cycle

23. Biogeochemical cycling refers to cycling of
(a) energy in the ecosystem.
(b) nutrients in the ecosystem.
(c) water in the ecosystem.
(d) gases between plants and animals.

24. The circulation or cycling of elements in an ecosystem is known as
(a) Geological cycling
(b) Geo-chemical cycling
(c) Bio-geochemical cycling
(d) Nutrient cycling

RESPONSE GRID					
	15. ⓐⓑⓒⓓ	16. ⓐⓑⓒⓓ	17. ⓐⓑⓒⓓ	18. ⓐⓑⓒⓓ	19. ⓐⓑⓒⓓ
	20. ⓐⓑⓒⓓ	21. ⓐⓑⓒⓓ	22. ⓐⓑⓒⓓ	23. ⓐⓑⓒⓓ	24. ⓐⓑⓒⓓ

Space for Rough Work

25. Maximum contribution of O_2 is from
(a) phytoplankton (b) grasslands
(c) herbs and shrubs (d) dense forest

26. Which of the following cycle would be affected if decomposers of an ecosystem vanish?
(a) Producer's cycle (b) Consumer's cycle
(c) Decomposer's cycle (d) Biogeochemical cycle

27. Role of bacteria in carbon cycle is
(a) photosynthesis
(b) chemosynthesis
(c) breakdown of organic compounds
(d) assimilation of nitrogen compounds

28. *Azotobacter* and *Clostridium* are nitrogen fixing bacteria and found in
(a) nodulated roots (b) leaves of certain plant
(c) free in soil (d) none of these

29. Crystalline rocks are natural source of which biogenetic element?
(a) Calcium (b) Phosphorus
(c) Magnesium (d) Sodium

30. Fishing and Guano birds assist the cycling of
(a) nitrogen (b) phosphorus
(c) carbon (d) N_2 and carbon both

RESPONSE GRID	25. ⓐⓑⓒⓓ 26. ⓐⓑⓒⓓ 27. ⓐⓑⓒⓓ 28. ⓐⓑⓒⓓ 29. ⓐⓑⓒⓓ 30. ⓐⓑⓒⓓ

DAILY PRACTICE PROBLEM SHEET 58 - BIOLOGY			
Total Questions	30	Total Marks	120
Attempted		Correct	
Incorrect		Net Score	
Cut-off Score	32	Qualifying Score	56
Success Gap = Net Score – Qualifying Score			
Net Score = (Correct × 4) – (Incorrect × 1)			

Space for Rough Work

DPP - Daily Practice Problems

Name : | Date :

Start Time : | End Time :

BIOLOGY 59

SYLLABUS : Biodiversity and Conservation

Max. Marks : 120 **Time : 60 min.**

GENERAL INSTRUCTIONS

- The Daily Practice Problem Sheet contains 30 MCQ's. For each question only one option is correct. Darken the correct circle/ bubble in the Response Grid provided on each page.
- You have to evaluate your Response Grids yourself with the help of solution booklet.
- Each correct answer will get you 4 marks and 1 mark shall be deduced for each incorrect answer. No mark will be given/ deducted if no bubble is filled. Keep a timer in front of you and stop immediately at the end of 60 min.
- The sheet follows a particular syllabus. Do not attempt the sheet before you have completed your preparation for that syllabus. Refer syllabus sheet in the starting of the book for the syllabus of all the DPP sheets.
- After completing the sheet check your answers with the solution booklet and complete the Result Grid. Finally spend time to analyse your performance and revise the areas which emerge out as weak in your evaluation.

DIRECTIONS (Q.1-Q.30) : There are 30 multiple choice questions. Each question has 4 choices (a), (b), (c) and (d), out of which ONLY ONE choice is correct.

1. Kanha National Park is famous for
   (a) birds (b) rhino
   (c) tiger (d) crocodile

2. "Chipko movement is concerned with
   (a) animal breeding
   (b) plant breeding
   (c) tiger project
   (d) plant conservation

3. Number of ecological hot spots in the world is
   (a) 14
   (b) 16
   (c) 20
   (d) 25

4. Convention on biodiversity came into force on
   (a) 29 Dec. 1993
   (b) 6 June 1992
   (c) 3 October 1994
   (d) 4 October 1993

RESPONSE GRID	1. ⓐⓑⓒⓓ	2. ⓐⓑⓒⓓ	3. ⓐⓑⓒⓓ	4. ⓐⓑⓒⓓ

Space for Rough Work

5. Number of biosphere reserves in May 2002 was
(a) 408 in 94 countries (b) 217 in 65 countries
(c) 381 in 74 countries (d) 449 in 103 countries

6. Which of the following does not occur in wild life conservation?
(a) Fauna conservation
(b) Flora conservation
(c) Use of soil & flora
(d) Prohibition of hunting

7. Biodiversity is determined by
(a) number of individuals in an area
(b) species richness
(c) species evenness
(d) both (b) and (c)

8. Ranthambore National Park is situated in
(a) Maharashtra
(b) Rajasthan
(c) Gujarat
(d) U.P.

9. Identify the correct match for tiger reserve and its state.
(a) Palamau-Orissa
(b) Bandipur-T.N.
(c) Manas-Assam
(d) Corbett-MP.

10. Which of the following species is endangered ?
(a) *Panthera leo persica*
(b) *Pythan molurus*
(c) *Chelonia mydas*
(d) *Pallas siberian crane*

11. New approach of conservation is the establishment of
(a) sancturies (b) reserve forests
(c) national parks (d) biosphere reserves

12. One of the *ex-situ* conservation methods for endangered species is
(a) wildlife sanctuaries
(b) biosphere reserves
(c) cryopreservation
(d) national parks

13. One of the most important functions of botanical gardens is that
(a) they provide a beautiful area for recreation.
(b) one can observe tropical plants there.
(c) they allow *ex-situ* conservation of germplasm.
(d) they provide the natural habitat for wild life.

14. Decrease in species diversity in tropical countries is mainly due to
(a) urbanisation (b) pollution
(c) deforestation (d) soil erosion

RESPONSE GRID					
	5. ⓐⓑⓒⓓ	6. 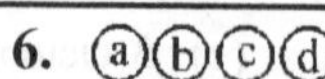	7. ⓐⓑⓒⓓ	8. ⓐⓑⓒⓓ	9. ⓐⓑⓒⓓ
	10. ⓐⓑⓒⓓ	11. ⓐⓑⓒⓓ	12. ⓐⓑⓒⓓ	13. ⓐⓑⓒⓓ	14. ⓐⓑⓒⓓ

Space for Rough Work

15. If at high altitude, birds become rare, the plants likely to disappear are
(a) Pine
(b) Orchids
(c) Oak
(d) *Rhododendrons*

16. Wild life conservation means the protection and preservation of
(a) ferocious wild animals only.
(b) wild plants only.
(c) non-cultivated plants and non-domesticated animals.
(d) all the above living in natural habitat.

17. If the Bengal tiger becomes extinct,
(a) *Hyenas* and wolves will become scarse.
(b) the wild area will be safe for man and domestic animals.
(c) its gene pool will be lost for ever.
(d) the population of beautiful animals like deers will be stabilized.

18. According to IUCN Red List, what is the status of Red Panda (*Ailurus fulgens*)?
(a) Vulnerable species
(b) Critically endangered species
(c) Extinct species
(d) Endangered species

19. Arboretum is
(a) a forest patch.
(b) a piece of social forestry.
(c) botanical garden with trees and shrubs.
(d) botanical garden with cacti and herbs.

20. Those species whose populations have been seriously depleted and whose ultimate security is not assured are known as
(a) threatened species (b) endangered species
(c) vulnerable species (d) rare species

21. Which of the following has become extinct in India ?
(a) Lion (b) Tiger
(c) Two horned Rhino (d) Dodo

22. Khaziranga wild life sanctuary is famous for
(a) tiger (b) musk deer
(c) elephant (d) rhino

23. Which one of the following pair is correctly matched to the wild life and national park ?
(a) Rann of Kutch-Wild ass
(b) Girforest - Tiger
(c) Manas - Elephant
(d) Corbett park - Asiatic lion

24. *Ex situ* conservation is carried out through
(a) funneling the threatened species into trade.
(b) offsite collections.
(c) gene banks.
(d) all of the above.

RESPONSE GRID	15. ⓐⓑⓒⓓ	16. ⓐⓑⓒⓓ	17. ⓐⓑⓒⓓ	18. ⓐⓑⓒⓓ	19. ⓐⓑⓒⓓ
	20. ⓐⓑⓒⓓ	21. ⓐⓑⓒⓓ	22. ⓐⓑⓒⓓ	23. ⓐⓑⓒⓓ	24. ⓐⓑⓒⓓ

Space for Rough Work

25. Gene banks comprise
   (a) seed banks, orchards, tissue culture and cryopreservaton.
   (b) seed banks, tissue culture and cryopreservation.
   (c) seed banks and cryopreservation.
   (d) tissue culture and cryopreservation.
26. Orthodox seeds can be preserved for long periods at a temperature of ?
   (a) $-196°C$ (b) $-10°$ to $-20°C$
   (c) $0° - 4°C$ (d) $10°$ to $15°C$
27. Endemic plants are those, which are
   (a) cosmopolitan in distribution.
   (b) restricted to a certain area.
   (c) gregarious in habit.
   (d) found in alpine region.
28. A high density of a protected animal in National Park can result into
   (a) predation
   (b) emigration
   (c) mutualism
   (d) intraspecific competition
29. An orthodox seed is one which can
   (a) Tolerate dehydration and low temperature.
   (b) Tolerate low temperature but not dehydration.
   (c) Tolerate dehydration but not low temperature.
   (d) None of the above.
30. Which of the following was the first established National Park in India ?
   (a) Bandipur (b) Corbett
   (c) Kanha (d) Periyar

RESPONSE GRID	25. ⓐⓑⓒⓓ	26. ⓐⓑⓒⓓ	27. ⓐⓑⓒⓓ	28. ⓐⓑⓒⓓ	29. ⓐⓑⓒⓓ
	30. ⓐⓑⓒⓓ				

DAILY PRACTICE PROBLEM SHEET 59 - BIOLOGY			
Total Questions	30	Total Marks	120
Attempted		Correct	
Incorrect		Net Score	
Cut-off Score	40	Qualifying Score	60
Success Gap = Net Score – Qualifying Score			
Net Score = (Correct × 4) – (Incorrect × 1)			

Space for Rough Work

DPP - Daily Practice Problems

Name : Date :

Start Time : End Time :

BIOLOGY 60

SYLLABUS : Environmental Issues

Max. Marks : 120 **Time : 60 min.**

GENERAL INSTRUCTIONS

- The Daily Practice Problem Sheet contains 30 MCQ's. For each question only one option is correct. Darken the correct circle/ bubble in the Response Grid provided on each page.
- You have to evaluate your Response Grids yourself with the help of solution booklet.
- Each correct answer will get you 4 marks and 1 mark shall be deduced for each incorrect answer. No mark will be given/ deducted if no bubble is filled. Keep a timer in front of you and stop immediately at the end of 60 min.
- The sheet follows a particular syllabus. Do not attempt the sheet before you have completed your preparation for that syllabus. Refer syllabus sheet in the starting of the book for the syllabus of all the DPP sheets.
- After completing the sheet check your answers with the solution booklet and complete the Result Grid. Finally spend time to analyse your performance and revise the areas which emerge out as weak in your evaluation.

DIRECTIONS (Q.1-Q.30) : There are 30 multiple choice questions. Each question has 4 choices (a), (b), (c) and (d), out of which ONLY ONE choice is correct.

1. Ozone layer depletion or hole in ozone layer is being found in
   (a) North pole (b) South pole
   (c) Russia (d) none of these
2. Eutrophication leads to death of fish due to
   (a) increased O_2 content
   (b) increased algae content.
   (c) decreased algae content.
   (d) decreased O_2 content.
3. What percentage of solar radiation is reflected in the outer space by troposphere ?
   (a) 10%
   (b) 17%
   (c) 34%
   (d) 90%
4. All organic substance found in the biosphere necessarily contain
   (a) silica
   (b) carbon
   (c) oxygen
   (d) nitrogen

RESPONSE GRID	1. ⓐⓑⓒⓓ	2. ⓐⓑⓒⓓ	3. ⓐⓑⓒⓓ	4. ⓐⓑⓒⓓ

Space for Rough Work

5. The region that lies up to 50 km from the Earth above the troposphere is called
   (a) mesosphere
   (b) stratosphere
   (c) troposphere
   (d) thermosphere

6. Zone of atmosphere near the ground is called
   (a) troposphere
   (b) stratosphere
   (c) mesosphere
   (d) homosphere

7. In case CO_2 of earth's atmosphere disappears, the temperature of earth's surface would
   (a) increase.
   (b) decrease.
   (c) depend on oxygen concentration.
   (d) remain the same.

8. Which one of the following has the least green house effect?
   (a) Carbon dioxide (b) Nitrogen gas
   (c) Water vapours (d) Ozone

9. Gas being produced by paddy fields and involved in global warming is
   (a) chlorine (b) methane
   (c) CO_2 (d) H_2S

10. Green house gases include
    (a) CO_2, CFC, CH_4 & NO_2
    (b) CO_2, O_2, N_2, NO_2 & NH_3
    (c) CH_4, N_2, CO_2 & NH_3
    (d) CFC, CO_2, NH_3 & N_2

11. Which gas of the atmosphere holds up ultraviolet rays?
    (a) O_2 (b) O_3
    (c) N_2 (d) CH_4

12. Checking of re-radiating heat by atmospheric dust, O_3, CO_2 and water vapours is called
    (a) ozone layer effect
    (b) solar effect
    (c) radioactive effect
    (d) green house effect

13. Which of the following statement is incorrect?
    (a) Carbon monoxide is the major environment pollutant.
    (b) All pollutants are not waste material.
    (c) N_2 and Mg can pollute water.
    (d) Lichens are affected by sulphur dioxide.

14. Lichens are important in the studies on atmospheric pollution because they
    (a) can grow in polluted atmosphere.
    (b) can readily multiply in polluted atmosphere.
    (c) efficiently purify the atmosphere.
    (d) are very sensitive to pollutants like sulphur dioxide.

RESPONSE GRID					
	5. ⓐⓑⓒⓓ	6.  ⓐⓑⓒⓓ	7. ⓐⓑⓒⓓ	8. ⓐⓑⓒⓓ	9. ⓐⓑⓒⓓ
	10. ⓐⓑⓒⓓ	11. ⓐⓑⓒⓓ	12. ⓐⓑⓒⓓ	13. ⓐⓑⓒⓓ	14. ⓐⓑⓒⓓ

Space for Rough Work

15. Which one of the following is the main air pollutant ?
(a) CO_2 (b) CO
(c) N_2 (d) sulphur

16. Cyanosis is caused by
(a) carbon dioxide
(b) nitrate concentration
(c) chlorine
(d) chromium

17. Gastro-intestinal disturbances is caused by
(a) chromium
(b) cadmium
(c) sodium
(d) selenium

18. Decrease in rate of mitosis in root tips is caused by
(a) SO_2
(b) CO_2
(c) Cu
(d) Pb

19. B.O.D. test (biological oxygen demand test) is made for measuring
(a) air pollution
(b) water pollution
(c) noise pollution
(d) soil pollution

20. In B.O.D. test, oxygen plays an important role to
(a) destroy inorganic matter.
(b) destroy pollution.
(c) destroy waste organic matter.
(d) none of the above

21. Noise pollution is created if noise is in excess to
(a) 70-75 dB
(b) 50-60 dB
(c) 80-99 dB
(d) 40-65 dB

22. Which is the most SO_2 polluted city ?
(a) Madras
(b) calcutta
(c) Delhi
(d) Mumbai

23. Accumulation of fluoride in plant leaves causes
(a) chlorosis
(b) stem rot
(c) tip burn
(d) none of these

24. Excessive inhalation of manganese causes
(a) pneumonia
(b) gout
(c) anaemia
(d) diphtheria

RESPONSE GRID					
	15. ⓐⓑⓒⓓ	16. ⓐⓑⓒⓓ	17. ⓐⓑⓒⓓ	18. ⓐⓑⓒⓓ	19. ⓐⓑⓒⓓ
	20. ⓐⓑⓒⓓ	21. ⓐⓑⓒⓓ	22. ⓐⓑⓒⓓ	23. ⓐⓑⓒⓓ	24. ⓐⓑⓒⓓ

Space for Rough Work

25. Water pollution can be stopped best by
(a) treating effluents to remove injurious chemicals
(b) rearing more fishes
(c) cultivating useful water plants
(d) spraying with DDT

26. Photochemical smog is related to the pollution of
(a) soil
(b) water
(c) air
(d) noise

27. The major source of pollution upto 80% of total air pollution in Metropolitan cities is due to
(a) pesticide
(b) radioactivity
(c) loudspeaker
(d) automobiles

28. Minamata disease is caused by
(a) air pollution
(b) soil pollution
(c) water pollution
(d) noise pollution

29. Bhopal gas accident occurred in which year and due to which gas?
(a) 1986/Methyl isocyanate
(b) 1984/Methyl isocyanate
(c) 1985/CO
(d) 1985/Ethyl isocyanate

30. Which of the following does not cause pollution?
(a) Thermal power project
(b) Nuclear energy project
(c) Automobile
(d) Hydroelectric project

RESPONSE GRID	
	25. ⓐⓑⓒⓓ 26. ⓐⓑⓒⓓ 27. ⓐⓑⓒⓓ 28. ⓐⓑⓒⓓ 29. ⓐⓑⓒⓓ 30. ⓐⓑⓒⓓ

DAILY PRACTICE PROBLEM SHEET 60 - BIOLOGY			
Total Questions	30	Total Marks	120
Attempted		Correct	
Incorrect		Net Score	
Cut-off Score	44	Qualifying Score	64
Success Gap = Net Score – Qualifying Score			
Net Score = (Correct × 4) – (Incorrect × 1)			

Space for Rough Work

DAILY PRACTICE PROBLEMS

BIOLOGY ANSWER KEY

Ch-1 Answer

1. (a)	2. (b)	3. (b)	4. (b)
5. (a)	6. (c)	7. (b)	8. (a)
9. (c)	10. (b)	11. (a)	12. (a)
13. (c)	14. (d)	15. (d)	16. (a)
17. (d)	18. (a)	19. (c)	20. (c)
21. (c)	22. (c)	23. (b)	24. (a)
25. (b)	26. (d)	27. (d)	28. (a)
29. (b)	30. (c)		

Ch-2 Answer

1. (b)	2. (a)	3. (b)	4. (c)
5. (a)	6. (a)	7. (d)	8. (a)
9. (c)	10. (d)	11. (b)	12. (d)
13. (c)	14. (c)	15. (a)	16. (d)
17. (a)	18. (b)	19. (c)	20. (b)
21. (a)	22. (d)	23. (c)	24. (b)
25. (c)	26. (b)	27. (c)	28. (a)
29. (c)	30. (c)		

Ch-3 Answer

1. (a)	2. (b)	3. (d)	4. (c)
5. (d)	6. (c)	7. (b)	8. (c)
9. (c)	10. (a)	11. (d)	12. (d)
13. (a)	14. (b)	15. (b)	16. (d)
17. (a)	18. (b)	19. (d)	20. (b)
21. (c)	22. (c)	23. (a)	24. (d)
25. (b)	26. (d)	27. (d)	28. (c)
29. (a)	30. (c)		

Ch-4 Answer

1. (d)	2. (b)	3. (b)	4. (c)
5. (b)	6. (a)	7. (c)	8. (b)
9. (a)	10. (b)	11. (c)	12. (d)
13. (a)	14. (b)	15. (b)	16. (b)
17. (a)	18. (c)	19. (d)	20. (c)
21. (c)	22. (a)	23. (b)	24. (b)
25. (d)	26. (b)	27. (b)	28. (b)
29. (d)	30. (a)		

Ch-5 Answer

1. (a)	2. (b)	3. (d)
4. (a)	5. (a)	6. (b)
7. (c)	8. (a)	9. (c)
10. (c)	11. (a)	12. (c)
13. (a)	14. (d)	15. (d)
16. (a)	17. (d)	18. (a)
19. (a)	20. (d)	21. (a)
22. (a)	23. (a)	24. (c)
25. (a)	26. (b)	27. (a)
28. (d)	29. (c)	30. (d)

Ch-6 Answer

1. (d)	2. (c)	3. (b)
4. (d)	5. (d)	6. (b)
7. (b)	8. (c)	9. (c)
10. (d)	11. (d)	12. (c)
13. (c)	14. (c)	15. (c)
16. (d)	17. (b)	18. (b)
19. (a)	20. (c)	21. (c)
22. (c)	23. (d)	24. (a)
25. (c)	26. (d)	27. (d)
28. (a)	29. (c)	30. (b)

Ch-7 Answer

1. (b)	2. (b)	3. (c)
4. (d)	5. (b)	6. (d)
7. (a)	8. (a)	9. (c)
10. (a)	11. (c)	12. (c)
13. (d)	14. (a)	15. (a)
16. (a)	17. (c)	18. (b)
19. (a)	20. (d)	21. (c)
22. (a)	23. (b)	24. (a)
25. (b)	26. (b)	27. (b)
28. (a)	29. (c)	30. (a)

Ch-8 Answer

1. (d)	2. (d)	3. (c)
4. (d)	5. (b)	6. (d)
7. (c)	8. (d)	9. (d)
10. (b)	11. (d)	12. (a)
13. (c)	14. (a)	15. (b)
16. (b)	17. (b)	18. (a)
19. (d)	20. (b)	21. (a)
22. (d)	23. (c)	24. (d)
25. (d)	26. (a)	27. (a)
28. (b)	29. (b)	30. (d)

DAILY PRACTICE PROBLEMS

BIOLOGY ANSWER KEY

Ch-9 Answer

1. (c) 2. (a) 3. (b)
4. (c) 5. (c) 6. (a)
7. (c) 8. (b) 9. (a)
10. (a) 11. (b) 12. (b)
13. (b) 14. (d) 15. (c)
16. (b) 17. (a) 18. (d)
19. (d) 20. (a) 21. (a)
22. (a) 23. (a) 24. (d)
25. (a) 26. (c) 27. (a)
28. (d) 29. (b) 30. (c)

Ch-10 Answer

1. (c) 2. (b) 3. (a)
4. (d) 5. (a) 6. (a)
7. (b) 8. (b) 9. (d)
10. (d) 11. (d) 12. (b)
13. (d) 14. (b) 15. (b)
16. (a) 17. (c) 18. (a)
19. (c) 20. (d) 21. (d)
22. (d) 23. (b) 24. (c)
25. (d) 26. (d) 27. (d)
28. (c) 29. (d) 30. (d)

Ch-11 Answer

1. (b) 2. (b) 3. (d)
4. (b) 5. (b) 6. (a)
7. (d) 8. (b) 9. (a)
10. (d) 11. (a) 12. (b)
13. (a) 14. (a) 15. (b)
16. (b) 17. (d) 18. (c)
19. (d) 20. (b) 21. (d)
22. (c) 23. (d) 24. (c)
25. (b) 26. (a) 27. (b)
28. (b) 29. (c) 30. (d)

Ch-12 Answer

1. (b) 2. (c) 3. (a)
4. (b) 5. (a) 6. (b)
7. (c) 8. (c) 9. (d)
10. (c) 11. (b) 12. (a)
13. (a) 14. (c) 15. (b)
16. (c) 17. (b) 18. (b)
19. (b) 20. (d) 21. (b)
22. (b) 23. (b) 24. (b)
25. (d) 26. (d) 27. (c)
28. (d) 29. (b) 30. (c)

Ch-13 Answer

1. (c) 2. (c) 3. (c)
4. (b) 5. (b) 6. (d)
7. (a) 8. (d) 9. (d)
10. (b) 11. (d) 12. (a)
13. (d) 14. (b) 15. (b)
16. (a) 17. (d) 18. (a)
19. (a) 20. (a) 21. (d)
22. (b) 23. (a) 24. (c)
25. (d) 26. (b) 27. (a)
28. (d) 29. (d) 30. (a)

Ch-14 Answer

1. (d) 2. (a) 3. (c)
4. (b) 5. (d) 6. (d)
7. (b) 8. (a) 9. (b)
10. (d) 11. (a) 12. (a)
13. (b) 14. (a) 15. (c)
16. (a) 17. (a) 18. (b)
19. (b) 20. (a) 21. (a)
22. (c) 23. (a) 24. (b)
25. (c) 26. (c) 27. (a)
28. (a) 29. (b) 30. (d)

Ch-15 Answer

1. (c) 2. (b) 3. (d)
4. (b) 5. (b) 6. (b)
7. (c) 8. (a) 9. (b)
10. (a) 11. (a) 12. (b)
13. (c) 14. (d) 15. (a)
16. (b) 17. (c) 18. (c)
19. (c) 20. (c) 21. (a)
22. (c) 23. (b) 24. (c)
25. (b) 26. (b) 27. (b)
28. (b) 29. (b) 30. (a)

Ch-16 Answer

1. (a) 2. (b) 3. (a)
4. (a) 5. (c) 6. (b)
7. (b) 8. (c) 9. (a)
10. (c) 11. (c) 12. (d)
13. (a) 14. (b) 15. (a)
16. (a) 17. (b) 18. (a)
19. (c) 20. (a) 21. (c)
22. (d) 23. (c) 24. (b)
25. (c) 26. (d) 27. (d)
28. (d) 29. (d) 30. (c)

DAILY PRACTICE PROBLEMS

BIOLOGY ANSWER KEY

Ch-17 Answer

1.	(a)	2.	(d)	3.	(b)
4.	(a)	5.	(a)	6.	(c)
7.	(d)	8.	(c)	9.	(a)
10.	(c)	11.	(b)	12.	(c)
13.	(c)	14.	(c)	15.	(b)
16.	(a)	17.	(a)	18.	(b)
19.	(c)	20.	(b)	21.	(a)
22.	(d)	23.	(d)	24.	(c)
25.	(c)	26.	(c)	27.	(b)
28.	(d)	29.	(d)	30.	(c)

Ch-18 Answer

1.	(d)	2.	(c)	3.	(c)
4.	(d)	5.	(c)	6.	(a)
7.	(b)	8.	(d)	9.	(a)
10.	(d)	11.	(b)	12.	(b)
13.	(a)	14.	(b)	15.	(b)
16.	(c)	17.	(b)	18.	(d)
19.	(c)	20.	(c)	21.	(d)
22.	(a)	23.	(b)	24.	(c)
25.	(b)	26.	(c)	27.	(b)
28.	(b)	29.	(c)	30.	(c)

Ch-19 Answer

1.	(a)	2.	(b)	3.	(b)
4.	(b)	5.	(b)	6.	(d)
7.	(d)	8.	(b)	9.	(b)
10.	(c)	11.	(b)	12.	(d)
13.	(c)	14.	(d)	15.	(a)
16.	(d)	17.	(a)	18.	(a)
19.	(c)	20.	(a)	21.	(d)
22.	(d)	23.	(c)	24.	(b)
25.	(c)	26.	(d)	27.	(c)
28.	(a)	29.	(d)	30.	(d)

Ch-20 Answer

1.	(b)	2.	(b)	3.	(d)
4.	(c)	5.	(b)	6.	(c)
7.	(a)	8.	(d)	9.	(b)
10.	(d)	11.	(c)	12.	(b)
13.	(d)	14.	(d)	15.	(c)
16.	(d)	17.	(d)	18.	(b)
19.	(b)	20.	(b)	21.	(a)
22.	(a)	23.	(d)	24.	(b)
25.	(b)	26.	(b)	27.	(c)
28.	(a)	29.	(d)	30.	(c)

Ch-21 Answer

1.	(a)	2.	(c)	3.	(a)
4.	(d)	5.	(c)	6.	(b)
7.	(a)	8.	(d)	9.	(a)
10.	(d)	11.	(c)	12.	(a)
13.	(c)	14.	(a)	15.	(d)
16.	(a)	17.	(a)	18.	(a)
19.	(b)	20.	(c)	21.	(b)
22.	(b)	23.	(a)	24.	(b)
25.	(b)	26.	(c)	27.	(a)
28.	(b)	29.	(a)	30.	(a)

Ch-22 Answer

1.	(d)	2.	(c)	3.	(a)
4.	(c)	5.	(b)	6.	(d)
7.	(a)	8.	(a)	9.	(d)
10.	(b)	11.	(c)	12.	(b)
13.	(d)	14.	(b)	15.	(a)
16.	(b)	17.	(c)	18.	(d)
19.	(a)	20.	(b)	21.	(d)
22.	(d)	23.	(c)	24.	(c)
25.	(b)	26.	(c)	27.	(a)
28.	(c)	29.	(a)	30.	(a)

Ch-23 Answer

1.	(a)	2.	(b)	3.	(d)
4.	(d)	5.	(b)	6.	(d)
7.	(c)	8.	(c)	9.	(d)
10.	(a)	11.	(c)	12.	(a)
13.	(a)	14.	(a)	15.	(d)
16.	(a)	17.	(a)	18.	(d)
19.	(d)	20.	(a)	21.	(a)
22.	(b)	23.	(b)	24.	(c)
25.	(b)	26.	(b)	27.	(a)
28.	(c)	29.	(c)	30.	(b)

Ch-24 Answer

1.	(c)	2.	(d)	3.	(d)
4.	(b)	5.	(c)	6.	(d)
7.	(c)	8.	(b)	9.	(d)
10.	(c)	11.	(b)	12.	(d)
13.	(c)	14.	(d)	15.	(a)
16.	(c)	17.	(a)	18.	(a)
19.	(c)	20.	(b)	21.	(c)
22.	(b)	23.	(c)	24.	(d)
25.	(a)	26.	(a)	27.	(a)
28.	(d)	29.	(b)	30.	(d)

DAILY PRACTICE PROBLEMS

BIOLOGY ANSWER KEY

Ch-25 Answer

1.	(c)	2.	(a)	3.	(b)
4.	(c)	5.	(d)	6.	(a)
7.	(d)	8.	(c)	9.	(d)
10.	(c)	11.	(d)	12.	(b)
13.	(b)	14.	(d)	15.	(d)
16.	(b)	17.	(a)	18.	(d)
19.	(a)	20.	(a)	21.	(b)
22.	(d)	23.	(c)	24.	(b)
25.	(d)	26.	(c)	27.	(c)
28.	(d)	29.	(b)	30.	(b)

Ch-26 Answer

1.	(a)	2.	(c)	3.	(b)
4.	(b)	5.	(a)	6.	(b)
7.	(c)	8.	(a)	9.	(d)
10.	(d)	11.	(d)	12.	(a)
13.	(d)	14.	(c)	15.	(b)
16.	(d)	17.	(b)	18.	(d)
19.	(a)	20.	(a)	21.	(a)
22.	(a)	23.	(b)	24.	(b)
25.	(d)	26.	(b)	27.	(b)
28.	(a)	29.	(a)	30.	(a)

Ch-27 Answer

1.	(a)	2.	(a)	3.	(d)
4.	(c)	5.	(b)	6.	(c)
7.	(b)	8.	(b)	9.	(a)
10.	(a)	11.	(c)	12.	(b)
13.	(d)	14.	(d)	15.	(a)
16.	(b)	17.	(c)	18.	(c)
19.	(d)	20.	(a)	21.	(d)
22.	(c)	23.	(b)	24.	(a)
25.	(c)	26.	(a)	27.	(c)
28.	(b)	29.	(a)	30.	(a)

Ch-28 Answer

1.	(a)	2.	(b)	3.	(d)
4.	(a)	5.	(b)	6.	(b)
7.	(b)	8.	(b)	9.	(a)
10.	(b)	11.	(c)	12.	(c)
13.	(b)	14.	(d)	15.	(d)
16.	(c)	17.	(b)	18.	(a)
19.	(a)	20.	(a)	21.	(c)
22.	(b)	23.	(c)	24.	(b)
25.	(c)	26.	(b)	27.	(b)
28.	(d)	29.	(d)	30.	(c)

Ch-29Answer

1.	(d)	2.	(c)	3.	(a)
4.	(c)	5.	(b)	6.	(b)
7.	(c)	8.	(c)	9.	(b)
10.	(b)	11.	(b)	12.	(d)
13.	(a)	14.	(d)	15.	(d)
16.	(a)	17.	(d)	18.	(a)
19.	(a)	20.	(c)	21.	(a)
22.	(a)	23.	(d)	24.	(d)
25.	(b)	26.	(a)	27.	(b)
28.	(d)	29.	(a)	30.	(d)

Ch-30 Answer

1.	(d)	2.	(a)	3.	(b)
4.	(b)	5.	(b)	6.	(a)
7.	(b)	8.	(a)	9.	(a)
10.	(a)	11.	(c)	12.	(b)
13.	(b)	14.	(c)	15.	(d)
16.	(b)	17.	(c)	18.	(c)
19.	(a)	20.	(b)	21.	(a)
22.	(d)	23.	(c)	24.	(d)
25.	(a)	26.	(d)	27.	(a)
28.	(b)	29.	(a)	30.	(c)

Ch-31 Answer

1.	(c)	2.	(b)	3.	(b)
4.	(d)	5.	(c)	6.	(a)
7.	(c)	8.	(c)	9.	(c)
10.	(b)	11.	(b)	12.	(a)
13.	(d)	14.	(a)	15.	(d)
16.	(b)	17.	(b)	18.	(c)
19.	(b)	20.	(a)	21.	(c)
22.	(b)	23.	(b)	24.	(d)
25.	(b)	26.	(a)	27.	(d)
28.	(d)	29.	(d)	30.	(a)

Ch-32 Answer

1.	(c)	2.	(d)	3.	(a)
4.	(d)	5.	(d)	6.	(c)
7.	(c)	8.	(c)	9.	(a)
10.	(c)	11.	(c)	12.	(d)
13.	(c)	14.	(a)	15.	(c)
16.	(a)	17.	(d)	18.	(c)
19.	(d)	20.	(a)	21.	(b)
22.	(c)	23.	(d)	24.	(c)
25.	(b)	26.	(c)	27.	(c)
28.	(a)	29.	(a)	30.	(c)

DAILY PRACTICE PROBLEMS

BIOLOGY ANSWER KEY

Ch-33 Answer

1.	(d)	2.	(c)	3.	(a)
4.	(d)	5.	(c)	6.	(c)
7.	(c)	8.	(d)	9.	(c)
10.	(b)	11.	(d)	12.	(a)
13.	(a)	14.	(a)	15.	(c)
16.	(b)	17.	(c)	18.	(d)
19.	(d)	20.	(b)	21.	(a)
22.	(a)	23.	(a)	24.	(d)
25.	(a)	26.	(d)	27.	(d)
28.	(b)	29.	(c)	30.	(c)

Ch-34 Answer

1.	(c)	2.	(c)	3.	(a)
4.	(b)	5.	(a)	6.	(a)
7.	(a)	8.	(c)	9.	(c)
10.	(d)	11.	(d)	12.	(d)
13.	(c)	14.	(a)	15.	(c)
16.	(d)	17.	(a)	18.	(d)
19.	(c)	20.	(d)	21.	(c)
22.	(c)	23.	(a)	24.	(d)
25.	(d)	26.	(d)	27.	(a)
28.	(b)	29.	(c)	30.	(b)

Ch-35 Answer

1.	(d)	2.	(b)	3.	(b)
4.	(b)	5.	(a)	6.	(b)
7.	(c)	8.	(d)	9.	(c)
10.	(b)	11.	(c)	12.	(d)
13.	(c)	14.	(a)	15.	(c)
16.	(b)	17.	(c)	18.	(c)
19.	(a)	20.	(c)	21.	(d)
22.	(c)	23.	(a)	24.	(a)
25.	(b)	26.	(d)	27.	(b)
28.	(a)	29.	(c)	30.	(b)

Ch-36 Answer

1.	(c)	2.	(a)	3.	(b)
4.	(b)	5.	(d)	6.	(d)
7.	(b)	8.	(b)	9.	(a)
10.	(a)	11.	(c)	12.	(a)
13.	(d)	14.	(b)	15.	(a)
16.	(c)	17.	(a)	18.	(a)
19.	(d)	20.	(c)	21.	(b)
22.	(c)	23.	(a)	24.	(c)
25.	(b)	26.	(b)	27.	(b)
28.	(a)	29.	(a)	30.	(d)

Ch-37 Answer

1.	(d)	2.	(c)	3.	(a)
4.	(d)	5.	(c)	6.	(c)
7.	(b)	8.	(c)	9.	(a)
10.	(a)	11.	(b)	12.	(c)
13.	(a)	14.	(c)	15.	(c)
16.	(a)	17.	(b)	18.	(d)
19.	(c)	20.	(a)	21.	(c)
22.	(a)	23.	(d)	24.	(a)
25.	(a)	26.	(b)	27.	(d)
28.	(d)	29.	(b)	30.	(d)

Ch-38 Answer

1.	(d)	2.	(d)	3.	(b)
4.	(a)	5.	(a)	6.	(d)
7.	(b)	8.	(a)	9.	(c)
10.	(d)	11.	(a)	12.	(a)
13.	(b)	14.	(d)	15.	(d)
16.	(a)	17.	(d)	18.	(b)
19.	(c)	20.	(a)	21.	(a)
22.	(d)	23.	(b)	24.	(b)
25.	(a)	26.	(b)	27.	(b)
28.	(a)	29.	(d)	30.	(c)

Ch-39 Answer

1.	(b)	2.	(c)	3.	(d)
4.	(a)	5.	(c)	6.	(c)
7.	(c)	8.	(d)	9.	(a)
10.	(d)	11.	(c)	12.	(b)
13.	(c)	14.	(c)	15.	(b)
16.	(b)	17.	(b)	18.	(a)
19.	(b)	20.	(b)	21.	(d)
22.	(d)	23.	(b)	24.	(b)
25.	(b)	26.	(d)	27.	(a)
28.	(a)	29.	(c)	30.	(c)

Ch-40 Answer

1.	(c)	2.	(d)	3.	(a)
4.	(a)	5.	(c)	6.	(a)
7.	(a)	8.	(c)	9.	(d)
10.	(d)	11.	(b)	12.	(d)
13.	(b)	14.	(d)	15.	(c)
16.	(c)	17.	(b)	18.	(a)
19.	(a)	20.	(d)	21.	(d)
22.	(d)	23.	(c)	24.	(d)
25.	(b)	26.	(a)	27.	(c)
28.	(a)	29.	(b)	30.	(d)

DAILY PRACTICE PROBLEMS

BIOLOGY ANSWER KEY

Ch-41 Answer

1.	(b)	2.	(b)	3.	(a)
4.	(a)	5.	(c)	6.	(c)
7.	(c)	8.	(d)	9.	(b)
10.	(a)	11.	(c)	12.	(d)
13.	(d)	14.	(c)	15.	(c)
16.	(c)	17.	(a)	18.	(a)
19.	(b)	20.	(d)	21.	(d)
22.	(a)	23.	(b)	24.	(c)
25.	(d)	26.	(c)	27.	(d)
28.	(d)	29.	(d)	30.	(b)

Ch-42 Answers

1.	(d)	2.	(b)	3.	(a)
4.	(d)	5.	(b)	6.	(d)
7.	(d)	8.	(b)	9.	(b)
10.	(a)	11.	(b)	12.	(c)
13.	(c)	14.	(b)	15.	(d)
16.	(b)	17.	(d)	18.	(b)
19.	(a)	20.	(c)	21.	(b)
22.	(c)	23.	(c)	24.	(a)
25.	(d)	26.	(c)	27.	(c)
28.	(b)	29.	(b)	30.	(c)

Ch-43 Answer

1.	(b)	2.	(c)	3.	(b)
4.	(b)	5.	(b)	6.	(a)
7.	(c)	8.	(d)	9.	(b)
10.	(d)	11.	(c)	12.	(d)
13.	(a)	14.	(d)	15.	(d)
16.	(c)	17.	(c)	18.	(b)
19.	(c)	20.	(c)	21.	(c)
22.	(a)	23.	(a)	24.	(a)
25.	(b)	26.	(c)	27.	(b)
28.	(d)	29.	(b)	30.	(b)

Ch-44 Answer

1.	(b)	2.	(b)	3.	(d)
4.	(c)	5.	(b)	6.	(b)
7.	(a)	8.	(b)	9.	(c)
10.	(d)	11.	(b)	12.	(b)
13.	(b)	14.	(b)	15.	(b)
16.	(d)	17.	(b)	18.	(a)
19.	(c)	20.	(a)	21.	(b)
22.	(d)	23.	(a)	24.	(b)
25.	(b)	26.	(d)	27.	(c)
28.	(b)	29.	(c)	30.	(a)

Ch-45 Answer

1.	(a)	2.	(a)	3.	(d)
4.	(b)	5.	(b)	6.	(a)
7.	(c)	8.	(b)	9.	(d)
10.	(b)	11.	(d)	12.	(d)
13.	(d)	14.	(d)	15.	(b)
16.	(b)	17.	(b)	18.	(c)
19.	(a)	20.	(a)	21.	(d)
22.	(d)	23.	(c)	24.	(a)
25.	(d)	26.	(c)	27.	(a)
28.	(b)	29.	(b)	30.	(c)

Ch-46 Answer

1	(c)	2	(a)	3	(a)
4	(a)	5	(a)	6	(c)
7	(a)	8	(b)	9	(a)
10	(a)	11	(b)	12	(c)
13	(d)	14	(b)	15	(b)
16	(c)	17	(b)	18	(d)
19	(a)	20	(c)	21	(d)
22	(c)	23	(a)	24	(c)
25	(b)	26	(b)	27	(d)
28	(b)	29	(a)	30	(a)

Ch-47 Answer

1	(d)	2	(c)	3	(c)
4	(b)	5	(c)	6	(d)
7	(c)	8	(a)	9	(a)
10	(a)	11	(a)	12	(d)
13	(d)	14	(b)	15	(d)
16	(b)	17	(c)	18	(a)
19	(c)	20	(a)	21	(b)
22	(c)	23	(c)	24	(c)
25	(a)	26	(b)	27	(b)
28	(d)	29	(c)	30	(b)

Ch-48 Answer

1	(c)	2	(b)	3	(b)
4	(b)	5	(d)	6	(d)
7	(c)	8	(d)	9	(b)
10	(a)	11	(a)	12	(c)
13	(b)	14	(b)	15	(b)
16	(d)	17	(b)	18	(c)
19	(d)	20	(b)	21	(d)
22	(d)	23	(a)	24	(d)
25	(a)	26	(a)	27	(b)
28	(b)	29	(c)	30	(b)

DAILY PRACTICE PROBLEMS

BIOLOGY ANSWER KEY

Ch-49 Answer

1.	(b)	2.	(a)	3.	(a)
4.	(a)	5.	(a)	6.	(a)
7.	(d)	8.	(c)	9.	(c)
10.	(c)	11.	(c)	12.	(a)
13.	(c)	14.	(b)	15.	(b)
16.	(a)	17.	(b)	18.	(d)
19.	(c)	20.	(a)	21.	(c)
22.	(c)	23.	(a)	24.	(c)
25.	(d)	26.	(a)	27.	(b)
28.	(b)	29.	(b)	30.	(a)

Ch-50 Answer

1.	(d)	2.	(d)	3.	(b)
4.	(a)	5.	(a)	6.	(a)
7.	(c)	8.	(a)	9.	(a)
10.	(b)	11.	(b)	12.	(b)
13.	(d)	14.	(a)	15.	(c)
16.	(b)	17.	(a)	18.	(b)
19.	(b)	20.	(b)	21.	(b)
22.	(d)	23.	(b)	24.	(b)
25.	(c)	26.	(d)	27.	(b)
28.	(d)	29.	(a)	30.	(a)

Ch-51 Answer

1.	(b)	2.	(d)	3.	(d)
4.	(a)	5.	(b)	6.	(a)
7.	(d)	8.	(c)	9.	(b)
10.	(a)	11.	(c)	12.	(d)
13.	(b)	14.	(b)	15.	(b)
16.	(d)	17.	(c)	18.	(d)
19.	(a)	20.	(c)	21.	(a)
22.	(d)	23.	(b)	24.	(c)
25.	(c)	26.	(b)	27.	(b)
28.	(d)	29.	(d)	30.	(a)

Ch-52 Answer

1.	(c)	2.	(b)	3.	(b)
4.	(d)	5.	(c)	6.	(c)
7.	(c)	8.	(b)	9.	(c)
10.	(b)	11.	(c)	12.	(d)
13.	(c)	14.	(c)	15.	(a)
16.	(c)	17.	(b)	18.	(a)
19.	(c)	20.	(d)	21.	(d)
22.	(a)	23.	(a)	24.	(a)
25.	(b)	26.	(a)	27.	(a)
28.	(b)	29.	(c)	30.	(a)

Ch-53 Answer

1.	(a)	2.	(a)	3.	(a)
4.	(c)	5.	(d)	6.	(b)
7.	(a)	8.	(d)	9.	(a)
10.	(a)	11.	(b)	12.	(a)
13.	(a)	14.	(d)	15.	(c)
16.	(a)	17.	(d)	18.	(c)
19.	(b)	20.	(d)	21.	(c)
22.	(b)	23.	(a)	24.	(d)
25.	(b)	26.	(b)	27.	(d)
28.	(a)	29.	(c)	30.	(d)

Ch-54 Answer

1.	(b)	2.	(a)	3.	(a)
4.	(b)	5.	(c)	6.	(a)
7.	(c)	8.	(a)	9.	(b)
10.	(a)	11.	(c)	12.	(c)
13.	(b)	14.	(a)	15.	(d)
16.	(a)	17.	(a)	18.	(c)
19.	(a)	20.	(d)	21.	(c)
22.	(b)	23.	(c)	24.	(a)
25.	(c)	26.	(c)	27.	(c)
28.	(d)	29.	(c)	30.	(c)

Ch-55 Answer

1.	(a)	2.	(c)	3.	(b)
4.	(c)	5.	(a)	6.	(a)
7.	(a)	8.	(c)	9.	(a)
10.	(d)	11.	(b)	12.	(d)
13.	(d)	14.	(a)	15.	(d)
16.	(c)	17.	(b)	18.	(b)
19.	(d)	20.	(b)	21.	(b)
22.	(b)	23.	(b)	24.	(d)
25.	(a)	26.	(a)	27.	(c)
28.	(b)	29.	(a)	30.	(a)

Ch-56 Answer

1.	(a)	2.	(b)	3.	(a)
4.	(d)	5.	(c)	6.	(b)
7.	(b)	8.	(b)	9.	(b)
10.	(d)	11.	(a)	12.	(b)
13.	(c)	14.	(d)	15.	(d)
16.	(b)	17.	(d)	18.	(d)
19.	(c)	20.	(c)	21.	(d)
22.	(a)	23.	(d)	24.	(b)
25.	(d)	26.	(a)	27.	(c)
28.	(b)	29.	(b)	30.	(b)

DAILY PRACTICE PROBLEMS

BIOLOGY ANSWER KEY

Ch-57 Answer

1.	(a)	2.	(a)	3.	(d)
4.	(d)	5.	(c)	6.	(d)
7.	(c)	8.	(a)	9.	(c)
10.	(a)	11.	(a)	12.	(d)
13.	(d)	14.	(b)	15.	(b)
16.	(c)	17.	(c)	18.	(b)
19.	(b)	20.	(a)	21.	(d)
22.	(b)	23.	(d)	24.	(b)
25.	(a)	26.	(c)	27.	(d)
28.	(c)	29.	(c)	30.	(c)

Ch-58 Answer

1.	(c)	2.	(a)	3.	(a)
4.	(d)	5.	(a)	6.	(c)
7.	(a)	8.	(b)	9.	(b)
10.	(d)	11.	(b)	12.	(b)
13.	(d)	14.	(c)	15.	(a)
16.	(b)	17.	(b)	18.	(b)
19.	(d)	20.	(c)	21.	(c)
22.	(b)	23.	(b)	24.	(c)
25.	(a)	26.	(d)	27.	(c)
28.	(c)	29.	(b)	30.	(b)

Ch-59 Answer

1.	(c)	2.	(d)	3.	(d)
4.	(a)	5.	(a)	6.	(c)
7.	(d)	8.	(b)	9.	(a)
10.	(b)	11.	(d)	12.	(c)
13.	(c)	14.	(c)	15.	(d)
16.	(d)	17.	(b)	18.	(c)
19.	(c)	20.	(b)	21.	(c)
22.	(d)	23.	(a)	24.	(d)
25.	(a)	26.	(b)	27.	(b)
28.	(d)	29.	(a)	30.	(b)

Ch-60 Answer

1.	(b)	2.	(d)	3.	(c)
4.	(b)	5.	(b)	6.	(a)
7.	(b)	8.	(b)	9.	(b)
10.	(a)	11.	(b)	12.	(d)
13.	(b)	14.	(d)	15.	(b)
16.	(b)	17.	(d)	18.	(d)
19.	(b)	20.	(c)	21.	(c)
22.	(b)	23.	(c)	24.	(a)
25.	(a)	26.	(c)	27.	(d)
28.	(c)	29.	(b)	30.	(b)

CHAPTER-WISE DPP SHEETS WITH SOLUTIONS

INDEX/CHAPTERS

DPP - Daily Practice Problems

Chapter-wise Sheets

Date : ____ **Start Time :** ____ **End Time :** ____

BIOLOGY

SYLLABUS : The Living World

Max. Marks : 180 **Marking Scheme :** + 4 for correct & (–1) for incorrect **Time : 60 min.**

INSTRUCTIONS : This Daily Practice Problem Sheet contains 45 MCQs. For each question only one option is correct. Darken the correct circle/ bubble in the Response Grid provided on each page.

1. The living organisms can be unexceptionally distinguished from the non-living things on the basis of their ability for
(a) interaction with the environment and progressive evolution
(b) reproduction
(c) growth and movement
(d) responsiveness to touch.

2. Which one of the following animals is correctly matched with its particular named taxonomic category ?
(a) Tiger - *tigris*, the species
(b) Cuttle fish - mollusca, a class
(c) Humans - primata, the family
(d) Housefly - musca, an order

3. Taxonomic hierarchy refers to
(a) Step-wise arrangement of all categories for classification of plants and animals
(b) A group of senior taxonomists who decide the nomenclature of plants and animals
(c) A list of botanists or zoologists who have worked on taxonomy of a species or group
(d) Classification of a species based on fossil record

4. Choose correct scientific name of mango.
(a) *Mangifera Indica*
(b) *Mangifera indica* Linn
(c) *Mangifera indica* Hook.
(d) *Mangifera indica* L

RESPONSE GRID	1. ⓐⓑⓒⓓ	2. ⓐⓑⓒⓓ	3. ⓐⓑⓒⓓ	4. ⓐⓑⓒⓓ

Space for Rough Work

5. Arrange the following taxonomic categories in increasing number of common characteristics w.r.t. plant mango
   A. Dicotyledonae
   B. Polymoniales
   C. *Mangifera*
   D. Angiospermae
   E. Anacardiaceae
   (a) A→D→B→E→C
   (b) A→D→C→B→E
   (c) D→A→B→E→C
   (d) D→A→C→E→B
6. The common characteristics between brinjal and wheat can be observed maximum at the level of their.
   (a) Division (b) Phylum
   (c) Kingdom (d) Both (b) and (c)
7. Study of number of chromosomes for resolving difficulties in classification is used in
   (a) Chemotaxonomy
   (b) Morphotaxonomy
   (c) Cytotaxonomy
   (d) Biochemical taxonomy
8. Most names in biological nomenclature of living organisms are taken from which language?
   (a) Hindi (b) Latin
   (c) German (d) French
9. The main objective of plant taxonomy is
   (a) to study the world's flora
   (b) to provide a method for identification and nomenclature
   (c) to provide Latin 'scientific' names for every group of plants in the world
   (d) all of these
10. Which one of the following has least similar characters?
   (a) Family (b) Class
   (c) Genus (d) Species
11. The ascending or descending arrangement of taxonomic categories is called as
   (a) classification (b) taxonomy
   (c) hierarchy (d) key
12. Select the false statement
   (a) Carolus Linnaeus described plants and classified them on the basis of their sexual parts
   (b) Some facts established by accurate and repeated observations do not require further verification
   (c) Study of the vestigeal organs is called teleology
   (d) 'White Revolution' resulted in enhanced production of milk
13. Animal taxonomists have named the animals according to:
   (a) International class for Zoology Nomenclature
   (b) Indian code for Zoology Nomenclature
   (c) International classification for Zoological Nomenclature
   (d) International code for Zoological Nomenclature
14. Which of the following statement is not true?
   (a) Homeostasis is a fundamental property of life
   (b) When the external temperature is warm, the superficial blood vessels constrict to prevent loss of body heat
   (c) Human beings are endothermic
   (d) Human beings are homeothermic
15. Species is :
   (a) population of individuals having same genotypes and phenotypes
   (b) a group of individuals inhabiting a geographical area
   (c) a group of interbreeding populations
   (d) population of one type
16. The usage of binomial names, for plant species was accepted by all after the publication of the work by :
   (a) Hooker (b) Linnaeus
   (c) Bentham (d) Darwin
17. Which of the following is less general in characters as compared to genus ?
   (a) Species (b) Division
   (c) Class (d) Family

RESPONSE GRID					
	5. ⓐⓑⓒⓓ	6. ⓐⓑⓒⓓ	7. ⓐⓑⓒⓓ	8. ⓐⓑⓒⓓ	9. ⓐⓑⓒⓓ
	10. ⓐⓑⓒⓓ	11. ⓐⓑⓒⓓ	12. ⓐⓑⓒⓓ	13. ⓐⓑⓒⓓ	14. ⓐⓑⓒⓓ
	15. ⓐⓑⓒⓓ	16. ⓐⓑⓒⓓ	17. ⓐⓑⓒⓓ		

Space for Rough Work

18. A taxon with reference to classification of living organisms can be defined as
(a) a group of similar genera
(b) a group of similar species
(c) a group of organisms based on chromosome numbers
(d) a group of any one rank of organisms

19. Species are considered as
(a) Real units of classification devised by taxonomists
(b) Real basic units of classification
(c) The lowest units of classification
(d) Artificial concept of human mind which cannot be defined in absolute terms

20. Linneaus system of classification is :
(a) Natural (b) Artificial
(c) Phylogenetic (d) Progressive

21. The book systema naturae was written by
(a) Carolus Von Linnaeus
(b) Hutchinson
(c) Engler and Prantl
(d) Bentham & Hooker

22. "Ordines Anomali" of Bentham and Hooker includes :
(a) Seed plants showing abnormal forms of growth and development
(b) Plants represented only in fossil state
(c) Plants described in the literature but which Bentham and Hooker did not see in original
(d) A few orders which could not be placed satisfactory in the classification.

23. What is true for individuals of same species?
(a) Live in same niche
(b) Live in same habitat
(c) Interbreeding
(d) Live in different habitat

24. Biosystematics aims at
(a) Identification and arrangement of organisms on the basis of their cytological characteristics
(b) The classification of organisms based on broad morphological characters
(c) Delimiting various taxa of organisms and establishing their relationships
(d) The classification of organisms based on their evolutionary history and establishing their phylogeny on the totality of various parameters from all fields of studies

25. An animal with same generic, specific and subspecific names is
(a) man (b) gorilla
(c) rabbit (d) elephant

26. Two similar holotypes are called
(a) Isotypes (b) Neotypes
(c) Syntypes (d) Mesotypes

27. Which of the following is not a taxon?
(a) Carnivora (b) Insectivora
(c) Herbivora (d) Mastigophora

28. The category which includes related families is
(a) Class (b) Phylum
(c) Order (d) Kingdom

29. NBRI is situated at
(a) Calcutta (b) Bombay
(c) Madras (d) Lucknow

30. A duplicate of nomenclature type is termed by a taxonomist as
(a) Syntype (b) Neotype
(c) Paratype (d) Isotype

31. Classification based on sequencing of DNA and chemical nature of protein is
(a) Chemotaxonomy
(b) Cytotaxonomy
(c) Adansonian taxonomy
(d) Karyotaxonomy

RESPONSE GRID					
	18. ⓐⓑⓒⓓ	19. ⓐⓑⓒⓓ	20. ⓐⓑⓒⓓ	21. ⓐⓑⓒⓓ	22. ⓐⓑⓒⓓ
	23. ⓐⓑⓒⓓ	24. ⓐⓑⓒⓓ	25. ⓐⓑⓒⓓ	26. ⓐⓑⓒⓓ	27. ⓐⓑⓒⓓ
	28. ⓐⓑⓒⓓ	29. ⓐⓑⓒⓓ	30. ⓐⓑⓒⓓ	31. ⓐⓑⓒⓓ	

Space for Rough Work

32. *Rattus rattus* scientific name is an example of
(a) Autonyms (b) Tautonyms
(c) Synonyms (d) Homonyms

33. Species belonging to different time period are called
(a) Syntype (b) Paratype
(c) Lectotype (d) Isotype

34. A new nomenclatural type submitted when the original material is missing is known as
(a) Holotype (b) Isotype
(c) Paratype (d) Neotype

35. Choose the incorrect statement regarding herbarium
(a) Dried specimens are poisoned by using DDT
(b) The collections are kept inside metallic vasculum
(c) Herbarium sheet is 41 × 29 cm
(d) It is used for alpha taxonomic research

36. No non-living object is capable of reproduction or replication by itself. Although reproduction can not be an all-inclusive defining characteristics of living organisms because
(a) Organisms like mules do not reproduce
(b) It brings about variation in offsprings
(c) Clones are blueprints of their parents
(d) Drones (male honey bees) are produced parthenogenetically

37. Which of the following have more characters in common ?
(a) Order (b) Class
(c) Phylum (d) Family

38. Mark the odd one in the following:
(a) Family (b) Class
(c) Taxon (d) Phylum

39. Which of the following is a name for a taxon that is identical to other such name?
(a) Autonym (b) Synonym
(c) Homonym (d) Tautonym

40. Which of the following is an automatically created name?
(a) Autonym
(b) Synonym
(c) Homonym
(d) Tautonym

41. When the specific epithet exactly repeats generic name, it is called
(a) Basionym (b) Synonym
(c) Homonym (d) Tautonym

42. Which of the following is most important for speciation ?
(a) Seasonal isolation
(b) Reproductive isolation
(c) Behavioural isolation
(d) Temporal isolation

43. Species occurring in different geographical areas are called as
(a) Sibling (b) Sympatric
(c) Allopatric (d) Neopatric

44. All members of different species of plants and animals present in particular area make up
(a) Population (b) Community
(c) Ecosystem (d) Biosphere

45. The study of the kind of life in outer space is known as :
(a) ecology (b) evolution
(c) anthropology (d) exobiology

RESPONSE GRID					
	32. ⓐⓑⓒⓓ	33. ⓐⓑⓒⓓ	34. ⓐⓑⓒⓓ	35. ⓐⓑⓒⓓ	36. ⓐⓑⓒⓓ
	37. ⓐⓑⓒⓓ	38. ⓐⓑⓒⓓ	39. ⓐⓑⓒⓓ	40. ⓐⓑⓒⓓ	41. ⓐⓑⓒⓓ
	42. ⓐⓑⓒⓓ	43. ⓐⓑⓒⓓ	44. ⓐⓑⓒⓓ	45. ⓐⓑⓒⓓ	

Space for Rough Work

DAILY PRACTICE PROBLEM DPP CHAPTERWISE 1 - BIOLOGY			
Total Questions	45	Total Marks	180
Attempted		Correct	
Incorrect		Net Score	
Cut-off Score	45	Qualifying Score	60
Success Gap = Net Score – Qualifying Score			
Net Score = (Correct × 4) – (Incorrect × 1)			

DPP - Daily Practice Problems

Chapter-wise Sheets

Date : | Start Time : | End Time :

BIOLOGY

SYLLABUS : Biological Classification

Max. Marks : 180 | **Marking Scheme : + 4 for correct & (–1) for incorrect** | **Time : 60 min.**

INSTRUCTIONS : This Daily Practice Problem Sheet contains 45 MCQs. For each question only one option is correct. Darken the correct circle/ bubble in the Response Grid provided on each page.

1. Which of the following processes are involved in the reproduction of protista?
   (a) Binary fission and fragmentation
   (b) Cell fusion and zygote formation
   (c) Spore formation and fragmentation
   (d) Budding and spore formation
2. In prokaryotes, the genetic material is
   (a) linear DNA with histones
   (b) circular DNA with histones
   (c) linear DNA without histones
   (d) circular DNA without histones
3. Which of the following bacteria carry out oxygenic photosynthesis by means of a photosynthetic apparatus similar to the eukaryotes?
   (a) Purple sulphur bacteria
   (b) Green sulphur bacteria
   (c) Cyanobacteria
   (d) More than one option is correct
4. Bacteria is a group of prokaryotic organisms which is characterised by
   (a) 70 S ribosomes
   (b) Peptidoglycan cell wall
   (c) Simple structure and complex behaviour
   (d) All of the above.
5. What may be a 'photosynthetic protistian' to one biologist may be 'a plant' to another? Which of the given below features of slime moulds shows linkage with plant?
   (a) Presence of holozoic nutrition
   (b) Presence of diverse sexual reproduction
   (c) Slime moulds have cellulosic spore wall
   (d) Formation of fruiting bodies

RESPONSE GRID	1. ⓐⓑⓒⓓ	2. ⓐⓑⓒⓓ	3. ⓐⓑⓒⓓ	4. ⓐⓑⓒⓓ	5. ⓐⓑⓒⓓ

Space for Rough Work

6. Choose the correct match
(a) *Gonyaulax* – Red sea
(b) *Euglena* – Chlorophyll a & c
(c) Desmids – Chrysophytes
(d) *Gymnodinium* – Hemicellulosic plates in wall

7. Nuclear dimorphism is shown by
(a) *Paramecium* (b) *Amoeba*
(c) *Plasmodium* (d) *Trypanosoma*

8. Most common type of genetic material present in bacteriophages is
(a) ds RNA (b) ss RNA
(c) ds DNA (d) ss DNA

9. Which of the following statement about Mycoplasma is true:
(a) They are smallest, disease causing thin walled organisms
(b) They differ from viruses in being cellular in organisation
(c) Insensitive to several antibiotics as they have 70S ribosomes
(d) They can survive without photosynthetic pigments and genetic material

10. Consider the following characters:
Non-motile spores, saprophytic unicellular eukaryotes, transfer of gametes by wind currents, Differentiation of plasmodium under suitable conditions.
How many of the characters given in box belong to slime moulds?
(a) Four (b) One
(c) Three (d) Two

11. Eubacteria can be differentiated from archaebacteria on the basis of
(a) Ribosomes (b) Gene of tRNA
(c) Cell wall (d) Nutrition

12. Select the pair that consists of plant or animal bacterial diseases.
(a) Cholera and typhoid
(b) Citrus canker and crown gall
(c) Malaria and dengue
(d) Both (a) and (b)

13. Cyanobacteria are classified under which of the following kingdoms?
(a) Monera (b) Protista
(c) Algae (d) Plantae

14. _______ are important decomposers that cause decay and decomposition of dead bodies of plants and animals.
(a) Saprotrophic bacteria
(b) Saprotrophic fungi
(c) Plants, like *Sarracenia*
(d) Both (a) and (b)

15. Chrysophytes are
(a) planktons (b) nektons
(c) benthic organisms (d) rooted submerged.

16. Eukaryotic, achlorophyllous and heterotrophic organisms are grouped under which of the following kingdoms?
(a) Monera (b) Protista
(c) Fungi (d) Plantae

17. Virion is
(a) nulceic acid of virus
(b) antiviral agent
(c) protein of virus
(d) completely assembled virus outside host.

18. In the five-kingdom system of classification, which single kingdom out of the following can include blue, green algae, nitrogen-fixing bacteria and methanogenic archaebacteria?
(a) Fungi (b) Plantae
(c) Protista (d) Monera

19. Viruses that infect bacteria, multiply and cause their lysis, are called
(a) Lysozymes (b) Lipolytic
(c) Lytic (d) Lysogenic

20. Phenetic classification of organisms is based on
(a) Observable characteristics of existing organisms
(b) The ancestral lineage of existing organisms
(c) Dendogram based on DNA characteristics
(d) Sexual characteristics

21. The practical purpose of classification of living organisms is to
(a) explain the origin of living organisms
(b) trace the evolution of living organisms
(c) name the living organisms
(d) facilitate identification of unknown organisms

RESPONSE GRID					
	6. ⓐⓑⓒⓓ	7. ⓐⓑⓒⓓ	8. ⓐⓑⓒⓓ	9. ⓐⓑⓒⓓ	10. ⓐⓑⓒⓓ
	11. ⓐⓑⓒⓓ	12. ⓐⓑⓒⓓ	13. ⓐⓑⓒⓓ	14. ⓐⓑⓒⓓ	15. ⓐⓑⓒⓓ
	16. ⓐⓑⓒⓓ	17. ⓐⓑⓒⓓ	18. ⓐⓑⓒⓓ	19. ⓐⓑⓒⓓ	20. ⓐⓑⓒⓓ
	21. ⓐⓑⓒⓓ				

Space for Rough Work

22. A system of classification in which a large number of traits are considered, is
(a) artificial system (b) synthetic system
(c) natural system (d) phylogenetic system

23. In five kingdom system, the main basis of classification is
(a) structure of nucleus (b) mode of nutrition
(c) structure of cell wall (d) asexual reproduction

24. Phenetic classification is based on
(a) Sexual characteristics
(b) The ancestral lineage of existing organisms
(c) Observable characteristics of existing organisms
(d) Dendograms based on DNA characteristics

25. In which kingdom would you classify the archaebecteria and nitrogen-fixing organisms. If the five-kingdom system of classification is used ?
(a) Monera (b) Plantae
(c) Fungi (d) Protista

26. Which of the following statements is not true for retroviruses?
(a) DNA is not present at any stage in the life cycle of retroviruses
(b) Retroviruses carry gene for RNA-dependent DNA polymerase
(c) The genetic material in mature retroviruses is RNA
(d) Retroviruses are causative agents for certain kinds of cancer in man

27. On how many criteria living organisms have been classified into five kingdoms ?
(a) Two (b) Four
(c) Five (d) Three

28. First true phylogenetic system of classification was given by
(a) Eichler (b) Engler and Prantl
(c) de Jussiaeu (d) de Candolle

29. Which of the following is not a protist ?
(a) *Taenia* (b) *Amoeba*
(c) *Paramecium* (d) *Euglena*

30. Which of the following is not correctly matched?
(a) Root knot disease - *Meloidogyne javanica*
(b) Smut of bajra - *Tolysporium penicillariae*
(c) Covered smut of barley - *Ustilago nuda*
(d) Late blight of potato - *Phytophthora infestans*

31. Which one of the following character was not used by R.H. Whittaker for biological classification ?
(a) Cell structure
(b) Physiological characters
(c) Thallus organisation
(d) Phylogenetic relationships

32. The first organisms to appear on earth were
(a) photoautotrophs (b) chemoautotrophs
(c) chemoheterotrophs (d) heterotrophs

33. 'Comma' shaped bacteria are known as
(a) coccus (b) spiral
(c) spirillum (d) vibrio

34. Slime moulds in the division myxomycota (true slime moulds) have
(a) pseudoplasmodia.
(b) spores that develop into free living amoeboid cells.
(c) spores that develop into flagellated gametes.
(d) feeding stages consisting of solitary individual cells.

35. Which one of the following statements about Mycoplasma is wrong ?
(a) They are pleomorphic.
(b) They are sensitive to penicillin.
(c) They cause diseases in plants.
(d) They are also called (Pleuro pneumonia like organisms) PPLO.

36. African sleeping sickness is due to
(a) *Plasmodium vivax* transmitted by Tse-tse fly
(b) *Trypanosoma lewsii* transmitted by Bed Bug
(c) *Trypanosoma gambiense* transmitted by *Glossina palpalis*
(d) *Entamoeba gingivalis* spread by Housefly

RESPONSE GRID				
22. ⓐⓑⓒⓓ	23. ⓐⓑⓒⓓ	24. ⓐⓑⓒⓓ	25. ⓐⓑⓒⓓ	26. ⓐⓑⓒⓓ
27. ⓐⓑⓒⓓ	28. ⓐⓑⓒⓓ	29. ⓐⓑⓒⓓ	30. ⓐⓑⓒⓓ	31. ⓐⓑⓒⓓ
32. ⓐⓑⓒⓓ	33. ⓐⓑⓒⓓ	34. ⓐⓑⓒⓓ	35. ⓐⓑⓒⓓ	36. ⓐⓑⓒⓓ

Space for Rough Work

37. Which one single organism or the pair of organisms is **correctly** assigned to its taxonomic group?
(a) *Paramecium* and *Plasmodium* belong to the same kingdom as that of *Penicillium*
(b) Lichen is a composite organism formed from the symbiotic association of an algae and a protozoan
(c) Yeast used in making bread and beer is a fungus
(d) *Nostoc* and *Anabaena* are examples of protista

38. Two animals which are the members of the same order must also be the members of the same :
(a) Class (b) Family
(c) Genus (d) Species

39. Bacteria lack alternation of generation because there is
(a) neither syngamy nor reduction division.
(b) distinct chromosomes are absent.
(c) no conjugation.
(d) no exchange of genetic material.

40. Capsid is
(a) genetic material of virus
(b) protein cover of virus
(c) extra genetic material of bacterium
(d) house keeping genome of bacterium

41. A group of fungi with septate mycelium in which sexual reproduction is either unknown or lacking are classified under
(a) phycomycetes (b) deuteromycetes
(c) ascomycetes (d) basidiomycetes

42. A few organisms are known to grow and multiply at temperatures of 100–105°C. They belong to
(a) marine archaebacteria
(b) thermophilic sulphur bacteria
(c) blue-green algae (cyanobacteria)
(d) thermophilic, subaerial fungi

43. Mycoplasma is pleuromorphic due to
(a) absence of cell wall
(b) presence of three layered cell membrane
(c) the presence of sterol
(d) None of these

44. Refer to the given figures of bacteria cell and *Nostoc* and choose the option which shows correct label for the structures marked as A, B, C, D and E ?

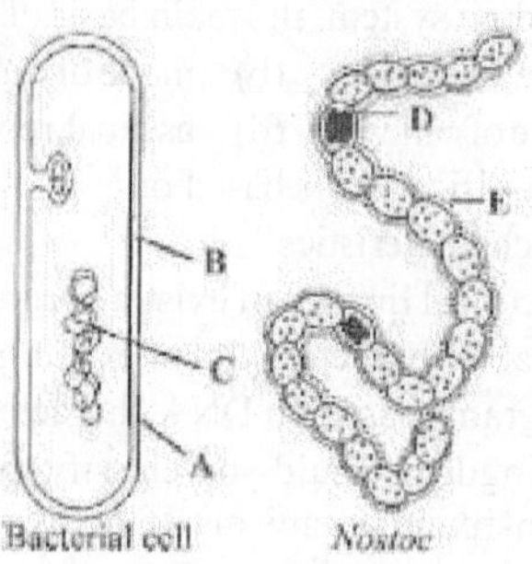

(a) A – Cell wall, B – Cell membrane, C – Heterocyst, D – DNA, E – Mucilagenous sheath
(b) A – Cell wall, B – Cell membrane, C – DNA, D – Heterocyst, E – Mucilagenous sheath
(c) A – Mucilagenous sheath, B – Cell membrane, C – DNA, D – Heterocyst, E – Cell wall
(d) A – Cell membrane, B – Cell wall, C – DNA, D – Heterocyst, E – Mucilagenous sheath

45. Choose the correct names of the different bacteria given below according to their shapes.

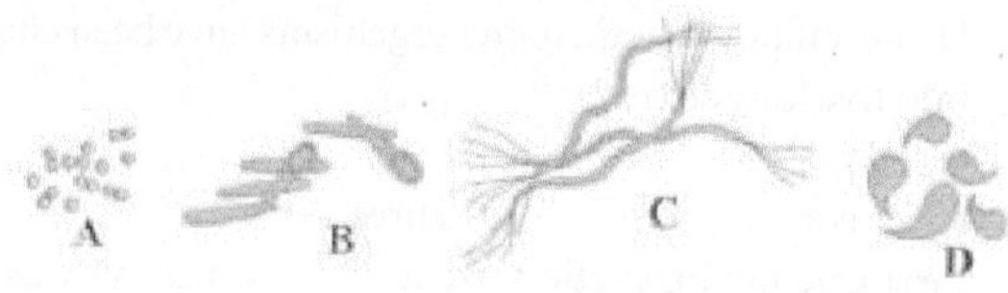

(a) A – Cocci, B – Bacilli, C – Spirilla, D – Vibrio
(b) A – Bacilli, B – Cocci, C – Spirilla, D – Vibrio
(c) A – Spirilla, B – Bacilli, C – Cocci, D – Vibrio
(d) A – Spirilla, B – Vibrio, C – Cocci, D – Bacilli

RESPONSE GRID					
	37. ⓐⓑⓒⓓ	38. ⓐⓑⓒⓓ	39. ⓐⓑⓒⓓ	40. ⓐⓑⓒⓓ	41. ⓐⓑⓒⓓ
	42. ⓐⓑⓒⓓ	43. ⓐⓑⓒⓓ	44. ⓐⓑⓒⓓ	45. ⓐⓑⓒⓓ	

Space for Rough Work

DAILY PRACTICE PROBLEM DPP CHAPTERWISE 2 - BIOLOGY			
Total Questions	45	Total Marks	180
Attempted		Correct	
Incorrect		Net Score	
Cut-off Score	45	Qualifying Score	60
Success Gap = Net Score – Qualifying Score			
Net Score = (Correct × 4) – (Incorrect × 1)			

DPP - Daily Practice Problems

Chapter-wise Sheets

Date : Start Time : End Time :

BIOLOGY (CB03)

SYLLABUS : Plant Kingdom

Max. Marks : 180 **Marking Scheme : + 4 for correct & (–1) for incorrect** **Time : 60 min.**

INSTRUCTIONS : This Daily Practice Problem Sheet contains 45 MCQs. For each question only one option is correct. Darken the correct circle/ bubble in the Response Grid provided on each page.

1. Floridean starch is found in
   (a) Chlorophyceae
   (b) Rhodophyceae
   (c) Phaeophyceae
   (d) Cyanophyceae
2. Peat moss is another name of
   (a) *Sphagnum* (b) *Marchantia*
   (c) *Riccia* (d) *Dryopteris*
3. Pteridophytes differ from mosses/bryophytes in possessing
   (a) independent gametophyte
   (b) well developed vascular system
   (c) archegonia structure
   (d) flagellate spermatozoids
4. Most plants are green in colour because
   (a) the atmosphere filters out all the colours of the visible light spectrum except green.
   (b) green light is the most effective wavelength region of the visible spectrum in sunlight for photosynthesis.
   (c) chlorophyll is least effective in absorbing green light.
   (d) green light allows maximum photosynthesis.
5. In Chlorophyceae, sexual reproduction occurs by
   (a) isogamy and anisogamy
   (b) isogamy, anisogamy and oogamy
   (c) oogamy only
   (d) anisogamy and oogamy

RESPONSE GRID	1. ⓐⓑⓒⓓ	2. ⓐⓑⓒⓓ	3. ⓐⓑⓒⓓ	4. ⓐⓑⓒⓓ	5. ⓐⓑⓒⓓ

Space for Rough Work

6. A water fern which is used as a green manure in rice fields is
(a) *Salvinia* (b) *Mucor*
(c) *Aspergillus* (d) *Azolla*

7. The largest flower found is known as
(a) *Rafflesia* (b) *Tecoma*
(c) *Musa* (d) Cauliflower

8. In fern, spores are formed in
(a) sporangium (b) oogonium
(c) archegonium (d) stomium

9. *Laminaria* (kelp) and *Fucus* (rock weed) are the examples of
(a) red algae
(b) brown algae
(c) green algae
(d) golden brown algae

10. People recovering from long illness are often advised to include the alga *Spirulina* in their diet because it
(a) makes the food easy to digest.
(b) is rich in proteins.
(c) has antibiotic properties.
(d) restores the intestinal microflora.

11. Which of the following cell organelle remains enveloped by a single unit membrane?
(a) Mitochondria (b) Lysosomes
(c) Nucleus (d) Chloroplast

12. Consider the following statements regarding the major pigments and stored food in the different groups of algae and choose the correct option.
(i) In Chlorophyceae, the stored food material is starch and the major pigments are chlorophyll-*a* and *d*.
(ii) In Phaeophyceae, laminarin is the stored food and major pigments are chlorophyll-*a* and *b*.
(iii) In Rhodophyceae, floridean starch is the stored food and the major pigments are chlorophyll-*a*, *d* and phycoerythrin.
(a) (i) is correct, but (ii) and (iii) are wrong.
(b) (i) and (ii) are correct, but (iii) is wrong.
(c) (i) and (iii) are correct, but (ii) is wrong.
(d) (iii) is correct, but (i) and (ii) are wrong.

13. Algae have cell wall made up of
(a) cellulose, galactans and mannans
(b) hemicellulose, pectins and proteins
(c) pectins, cellulose and proteins
(d) cellulose, hemicellulose and pectins.

14. Which plays an important role in the dispersal of spores in *Funaria*?
(a) Operculum
(b) Capsule
(c) Peristome and annulus
(d) Sporogonium

15. Read the following five statements (i – v) and answer the question.
(i) In *Equisetum* the female gametophyte is retained on the parent sporophyte.
(ii) In *Ginkgo* male gametophyte is not independent.
(iii) The sporophyte in *Riccia* is more developed than that in *Polytrichum.*
(iv) Sexual reproduction in *Volvox* is isogamous.
(v) The spores of slime molds lack cell walls.
How many of the above statements are correct?
(a) Two (b) Three
(c) Four (d) One

16. Which one of the following is common to multicellular fungi, filamentous algae and protonema of mosses?
(a) Diplontic life cycle
(b) Members of kingdom plantae
(c) Mode of Nutrition
(d) Multiplication by fragmentation

17. Which one of the following is a correct statement ?
(a) Pteridophyte gametophyte has a protonemal and leafy stage
(b) In gymnosperms female gametophyte is free-living
(c) Antheridiophores and archegoniophores are present in pteridophytes
(d) Origin of seed habit can be traced in pteridophytes

RESPONSE GRID					
	6. ⓐⓑⓒⓓ	7. ⓐⓑⓒⓓ	8. ⓐⓑⓒⓓ	9. ⓐⓑⓒⓓ	10. ⓐⓑⓒⓓ
	11. ⓐⓑⓒⓓ	12. ⓐⓑⓒⓓ	13. ⓐⓑⓒⓓ	14. ⓐⓑⓒⓓ	15. ⓐⓑⓒⓓ
	16. ⓐⓑⓒⓓ	17. ⓐⓑⓒⓓ			

Space for Rough Work

18. What is the similarity between gymnosperms and angiosperms ?
(a) Phloem of both have companian cells.
(b) Endosperm is formed before fertilization in both.
(c) Origin of ovule and seed is similar in both.
(d) Both have leaves, stem and roots.

19. In Chlorophyceae, sexual reproduction occurs by
(a) isogamy and anisogamy
(b) isogamy, anisogamy and oogamy
(c) oogamy only
(d) anisogamy and oogamy

20. In gymnosperms, the ovule is naked because
(a) ovary wall is absent (b) integuments are absent
(c) perianth is absent (d) nucellus is absent

21. How many meiotic division would be required to produce 101 female gametophytes in an angiosperm?
(a) 101 (b) 26
(c) 127 (d) None of these

22. Which one of the following is the major difference between mosses and ferns ?
(a) Ferns lack alternation of generation while mosses show the same.
(b) Mosses are facultative aerobes while ferns are obligate aerobes.
(c) Vascular bundles of ferns show xylem vessels while those of mosses lack it.
(d) Sporophytes of ferns live much longer as compared to the sporophytes of mosses.

23. Red snow causing alga is
(a) *Chlamydomonas nivalis*
(b) *Chlamydomonas reinhardtii*
(c) *Chlamydomonas debaryanum*
(d) *Chalmydomonas media*

24. Alginates (alginin), used as highly efficient gauze in internal operations are obtained from cell walls of
(a) Cyanophyceae (b) Phaeophyceae
(c) Rhodophyceae (d) All of these

25. Bryophytes resemble algae in the following aspects
(a) Filamentous body, presence of vascular tissues and autotrophic nutrition
(b) Differentiation of plant body into root, stem and leaves and autotrophic nutrition
(c) Thallus like plant body, presence of root and autotrophic nutrition
(d) Thallus like plant body, lack of vascular tissues and autotrophic nutrition

26. In sexual life cycle of *Agaricus*, dikaryotization (n + n) is brought about by
(a) Fusion of male and female sex organs
(b) Fusion of vegetative cells of different genotypes
(c) Somatogamy between basidiospores
(d) Fusion of motile gametes

27. Read the following features properly
A. Free living
B. Mostly photosynthetic
C. Mostly parasitic
D. Inconspicuous
E. Unicellular

How many of the given features are correct for prothallus of pteridophytes?
(a) Three (b) Five
(c) Four (d) Two

28. Identify the correctly matched pair:

	Class	Example	Feature
(a)	Psilopsida	*Lycopodium*	Seed habit
(b)	Sphenopsida	*Selaginella*	Strobilus
(c)	Lycopsida	*Psilotum*	Homosporous
(d)	Pteropsida	*Dryopteris*	Macrophylls

29. Angiosperms have dominated the land flora primarily because of their
(a) power of adaptability in diverse habitat
(b) property of producing large number of seeds
(c) nature of self pollination
(d) domestication by man

RESPONSE GRID					
	18. ⓐⓑⓒⓓ	19. ⓐⓑⓒⓓ	20. ⓐⓑⓒⓓ	21. ⓐⓑⓒⓓ	22. ⓐⓑⓒⓓ
	23. ⓐⓑⓒⓓ	24. ⓐⓑⓒⓓ	25. ⓐⓑⓒⓓ	26. ⓐⓑⓒⓓ	27. ⓐⓑⓒⓓ
	28. ⓐⓑⓒⓓ	29. ⓐⓑⓒⓓ			

Space for Rough Work

30. Which out of the following are included under tracheophyta i.e., vascular plants?
(a) Pteridophytes (b) Gymnosperms
(c) Angiosperms (d) All of these

31. At least a half of the total CO_2 fixation on earth is carried out through photosynthesis by
(a) angiosperms (b) gymnosperms
(c) algae (d) bryophytes

32. The embryonic development in bryophytes takes place in the
(a) protonema (b) sporangium
(c) antheridium (d) archegonium.

33. The spread of living pteridophytes is limited and is restricted to narrow geographical region because of
(a) gametophytic growth needs cool, damp and shady places
(b) requirement of water for fertilization
(c) absence of stomata in leaf and absence of vascular tissue
(d) both (a) and (b)

34. Gymnosperm called as a living fossil is
(a) *Cycas* (b) *Ginkgo*
(c) *Juniperus* (d) Both (a) and (b).

35. The sporophyte is the dominant phase in
(a) pteridophytes (b) gymnosperms
(c) angiosperms (d) all of these.

36. Which kind of life-cycle pattern is exhibited by seed-bearing plants?
(a) Haplontic (b) Diplontic
(c) Haplo-diplontic (d) All of these

37. Plants reproducing by spores such as mosses and ferns are grouped under the general term
(a) Thallophytes (b) Cryptogams
(c) Bryophytes (d) Sporophytes

38. Angiosperms have dominated the land flora primarily because of their
(a) Power of adaptability in diverse habitat
(b) Property of producing large number of seeds
(c) Nature of self pollination
(d) Domestication by man

39. Many blue-green algae occur in thermal springs (hot water springs). The temperature tolerance of these algae have been attributed to their
(a) cell wall structure
(b) mitochondrial structure
(c) modern cell organization
(d) importance of homopolar bonds in their proteins

40. Which of the following occurs both in fresh as well as in marine water ?
(a) Spirogyra (b) *Cladophora*
(c) Oedogonium (d) *Cephaleuros*

41. The pyrenoids are made up of
(a) proteinaceous centre and starchy sheath
(b) core of protein surrounded by fatty sheath
(c) core of starch surrounded by sheath of protein
(d) core of nucleic acid surrounded by protein sheath

42. Blue green algae have
(a) chlorophyll (b) xanthyophyll
(c) phycocyanin (d) fucoxanthin

43. Parasitic alga is
(a) *Volvox* (b) Ulothrix
(c) *Porphyra* (d) *Cephaleuros*

44. Which one of the following pairs of plants are not seed producers?
(a) *Funaria* and *Pinus* (b) Fern and *Funaria*
(c) *Funaria* and *Ficus* (d) *Ficus* and *Chlamydomonas*

45. Neck canal cells are absent in archegonia of –
(a) Bryophytes (b) Gymnosperms
(c) Pteridophytes (d) All of these

RESPONSE GRID					
	30. ⓐⓑⓒⓓ	31. ⓐⓑⓒⓓ	32. ⓐⓑⓒⓓ	33. ⓐⓑⓒⓓ	34. ⓐⓑⓒⓓ
	35. ⓐⓑⓒⓓ	36. ⓐⓑⓒⓓ	37. ⓐⓑⓒⓓ	38. ⓐⓑⓒⓓ	39. ⓐⓑⓒⓓ
	40. ⓐⓑⓒⓓ	41. ⓐⓑⓒⓓ	42. ⓐⓑⓒⓓ	43. ⓐⓑⓒⓓ	44. ⓐⓑⓒⓓ
	45. ⓐⓑⓒⓓ				

Space for Rough Work

DAILY PRACTICE PROBLEM DPP CHAPTERWISE 3 - BIOLOGY			
Total Questions	45	Total Marks	180
Attempted		Correct	
Incorrect		Net Score	
Cut-off Score	35	Qualifying Score	50
Success Gap = Net Score – Qualifying Score			
Net Score = (Correct × 4) – (Incorrect × 1)			

DPP - Daily Practice Problems

Chapter-wise Sheets

Date : | Start Time : | End Time :

BIOLOGY CB04

SYLLABUS : Animal Kingdom

Max. Marks : 180 **Marking Scheme : + 4 for correct & (–1) for incorrect** **Time : 60 min.**

INSTRUCTIONS : This Daily Practice Problem Sheet contains 45 MCQs. For each question only one option is correct. Darken the correct circle/ bubble in the Response Grid provided on each page.

1. Classification of Porifera is based on
(a) branching (b) spicules
(c) reproduction (d) symmetry

2. A chordate character is
(a) gills (b) spiracles
(c) postanal tail (d) chitinous exoskeleton

3. Which one of the following pairs of animals comprises 'jawless fishes'?
(a) Mackerals and Rohu (b) Lampreys and hag fishes
(c) Guppies and hag fishes (d) Lampreys and eels

4. Leech is
(a) carnivorous (b) sanguivorous
(c) ectoparasite (d) Both (b) and (c)

5. Which one of the following groups of animals is bilaterally symmetrical and triploblastic?
(a) Aschelminthes (round worms)
(b) Ctenophores
(c) Sponges
(d) Coelenterates (Cnidarians)

6. Which of the following animal is cold blooded and has 4 - chambered heart?
(a) Salamander (b) *Ornithorhynchus*
(c) Crocodile (d) *Calotes*

7. Which one of the following is NOT a characteristic of phylum Annelida?
(a) Closed circulatory system
(b) Segmentation
(c) Pseudocoelom
(d) Ventral nerve cord

RESPONSE GRID					
	1. ⓐⓑⓒⓓ	2. ⓐⓑⓒⓓ	3. ⓐⓑⓒⓓ	4. ⓐⓑⓒⓓ	5. ⓐⓑⓒⓓ
	6. ⓐⓑⓒⓓ	7. ⓐⓑⓒⓓ			

Space for Rough Work

8. Which one of the following characters is not typical of the class mammalia?
(a) Thecodont dentition
(b) Alveolar lungs
(c) Ten pairs of cranial nerves
(d) Seven cervical vertebrae

9. Which one of following feature is possessed by Crustaceans and not by insects?
(a) Paired limbs
(b) Two pairs of antenna
(c) Chitinous exoskeleton
(d) Bilateral symmetry

10. Poison glands of snake are modified
(a) Linguals (b) Sublinguals
(c) Maxillaries (d) Parotids

11. The adults are radially symmetrical but larvae exhibit bilateral symmetry in
(a) Mollusca (b) Hemichordata
(c) Echinodermata (d) Cephalochordata

12. Which one of the following categories of animals, is correctly described with no single exception in it?
(a) All sponges are marine and have collared cells.
(b) All mammals are viviparous and possess diaphragm for breathing.
(c) All bony fishes have four pairs of gills and an operculum on each side.
(d) All reptiles possess scales, have a three chambered heart and are cold blooded (poikilothermal).

13. Solenocytes and metanephridia are excretory organs of
(a) Annelida and Arthropoda
(b) Platyhelminthes and Annelida
(c) Coelenterata and Mollusca
(d) Aschelminthes and Annelida

14. Radial symmetry occurs in
(a) Porifera and Coelenterata
(b) Coelenterata and Echinodermata
(c) Coelenterata and Platyhelminthes
(d) Arthropoda and Mollusca

15. Which of the following statements is/are not true?
(i) In Urochordata, notochord is present in larval tail.
(ii) In Cephalochordata, notochord extends from head to tail region.
(iii) *Branchiostoma* belongs to hemichordata.
(iv) Only one class of living members, class Cyclostomata represents the super class agnatha
(a) (ii) and (iv) (b) (i), (iii) and (iv)
(c) (iii) only (d) (i) and (iv)

16. In *Amoeba* and *Paramecium* osmoregulation occurs through
(a) pseudopodia (b) nucleus
(c) contractile vacuole (d) general surface

17. Animals with metameric segmentation, bilateral symmetry and closed circulatory system belong to phylum
(a) Annelida (b) Echinodermata
(c) Arthropoda (d) Mollusca

18. Which one of the following characters is **not** typical of the class Mammalia?
(a) Thecodont dentition
(b) Alveolar lungs
(c) Ten pairs of cranial nerves
(d) Seven cervical vertebrae

19. The segments of earthworms are
(a) Apparent in the embryo but not in the adult
(b) Specialised for different functions and are present in endoderm only
(c) Present in mesoderm but not in the ectoderm
(d) Repetitive, with serial repetition of at least some organs

20. The evolution of an internal body cavity/coelom offered an advantage in animal body design in all areas, except
(a) Evolution of effecient organ systems
(b) Provides space within which the gonads can expand and large number of gametes stored
(c) Circulation
(d) Greater freedom of movement

RESPONSE GRID					
	8. ⓐⓑⓒⓓ	9. ⓐⓑⓒⓓ	10. ⓐⓑⓒⓓ	11. ⓐⓑⓒⓓ	12. ⓐⓑⓒⓓ
	13. ⓐⓑⓒⓓ	14. ⓐⓑⓒⓓ	15. ⓐⓑⓒⓓ	16. ⓐⓑⓒⓓ	17. ⓐⓑⓒⓓ
	18. ⓐⓑⓒⓓ	19. ⓐⓑⓒⓓ	20. ⓐⓑⓒⓓ		

Space for Rough Work

21. Animals like bats have/are
(a) Hollow skeleton
(b) Feathers which are modified reptilian scales
(c) Endothermic
(d) Efficient respiration. Non-vascular air sacs are connected to lungs to supplement respiration

22. Which of the following is incorrect match of animal group/life style/structure/function?

	Animal	Lifestyle	Structure/functions group
(a)	Sponges	Sessile filter feeders	Amoebocytes/carry food and wastes Spicules/support and protection spongin / support
(b)	Cnidarians	Free floating or attached	Gastrovascular cavity/digestion Cnidocytes/ protection and food getting
(c)	Flatworms	Free living or parasite	Flame cells/ excretion Tegument/ protection
(d)	Molluscs	Terrestrial, marine, fresh water inhabitants	Radula/feeding mantle/motility

23. Complete the following analogy : Pigeon's milk : Crop :: Song :
(a) Syrinx (b) Trachea
(c) Proventriculus (d) Anterior air sacs

24. Which of the following is not correct matching of phylum and its three examples?
(a) Annelida : *Aphrodite, Chaetopterus, Bonnelia*
(b) Mollusca : *Teredo, Aplysia, Chaetopleura*
(c) Aschelminthes : *Ancylostoma, Enterobius, Tubifex*
(d) Arthropoda : *Buthus, Lepisma, Leptocorisa*

25. Hemichordates differ from chordates, in that hemichordates
(a) Are gill breathers
(b) Fertilisation is external and development is indirect
(c) Do not possess notochord
(d) Are bilaterally symmetrical, triploblastic and coelomate animals

26. All birds have
(a) Oil gland at the base of tail
(b) Feather on their body and can fly
(c) Nests to care their babies
(d) Internal fertilization, are oviparous and eggs are covered with calcareous shell

27. Which one of the following is a coelenterate?
(a) Sea mouse (b) Sea anemone
(c) Sea urchin (d) Sea cucumber

28. Presence of external ear pinna, body hairs, four chambered heart are the characters of
(a) *Macropus* (b) *Balaenoptera*
(c) *Psittacula* (d) *Aptenodytes*

29. Which one of the following statements about certain given animals is correct?
(a) Round worms are pseudo-coelomates
(b) Molluscs are acoelomates
(c) Insects are pseudo-coelomates
(d) Flatworms are coelomates

30. Which one of the following phyla is correctly matched with its two general characteristics?
(a) Echinodermata – pentamerous radial symmetry and mostly internal fertilization
(b) Mollusca – normally oviparous and development through a trochophore or veliger larva
(c) Arthropoda – body divided into head, thorax and abdomen and respiration by mouth
(d) Chordata – notochord persists throughout and separate anal and urinary openings to the outside

31. Amphibians share with reptiles all of the following characters expect
(a) ventral heart
(b) external fertilization and indirect development
(c) dioecious, oviparous
(d) cold blooded or poikilotherms

RESPONSE GRID					
	21. ⓐⓑⓒⓓ	22. ⓐⓑⓒⓓ	23. ⓐⓑⓒⓓ	24. ⓐⓑⓒⓓ	25. ⓐⓑⓒⓓ
	26. ⓐⓑⓒⓓ	27. ⓐⓑⓒⓓ	28. ⓐⓑⓒⓓ	29. ⓐⓑⓒⓓ	30. ⓐⓑⓒⓓ
	31. ⓐⓑⓒⓓ				

Space for Rough Work

32. Which of the following statements about the body cavity of animals is true ?
(a) The body cavity of coelomates develops from the embryonic ectoderm.
(b) The acoelomates' body cavity is filled with liquid.
(c) The pseudocoel of the pseudocoelomates have a peritoneum.
(d) The acoelomates do not have an enclosed body cavity.

33. Sponges have a very simple body plan. Which of the following statements about sponge structure or function is false ?
(a) Choanocytes are flagellated cells that play a role in feeding.
(b) Large species are found in areas of heavy wave action, where food is most abundant.
(c) Individual sponges are both male and female.
(d) Water enters a sponge through pores and exits via one or more oscula.

34. Which of the following traits is not shared by the Ctenophora and the Cnidaria ?
(a) Both are diploblastic
(b) Both have radial symmetry
(c) Both have complete guts.
(d) Both have feeding tentacles.

35. Earthworm has
(a) Two eyes (b) Many eyes
(c) No eyes (d) One eye.

36. Which of the following statements is not true of the Rotifera ?
(a) They have a complete gut with an anterior mouth and posterior anus.
(b) They are coelomates
(c) The corona is a ciliated organ used in acquiring food.
(d) They use a hydrostatic skeleton.

37. The combination of a true coelom and repeating body segmentation allows the annelids (unlike the anatomically "simpler" worms) to do which of the following ?
(a) Attain complex body shapes and thus locomote more precisely
(b) Move through loose marine sediments
(c) Be hermaphroditic
(d) Inject paralytic poisons into their prey

38. An animal is divided along its main body axis to produce similar halves. Which of the following types of symmetry could apply ?
(a) Spherical (b) Radial
(c) Radial or biradial (d) Bilateral

39. Cephalization is a characteristic mainly associated with which of the following types of body symmetry in animals ?
(a) Asymmetric (b) Radial
(c) Biradial (d) Bilateral

40. Amphids present on ventrolateral lips of *Ascaris* are
(a) Chemoreceptors (b) Olfactoreceptors
(c) Tactoreceptors (d) Gustatoreceptors

41. Corals are common representatives of the cnidarian class______.
(a) Scyphozoa (b) Anthozoa
(c) Porifera (d) Placozoa

42. Which of the following characteristics is unique to the phylum Cnidaria ?
(a) Sexual reproduction
(b) Symbiotic associations with other organisms
(c) Sedentary body forms
(d) Nematocysts

43. Which of the following statements is true of all flatworms ?
(a) Flatworms are biradially symmetric
(b) Flatworms have a complete digestive system
(c) Flatworms tend to have large, thickened bodies
(d) Flatworms are triploblastic

44. Which of the following structures is absent from a typical gastropod mollusk (e.g., a garden snail) ?
(a) Protective shell (b) Head
(c) Radula (d) None of these

45. Which two of the following are found in the mesophyl or protein matrix and serve as structural support for a sponge ?
(a) spicule, spongin (b) osculum, spicule
(c) medusa, polyp (d) polyp, osculum

RESPONSE GRID					
	32. ⓐⓑⓒⓓ	33. ⓐⓑⓒⓓ	34. ⓐⓑⓒⓓ	35. ⓐⓑⓒⓓ	36. ⓐⓑⓒⓓ
	37. ⓐⓑⓒⓓ	38. ⓐⓑⓒⓓ	39. ⓐⓑⓒⓓ	40. ⓐⓑⓒⓓ	41. ⓐⓑⓒⓓ
	42. ⓐⓑⓒⓓ	43. ⓐⓑⓒⓓ	44. ⓐⓑⓒⓓ	45. ⓐⓑⓒⓓ	

Space for Rough Work

DAILY PRACTICE PROBLEM DPP CHAPTERWISE 4 - BIOLOGY			
Total Questions	45	Total Marks	180
Attempted		Correct	
Incorrect		Net Score	
Cut-off Score	40	Qualifying Score	65
Success Gap = Net Score – Qualifying Score			
Net Score = (Correct × 4) – (Incorrect × 1)			

DPP - Daily Practice Problems

Chapter-wise Sheets

Date : ____ **Start Time :** ____ **End Time :** ____

BIOLOGY

SYLLABUS : Morphology of Flowering Plants

Max. Marks : 180 **Marking Scheme : + 4 for correct & (–1) for incorrect** **Time : 60 min.**

INSTRUCTIONS : This Daily Practice Problem Sheet contains 45 MCQs. For each question only one option is correct. Darken the correct circle/ bubble in the Response Grid provided on each page.

1. Which one of the following is a true fruit?
   (a) Apple (b) Pear
   (c) Cashew nut (d) Coconut
2. Pulses are belong to the family
   (a) fabaceae (b) asteraceae
   (c) poaceae (d) solanaceae
3. In a cereal grain the single cotyledon of embryo is represented by
   (a) scutellum (b) prophyll
   (c) coleoptile (d) coleorrhiza
4. Perisperm is
   (a) remnant of endosperm
   (b) persistant nucellus
   (c) remnant of embryo
   (d) part of endosperm
5. The mode of catching insects in *Drosera* plants is by means of
   (a) sensitive glandular hairs which secrete a sweet, viscous, shining substance.
   (b) specially sensitive trigger hairs.
   (c) leaves which are modified into pitcher.
   (d) leaf segments modified into bladder.
6. Insectivorous plants grow in
   (a) calcium deficient soil
   (b) carbon deficient soil
   (c) magnesium deficient soil
   (d) nitrogen deficient soil
7. Which part of the coconut produces coir?
   (a) Seed coat (b) Mesocarp
   (c) Epicarp (d) Pericarp

RESPONSE GRID					
	1. ⓐⓑⓒⓓ	2. ⓐⓑⓒⓓ	3. ⓐⓑⓒⓓ	4. ⓐⓑⓒⓓ	5. ⓐⓑⓒⓓ
	6. ⓐⓑⓒⓓ	7. ⓐⓑⓒⓓ			

Space for Rough Work

8. Pineapple (ananas) fruit develops from
   (a) a multipistillate syncarpous flower
   (b) a cluster of compactly borne flowers on a common axis
   (c) a multilocular monocarpellary flower
   (d) a unilocular polycarpellary flower

9. Scutellum is a/an
   (a) protective covering of radicle
   (b) protective covering of plumule
   (c) endosperm of gymnosperms
   (d) shield-shaped cotyledon

10. Fibrous root system is better adopted than tap root system for
   (a) transport of organic matter
   (b) absorption of water and minerals
   (c) storage of food
   (d) anchorage of plant to soil

11. Velamen is found in
   (a) roots of screwpine
   (b) aerial and terrestrial roots of orchids
   (c) leaves of *Ficus elastica*
   (d) only aerial roots of orchids

12. Hypanthodium is
   (a) thalamus (b) fruit
   (c) inflorescence (d) ovary

13. Which of the following statement (s) is/are incorrect?
   (i) Calyx and corolla are reproductive organs of a flower.
   (ii) Zygomorphic flower can be divided into two equal radial halves in any radial plane.
   (iii) Flowers without bracts are termed as bracteate.
   (iv) Parthenocarpic fruit is formed after fertilization of the ovary.
   (v) In legumes, seed is non-endospermic.
   (vi) Radical buds develop on roots.
   (a) (i), (ii), (iii) and (iv) (b) (i), (ii) and (v)
   (c) (iii), (iv) and (vi) (d) (i), (iv) and (v)

14. Milky water of green coconut is
   (a) liquid nucellus
   (b) liquid of female gametophyte
   (c) liquid endosperm
   (d) liquid embryo

15. Clove is
   (a) flower bud (b) axillary bud
   (c) thalamus (d) ovule

16. When gynoecium is present in the top most position of thalamus, the flower is known as
   (a) inferior (b) epigynous
   (c) perigynous (d) hypogynous

17. Which is not a stem modification ?
   (a) Rhizome of ginger (b) Corm of *Colocasia*
   (c) Pitcher of *Nepenthes* (d) Tuber of potato

18. Which option is correctly matched with the diagrams?

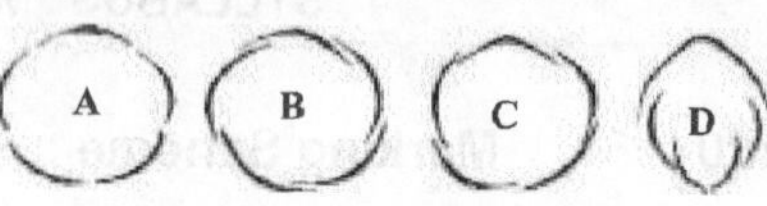

   (a) A-Valvate B-Twisted, C-Imbricate, D-Vexillary
   (b) A-Vexillary, B-Valvate, C-Twisted, D-Imbricate
   (c) A-Imbricate, B-Vexillary, C-Valvate, D-Twisted
   (d) A-Twisted, B-Imbricate, C-Vexillary, D-Valvate

RESPONSE GRID

8. ⓐⓑⓒⓓ 9. ⓐⓑⓒⓓ 10. ⓐⓑⓒⓓ 11. ⓐⓑⓒⓓ 12. ⓐⓑⓒⓓ
13. ⓐⓑⓒⓓ 14. ⓐⓑⓒⓓ 15. ⓐⓑⓒⓓ 16. ⓐⓑⓒⓓ 17. ⓐⓑⓒⓓ
18. ⓐⓑⓒⓓ

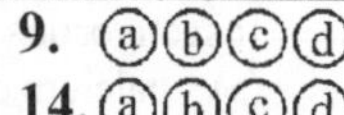

Space for Rough Work

19. Given below are the diagrammatic representation of position of floral parts on thalamus, condition of ovary and example. Find the correctly matched combination?

	Position of floral parts on thalamus	Condition of ovary	Example
(a)		$\overline{G}$	Cucumber
(b)		G–	Brinjal
(c)		$\underline{G}$	Plum
(d)		$\underline{G}$	Rose

20. Match Column-I with Column-II and select the correct option from the codes given below.

Column-I	Column-II
A. Thorns	I. Vegetative propagation
B. Phylloclades	II. Defensive mechanism
C. Runners	III. Mechanical support
D. Stilt roots	IV. Absorption of nutrition
E. Haustoria	V. Photosynthesis

(a) A-I; B-IV; C-III; D-II; E-I
(b) A-II; B-V; C-III; D-I; E-IV
(c) A-II; B-V; C-I; D-III; E-IV
(d) A-III; B-V; C-IV; D-I; E-II

21. Maize grain is a fruit known as
(a) cypsela (b) caryopsis
(c) legume (d) achene

22. Monocotyledonous root differs from dicot root in which of the following internal features (a to d)?
(a) Presence of parenchymatous pericycle.
(b) Absence of fewre xylem bundle.
(c) Presence of large and well-developed pith.
(d) Presence of parenchymatous cortex without intercellular spaces.

23. Select correct match w.r.t column I & II.

Column I	Column II
A. Modified tap root for respiration	I. Zea mays
B. Storage tap root	II. ipomoea
C. Modified aventitious root for mechanical support	III. Rhizophora
D. Modified adventitious root for food storge	IV. Turnip

(a) A-III; B-IV; C-I; D-II
(b) A-III; B-IV; C-II; D-I
(c) A-IV; B-II; C-I; D-III
(d) A-III; B-II; C-I; D-IV

24. The modified stem in some plants of arid region is
(a) Tendril for climbing as in *Passiflora*
(b) Spines for defence mechanism
(c) Phylloclade for food synthesis
(d) Phyllode for food synthesis

25. The modified stem in grasses, strawberry and *Crysanthemum* is concerned with special functions i.e.,
i. Food storage
ii. Vegetative propagation
iii. Assimilation
iv. Spread to new niches
v. Perennation
(a) ii, iv (b) i, ii, v
(c) ii, iv, v (d) iii, iv, v

RESPONSE GRID					
	19. ⓐⓑⓒⓓ	20. ⓐⓑⓒⓓ 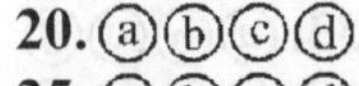	21. ⓐⓑⓒⓓ	22. ⓐⓑⓒⓓ	23. ⓐⓑⓒⓓ
	24. ⓐⓑⓒⓓ	25. ⓐⓑⓒⓓ			

Space for Rough Work

26. In which of the following type of flowers stamens are superior in position?
(a) Hypogynous (b) Perigynous
(c) Epigynous (d) Protogynous

27. Inner layer of pericarp is hard and stony in
(a) Dateplam, Almond (b) Wood, apple, Pea
(c) Mango, Coconut (d) Pear, Litchi

28. Find out the incorrect match.
(a) Sterile stamen – Staminode
(b) Stamens attached to petals – Epipetalous
(c) Stamens attached to perianth – Episepalous
(d) Free stamens – Polyandrous

29. Ovary is said to be half inferior in which of the following conditions?
(a) Hypogynous (b) Perigynous
(c) Epigynous (d) Both (b) and (c)

30. Identify the family which shows the following diagnostic features.
Flowers pentamerous, gynoecium-bicarpellary, syncarpous, ovary placed obliquely, placentation axile, placenta swollen.
(a) Solanaceae (b) Leguminosae
(c) Papilionaceae (d) Liliaceae

31. Select the pair which contains monocotyledonous families.
(a) Solanaceae and Brassicaceae
(b) Fabaceae and Asteraceae
(c) Liliaceae and Poaceae
(d) None of these

32. In Nepenthes (pitcher plant), the pitcher is formed due to modification of
(a) leaf petiole (b) leaf lamina
(c) tendril (d) leaflet

33. Example for tuberous adventitious roots
(a) Dahlia (b) Carrot
(c) Radish (d) Beet

34. A root-cap is usually absent in the roots of
(a) Hydrophytes (b) Epiphytes
(c) Parasites (d) All of the above

35. An example of negatively geotropic root
(a) Coralloid root of *Cycas*
(b) Pneumatophore of mangroves
(c) Assimilatory roots of *Trapa*
(d) More than one of the above.

36. *Santalum album* is normally considered as a
(a) Complete root parasite
(b) Partial root parasite
(c) Complete stem parasite
(d) Partial stem parasite

37. An example of tuberous root that is a modification of tap foot
(a) Radish (b) *Mirabilis*
(c) Sweet Potato (d) *Ipomoea*

38. Ginger is a stem and not a root because
(a) It stores food
(b) It is bitter in taste
(c) It has nodes and internodes
(d) It is non-green in colour.

39. In *Allium*, the leafless part of the stem which bears flower is called
(a) Culm (b) Scape (c) Caudex (d) Bulb

40. Sweet Potato is a modification of
(a) Root (b) Stem
(c) Bud (d) Flowering axis

41. Epiphyllous buds serve the function of
(a) Respiration (b) Nutrition
(c) Reproduction (d) Absorption

42. In a potato plant the tubers develop on
(a) Primary root (b) Secondary root
(c) Tertiary root (d) Stolon

43. Root is the prolongation of
(a) Plumule (b) Radicle
(c) Stem (d) Branches

44. Food stored in a bulb is within
(a) A swollen stem (b) Swollen leaf-bases
(c) Enlarged roots (d) In the inflorescence

45. Cladode is the modification of
(a) Whole stem (b) Axillary bud
(c) Leaf (d) Leaflets.

RESPONSE GRID					
	26. ⓐⓑⓒⓓ	27. ⓐⓑⓒⓓ	28. ⓐⓑⓒⓓ	29. ⓐⓑⓒⓓ	30. ⓐⓑⓒⓓ
	31. ⓐⓑⓒⓓ	32. ⓐⓑⓒⓓ	33. ⓐⓑⓒⓓ	34. ⓐⓑⓒⓓ	35. ⓐⓑⓒⓓ
	36. ⓐⓑⓒⓓ	37. ⓐⓑⓒⓓ	38. ⓐⓑⓒⓓ	39. ⓐⓑⓒⓓ	40. ⓐⓑⓒⓓ
	41. ⓐⓑⓒⓓ	42. ⓐⓑⓒⓓ	43. ⓐⓑⓒⓓ	44. ⓐⓑⓒⓓ	45. ⓐⓑⓒⓓ

Space for Rough Work

DAILY PRACTICE PROBLEM DPP CHAPTERWISE 5 - BIOLOGY			
Total Questions	45	Total Marks	180
Attempted		Correct	
Incorrect		Net Score	
Cut-off Score	45	Qualifying Score	60
Success Gap = Net Score – Qualifying Score			
Net Score = (Correct × 4) – (Incorrect × 1)			

DPP - Daily Practice Problems

Chapter-wise Sheets

Date : ______ **Start Time :** ______ **End Time :** ______

BIOLOGY

SYLLABUS : Anatomy of Flowering Plants

Max. Marks : 180 **Marking Scheme : + 4 for correct & (–1) for incorrect** **Time : 60 min.**

INSTRUCTIONS : This Daily Practice Problem Sheet contains 45 MCQs. For each question only one option is correct. Darken the correct circle/ bubble in the Response Grid provided on each page.

1. During formation of leaves and elongation of stem, some cells ' left behind' from the shoot apical meristem, constitute ____
   (a) Lateral meristem (b) Axillary bud
   (c) Cork cambium (d) Fascicular cambium
2. Function of companion cells is
   (a) providing energy to sieve elements for active transport.
   (b) providing water to phloem.
   (c) loading of sucrose into sieve elements by passive transport.
   (d) loading of sucrose into sieve elements.
3. The cork cambium, cork and secondary cortex are collectively called
   (a) phelloderm (b) phellogen
   (c) periderm (d) phellem
4. Which of the following does not have stomata ?
   (a) Hydrophytes (b) Mesophytes
   (c) Xerophytes (d) Submerged hydrophytes
5. The chief water conducting elements of xylem in gymnosperms are :
   (a) vessels (b) fibres
   (c) transfusion tissue (d) tracheids
6. Which is correct about transport or conduction of substances?
   (a) Organic food moves up through phloem.
   (b) Organic food moves up through phloem
   (c) Inorganic food moves upwardly and downwardly through xylem
   (d) Organic food moves upwardly and downwardly through phloem

RESPONSE GRID	1. ⓐⓑⓒⓓ 2. ⓐⓑⓒⓓ 3. ⓐⓑⓒⓓ 4. ⓐⓑⓒⓓ 5. ⓐⓑⓒⓓ 6. ⓐⓑⓒⓓ

Space for Rough Work

7. A narrow layer of thin walled cells found between phloem/ bark and wood of a dicot is
(a) cork cambium (b) vascular cambium
(c) endodermis (d) pericycle

8. Which of the following statements is true?
(a) Vessels are multicellular with narrow lumen.
(b) Tracheids are multicellular with narrow lumen.
(c) Vessels are unicellular with wide lumen.
(d) Tracheids are unicellular with wide lumen.

9. The quiescent centre in root meristem serves as a
(a) site for storage of food which is utilized during maturation.
(b) reservoir of growth hormones.
(c) reserve for replenishment of damaged cells of the meristem.
(d) region for absorption of water.

10. Root hair arises from
(a) pericycle (b) endodermis
(c) cortex (d) epiblema

11. Phellogen and phellem respectively denote
(a) cork and cork cambium
(b) cork cambium and cork
(c) secondary cortex and cork
(d) cork and secondary cortex

12. The common bottle cork is a product of :
(a) Dermatogen (b) Phellogen
(c) Xylem (d) Vascular cambium

13. A vascular bundle in which the protoxylem is pointing to the periphery is called
(a) endarch (b) exarch
(c) radial (d) closed

14. Which of the following does not have stomata?
(a) Hydrophytes
(b) Mesophytes
(c) Xerophytes
(d) Submerged hydrophytes

15. Collenchyma differs from parenchyma in having
(a) living protoplasm
(b) cellulose walls
(c) vacuoles
(d) pectin deposits at corners

16. Which of the following plant shows multiple epidermis?
(a) *Croton* (b) *Allium*
(c) *Nerium* (d) *Cucurbita*

17. Epidermis is absent in
(a) root tip and shoot tip (b) shoot bud and floral bud
(c) ovule and seed (d) petiole and pedicel

18. Which of the following layer is present nearest of plasma membrane in plant cell?
(a) Secondary wall (b) Middle lamella
(c) Primary wall (d) Tonoplast

19. Which one of the following statement is false?
(i) Epidermal cell has small amount of cytoplasm and a large vacuole.
(ii) Waxy cuticle layer is absent in roots.
(iii) Root hairs are unicellular, while stem hairs / trichomes are multicellular.
(iv) Trichomes may be branched or unbranched, soft or stiff and prevent transpiration.
(v) Guard cells are dumbell shaped in dicots and bean-shaped in monocots.
(a) (i) only (b) (iv) only
(c) (iii) only (d) (v) only

20. Which option is true about heart wood/duramen?
(i) It does not help in water and mineral conduction.
(ii) It is dark coloured but soft.
(iii) It has tracheary elements filled with tannins, resins, gums, oil, etc.
(iv) It is a peripheral part.
(v) Sensitive to microbes and insects, hence least durable.
(a) (i) and (iii) (b) (ii) and (iii)
(c) (iv) and (v) (d) (iii) and (iv)

RESPONSE GRID					
	7. ⓐⓑⓒⓓ	8. ⓐⓑⓒⓓ	9. ⓐⓑⓒⓓ	10. ⓐⓑⓒⓓ	11. ⓐⓑⓒⓓ
	12. ⓐⓑⓒⓓ	13. ⓐⓑⓒⓓ	14. ⓐⓑⓒⓓ	15. ⓐⓑⓒⓓ	16. ⓐⓑⓒⓓ
	17. ⓐⓑⓒⓓ	18. ⓐⓑⓒⓓ	19. ⓐⓑⓒⓓ	20. ⓐⓑⓒⓓ	

Space for Rough Work

21. In land plants, the guard cells differ from other epidermal cells in having
(a) cytoskeleton (b) mitochondria
(c) endoplasmic reticulum (d) chloroplasts

22. Which of the following statement(s) is/are true?
(i) Uneven thickening of cell wall is characteristic of sclerenchyma.
(ii) Periblem forms cortex of the stem and the root.
(iii) Tracheids are the chief water transporting elements in gymnosperms.
(iv) Companion cell is devoid of nucleus at maturity.
(v) The Commercial cork is obtained from *Quercus suber.*
(a) (i) and (iv) only (b) (ii) and (v) only
(c) (iii) and (iv) only (d) (ii), (iii) and (v) only

23. In endarch condition of xylem, protoxylem lies ______ of metaxylem.
(a) on inner side
(b) on outer side
(c) both on inner and outer side
(d) in centre

24. Y-shaped arrangement of xylem vessels is found in
(a) monocot stem (b) dicotstem
(c) monocot root (d) dicot root

25. Study carefully the following statements and select the incorrect one(s).
(i) Lateral roots develop from pericycle.
(ii) Endodermis is the innermost layer of cortex.
(iii) Sap wood is the central, dark coloured, non-conducting part of secondary xylem
(a) (i) and (ii) (b) (ii) and (iii)
(c) (i) only (d) (iii) only

26. Removed parts of grasses by the grazing herbivores are regenerated fast by
(a) Both apical and intercalary meristems
(b) Lateral meristems only
(c) Secondary meristems
(d) Intercallary meristems

27. Which of the following statement for early wood is correct?
(a) Xylary elements having narrow vessels
(b) Darker with higher density
(c) Lighter in colour and has high density
(d) Having vessels with wider cavities

28. Choose correct option w.r.t origin and position of meristem reponsible for the regeneration of parts removed by the grazing herbivores.

	Origin	Position
(a)	Secondary	Lateral
(b)	Primary	Apical
(c)	Secondary	Apical
(d)	Primary	Intercalated

29. Companion cells are
(a) Thin, parrenchymatous, enucleated cells
(b) Thick, sclerenchymatous, nucleated cells
(c) Thin, parenchymatous, nucleated cells
(d) Thick, collenchymatous, enucleated cells

30. Conjoint type of vascular bundle are present in
(a) Root, stem (b) Root, leaves
(c) Stem, leaves (d) Root only

31. Activity of cambium is controlled by
(a) Physiological and environmental factors
(b) Hormonal and external factors
(c) Internal and external factors
(d) Move than one option is correct

32. Epidermal and ground tissue system are made of
(a) Meristem and simple permanet tissue
(b) Primary and secondary meristem
(c) Simple and complex permanent tissue
(d) Simple permanent tissues

RESPONSE GRID					
	21. ⓐⓑⓒⓓ	22. ⓐⓑⓒⓓ	23. ⓐⓑⓒⓓ	24. ⓐⓑⓒⓓ	25. ⓐⓑⓒⓓ
	26. ⓐⓑⓒⓓ	27. ⓐⓑⓒⓓ	28. ⓐⓑⓒⓓ	29. ⓐⓑⓒⓓ	30. ⓐⓑⓒⓓ
	31. ⓐⓑⓒⓓ	32. ⓐⓑⓒⓓ			

Space for Rough Work

33. Root region with thin-walled, small cells having dense cytoplasm is
(a) Region of cell elongation
(b) Maturation region
(c) Proximal to cell elongation zone
(d) Distal to cell elongation zone

34. Periderm includes
(a) Cork and secondary cortex
(b) Only cork cambium
(c) Cork cambium and cork only
(d) Cork, phellogen and secondary cortex

35. Choose incorrect statement from the given below
I. In dicot stem, the innemost layer of cortex is endodermis.
II. Endodermis with casparian strips is found in dicot stem.
III. In dicot stem, the cells of endodermis are rich in starch grain.
(a) I & II (b) II & III
(c) I & III (d) Only II

36. The cells which lie between xylem and phloem in dicot root are
(a) Pith rays
(b) Conjuctive tissue
(c) Interfascicular cambium
(d) Intrafascicular cambium

37. Which tissue of the dicot root is involved in the formation of vascular cambium?
(a) Pericyle part above primary phloem
(b) Whole pericycle
(c) Thich walled parenchymatous cells of pith
(d) Portion of pericycle above the protoxylem

38. Vascular bundles in dicot leaves are
(a) Conjoint, bicollateral, closed
(b) Radia, open
(c) Conjoint, collateral, closed
(d) Conjoint, collaterl, open

39. Which of the following represents the functions of veins in the leaves?
(a) Transport of water and minerals
(b) Mechanical support
(c) Transport of organic food material
(d) All of these

40. In *Cuscuta* the nodes give rise to special roots which penetrate the host tissue upto
(a) Cortex (b) Phloem
(c) Epidermis (d) Pericycle

41. A tylose is formed form :
(a) ray parenchyma
(b) inner parenchyma
(c) paratracheal parenchyma
(d) metatracheal parenchyma

42. Which of the following plant shows multiple epidermis?
(a) *Croton* (b) *Allium*
(c) *Nerium* (d) *Cucurbita*

43. In the monocot root, we obseve
(a) suberized exodermis, polyarch xylem , pith
(b) exodermis, endarch, tetarch closed bundles
(c) conjoint, collateral, open, polyarch vascular bundle
(d) suberized exodermis, casparian strip, passage cells, cambium

44. In a dicotyledonous stem, the sequence of tissues from the outside to the inside is :
(a) phellem-pericycle-endodermis-phloem
(b) phellem-phloem-endodermis-pericycle
(c) phellem-endodermis-pericycle-phloem
(d) pericycle-phellem-endodermis-phloem

45. The tunica corpus theory was proposed by :
(a) Nageli
(b) Hanstein
(c) Schmidt
(d) Haberlandt

RESPONSE GRID					
	33. ⓐⓑⓒⓓ	34. ⓐⓑⓒⓓ	35. ⓐⓑⓒⓓ	36. ⓐⓑⓒⓓ	37. ⓐⓑⓒⓓ
	38. ⓐⓑⓒⓓ	39. ⓐⓑⓒⓓ	40. ⓐⓑⓒⓓ	41. ⓐⓑⓒⓓ	42. ⓐⓑⓒⓓ
	43. ⓐⓑⓒⓓ	44. ⓐⓑⓒⓓ	45. ⓐⓑⓒⓓ		

Space for Rough Work

DAILY PRACTICE PROBLEM DPP CHAPTERWISE 6 - BIOLOGY			
Total Questions	45	Total Marks	180
Attempted		Correct	
Incorrect		Net Score	
Cut-off Score	45	Qualifying Score	60
Success Gap = Net Score – Qualifying Score			
Net Score = (Correct × 4) – (Incorrect × 1)			

DPP - Daily Practice Problems

Chapter-wise Sheets

Date :   **Start Time :**   **End Time :**

BIOLOGY

SYLLABUS : Structural Organisation in Animals

Max. Marks : 180   **Marking Scheme : + 4 for correct & (–1) for incorrect**   **Time : 60 min.**

INSTRUCTIONS : This Daily Practice Problem Sheet contains 45 MCQs. For each question only one option is correct. Darken the correct circle/ bubble in the Response Grid provided on each page.

1. The kind of epithelium which forms the inner walls of blood vessels is
(a) cuboidal epithelium
(b) columnar epithelium
(c) ciliated columnar epithelium
(d) squamous epithelium

2. The cell junctions called tight, adhering and gap junctions are found in
(a) connective tissue   (b) epithelial tissue
(c) neural tissue   (d) muscular tissue

3. Housefly possesses
(a) two pairs of wings   (b) one pair of wings
(c) three pairs of wings   (d) four pair of wings

4. Excretory organs of Cockroach are
(a) flame cells
(b) nephridia
(c) hreen glands
(d) malpighian tubules

5. Myoglobin is present in
(a) all muscle fibres
(b) white muscle fibres only
(c) red muscle fibres only
(d) both white and red muscle fibres

6. Intercalated discs are the communication junctions between the cells of
(a) cardiac muscles
(b) striped muscles
(c) adipose tissue
(d) nerve and striated muscles

RESPONSE GRID	1. ⓐⓑⓒⓓ	2. ⓐⓑⓒⓓ	3. ⓐⓑⓒⓓ	4. ⓐⓑⓒⓓ	5. ⓐⓑⓒⓓ
	6. ⓐⓑⓒⓓ				

Space for Rough Work

7. What is common among silverfish, scorpion, crab and honey bee?
(a) Compound eyes (b) Poison glands
(c) Jointed legs (d) Metamorphosis

8. Mast cells are associated with
(a) exocrine glands
(b) endocrine glands
(c) areolar connective tissue
(d) neural tissue

9. Lymph differs from blood in possessing
(a) only WBCs.
(b) more RBCs and WBCs.
(c) more RBCs and few WBCs.
(d) more WBCs and few RBCs.

10. Pseudostratified epithelium is found in
(a) oesophagus (b) respiratory tract
(c) urinary tract (d) kidney

11. Given below are four matchings of an animal and its kind of respiratory organ :
(i) Silver fish – trachea
(ii) Scorpion – book lung
(iii) Sea squirt – pharyngeal gills
(iv) Dolphin – skin
The correct matchings are
(a) (iii) and (iv) (b) (i) and (iv)
(c) (i), (ii) and (iii) (d) (ii) and (iv)

12. Tendons and ligaments are the examples of
(a) areolar connective tissue
(b) adipose tissue
(c) dense regular connective tissue
(d) loose connective tissue

13. Major inorganic component of vertebrate bone is
(a) calcium carbonate
(b) calcium phosphate
(c) sodium hydroxide
(d) potassium hydroxide

14. Histamine is secreted by
(a) mast cells (b) fibroblast
(c) histocytes (d) melanocytes

15. Which of the following is not a granulocyte?
(a) Lymphocyte (b) Neutrophil
(c) Basophil (d) Eosinophil

16. The type of epithelial cells which line inner surface of fallopian tubes, bronchioles and small bronchi are known as
(a) squamous epithelium (b) columnar epithelium
(c) ciliated epithelium (d) cubical epithelium

17. Male and female cockroaches can be distinguished externally through
(a) anal styles in male
(b) anal cerci in female
(c) anal style and antennae in females
(d) both (b) and (c)

18. The long bones are hollow and connected by air passage. They are characteristic of
(a) aves (b) mammalia
(c) reptilia (d) sponges

19. The most active phagocytic white blood cells are
(a) neutrophils and monocytes
(b) neutrophils and eosinophils
(c) lymphocytes and macrophages
(d) eosinophils and lymphocytes

20. In the mouthparts of the cockroach, the organ of mastication is
(a) labium (b) maxillae
(c) mandibles (d) labrum

21. Which one of the following groups of structures/organs have similar function?
(a) Typhlosole in earthworm, intestinal villi in rat and contractile vacuole in *Amoeba*.
(b) Nephridia in earthworm, malpighian tubules in cockroach and urinary tubules in rat.
(c) Antennae of cockroach, tympanum of frog and clitellum of earthworm.
(d) Incisors of rat, gizzard (proventriculus) of cockroach and tube feet of starfish.

RESPONSE GRID					
	7. ⓐⓑⓒⓓ	8. ⓐⓑⓒⓓ	9. ⓐⓑⓒⓓ	10. ⓐⓑⓒⓓ	11. ⓐⓑⓒⓓ
	12. ⓐⓑⓒⓓ	13. ⓐⓑⓒⓓ	14. ⓐⓑⓒⓓ	15. ⓐⓑⓒⓓ	16. ⓐⓑⓒⓓ
	17. ⓐⓑⓒⓓ	18. ⓐⓑⓒⓓ	19. ⓐⓑⓒⓓ	20. ⓐⓑⓒⓓ	21. ⓐⓑⓒⓓ

Space for Rough Work

22. Which of the following statements are wrong?
(i) Leucocytes disintegrate in the spleen and liver.
(ii) RBC, WBC and blood platelets are produced by bone marrow.
(iii) Neutrophils bring about destruction and detoxification of toxins of protein origin.
(iv) The important function of lymphocytes is to produce antibodies.
(a) (i) and (ii) only (b) (i) and (iv) only
(c) (i) and (iii) only (d) (ii) and (iii) only

23. Malpighian tubules are the excretory organs in
(a) Cockroach (b) Platyhelminthes
(c) *Ascaris* (d) *Pila*

24. Young one of the cockroach is called
(a) Naid (b) Grub
(c) Nymph (d) Maggot

25. Cilia, Flagella and microvilli are associated with the
(a) Connective tissue (b) Epithelial tissue
(c) Nervous tissue (d) Muscular tissue

26. In which of the following animals, respiration occurs without any respiratory organ?
(a) Frog (b) Fish
(c) Cockroach (d) Earthworm

27. Which of the following type of cell junction is not found in animal tissues ?
(a) Desmosome (b) Tight junction
(c) Gap junction (d) Plasmodesmata

28. Which one of the following cellular components of the blood is responsible for the production of antibodies?
(a) Thrombocyte (b) Lymphocyte
(c) Monocyte (d) Erythrocyte

29. Sprain is caused due to the excessive pulling of
(a) Muscles (b) Tendons
(c) Ligaments (d) All of the above

30. In cockroach, the corpora allata is
(a) A sense organ (b) A tactile organ
(c) An endocrine organ (d) A digestive gland

31. Which of the following is incorrect w.r.t stucture/gland of cockroach and its total number in body?
(a) Collateral glands in female cockroach - 2
(b) Malpighian tubules - 100- 150
(c) Spiracles - 10
(d) Ommatidia in each compound eye - 2000

32. The function of basement membrane in epithelial tissue is to
(a) Facilitate communication of cells with each other
(b) Allow rapid ion transfer among cells
(c) Separate the epithellal tissue from underlying connective tissue
(d) Produce multiple layers of cells in a compound epithelium

33. All among the following are function of connective tissue except
(a) Binding support
(b) Fat storage
(c) Body defense and transport
(d) Propelling mucus towards the pharynx

34. Mark the statement that is incorrect for smooth muscles.
(a) Smooth muscles from the muscular component of the visceral organs such as blood vessels, urinary bladder, ciliary body etc.
(b) Smooth muscles cells do not have the troponin-tropomyosin mechanism of controlling contraction
(c) Irritability is shown by all smooth muscles except ciliary muscles
(d) Hair root muscles and muscles on the large blood vessels are in example of multi-unit smooth muscles

35. The avascular nature of cartilage is responsible for its
(a) Flexibility
(b) Slow repair
(c) Suitability for embryonic endoskeleton
(d) Transparent consistency

RESPONSE GRID					
	22. ⓐⓑⓒⓓ	23. ⓐⓑⓒⓓ	24. ⓐⓑⓒⓓ	25. ⓐⓑⓒⓓ	26. ⓐⓑⓒⓓ
	27. ⓐⓑⓒⓓ	28. ⓐⓑⓒⓓ	29. ⓐⓑⓒⓓ	30. ⓐⓑⓒⓓ	31. ⓐⓑⓒⓓ
	32. ⓐⓑⓒⓓ	33. ⓐⓑⓒⓓ	34. ⓐⓑⓒⓓ	35. ⓐⓑⓒⓓ	

Space for Rough Work

36. Match Column-I with Column-II and select the correct option from the codes given below.

	Column-I		Column-II
A.	Hyaline cartilage	I.	Pectoral girdle of frog
B.	Fibrous cartilage	II.	Long bones; sternum; ribs
C.	Elastic cartilage	III.	Pubic symphysis
D.	Calcified cartilage	IV.	Eustachain tube; epiglottis

(a) A-I; B-II; C-III; D-IV
(b) A-II; B-III; C-IV; D-I
(c) A-II; B-IV; C-III; D-I
(d) A-IV; B-III; C-II; D-I

37. The type of epithelial cells which line the inner surface of Fallopian tubes, bronchioles and small bronchi are known as
(a) squamous epithelium
(b) columnar epithelium
(c) ciliated epithelium
(d) cubical epithelium

38. The wall of heart is thick due to
(a) Pericardium (b) Epicardium
(c) Myocardium (d) Endocardium

39. The strongest cartilage is
(a) Hyaline cartilage
(b) Elastic cartilage
(c) Fibrous cartilage
(d) None of the above

40. Nissl's granules are found in cyton of nerve cells. These have affinity for basic dyes. The granules are made up of
(a) Protein (b) DNA
(c) Amino acids (d) RNA

41. The inability to absorb digested nutrients may be due to damage of which type of epithelium?
(a) Ciliated columnar
(b) Simple columnar
(c) Simple squamous
(d) Simple cuboidal

42. The presence of a basement membrane is typical of most
(a) Epithelial tissues
(b) Connective tissues
(c) Nervous tissues
(d) Muscle tissues

43. The most appropriate definition of Neuroglial cells are that they are
(a) Nonsensory supporting cells
(b) Secretory cells
(c) Sensory cells
(d) Sensory and supporting cells

44. Covering around bone is called
(a) Perichondrion
(b) Periosteum
(c) Epiosteum
(d) Endosteum

45. The figure given below shows the head region of cockroach. Identify A to F.

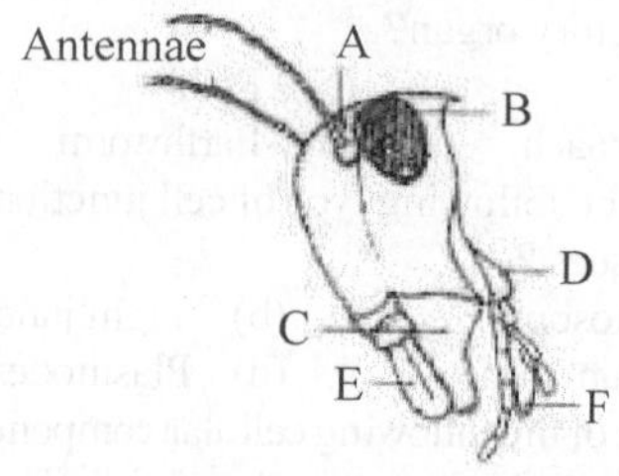

(a) A- Compound eye, B-Ocellus, C-Maxilla, D-Mandible, E-Labrum, F-Labium
(b) A- Ocellus, B-Compound eye, C-Mandible, D-Maxilla, E-Labrum, F-Labium
(c) A- Ocellus, B-Compound eye, C-Mandible, D-Maxilla, E-Labium, F-Labrum
(d) A- Ocellus, B-Compound eye, C-Maxilla, D-Mandible, E-Labrum, F-Labium

RESPONSE GRID					
	36. ⓐⓑⓒⓓ	37. ⓐⓑⓒⓓ	38. ⓐⓑⓒⓓ	39. ⓐⓑⓒⓓ	40. ⓐⓑⓒⓓ
	41. ⓐⓑⓒⓓ	42. ⓐⓑⓒⓓ	43. ⓐⓑⓒⓓ	44. ⓐⓑⓒⓓ	45. ⓐⓑⓒⓓ

Space for Rough Work

DAILY PRACTICE PROBLEM DPP CHAPTERWISE 7 - BIOLOGY			
Total Questions	45	Total Marks	180
Attempted		Correct	
Incorrect		Net Score	
Cut-off Score	40	Qualifying Score	55
Success Gap = Net Score – Qualifying Score			
Net Score = (Correct × 4) – (Incorrect × 1)			

DPP - Daily Practice Problems

Chapter-wise Sheets

Date : **Start Time :** **End Time :**

BIOLOGY CB08

SYLLABUS : Cell: The Unit of Life

Max. Marks : 180 **Marking Scheme : + 4 for correct & (–1) for incorrect** **Time : 60 min.**

INSTRUCTIONS : This Daily Practice Problem Sheet contains 45 MCQs. For each question only one option is correct. Darken the correct circle/ bubble in the Response Grid provided on each page.

1. The cell organelle involved in glycosylation of protein is
   (a) ribosome
   (b) peroxisome
   (c) endoplasmic reticulum
   (d) mitochondria
2. The outer layer of vacuole is called
   (a) cell wall (b) tonoplast
   (c) plasmalayer (d) leucoplast
3. Which of the following cell organelle remains enveloped by a single unit membrane?
   (a) Mitochondria (b) Lysosomes
   (c) Nucleus (d) Chloroplast
4. Choose the correct option.
   (i) Lysosomes are double membranous vesicles budded off from Golgi apparatus and contain digestive enzymes.
   (ii) Endoplasmic reticulum consists of a network of membranous tubule and helps in transport, synthesis and secretion.
   (iii) Leucoplasts are bound by two membranes, lack pigment but contain their own DNA and protein synthesising machinery.
   (iv) Sphaerosomes are single membrane bound organelle which are associated with synthesis and storage of lipids.
   (a) (i) only (b) (i) and (ii)
   (c) (ii), (iii) and (iv) (d) All of these
5. The nucleolus is the site of formation of
   (a) spindle fibres (b) chromosomes
   (c) ribosomes (d) peroxisomes

RESPONSE GRID	1. ⓐⓑⓒⓓ	2. ⓐⓑⓒⓓ	3. ⓐⓑⓒⓓ	4. ⓐⓑⓒⓓ	5. ⓐⓑⓒⓓ

Space for Rough Work

6. Which one of the following combination is mismatched?
(a) Glycocalyx - may be capsule or slime layer
(b) Pili - Reproduction
(c) Cell wall - Protective, determines shape, prevents from bursting
(d) Flagella, Pili and Fimbriae - Surface structures of bacterial cell

7. The fluidity of membranes in a plant in cold weather may be maintained by
(a) increasing the number of phospholipids with unsaturated hydrocarbon tails
(b) increasing the proportion of integral proteins
(c) increasing concentration of cholesterol in membrane
(d) increasing the number of phospholipids with saturated hydrocarbon tail

8. The cell as a basic unit of structure of living beings was discovered by
(a) Aristotle
(b) Robert Hooke
(c) Schleiden and Schwann
(d) Gregore Mendel

9. Which pair of structures are usually found in both plant and animal cells?
(a) Cell membrane and nucleolus
(b) Cell membrane and cell wall
(c) Nucleolus and chloroplast
(d) Nucleus and cell wall

10. Most abundant lipid in the cell membrane is
(a) cholesterol (b) phospholipids
(b) glycolipids (d) cerebrosides

11. If you remove the fimbriae from the bacterial cell, which of the following would you expect to happen?
(a) The bacteria could no longer swim
(b) The bacteria would not adhere to the host tissue
(c) Transportation of molecules across the membrane would stop
(d) The shape of bacteria would change

12. Cell recognition and adhesion are facilitated by components of plasma membrane. These components are generally
(a) protein molecules alone
(b) lipids alone
(c) both lipids and proteins
(d) glycolipids and glycoproteins

13. Smooth endoplasmic reticulum is well developed in the cells which synthesize
(a) steorids (b) proteins
(c) carbohydrates (d) all of these.

14. Select the option with correct labelling of given structure of Golgi apparatus.

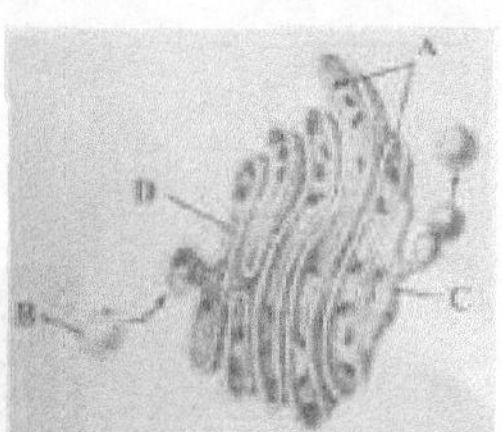

	A	B	C	D
(a)	Cisternae	Vesicle	*trans* face	*cis* face
(b)	Cisternae	Vesicle	*cis* face	*trans* face
(c)	Vesicle	Cisternae	*cis* face	*trans* face
(d)	Tubules	Vesicle	*trans* face	*cis* face

15. The molecules in the membrane that limit its permeability are the
(a) carbohydrates (b) phospholipids
(c) proteins (d) water

16. pH of vacuolar cell sap is
(a) neutral and isotonic.
(b) alkaline and isotonic.
(c) acidic and hypertonic.
(d) equal to cytoplasm and isotonic.

17. All plastids have essentially the same structure because
(a) they have to perform the same function
(b) they are localised in the aerial parts of plants

RESPONSE GRID				
6. ⓐⓑⓒⓓ	7. ⓐⓑⓒⓓ	8. ⓐⓑⓒⓓ	9. ⓐⓑⓒⓓ	10. ⓐⓑⓒⓓ
11. ⓐⓑⓒⓓ	12. ⓐⓑⓒⓓ	13. ⓐⓑⓒⓓ	14. ⓐⓑⓒⓓ	15. ⓐⓑⓒⓓ
16. ⓐⓑⓒⓓ				

Space for Rough Work

(c) one type of plastids can differentiate into another type of plastids depending upon the cell requirements
(d) all plastids have to store starch, lipids and proteins

18. A Semi-autonomous organelle.
B Have linear DNA as well as RNA
C Carry out ATP synthesis.
D Have quantasomes embedded in thylakoid membrane
E. Occurs in all photosynthetic organisms.
Find the correct number of statements w.r.t. plastids.
(a) One (b) Two
(c) Three (d) Four

19. Select incorrect matching
(a) Elaioplasts - Oils
(b) Chromoplasts - Fat soluble anthocyanin pigments
(c) Mitochondria - Fission in G_2 phase
(d) Contractile vacuole - Excretion

20. Quasi-fluid nature of membrane is due to
(a) Phospholipid (b) Integral protein
(c) Peripheral protein (d) Sugar moiety

21. Gas vacuole is present in
(a) Blue green algae
(b) Purple photosynthetic bacteria
(c) Green photosynthetic bacteria
(d) All of the above

22. Which of the following feature is not associated with centrosome?
(a) Pericentriolar material
(b) Two cylindrical structures
(c) Two centriole
(d) Lipid bilayer covering

23. What is the site of DNA and centriole duplication respectively?
(a) Nucleus, nucleus
(b) Nucleus, cytoplasm
(c) Cytoplasm, nucleus
(d) Nucleus, nucleolous

24. Cell wall
(a) Helps in cell to cell interaction
(b) Protects the cell from infection
(c) Contains minerals like calcium carbonate in certain algae
(d) All are correct

25. Read the following statements carefully and choose the correct options w.r.t. eukaryotic cell.
I. All eukaryotic cells are identical in structure
II. Mitochondria and plastids are semi-autonomous organelles
III. Ribosomes are associated with plasma membrane
IV. There is an extensive compartmentalization of cytoplasm through the presence of membrane bound organelles
(a) I & IV (b) II & IV
(c) I & III (d) II & III

26. Golgi bodies are involved in
(a) Recycling of broken plasma membrane during endocytosis
(b) Synthesis of glycolipids
(c) Modification of proteins
(d) All of the above

27. Which of the following organelles lack membrane in eukaryotic cell?
A. Cilia B. Lysosome
C. RER D. Ribosomes
E. Flagella F. Centrioles
(a) D & F (b) C & D
(c) A & D (d) A & E

28. Aleuroplasts, amyloplasts and elaioplasts
(a) Divide by multiple fission
(b) Store protein, starch and fat respectively
(c) Help in photolysis of water
(d) Store reserve food and pigments

29. Reformation of nucleolus, golgi complex and ER occurs in
(a) Telophase (b) Metaphase
(c) Prophase (d) Anaphase

30. Ribosomes of the cytoplasm, chloroplast and mitochondrion are respectively
(a) 80S, 80S and 70S (b) 80S, 70S and 70S
(c) 70S in all (d) 80S in all

31. Integral cell membrane proteins
(a) are partially embedded in lipid layers
(b) are completely embedded in lipid layers

RESPONSE GRID					
	17. ⓐⓑⓒⓓ	18. ⓐⓑⓒⓓ	19. ⓐⓑⓒⓓ	20. ⓐⓑⓒⓓ	21. ⓐⓑⓒⓓ
	22. ⓐⓑⓒⓓ	23. ⓐⓑⓒⓓ	24. ⓐⓑⓒⓓ	25. ⓐⓑⓒⓓ	26. ⓐⓑⓒⓓ
	27. ⓐⓑⓒⓓ	28. ⓐⓑⓒⓓ	29. ⓐⓑⓒⓓ	30. ⓐⓑⓒⓓ	

Space for Rough Work

(c) show lateral but not vertical movements within bilayer of lipid
(d) All of these

32. Which group of organelles is involved in synthesis of substances needed by cell?
(a) Lysosome, vacuole, ribosome
(b) Vacuole, RER, SER
(c) Ribosome, RER, SER
(d) RER, lysosome, vacuole

33. Who gave the lamellar or sandwich model of cell membrane?
(a) Singer and Nicolson
(b) Danielle and Davson
(c) J. Robertson
(d) None of these

34. Microtubules are absent in
(a) mitochondria (b) flagella
(c) spindle fibres (d) centriole

35. Which of the following contributes to differences in the two sides of the cell membrane?
(a) Differences in peripheral proteins
(b) Different domains expressed on the ends of integral proteins
(c) Differences in phospholipid types
(d) All of the above

36. Which of the following cell membrane components serve as recognition signals for interactions between cells?
(a) Recognition proteins
(b) Glycolipids or glycoproteins
(c) Phospholipids
(d) Integral proteins

37. Channel proteins allow ions that would not normally pass through the cell membrane to go through the channel. What properties of the proteins are responsible for this?
(a) The channels are often composed of charged or polar R groups.
(b) The channels are often composed of hydrophobic R groups.
(c) a and b
(d) None of the above

38. Which of the following is present in both prokaryotes and eukaryotes?
(a) Lysosome (b) Vesicles
(c) Chloroplast (d) Plasma membrane

39. Both chloroplasts and mitochondria
(a) have multiple membranes.
(b) have highly structured innermost membranes.
(c) are found only in eukaryotic cells.
(d) All of the above

40. Microtubules, motor proteins, and actin filaments are all part of the
(a) mechanism of photosynthesis that occurs in chloroplasts.
(b) rough ER in prokaryotic cells.
(c) cytoskeleton of eukaryotic cells.
(d) process that moves small molecules across cell membranes.

41. The cell wall of both bacteria and cyanobacteria contains
(a) Lipid (b) Pectin
(c) Protein (d) Muramic acid

42. Mesosomes were taken as
(a) Golgi bodies (b) Plastids
(c) Mitochondria (d) Endoplasmic reticulum

43. Pit membrane of simple pit is formed by :
(a) Secondary cell wall (b) Middle lamella
(c) Primary cell wall (d) Plasma

44. Which one of the following cell organelles found only in plants?
(a) Golgi complex (b) Mitochondria
(c) Plastids (d) Ribosomes

45. Peroxisomes are rich in
(a) DNA (b) RNA
(c) Catalytic enzymes (d) Oxidative enzymes

RESPONSE GRID					
	31. ⓐⓑⓒⓓ	32. ⓐⓑⓒⓓ	33. ⓐⓑⓒⓓ	34. ⓐⓑⓒⓓ	35. ⓐⓑⓒⓓ
	36. ⓐⓑⓒⓓ	37. ⓐⓑⓒⓓ	38. ⓐⓑⓒⓓ	39. ⓐⓑⓒⓓ	40. ⓐⓑⓒⓓ
	41. ⓐⓑⓒⓓ	42. ⓐⓑⓒⓓ	43. ⓐⓑⓒⓓ	44. ⓐⓑⓒⓓ	45. ⓐⓑⓒⓓ

Space for Rough Work

DAILY PRACTICE PROBLEM DPP CHAPTERWISE 8 - BIOLOGY			
Total Questions	45	Total Marks	180
Attempted		Correct	
Incorrect		Net Score	
Cut-off Score	45	Qualifying Score	60
Success Gap = Net Score – Qualifying Score			
Net Score = (Correct × 4) – (Incorrect × 1)			

DPP - Daily Practice Problems

Chapter-wise Sheets

Date : | Start Time : | End Time :

BIOLOGY CB09

SYLLABUS : Biomolecules

Max. Marks : 180 **Marking Scheme : + 4 for correct & (–1) for incorrect** **Time : 60 min.**

INSTRUCTIONS : This Daily Practice Problem Sheet contains 45 MCQs. For each question only one option is correct. Darken the correct circle/ bubble in the Response Grid provided on each page.

1. A nucleotide is formed of
   (a) Purine, pyrimidine and phosphate
   (b) Purine, sugar and phosphate
   (c) Nitrogen base, sugar and phosphate
   (d) Pyrimidine, sugar and phosphate
2. Glycogen is a polymer of
   (a) galactose (b) glucose
   (c) fructose (d) sucrose
3. Most abundant organic compound on earth is
   (a) Protein (b) Cellulose
   (c) Lipids (d) Steroids
4. Protein synthesis in a cell takes place
   (a) only in the cytoplasm
   (b) in the nucleolus as well as in cytoplasm
   (c) in cytoplasm as well as in mitochondria
   (d) only on ribosomes attached to the nuclear envelope
5. In RNA, thymine is replaced by
   (a) Adenine (b) Guanine
   (c) Cytosine (d) Uracil
6. Which of the following is a reducing sugar?
   (a) Galactose (b) Gluconic acid
   (c) β-methyl galactoside (d) Sucrose
7. An enzyme that can stimulate germination of barley seeds is
   (a) lipase (b) protease
   (c) invertase (d) α-amylase
8. Carrier ions like Na^+ facilitate the absorption of substances like:
   (a) amino acids and glucose
   (b) glucose and fatty acids
   (c) fatty acids and glycerol
   (d) fructose and some amino acids

RESPONSE GRID					
	1. ⓐⓑⓒⓓ	2. ⓐⓑⓒⓓ	3. ⓐⓑⓒⓓ	4. ⓐⓑⓒⓓ	5. ⓐⓑⓒⓓ
	6. ⓐⓑⓒⓓ	7. ⓐⓑⓒⓓ	8. ⓐⓑⓒⓓ		

Space for Rough Work

9. An enzyme brings about
   (a) decrease in reaction time
   (b) increase in reaction time
   (c) increase in activation energy
   (d) reduction in activation energy
10. The enzymes hexokinase which catalyses glucose to glucose-6-phosphate in glycolysis is inhibited by glucose-6-phosphate. This is an example of
   (a) competitive inhibition
   (b) non-competitive inhibition
   (c) feedback allosteric inhibition
   (d) positive feedback
11. The enormous diversity of protein molecules is due mainly to the diversity of
   (a) amino groups on the amino acids
   (b) R groups on the amino acids
   (c) amino acid sequences within the protein molecule
   (d) peptide bonds
12. Length of one turn of the helix in a B-form DNA is approximately
   (a) 3.4 nm (b) 2 nm
   (c) 0.34 nm (d) 20 nm
13. The catalytic efficiency of two different enzymes can be compared by the
   (a) formation of the product
   (b) pH optimum value
   (c) K_m value
   (d) molecular size of the enzyme
14. Which of the following sets contains polysaccharides?
   (a) Glucose, fructose, lactose
   (b) Starch, glycogen, cellulose
   (c) Sucrose, maltose, cellulose
   (d) Galactose, starch, sucrose
15. Sigmoid growth curve is represented by
   (a) $dN/dt = rN$
   (b) $dN/dt = rN(1 - N/K)$
   (c) $Nt = N_0 + B + I - D - E$
   (d) $dN/dt = 1 - N/K$
16. A coenzyme is
   (a) Same enzyme that occurs in different tissues such as heart and muscle
   (b) One that shares the function of other enzyme
   (c) Organic or inorganic in nature and helps activate metabolic enzymes
   (d) Organic non-protein in nature and helps to activate metabolic enzymes
17. The most basic amino acid is
   (a) Arginine (b) Histidine
   (c) Glycine (d) Glutamine
18. Which of the following is not an aromatic amino acid?
   (a) Valine (b) Tyrosine
   (c) Phenylalanine (d) Tryptophan
19. The class of enzymes that catalyze the removal of a group from a substrate without addition of water, leaving double bonds, is
   (a) Transferases (b) Dehydrogenases
   (c) Hydrolases (d) Lyases
20. Which of the following is a component of vitamin (thiamine, biotin), Acetyl CoA, cysteine, methionine and ferredoxin?
   (a) Fe (b) S
   (c) Co (d) K
21. The helical structure of protein is stabilized by
   (a) Ester (b) Peptide bonds
   (c) Disulphide bonds (d) Hydrogen bonds
22. Cellulose is a polymer of
   (a) α-glucose (b) α-D-glucose
   (c) β-D-glucose (d) β-L-glucose

RESPONSE GRID					
	9. ⓐⓑⓒⓓ	10. ⓐⓑⓒⓓ	11. ⓐⓑⓒⓓ	12. ⓐⓑⓒⓓ	13. ⓐⓑⓒⓓ
	14. ⓐⓑⓒⓓ	15. ⓐⓑⓒⓓ	16. ⓐⓑⓒⓓ	17. ⓐⓑⓒⓓ	18. ⓐⓑⓒⓓ
	19. ⓐⓑⓒⓓ	20. ⓐⓑⓒⓓ	21. ⓐⓑⓒⓓ	22. ⓐⓑⓒⓓ	

Space for Rough Work

23. The given graph shows the effect of substrate concentration on the rate of reaction of the enzyme green -gram - phosphatase. What does the graph indicate ?

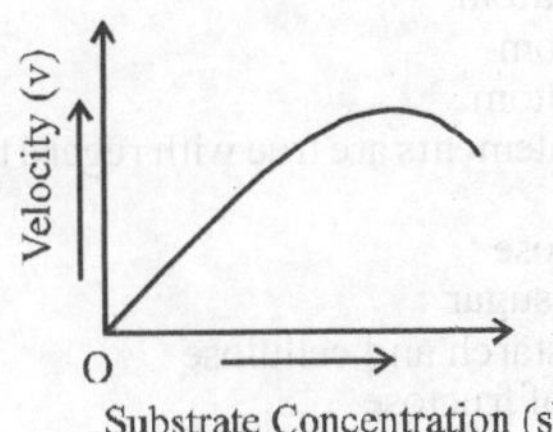

(a) The rate of enzyme reaction is directly proportional to the substrate concentration
(b) Presence of an enzyme inhibitor in the reaction mixture
(c) Formation of an enzyme-substrate complex
(d) At higher substrate-concentration the pH increases.

24. The Km value of the enzyme is the value of the substrate concentration at which the reaction reaches to
(a) Zero (b) 2 Vmax
(c) ½ Vmax (d) ¼ Vmax

25. Which group contains biocatalysts?
(a) Myosin, oxytocin, adrenalin
(b) Peptidase, amylase, rennin
(c) Glucose, amino acids, fatty acids
(d) Rhodopsin, pepsin, steapsin

26.

```
                          O
                          ||
      O       CH2—O—C—R1
      ||         |
R2—C—O—CH        O
                 |          ||
              CH2—O—P—O—CH2—CH2
                          |              |
                          OH            N+
                                     /  |  \
                                  CH3 CH3 CH3
```

Given structural formula is correctly identified alongwith its related function by which of the following options?
(a) Cholesterol – A component of animal cell membrane
(b) Lecithin – A component of cell membrane
(c) Triglyceride – An energy source
(d) Adenosine – A component of nucleic acids

27. Select the option that correctly identifies the chemical bonds present in the given biomolecules.
Polysaccharides – A, Proteins – B, Fats – C, Water – D

	A	B	C	D
(a)	Ester	Peptide	Glycosidic	Hydrogen
(b)	Glycosidic	Peptide	Ester	Hydrogen
(c)	Glycosidic	Peptide	Hydrogen	Ester
(d)	Hydrogen	Ester	Peptide	Glycosidic

28. Which of the following statements is not correct regarding chitin?
(a) It is a storage polysaccharide
(b) It is a homopolysccharide
(c) It is a constituent of arthropod exoskeleton and fungal cell wall
(d) It is the second most abundant carbohydrate on earth

29. Which of the following secondary metabolites are used as drugs?
(a) Abrin and ricin
(b) Vinblastin and curcumin
(c) Anthocyanins
(d) Gums and cellulose

30. Enzymes that catalyse removal of groups from substrates by mechanisms other than hydrolysis, and addition of groups to double bonds, are called
(a) ligases (b) lyases
(c) hydrolases (d) dehydrogenases

31. Which of the following statement is incorrect w.r.t. starch?
(a) Starch consists of unbranched amylose and branched amylopectin
(b) It is a polymer of α-D-glucose
(c) Successive glucose units are linked together by α 1 → 6 linkage and at branching α 1 → 4 linkage is found
(d) Starch turns black with iodine

RESPONSE GRID	23. ⓐⓑⓒⓓ	24. ⓐⓑⓒⓓ	25. ⓐⓑⓒⓓ	26. ⓐⓑⓒⓓ	27. ⓐⓑⓒⓓ
	28. ⓐⓑⓒⓓ	29. ⓐⓑⓒⓓ	30. ⓐⓑⓒⓓ	31. ⓐⓑⓒⓓ	

Space for Rough Work

32. Cellulose differs from chitin in
(a) Branching
(b) Type of glycosidic bond
(c) Type of monomer unit
(d) Abundancy in biosphere

33. Oils are rich in
(a) Glycerol that possesses three hydroxyl groups
(b) Saturated fatty acids
(c) Fats that are generally liquid at room temperature
(d) Esters of fatty acids

34. Bonds that do not exist in tertiary structure of proteins
(a) Covalent bonds
(b) Phosphodiester bonds
(c) Hydrophobic etnractions
(d) Ionic bonds

35. Cleavage of specific covalent bonds and removal of groups without hydrolysis is the property of
(a) Isomerases (b) Lyases
(c) Hydrolases (d) Transferases

36. The catalytic efficiency of two different enzymes can be compared by the
(a) K_m value
(b) pH value
(c) Formation of the product
(d) Molecular size of the enzyme

37. Which of the following is an example of isozyme?
(a) α-Amylase
(b) Glucokinase
(c) Lactate dehydrogenase
(d) All of these

38. Essential amino acids include
(a) leucine (b) valine
(c) tryptophan (d) all of these

39. Acidic amino acids have two –COOH groups and one –NH_2 group per molecule. Select the pair that consists of acidic amino acids.
(a) Aspartic acid, glutamic acid
(b) Lysine, arginine
(c) Glycine, alanine
(d) Both (a) and (b)

40. The structure of glucose and galactose are same except with regard to
(a) First carbon atom
(b) Second carbon atom
(c) Third carbon atom
(d) Fourth carbon atom

41. All the following statements are true with regard to glucose, except :
(a) It is an aldohexose
(b) It is a reducing sugar
(c) It is present in starch and cellulose
(d) It is an epimer of fructose

42. Guanylic acid is a
(a) Nucleoside of purine
(b) Nucleoside of pyrimidine
(c) Nucleotide of purine
(d) Nucleotide of pyrimidine

43. Long chain molecules of fatty acids are formed by
(a) Polymerisation of 2 carbon compounds
(b) Decomposition of fats
(c) Polymerisation of glycogen
(d) Conversion of glycogen

44. Read the following four statements (A to D) and mark the option that has both correct statements
A. Blood conc. of glucose in a normal healthy man is 4.5 to 5.0 mM
B. In proteins only left handed helices are observed
C. The pitch of B- DNA is 3.4 Angstrom
D. At each step of ascent the strand of DNA turns 36 degree
(a) A, B (b) B, C (c) C, D (d) A, D

45. Which one of the following biomolecules is correctly characterized?
(a) Lecithin - A phosphorylated glyceride found in cell membrane.
(b) Palmitic acid - An unsaturated fatty acid with 18 carbon atoms.
(c) Adenylic acid - Adenosine with a glucose phosphate molecule.
(d) Alanine amino acid - Contains an amino group and an acidic group anywhere in the molecule.

RESPONSE GRID					
	32. ⓐⓑⓒⓓ	33. ⓐⓑⓒⓓ	34. ⓐⓑⓒⓓ	35. ⓐⓑⓒⓓ	36. ⓐⓑⓒⓓ
	37. ⓐⓑⓒⓓ	38. ⓐⓑⓒⓓ	39. ⓐⓑⓒⓓ	40. ⓐⓑⓒⓓ	41. ⓐⓑⓒⓓ
	42. ⓐⓑⓒⓓ	43. ⓐⓑⓒⓓ	44. ⓐⓑⓒⓓ	45. ⓐⓑⓒⓓ	

Space for Rough Work

DAILY PRACTICE PROBLEM DPP CHAPTERWISE 9 - BIOLOGY			
Total Questions	45	Total Marks	180
Attempted		Correct	
Incorrect		Net Score	
Cut-off Score	45	Qualifying Score	55
Success Gap = Net Score – Qualifying Score			
Net Score = (Correct × 4) – (Incorrect × 1)			

DPP - Daily Practice Problems

Chapter-wise Sheets

Date :   **Start Time :**   **End Time :**

BIOLOGY

SYLLABUS : Cell Cycle and Cell Division

Max. Marks : 180   **Marking Scheme :** + 4 for correct & (–1) for incorrect   **Time : 60 min.**

INSTRUCTIONS : This Daily Practice Problem Sheet contains 45 MCQs. For each question only one option is correct. Darken the correct circle/ bubble in the Response Grid provided on each page.

1. In cell cycle, DNA replication takes place in
   (a) G_1 phase (b) G_2 phase
   (c) mitotic metaphase (d) S phase
2. During cell division, the spindle fibres attach to the chromosome at a region called
   (a) chromocentre (b) kinetochore
   (c) centriole (d) chromomere
3. Chromosome duplication without nuclear division refers to
   (a) meiosis (b) mitosis
   (c) androgenesis (d) endomitosis
4. During which stages (or prophase I substages) of meiosis do you expect to find the bivalents and DNA replication respectively?
   (a) Pachytene and interphase (between two meiotic divisions)
   (b) Pachytene and interphase (just prior to prophase I)
   (c) Pachytene and S phase (of interphase just prior to prophase I)
   (d) Zygotene and S phase (of interphase prior to prophase I)
5. The two chromatids of a metaphase chromosome represent
   (a) replicated chromosomes to be separated at anaphase
   (b) homologous chromosomes of a diploid set
   (c) non-homologous chromosomes joined at the centromere
   (d) maternal and paternal chromosomes joined at the centromere
6. Recombination of genes occur at
   (a) prophase in mitosis
   (b) prophase I in meiosis
   (c) prophase II in meiosis
   (d) metaphase II in meiosis

RESPONSE GRID	1. ⓐⓑⓒⓓ	2. ⓐⓑⓒⓓ	3. ⓐⓑⓒⓓ	4. ⓐⓑⓒⓓ	5. ⓐⓑⓒⓓ
	6. ⓐⓑⓒⓓ				

Space for Rough Work

7. Four daughter cells formed after meiosis are
(a) genetically similar (b) genetically different
(c) anucleate (d) multinucleate

8. In animal cells, cytokinesis involves
(a) the separation of sister chromatids
(b) the contraction of the contractile ring of microfilament
(c) depolymerisation of kinetochore microtubules
(d) a protein kinase that phosphorylates other enzymes

9. The number of chromatids in a chromosome at anaphase is
(a) 2 in mitosis and 1 in meiosis
(b) 1 in mitosis and 2 in meiosis
(c) 2 each in mitosis and meiosis
(d) 2 in mitosis and 4 in meiosis

10. During cell division, sometimes there will be failure of separation of sister chromatids. This event is called
(a) interference (b) complementation
(c) non-disjunction (d) coincidence

11. In which stage of the cell cycle histone proteins synthesized in a eukaryotic cell?
(a) During G_2 stage of prophase
(b) During S-phase
(c) During entire prophase
(d) During telophase

12. What is true about telophase stage?
(a) Chromosomes lose their identity as discrete elements
(b) Chromosomes cluster at opposite spindle poles
(c) Nuclear envelope, nucleolus, Golgi complex and ER reform
(d) All of these

13. Match Column-I with Column-II and select the correct option from the codes given below.

	Column-I		Column-II
A.	Disintegration of nuclear membrane	I.	Anaphase
B.	Appearance of nucleolus	II.	Prophase
C.	Division of centromere	III.	Telophase
D.	Replication of DNA	IV.	S-phase

(a) A-II; B-III; C-I; D-IV
(b) A-II; B-III; C-IV; D-I
(c) A-III; B-II; C-I; D-IV
(d) A-III; B-II; C-IV; D-I

14. The centromere is situated close to its ends and forming one extremely short and one very long arm in
(a) Metacentric chromosome
(b) Sub-metacentric chromosome
(c) Acrocentric chromosome
(d) Telocentric chromosome

15. Select the events that do not occur in interphase stage of cell-cycle
A. RNA and protein synthesis.
B. Cytoplasmic growth.
C. Polymerisation of spindle fibres protein.
D. Disappearance of Golgi bodies and ER.
E. DNA molecules in highly supercoiled stage.
(a) C, D & E (b) D & E only
(c) B, C & D (d) C & D only

16. The stage between two meiotic divisions is called interkinesis and
(a) Is long lived
(b) Is followed by prophase I
(c) Is generally short lived and followed by prophase II
(d) Involves duplication of genes and centrioles

17. Which is correct w.r.t. anaphase?
(a) Centromeres split and chromatids separate
(b) Spindle fibres attach to kinetochores
(c) Chromosomes are moved to spindle equator
(d) Chromatid splits by recombinase activity

18. Maximum cytoplasmic growth occurs in
(a) G_1-phase (b) S-phase
(c) G_2-phase (d) M-phase

19. Diagrammatic representation of chromosomes of a species, is called
(a) Karyotype (b) Crytogram
(c) Cladogram (d) Idiogram

20. Which one of the following events is incorrect for cell cycle?
(a) All events are under genetic control
(b) Maximum cell growth occurs in M-phase
(c) DNA synthesis occurs only during one specific
(d) Centriole duplication occurs in S-phase

RESPONSE GRID

7. ⓐⓑⓒⓓ 8. ⓐⓑⓒⓓ 9. ⓐⓑⓒⓓ 10. ⓐⓑⓒⓓ 11. ⓐⓑⓒⓓ
12. ⓐⓑⓒⓓ 13. ⓐⓑⓒⓓ 14. ⓐⓑⓒⓓ 15. ⓐⓑⓒⓓ 16. ⓐⓑⓒⓓ
17. ⓐⓑⓒⓓ 18. ⓐⓑⓒⓓ 19. ⓐⓑⓒⓓ 20. ⓐⓑⓒⓓ

Space for Rough Work

21. The sequence of events by which a cell duplicates its genome, synthesises other constituents of the cell and eventually divide into two daughter cell is termed as
(a) Karyochorists (b) I-phase
(c) Cell cycle (d) M-phase

22. Lampbrush chromosomes are seen in which typical stage?
(a) Mitotic anaphase (b) Mitotic prophase
(c) Mitotic metaphase (d) Meiotic prophase

23. In telophase of mitosis, the mitotic spindle breaks down and nuclear membranes form. This is essentially the opposite of what happens in
(a) prophase. (b) interphase.
(c) metaphase. (d) S phase.

24. Most cells divide if they receive the proper signal at a checkpoint in the ____ phase of the cell cycle.
(a) M (b) G_1 (c) S (d) G_2

25. Which of the following carry the same genetic information?
(a) sister chromatids
(b) X and Y chromosomes
(c) all autosomes
(d) homologous chromosomes

26. A zoologist examined an intestine cell from a crayfish and counted 200 chromosomes, each consisting of 2 chromatids, at prophase I of mitosis. What would he expect to see in each of the four cells at telophase II of meiosis if he looked in the crayfish ovary?
(a) 50 chromosomes, each consisting of 2 chromatids
(b) 50 chromosomes, each consisting of 1 chromatid
(c) 100 chromosomes, each consisting of 2 chromatids
(d) 100 chromosomes, each consisting of 1 chromatid

27. Which of the following is true of kineto-chores?
(a) They are localized at the centromere of each chromosome.
(b) They are the sites where microtubules attach to separate the chromosomes.
(c) They are organized so that there is one per sister chromatid in meiosis.
(d) All of the above

28. Chromosome movement during anaphase is the result of :
(a) the molecular motors at the kinetochores that move the chromosomes toward the poles.
(b) molecular motors at the centrosome that pull the microtubules toward the poles.
(c) shortening of the microtubules at the centrosome that pull the chromosomes toward the poles.
(d) a and c

29. Programmed cell death (apoptosis) :
(a) occurs in cells that have been deprived of essential nutrients.
(b) occurs only in cells that have damaged DNA
(c) is a natural process during development.
(d) is signaled by the initiated of mitosis.

30. The following graph represents the changes in the quantity of DNA present in the cell cycle at different stages.

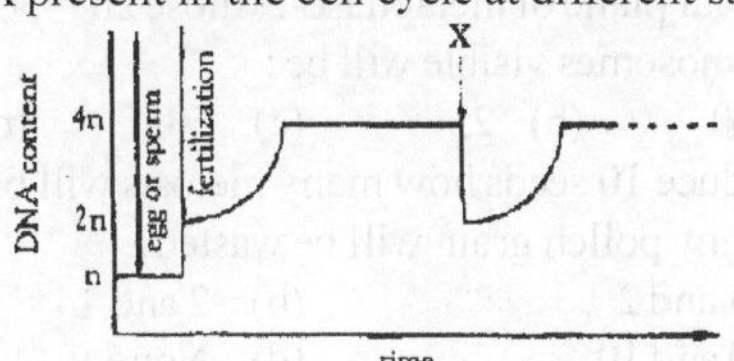

What stage takes place at X?
(a) anaphase (b) cytokinesis
(c) interphase (d) metaphase

31. The absence of centrioles from higher plant cells means that during somatic cell nuclear division there is :
(a) no apparent organiser of mitotic spindles
(b) no equatorial arrangement of chromo-somes at metaphase.
(c) no new cell wall laid down at telophase.
(d) no spindle formed.

32. The amount of DNA in a mammalian cell in early prophase I is x.
What is the amount of DNA in the same cell in anaphase I of meiosis?
(a) $\frac{x}{4}$ (b) $\frac{x}{2}$ (c) x (d) 2x

RESPONSE GRID					
	21. ⓐⓑⓒⓓ	22. ⓐⓑⓒⓓ	23. ⓐⓑⓒⓓ	24. ⓐⓑⓒⓓ	25. ⓐⓑⓒⓓ
	26. ⓐⓑⓒⓓ	27. ⓐⓑⓒⓓ	28. ⓐⓑⓒⓓ	29. ⓐⓑⓒⓓ	30. ⓐⓑⓒⓓ
	31. ⓐⓑⓒⓓ	32. ⓐⓑⓒⓓ			

Space for Rough Work

33. Fully differentiated cells which do not divide are supposed to be in :
(a) G_1 - phase (b) G_2 - phase
(c) S - phase (d) G_0 - phase

34. A cell can not divide if it does not cross :
(a) Hayflick limit (b) cytokinesis
(c) restriction point (d) G_0 - phase

35. Synapsis is of ______ kind(s) :
(a) one (b) two (c) three (d) four

36. If a stock has 2n = 48 and scion microspore mother cell has 2n = 24; then root cell and the microspores will have ______ chromosomes respectively.
(a) 12, 48 (b) 48, 12 (c) 24, 12 (d) 24, 96

37. A plant has number of chromosome groups arranged at equatorial plane of metaphase-I whose 2n = 50; the number of chromosomes visible will be :
(a) 100 (b) 25 (c) 50 (d) 75

38. To produce 10 seeds how many meiosis will be needed and how many pollen grain will be wasted?
(a) 13 and 2 (b) 2 and 14
(c) 10 and 10 (d) None

39. Which one of the following list contain the correct order of meiotic events ?
(a) Separation of sister chromatids, recombination, formation of the synaptonemal complex, separation of homologous chromosomes
(b) Separation of homologous chromosomes, formation of the synaptonemal complex, recombination, separation of sister chromatids
(c) Formation of synaptonemal complex, recombination, separation of sister chromatids, separation of homologous chromosomes
(d) Formation of the synaptonemal complex, recombination, separation of homologous chromosomes, separation of sister chromatids.

40. In meiosis, actual haploidy in terms of DNA comes in
(a) Metaphase-I (b) Anaphase-II
(c) Anaphase-I (d) Interkinesis

41. Which of the following can not be considered as mitogen ?
(a) Cytokinin (b) Insulin
(c) Cholchicine (d) Auxin

42. G_2 phase is not assiciated with
(a) Synthesis of some non-histone proteins
(b) Synthesis of tubulin proteins for spindle fibres
(c) DNA synthesis
(d) Duplication of centrioles

43. Replication of centriole occurs during
(a) Early anaphase (b) Mid metaphase
(c) Late metaphase (d) Interphase

44. A cell in mitotic prophase can be distinguished from a cell in meiotic prophase by
(a) Formation of tetrad in a meiotic cell
(b) The terminalization of chiasmata in late prophase of mitosis
(c) Zipping in early prophase of mitosis
(d) Presence of only half as many chromosomes in the meiotic cell

45. The diagram shows a cell whose diploid chromosome number is four. Which one of the following option shows correct stage of cell ?

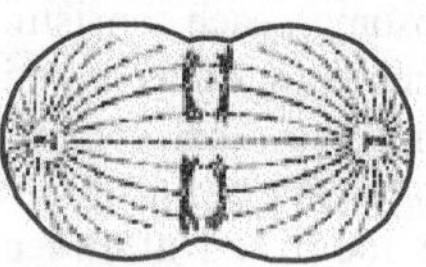

(a) Metaphase
(b) Anaphase of mitosis
(c) Anaphase I of meiosis
(d) Anaphase II of meiosis

RESPONSE GRID					
	33. ⓐⓑⓒⓓ	34. ⓐⓑⓒⓓ	35. ⓐⓑⓒⓓ	36. ⓐⓑⓒⓓ	37. ⓐⓑⓒⓓ
	38. ⓐⓑⓒⓓ	39. ⓐⓑⓒⓓ	40. ⓐⓑⓒⓓ	41. ⓐⓑⓒⓓ	42. ⓐⓑⓒⓓ
	43. ⓐⓑⓒⓓ	44. ⓐⓑⓒⓓ	45. ⓐⓑⓒⓓ		

Space for Rough Work

DAILY PRACTICE PROBLEM DPP CHAPTERWISE 10 - BIOLOGY			
Total Questions	45	Total Marks	180
Attempted		Correct	
Incorrect		Net Score	
Cut-off Score	40	Qualifying Score	50
Success Gap = Net Score – Qualifying Score			
Net Score = (Correct × 4) – (Incorrect × 1)			

DPP - Daily Practice Problems

Chapter-wise Sheets

Date : | Start Time : | End Time :

BIOLOGY CB11

SYLLABUS : Transport in Plants

Max. Marks : 180 **Marking Scheme : + 4 for correct & (–1) for incorrect** **Time : 60 min.**

INSTRUCTIONS : This Daily Practice Problem Sheet contains 45 MCQs. For each question only one option is correct. Darken the correct circle/ bubble in the Response Grid provided on each page.

1. Water potential is equal to
(a) $\Psi_s + O.P.$ (b) $\Psi_s + T.P.$
(c) $\Psi_p + \Psi_w$ (d) $\Psi_s + \Psi_p$

2. The movement of mineral ions into plant root cells as a result of diffusion is called
(a) osmosis
(b) active absorption
(c) passive absorption
(d) endocytosis

3. In soil, the water available for root absorption is
(a) gravitational water (b) capillary water
(c) hygroscopic water (d) combined water

4. Plasmolysis occurs when the cell is placed in
(a) isotonic solution (b) hypotonic solution
(c) hypertonic solution (d) None of the above

5. The rate of transpiration in plants is dependent upon
(a) temperature and soil
(b) light and temperature
(c) wind, temperature and light
(d) light, temperature, atmospheric humidity and wind

6. The water potential and osmotic potential of pure water are
(a) zero and zero (b) 100 and 100
(c) zero and 100 (d) 100 and zero

7. Stomata open and close due to
(a) circadian rhythm
(b) genetic clock
(c) pressure of gases inside the leaves
(d) turgor pressure of guard cells

8. Movement of ions or molecules in a direction opposite to that of prevailing electrochemical gradient is known as
(a) diffusion (b) active transport
(c) pinocytosis (d) brownian movement

9. Osmosis means movement of
(a) solute from lower concentration to higher concentration.
(b) solute from higher concentration to lower concentration.
(c) solvent from low concentration of solution to higher concentration of solution.
(d) solvent from higher concentration of solution to low concentration of solution.

RESPONSE GRID					
	1. ⓐⓑⓒⓓ	2. ⓐⓑⓒⓓ	3. ⓐⓑⓒⓓ	4. ⓐⓑⓒⓓ	5. ⓐⓑⓒⓓ
	6. ⓐⓑⓒⓓ	7. ⓐⓑⓒⓓ	8. ⓐⓑⓒⓓ	9. ⓐⓑⓒⓓ	

Space for Rough Work

10. Solution outside a cell has higher concentration than cytoplasm, then the solution is
(a) isotonic (b) acidic
(c) hypotonic (d) hypertonic

11. Which of the following is used as antitranspirant?
(a) Calcium carbonate
(b) Phenyl mercuric acetate
(c) Cobalt chloride
(d) Naphthol acetic acid

12. During $Na^+ - K^+$ pump
(a) $3Na^+$ and $2K^+$ are transported
(b) $1Na^+$ and $2K^+$ are transported
(c) $3Na^+$ and $3K^+$ are transported
(d) Depends on requirement of cell

13. In rainy season, door gets swelled due to
(a) imbibition (b) diffusion
(c) transpiration (d) respiration

14. Turgor pressure become equal to the wall pressure when
(a) water leaves the cell
(b) water enter the cell
(c) no exchange of water takes place
(d) solute goes from cell into water

15. Addition of a solute to pure water causes
(a) negative water potential.
(b) more negative water potential.
(c) positive water potential.
(d) more positive water potential.

16. Which one of the following process help the water- soluble inorganic nutrients go down into the soil horizon and get precipitated as unavailable salts?
(a) Fragmentation (b) Leaching
(c) Catabolism (d) Humification

17. Water logging of soil makes it physiologically dry because:
(a) this condition does not allow the capillary force to work
(b) this condition does not allow oxygen to enter the soil
(c) Both (a) and (b)
(d) None of the above

18. The productivity of a crop declines when leaves begin to wilt mainly because
(a) the chlorophyll of wilting leaves decomposes.
(b) flaccid mesophyll cells are incapable of photosynthesis.
(c) stomata close, preventing CO_2 from entering the leaf.
(d) photolysis, the water-splitting step of photosynthesis, cannot occur when there is a water deficiency.

19. The phenomenon of plasmolysis is evident when cells are kept in
(a) hypotonic solution
(b) hypertonic solution
(c) isotonic solution
(d) None of the above

20. What mechanism explains the movement of sucrose from source to sink ?
(a) Evaporation of water and active transport of sucrose from sink.
(b) Osmotic movement of water into the sucrose loaded sieve tube cells creating a higher hydrostatic pressure into the source than in the sink.
(c) Tension created by differences in hydrostatic pressure in the source and sink.
(d) Active transport of sucrose through the sieve tube membranes driven by proton pump.

21. Water will be absorbed by root hairs when
(a) concentration of salt in the soil is high.
(b) concentration of solutes in the cell sap is high.
(c) plant is rapidly respiring.
(d) they are separated from soil by a permeable membrane.

22. Proton pumps in the plasma membranes of plant cells may
(a) generate a membrane potential that helps drive cations into the cell through their specific carriers.
(b) be coupled to the movement of K^+ into guard cells.
(c) drive the accumulation of sucrose in sieve-tube members.
(d) be involved in all of the above

23. Which one of the following graphs most closely represents the relationship between the rate of transpiration of a mesophytic leaf and the atmospheric humidity ?

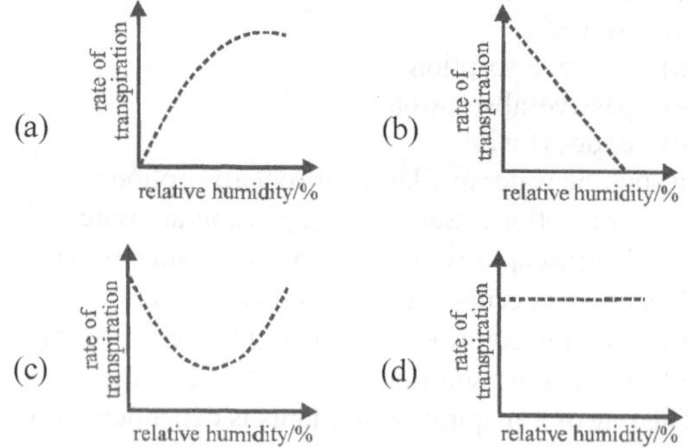

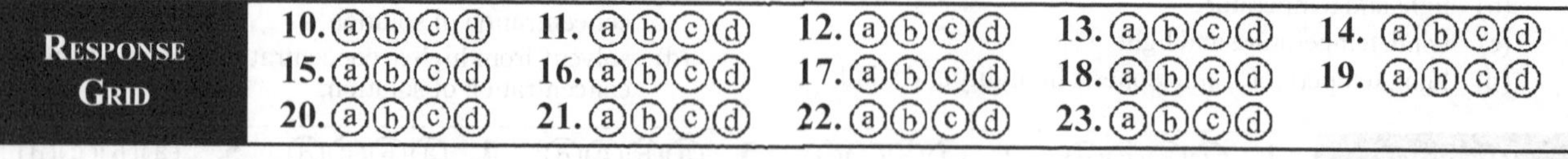

Space for Rough Work

24. A plant cell placed in pure water will
(a) expand until the osmotic potential or solute potential reaches that of water.
(b) becomes more turgid until the pressure potential of cell reaches its osmotic potential.
(c) become more turgid until the osmotic potential reaches that of pure water.
(d) becomes less turgid until the osmotic potential reaches that of pure water.

25. The casparian strip prevents water and minerals from entering the stele through the
(a) plasmodesmata (b) apoplast
(c) symplast (d) xylem vessel

26. If a soft stemmed plant, is cut horizontally near the base of its stem with a sharp blade on early morning of a humid day, drops of solution ooze through cut stem. This is due to
(a) guttation (b) bleeding
(c) transpiration pull (d) root pressure

27. Which of the following statements is incorrect?
(a) Endodermis is impervious to water due to the presence of suberized Casparian stips.
(b) Xylem vessels and tracheids, being non-living, are parts of the apoplast
(c) Ascent of sap is best explained by root pressure theory.
(d) None of these

28. In part A of a plant, sugars are actively transported into the phloem tissue. In part B, sugars are actively transported out of the phloem. Which way will the phloem sap move under these conditions?
(a) From A to B
(b) From B to A
(c) First from A to B; then, once the pressure builds up, from B to A
(d) First from B to A; then, once the pressure builds up, from A to B

29. If the pressure potential is +0.16 megapascals (mPa) and the osmotic potential is –0.24 megapascals, then the water potential would be
(a) +0.4 mPa (b) +0.08 mPa
(c) –0.08 mPa (d) + 0.16 mPa.

30. The movement of water up the stems of tall plants is least dependent on which of the following factors ?
(a) Guttation
(b) Transpiration
(c) Cohesiveness of water molecules
(d) Tension within column of water molecules

31. As a leaf loses water, a series of events typically occurs leading to the closing of stomata. Which of the following shows the correct sequence of these events ?
(a) Potassium ions enter guard cells, water passively enters guard cells, abscisic acid is released by leaf cells, pressure potential in guard cells decreases
(b) Potassium ions enter guard cells, water passively leaves guard cells, pressure potential in guard cells decreases
(c) Water passively leaves guard cells, pressure potential in guard cells decreases, abscisic acid is released by leaf cells, potassium ions leave guard cells
(d) Abscisic acid is released by leaf cells, potassium ions leave guard cells, water passively leaves guard cells, pressure potential in guard cells decreases

32. Which of the following explains why the leaf cells of plants within the family Crassulaceae have a lower pH at night than during the day ?
(a) These plants keep their stomata open at night but clossed during the day.
(b) Carbon dioxide in these plants is chemically fixed in orgnanic acids during the night, but these acids are metabolized during the day.
(c) These plants have a buildup of abscisic acid at night.
(d) Because their stomata are closed during the day, excess carbon dioxide within the leaf forms carbonic acid.

33. Which of the following characteristics applies to both xylem and phloem transport ?
(a) Follows a water potential gradient
(b) Involves only living cells
(c) Can occur in both directions
(d) Involves active transport of solute with passive movement of water

34. Which of the following cell types would be the third cell type a sugar molecule is likely to encounter on its route from its site of production in a chloroplast to its site of storage in the root ?
(a) Mesophyll cell
(b) Companion cell
(c) Sieve tube element
(d) Parenchyma

35. The primary difference between the apoplast and the symplast is that the
(a) apoplast is nonliving spaces and cell walls.
(b) apoplast relies on active transport.
(c) symplast is nonliving spaces and cell walls.
(d) apoplast prevents passive diffusion.

RESPONSE GRID					
	24. ⓐⓑⓒⓓ	25. ⓐⓑⓒⓓ	26. ⓐⓑⓒⓓ	27. ⓐⓑⓒⓓ	28. ⓐⓑⓒⓓ
	29. ⓐⓑⓒⓓ	30. ⓐⓑⓒⓓ	31. ⓐⓑⓒⓓ	32. ⓐⓑⓒⓓ	33. ⓐⓑⓒⓓ
	34. ⓐⓑⓒⓓ	35. ⓐⓑⓒⓓ			

Space for Rough Work

36. Tension is a result of which of the following?
(a) Transpiration at the leaf surface
(b) The cohesive nature of water
(c) The narrowness of the xylem tube
(d) All of the above

37. Which of the following regulates stomatal opening and closing ?
(a) Abscisic acid levels
(b) Light levels
(c) Carbon dioxide concentrations
(d) All of the above

38. Regulators of stomatal opening and closing work by activating the
(a) proton pump in guard cells.
(b) proton pump in stomata
(c) sodium-potassium pump
(d) all of the above

39. Which of the following would not contribute to water uptake by a plant cell ?
(a) an increase in the water potential (ψ) of the surrounding solution
(b) a decrease in pressure on the cell exerted by the wall
(c) the uptake of solutes by the cell
(d) an increase in tension on the surrounding solution.

40. Which of the following does not appear to involve active transport across membranes?
(a) the movement of mineral nutrients from the apoplast to the symplast
(b) the movement of sugar from mesophyll cells into sieve-tube members in corn
(c) the movement of sugar from one sieve-tube member to the next
(d) K^+ uptake by guard cells during stomatal opening.

41. The productivity of a crop declines when leaves begin to wilt mainly because
(a) the chlorophyll of wilting leaves decomposes.
(b) flaccid mesophyll cells are incapable of photosynthesis
(c) stomata close, preventing CO_2 from entering the leaf.
(d) photolysis, the water-splitting step of photosynthesis, cannot occur when there is a water deficiency

42. Imagine cutting a live twig from a tree and examining the cut surface of the twig with a magnifying glass. You locate the vascular tissue and observe a growing droplet of fluid exuding from the cut surface. This fluid is probably
(a) phloem sap
(b) xylem sap
(c) guttation fluid
(d) fluid of the transpiration stream

43. Xylem moves its materials
(a) from the leaves to the roots.
(b) from the roots to the leaves.
(c) when the plant is actively photosynthesizing.
(d) if the proper pressure is maintained.

44. Primary and secondary active transport both
(a) Generate ATP
(b) Use ATP directly
(c) Can move solutes against their concentration gradient
(d) Include the passive movement of glucose molecule

45. Aquaporins are
(a) cytoplasmic connections between cortex cells.
(b) openings in the lower epidermis of leaves through which water vapour escapes.
(c) openings into root hairs through which water enters.
(d) water specific channels in membranes that may regulate the rate of osmosis.

RESPONSE GRID					
	36. ⓐⓑⓒⓓ	37. ⓐⓑⓒⓓ	38. ⓐⓑⓒⓓ	39. ⓐⓑⓒⓓ	40. ⓐⓑⓒⓓ
	41. ⓐⓑⓒⓓ	42. ⓐⓑⓒⓓ	43. ⓐⓑⓒⓓ	44. ⓐⓑⓒⓓ	45. ⓐⓑⓒⓓ

Space for Rough Work

DAILY PRACTICE PROBLEM DPP CHAPTERWISE 11 - BIOLOGY			
Total Questions	45	Total Marks	180
Attempted		Correct	
Incorrect		Net Score	
Cut-off Score	40	Qualifying Score	55
Success Gap = Net Score – Qualifying Score			
Net Score = (Correct × 4) – (Incorrect × 1)			

DPP - Daily Practice Problems

Chapter-wise Sheets

Date : Start Time : End Time :

BIOLOGY CB12

SYLLABUS : Mineral Nutrition

Max. Marks : 180 **Marking Scheme : + 4 for correct & (–1) for incorrect** **Time : 60 min.**

INSTRUCTIONS : This Daily Practice Problem Sheet contains 45 MCQs. For each question only one option is correct. Darken the correct circle/ bubble in the Response Grid provided on each page.

1. Which of the following is a free living aerobic non-photosynthetic nitrogen-fixer?
   (a) *Rhizobium* (b) *Azotobacter*
   (c) *Azospirillum* (d) *Nostoc*
2. Which element is located at the centre of the porphyrin ring in chlorophyll ?
   (a) Manganese (b) Calcium
   (c) Magnesium (d) Potassium
3. Which one of the following mineral elements plays an important role in biological nitrogen fixation ?
   (a) Molybdenum (b) Copper
   (c) Manganese (d) Zinc
4. Leghaemoglobin helps in
   (a) nitrogen fixation
   (b) protecting nitrogenase from O_2
   (c) destroys bacteria
   (d) transport of food in plants
5. 'Whip-tail' disease in cauliflower is noted due to deficiency of
   (a) manganese (b) magnesium
   (c) molybdenum (d) nitrogen
6. Which of the following is correct set of micronutrient for plants?
   (a) Mg, Si, Fe, Cu, Ca (b) Cu, Fe, Zn, B, Mn
   (c) Mg, Fe, Zn, B, Mn (d) Mo, Zn, Cl, Mg, Ca

RESPONSE GRID	1. ⓐⓑⓒⓓ	2. ⓐⓑⓒⓓ	3. ⓐⓑⓒⓓ	4. ⓐⓑⓒⓓ	5. ⓐⓑⓒⓓ
	6. ⓐⓑⓒⓓ				

Space for Rough Work

7. Which of the following statements about nitrification is not true ?
(a) *Nitrobacter* oxidizes nitrite to nitrate.
(b) *Nitrosomonas* and *Nitrosococcus* convert ammonium ions to nitrite.
(c) Nitrification reactions are energy-producing (exergonic) reactions.
(d) Heterotrophic plants are more directly dependent on the nitrifying bacteria for usable nitrogen than autotrophic plants.

8. A free living nitrogen-fixing cyanobacterium which can also form symbiotic association with the water fern *Azolla* is
(a) *Tolypothrix* (b) *Chlorella*
(c) *Nostoc* (d) *Anabaena*

9. *Azotobacter* and *Beijerinckia* are the examples of
(a) symbiotic nitrogen-fixers
(b) non-symbiotic nitrogen-fixers
(c) ammonifying bacteria
(d) disease causing bacteria

10. Which of the following is not performed by root hairs ?
(a) Water uptake (b) Oxygen uptake
(c) Mineral uptake (d) CO_2 uptake

11. The plant ash is an indication of.
(a) organic matter of plant.
(b) waste product.
(c) mineral salts absorbed by plants.
(d) None of the above

12. Plant obtains nitrogen from soil in the form of
(a) nitrite (b) nitrate
(c) ammonia (d) hydroxylamine

13. Hydroponics is
(a) nutrient less culture (b) water less culture
(c) soilless culture (d) none of these

14. Which element is required in the germination of pollen grain?
(a) Chlorine (b) Potassium
(c) Boron (d) Calcium

15. $N_2 + 8e^- + 8H^+ + 16\,ATP \rightarrow 2NH_4 + H_2 + 16ADP + 16Pi$
The above equation refers to
(a) ammonification (b) nitrification
(c) nitrogen fixation (d) denitrification

16. A trace element essential for plant growth and radio-isotope, which is used in cancer therapy is
(a) cobalt (b) calcium
(c) sodium (d) iron

17. Which one of the following elements in plants is not remobilised?
(a) Phosphorus (b) Calcium
(c) Potassium (d) Sulphur

18. Element necessary for middle lamella is
(a) Ca (b) Zn
(c) K (d) Cu

19. Which of the following element is necessary for translocation of sugars in plants ?
(a) Boron (b) Molybdenum
(c) Manganese (d) Iron

20. Plants die from prolonged water-logging because
(a) soil nutrients become very dilute.
(b) root respiration stops.
(c) cell sap in the plants becomes too dilute.
(d) nutrients leach down due to excess water.

21. Conversion of ammonia into nitrates through *Nitrosomonas* is called
(a) nitrogen fixation (b) nitrification
(c) denitrification (d) ammonification

RESPONSE GRID				
7. ⓐⓑⓒⓓ	8. ⓐⓑⓒⓓ	9. ⓐⓑⓒⓓ	10. ⓐⓑⓒⓓ	11. ⓐⓑⓒⓓ
12. ⓐⓑⓒⓓ	13. ⓐⓑⓒⓓ	14. ⓐⓑⓒⓓ	15. ⓐⓑⓒⓓ	16. ⓐⓑⓒⓓ
17. ⓐⓑⓒⓓ	18. ⓐⓑⓒⓓ	19. ⓐⓑⓒⓓ	20. ⓐⓑⓒⓓ	21. ⓐⓑⓒⓓ

Space for Rough Work

22. Which of the following is not a correct pairing of a macronutrient and the major functions it performs in the life of a plant ?
   (a) Potassium-enzyme activation, water balance, ion balance.
   (b) Calcium-activity of membranes and cytoskeleton, second messenger.
   (c) Sulphur-in proteins and coenzymes.
   (d) Iron-in active sites of many redox enzymes and electron carriers.
23. Which of the following minerals activate the enzymes involved in respiration?
   (a) nitrogen and phosphorus
   (b) magnesium and manganese
   (c) potassium and calcium
   (d) sulpur and iron
24. Soil can easily become deficient in______ because these ions are negatively charged and do not stick to negatively charged clay particles.
   (a) Nitrate (b) Calcium
   (c) Ammonium (d) Magnesium
25. Legumes' roots have swellings called nodules that
   (a) produce antibiotics that protect the plant from soil bacteria
   (b) provide a steady supply of sugar to the host plant
   (c) increases the surface area for water uptake
   (d) contain nitrogen-fixing bacteria
26. Macronutrients are_____ than micronutrients.
   (a) Larger molecules
   (b) Needed in greater quantities
   (c) More essential
   (d) More important for growth
27. 'Hunger signs' in plants are :
   (a) Symptoms due to lesser water absorption in plants
   (b) Sympotms due to poor photosynthesis in plants
   (c) Deficiency symptoms of particular mineral nutrients
   (d) None of the above
28. Death of stem and root tips occur due to deficiency of :
   (a) Calcium (b) Nitrogen
   (c) Carbon (d) Phosphorus
29. Which of the following is not involved in the phenomenon called ion exchange ?
   (a) Negatively charged clay particles
   (b) pH
   (c) Cations like K^+, Mg^{2+}, and Ca^{2+}
   (d) H_2S
30. Tea yellow is a disease of tea plants produced due to the deficiency of-
   (a) Nitrogen (b) Sulphur
   (c) Potassium (d) Phosphorus
31. Nitrogenase
   (a) is insensitive to oxygen.
   (b) contains magnesium
   (c) releases two NH_3 molecules as products.
   (d) requires an aerobic environment.
32. Deficiency of iron causes :
   (a) interveinal chlorosis first on young leaves
   (b) decrease in protein synthesis
   (c) reduced leaves and stunted growth
   (d) bending of stem tip
33. You notice that the young leaves of your tomato plants are very yellow. What type of deficiency does this suggest ?
   (a) Nitrogen (b) Carbon
   (c) Water (d) Iron

RESPONSE GRID					
	22. ⓐⓑⓒⓓ	23. ⓐⓑⓒⓓ	24. ⓐⓑⓒⓓ	25. ⓐⓑⓒⓓ	26. ⓐⓑⓒⓓ
	27. ⓐⓑⓒⓓ	28. ⓐⓑⓒⓓ	29. ⓐⓑⓒⓓ	30. ⓐⓑⓒⓓ	31. ⓐⓑⓒⓓ
	32. ⓐⓑⓒⓓ	33. ⓐⓑⓒⓓ			

Space for Rough Work

34. Where does most nutrient uptake occur in roots ?
(a) At the root tip, where root tissue first encounters soil away from the zone of nutrient depletion.
(b) At the casparian strip, where ions must enter the symplast prior to entering xylem cells.
(c) In the symplastic and apoplastic pathways.
(d) In root hairs and in the zone of maturation.

35. It is possible to determine whether an element is essential by observing growth of plants
(a) On soil in which the particular element is removed
(b) On soil in which only the particular element is present
(c) On an inert medium to which solution of only the particular element is added
(d) On an inert medium to which a nutrient solution excluding that particular element, is added.

36. Terrestrial plants obtain all of the following nutrients from the soil except
(a) potassium (b) nitrogen
(c) phosphorus (d) carbon

37. For chlorophyll formation a plant needs :
(a) Fe, Ca & light (b) Fe, Mg & Light
(c) Ca, K & light (d) Mn & Cu

38. Which element essential for the stability of chromosome structure?
(a) Zn (b) Ca
(c) Mo (d) Fe

39. Which mineral nutrients are called critical element for crops?
(a) N, P, K (b) C, H, O
(c) N, S, Mg (d) K, Ca, Fe

40. Nitrogenase enzymes are extremely sensitive to ________ molecules -
(a) Hydrogen (b) Oxygen
(c) Water (d) CO_2

41. Gray spots of Oat are caused by deficiency of :
(a) Cu (b) Zn
(c) Mn (d) Fe

42. Which element participates in photolysis of water, ionic balance maintenance and solute concentration?
(a) Cl^- (b) B
(c) Na^+ (d) Mg^{2+}

43. "Reclamation" and "Little leaf" disease, caused by deficiency of
(a) Zn and Mo (b) Cu and Zn
(c) Cu and B (d) Mn and Cu

44. Brown heart rot of beets is due to deficiency of
(a) B (b) P
(c) Mg (d) Mo

45. Which element is related with Khaira disease of paddy & auxin synthesis?
(a) Fe (b) Zn
(c) B (d) Cu

RESPONSE GRID					
	34. ⓐⓑⓒⓓ	35. ⓐⓑⓒⓓ	36. ⓐⓑⓒⓓ	37. ⓐⓑⓒⓓ	38. ⓐⓑⓒⓓ
	39. ⓐⓑⓒⓓ	40. ⓐⓑⓒⓓ	41. ⓐⓑⓒⓓ	42. ⓐⓑⓒⓓ	43. ⓐⓑⓒⓓ
	44. ⓐⓑⓒⓓ	45. ⓐⓑⓒⓓ			

Space for Rough Work

DAILY PRACTICE PROBLEM DPP CHAPTERWISE 12 - BIOLOGY			
Total Questions	45	Total Marks	180
Attempted		Correct	
Incorrect		Net Score	
Cut-off Score	40	Qualifying Score	55
Success Gap = Net Score – Qualifying Score			
Net Score = (Correct × 4) – (Incorrect × 1)			

DPP - Daily Practice Problems

Chapter-wise Sheets

Date : | **Start Time :** | **End Time :**

BIOLOGY CB13

SYLLABUS : Photosynthesis in Higher Plants

Max. Marks : 180 | **Marking Scheme :** + 4 for correct & (–1) for incorrect | **Time : 60 min.**

INSTRUCTIONS : This Daily Practice Problem Sheet contains 45 MCQs. For each question only one option is correct. Darken the correct circle/ bubble in the Response Grid provided on each page.

1. Quantasomes are present in
(a) chloroplast (b) mitochondria
(c) golgi body (d) lysosome

2. The C_4 plants are photosynthetically more efficient than C_3 plants because:
(a) the CO_2 compensation point is more
(b) CO_2 generated during photorespiration is trapped and recycled through PEP carboxylase
(c) the CO_2 efflux is not prevented
(d) they have more chloroplasts

3. Photorespiration is favoured by
(a) high O_2 and low CO_2
(b) low light and high O_2
(c) low temperature and high O_2
(d) low O_2 and high CO_2

4. Which one is the correct summary equation of photosynthesis?
(a) $C_6H_{12}O_6 + 6O_2 \longrightarrow 6CO_2 + 6H_2O + \text{energy}$
(b) $C_6H_{12}O_6 + 6O_2 + 6H_2O \longrightarrow 6CO_2 + 12H_2O + \text{energy}$
(c) $6CO_2 + 6H_2O \longrightarrow 6H_2O + C_6H_{12}O_6$
(d) $6CO_2 + 12H_2O \xrightarrow[\text{Chlorophyll}]{\text{Light}} 6O_2 + C_6H_{12}O_6 + 6H_2O$

5. C_4 acid, formed in the mesophyll of C_4 plants leaf during photosynthesis is
(a) OAA or malic acid or aspartic acid
(b) pyruvic acid
(c) succinic acid
(d) fumaric acid

6. The organelles which take part in photo-respiration are
(a) chloroplast, mitochondria, nucleus
(b) chloroplast, mitochondria, lysosome
(c) mitochondria, chloroplast, peroxisome
(d) mitochondria, lysosomes, peroxisome

RESPONSE GRID	1. ⓐⓑⓒⓓ	2. ⓐⓑⓒⓓ	3. ⓐⓑⓒⓓ	4. ⓐⓑⓒⓓ	5. ⓐⓑⓒⓓ
	6. ⓐⓑⓒⓓ				

Space for Rough Work

7. Grana are present inside the :
(a) mitochondria (b) chloroplast
(c) endoplasmic reticulum (d) ribosome

8. Photosynthetically active radiation is represented by the range of wavelength of
(a) 340 - 450 nm (b) 400 - 700 nm
(c) 500 - 600 nm (d) 400 - 950 nm

9. Which of the following statement is incorrect ?
(a) C_3 plants respond to higher temperature, show higher photosynthetic rate while C_4 plants have lower optimum temperature.
(b) Tropical plants have higher temperature optimum than the plants adapted to temperate climate.
(c) Light reaction is less temperature sensitive than dark reaction.
(d) The effect of water as a factor is more through its effect on plant, rather than directly on photosynthesis.

10. Chemiosmotic hypothesis given by Peter Mitchell proposes the mechanism of
(a) synthesis of ATP (b) synthesis of $FADH_2$
(c) synthesis of NADH (d) synthesis of NADPH

11. RuBisCO stands for
(a) Ribulose Biphosphate Carboxylase Oxygenase
(b) Ribulose Phosphate Carboxylase Oxygenase
(c) Ribulose Phosphate Carboxylic Oxygenase
(d) None of the above

12. Which enzyme is most abundantly found on earth?
(a) Catalase (b) RuBisCo
(c) Nitrogenase (d) Invertase

13. The first carbon dioxide acceptor in C_4-plants is
(a) Phosphoenol-pyruvate
(b) Ribulose 1, 5-diphosphate
(c) Oxalo-acetic acid
(d) Phosphoglyceric acid

14. Which one of the following is essential for photolysis of water ?
(a) Manganese (b) Zinc
(c) Copper (d) Boron

15. Which element is required in the germination of pollen grain?
(a) Chlorine (b) Potassium
(c) Boron (d) Calcium

16. In a CAM plant, the concentration of organic acid
(a) increases during the day.
(b) decreases during the day.
(c) increases during night.
(d) decreases or increases during day.

17. Which of the following statements regarding photorespiration are **true**?
(a) Photorespiration is a metabolically expensive pathway.
(b) Photorespiration is avoided when CO_2 is abundant.
(c) Photorespiration results in a loss of usable carbon dioxide.
(d) All of the above

18. The Z-scheme refers to
(a) the type of photosynthesis occurs in plants found in areas with minimal precipitation.
(b) the pattern of grana within the chloroplasts of photosynthetic plants.
(c) the carbon-fixation process which is also known as the Calvin cycle.
(d) an energy diagram for the transfer of electrons in the light reactions of photosynthesis in plants.

19. Which one is a C_4-plant?
(a) Papaya (b) Pea
(c) Potato (d) Maize/Corn

20. In an experiment, mature leaves on the plant were enclosed for a fixed amount of time in a transparent bag that had radioactive CO_2. In which part of the plant will maximum radioactivity be found after some time?
(a) Actively growing leaves.
(b) Guard cells of all the leaves.
(c) In mature leaves.
(d) Senescing leaves and roots.

21. When a photosynthetic plant is transferred to an atmosphere of enriched O_2, its rate of
(a) photosynthesis would increase.
(b) photosynthesis would decrease.
(c) respiration would decrease.
(d) osmosis would increase.

RESPONSE GRID					
	7. ⓐⓑⓒⓓ	8. ⓐⓑⓒⓓ	9. ⓐⓑⓒⓓ	10. ⓐⓑⓒⓓ	11. ⓐⓑⓒⓓ
	12. ⓐⓑⓒⓓ	13. ⓐⓑⓒⓓ	14. ⓐⓑⓒⓓ	15. ⓐⓑⓒⓓ	16. ⓐⓑⓒⓓ
	17. ⓐⓑⓒⓓ	18. ⓐⓑⓒⓓ	19. ⓐⓑⓒⓓ	20. ⓐⓑⓒⓓ	21. ⓐⓑⓒⓓ

Space for Rough Work

22. In C_4 plants, agranal chloroplasts are found in
(a) mesophyll cells
(b) epidermal cell chloroplasts of green stem
(c) bundle sheath cells
(d) chloroplasts of guard cells

23. Oxysomes or $F_0 - F_1$ particles occur on
(a) thylakoids
(b) mitochondrial surface
(c) inner mitochondrial membrane
(d) chloroplast surface

24. Photorespiration is favoured by
(a) high O_2 and low CO_2
(b) low light and high O_2
(c) low temperature and high O_2
(d) low O_2 and high CO_2

25. In the leaves of C_4 plants, malic acid formation during CO_2 fixation occurs in the cells of
(a) bundle sheath (b) phloem
(c) epidermis (d) mesophyll

26. The graph shows the relation between light intensity and the giving off and taking up of carbon dioxide by the leaves of a plant. Why is most carbon dioxide given off when the light intensity is zero units ?

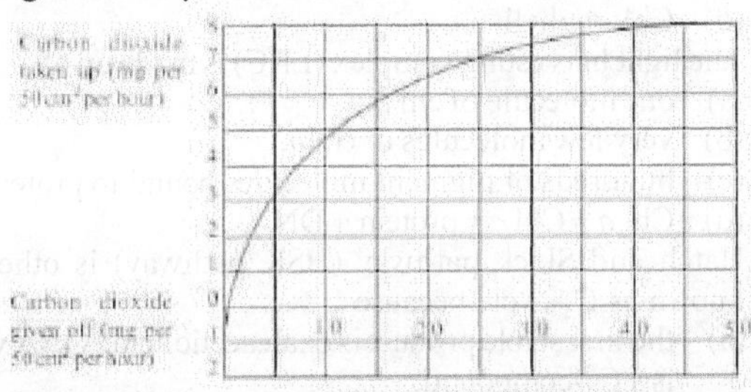

(a) Because it is just the start of the experiment.
(b) Only respiration is taking place at this intensity of light.
(c) Only photosynthesis is taking place at this intensity of light.
(d) The rate of photosynthesis is equivalent to the rate of respiration.

27. In photosystem-I the first electron acceptor is
(a) cytochrome (b) plastocyanin
(c) an iron-sulphur protein (d) ferredoxin

28. Which of the following statement is false?
(a) H_2S, not H_2O, is involved in photosynthesis of purple sulphur bacteria.
(b) Light and dark reactions are stopped in the absence of light.
(c) Calvin cycle occurs in the grana of chloroplast.
(d) ATP is produced during light reaction *via* chemiosmosis.

29. Given below is the pathway of light reaction. Identify the given blanks indicated by A, B, C, D and E.

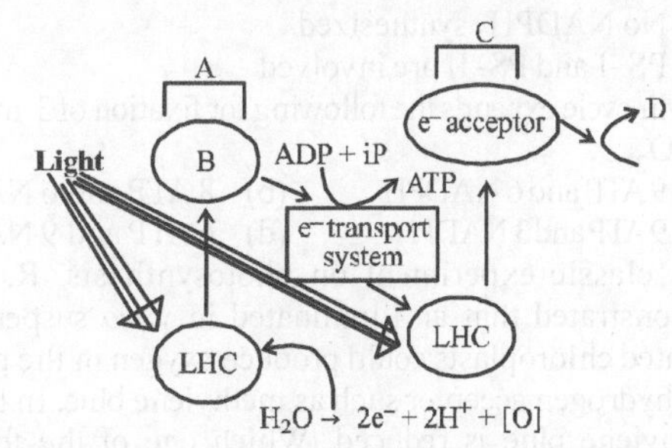

	A	B	C	D
(a)	P 700	H^+ acceptor	P680	$NADP^+$
(b)	Photosystem I	e^- acceptor	Photosystem II	$NADPH_2^+$ ATP
(c)	Photosystem II	H^+ acceptor	P700	NADPH
(d)	Photosystem II	e^- acceptor	Photosystem I	NADPH + H^+

30. Warburg effect refers to
(a) decreased photosynthetic rate at very high O_2 concentration
(b) increased photosynthetic rate at very high O_2 concentration
(c) decreased photosynthetic rate at very low O_2 concentration
(d) increased photosynthetic rate at very low O_2 concentration

RESPONSE GRID					
	22. ⓐⓑⓒⓓ	23. ⓐⓑⓒⓓ	24. ⓐⓑⓒⓓ	25. ⓐⓑⓒⓓ	26. ⓐⓑⓒⓓ
	27. ⓐⓑⓒⓓ	28. ⓐⓑⓒⓓ	29. ⓐⓑⓒⓓ	30. ⓐⓑⓒⓓ	

Space for Rough Work

31. Read the following statements and select the correct ones.
(i) PS I is involved in non-cyclic photophosphorylation only.
(ii) PS II is involved in both cyclic and non-cyclic photophosphorylation
(iii) Stroma lamellae membranes possess PS I only, whereas grana lamellae membranes possess both PS I and PS II.
(a) (i) only (b) (ii) only
(c) (iii) only (d) (i), (ii) and (iii)

32. Which is not correct for cyclic photophosporylation ?
(a) No O_2 given off
(b) No water consumed
(c) No $NADPH_2$ synthesized
(d) PS–I and PS–II are involved

33. Calvin cycle expends the following for fixation of 3-molecules of CO_2
(a) 9 ATP and 6 $NADPH_2$ (b) 8 ATP and 6 $NADPH_2$
(c) 9 ATP and 3 $NADPH_2$ (d) 6 ATP and 9 $NADPH_2$

34. In a classic experiment on photosynthesis, R. L. Hill demonstrated that an illuminated in *vitro* suspension of isolated chloroplasts could produce oxygen in the presence of a hydrogen acceptor such as methylene blue. In this case methylene blue is reduced. Which one of the following compounds replaces methylene blue in the intact photosynthesising plant ?
(a) adenosine triphosphate (ATP)
(b) carbon dioxide
(c) nicotinamide adenine dinucleotide phosphate (NADP)
(d) phosphoglyceric acid (PGA)

35. The key compound of Calvin cycle is
(a) PGA (b) PGAL
(c) DHAP (d) DPGA

36. Which one of the following is a correct outline of the main events in photosynthesis?
(a) Oxygen reacts with a carbohydrate to produce water and carbon dioxide in the presence of light.
(b) Light joins carbon dioxide to an acceptor compound which is then reduced by hydrogen obtained from water.
(c) Light splits water and the resulting hydroxyl group combines with a compound which has incorporated carbon dioxide
(d) Carbon dioxide combines with an acceptor compound and this is reduced by hydrogen split from water by light.

37. Who used prism, white light, green alga, *Cladophora* and aerobic bacteria and plotted the action spectra for photosynthesis?
(a) Sachs (b) Arnon
(c) Arnold (d) Englemann

38. Which one of the following pigment does not occur in the chloroplast?
(a) Carotene (b) Xanthophyll
(c) Chlorophyll '*b*' (d) Anthocyanin

39. Which pigment is water soluble?
(a) Chlorophyll (b) Carotene
(c) Anthocyanin (d) Xanthophyll

40. Generally CO_2 is not limiting for hydrophytes –
(a) Mesophytes plants fix H_2S in their photosynthesis.
(b) These plants also CO_2 have from water in the form of HCO_3.
(c) Glucose is not required for their respiration
(d) All of the above

41. Suspension of isolated thylakoids in culture medium containing CO_2 and H_2O does not produce hexose due to absence of
(a) ATP (b) Enzyme
(c) Proteins (d) Hill reagent

42. Stroma in the chloroplasts of higher plant contains
(a) Light-independent reaction enzymes
(b) Light-dependent reaction enzymes
(c) Ribosomes
(d) Chlorophyll

43. The light harvesting complex (LHC) is made up of
(a) one molecule of chl *a*.
(b) very few molecules of chl *a*.
(c) hundreds of pigment molecules bound to proteins.
(d) Chl *a* + Chl *c* + protein + DNA.

44. Hatch and Slack pathway (HSK pathway) is otherwise known as C_4-cycle because
(a) the first stable product is oxaloacetic acid / OAA which is a C_4-compound.
(b) the primary CO_2 acceptor is OAA, a C_4-compounds.
(c) all intermediate metabolites are C_4-compound.
(d) at one time 4CO_2 molecules take part in carboxylation pathway.

45. The first acceptor of electrons from an excited chlorophyll molecule of photosystem II is –
(a) Quinone (b) Cytochrome
(c) Iron-sulphur protein (d) Ferredoxin

RESPONSE GRID					
	31. ⓐⓑⓒⓓ	32. ⓐⓑⓒⓓ	33. ⓐⓑⓒⓓ	34. ⓐⓑⓒⓓ	35. ⓐⓑⓒⓓ
	36. ⓐⓑⓒⓓ	37. ⓐⓑⓒⓓ	38. ⓐⓑⓒⓓ	39. ⓐⓑⓒⓓ	40. ⓐⓑⓒⓓ
	41. ⓐⓑⓒⓓ	42. ⓐⓑⓒⓓ	43. ⓐⓑⓒⓓ	44. ⓐⓑⓒⓓ	45. ⓐⓑⓒⓓ

Space for Rough Work

DAILY PRACTICE PROBLEM DPP CHAPTERWISE 13 - BIOLOGY			
Total Questions	45	Total Marks	180
Attempted		Correct	
Incorrect		Net Score	
Cut-off Score	45	Qualifying Score	60
Success Gap = Net Score – Qualifying Score			
Net Score = (Correct × 4) – (Incorrect × 1)			

DPP - Daily Practice Problems

Chapter-wise Sheets

Date : | **Start Time :** | **End Time :**

BIOLOGY (CB14)

SYLLABUS : Respiration in Plants

Max. Marks : 180 **Marking Scheme : + 4 for correct & (–1) for incorrect** **Time : 60 min.**

INSTRUCTIONS : This Daily Practice Problem Sheet contains 45 MCQs. For each question only one option is correct. Darken the correct circle/ bubble in the Response Grid provided on each page.

1. Respiratory Quotient (R.Q.) is defined as:
   (a) vol. of O_2/vol. of CO_2
   (b) vol. of CO_2 /vol. of O_2
   (c) vol. of O_2/vol. of N_2
   (d) vol. of N_2/vol. of CO_2
2. What is the end product of glycolysis of a glucose molecule?
   (a) Fructose 1, 6- diphosphate
   (b) Pyruvate and ATP
   (c) Phosphoglyceraldehyde
   (d) Lactic acid and ATP
3. Connecting link between glycolysis and Krebs cycle is/ before entering Krebs cycle pyruvate is changed to
   (a) oxaloacetate (b) phosphoenol pyruvate
   (c) pyruvate (d) acetyl CoA
4. Site of respiration in bacteria is
   (a) episome (b) ribosome
   (c) mesosome (d) microsome
5. Terminal cytochrome of respiratory chain which donates electrons to oxygen is
   (a) Cyt. b (b) Cyt. c
   (c) Cyt. a_1 (d) Cyt. a_3
6. Which of the following processes make direct use of oxygen ?
   (a) Glycolysis (b) Fermentation
   (c) Electron transport (d) citric acid cycle
7. Glycolysis is a
   (a) redox process (b) aerobic process
   (c) oxidative process (d) reductive process
8. In Kreb's cycle, the FAD participates as electron acceptor during the conversion of
   (a) succinyl CoA to succinic acid
   (b) α - ketoglutarate to succinyl CoA
   (c) fumaric acid to malic acid
   (d) succinic acid to fumaric acid
9. The major reason that glycolysis is not as energy productive as respiration is that
   (a) NAD^+ is regenerated by alcohol or lactate production, without the high-energy electrons passing through the electron transport chain.
   (b) it is the pathway common to fermentation and respiration.
   (c) it does not take place in a specialized membrane-bound organelle.
   (d) pyruvate is more reduced than CO_2; it still contains much of the energy from glucose.

RESPONSE GRID					
	1. ⓐⓑⓒⓓ	2. ⓐⓑⓒⓓ	3. ⓐⓑⓒⓓ	4. ⓐⓑⓒⓓ	5. ⓐⓑⓒⓓ
	6. ⓐⓑⓒⓓ	7. ⓐⓑⓒⓓ	8. ⓐⓑⓒⓓ	9. ⓐⓑⓒⓓ	

Space for Rough Work

10. Which of the following is a 4-carbon compound?
(a) Oxaloacetic acid (b) Citric acid
(c) Phosphoglyceric acid (d) Phosphoenol pyruvate

11. Which of the following is a biological uncoupler of oxidative phosphorylation ?
(a) Thermogenin
(b) 2, 4 – Dichlorophenoxy acetic acid
(c) 2, 4 – Dinitrophenol
(d) Ethylene diaminotetra acetic acid

12. Which of the following is not true for oxidative phosphorylation?
(a) It uses oxygen as the initial electron donor.
(b) It involves the redox reactions of electron transport chain.
(c) It involves an ATP synthase located in the inner mitochondrial membrane.
(d) It depends on chemiosmosis.

13. In which of the following reaction of glycolysis, a molecule of water is removed from the substrate ?
(a) Frucoste-6-phosphate $\rightarrow$ Fructose 1, 6-biphosphate
(b) 3-phosphate glyceraldehyde $\rightarrow$ 1, 3 biphosphoglyceric acid
(c) Phosphoenol-pyruvate $\rightarrow$ Pyruvic acid
(d) 2-phosphoglycerate $\rightarrow$ PEP

14. The first dicarboxylic acid in Krebs' cycle is
(a) isocitric acid (b) pyruvic acid
(c) oxalo acetic acid (d) α-ketoglutaric acid

15. Final electron acceptor in oxidative phosphorylation is
(a) hydrogen (b) dehydrogenase
(c) cytochrome (d) oxygen

16. Quantasomes are found in
(a) mitochondria (b) chloroplast
(c) lysosome (d) endoplasmic reticulum

17. Which of the following statement(s) is/are incorrect ?
(i) Proton channel of oxysome / complex V / ATP synthase is located in F_1.
(ii) Metabolic water is produced in terminal oxidation / produced in respiration.
(iii) CoQ accepts electron from NADH dehydrogenase (complex I) and also can accept electron from $FADH_2$/ succinate Q-reductase / complex II.
(iv) Cytochrome *c* is a small protein attached to outer surface of the inner mitochondrial membrane and acts as mobile carrier for transfer of electrons between complex I (Cyt *bc*, complex) and III.
(v) Complex IV refers to cytochrome *c* oxidase (cyt a, a_3 and 2 Cu per centre).
(vi) If a cell is treated with a drug that inhibits ATP synthase, the pH of mitochondrial matrix will increase.
(a) (i), (ii) and (iii) (b) (iii), (v) and (vi)
(c) (i) and (iv) (d) Only (iii)

18. Which of the following is an important intermediate found in all the types of respiration ?
(a) Acetyl CoA (b) Pyruvic acid
(c) Oxaloacetate (d) Tricarboxylic acid

19. Incomplete oxidation of glucose into pyruvic acid with several intermediate steps is known as
(a) TCA-pathway (b) Glycolysis
(c) HMS-pathway (d) Krebs cycle

20. Common enzyme in glycolysis and pentose phosphate pathways is
(a) hexokinase (b) aconitase
(c) fumarase (d) dehydrogenase

21. Out of 36 ATP molecules produced per glucose molecule during respiration
(a) 2 are produced outside glycolysis and 34 during respiratory chain.
(b) 2 are produced outside mitochondria and 34 inside mitochondria.
(c) 2 during glycolysis and 34 during Krebs cycle.
(d) all are formed inside mitochondria.

22. R.Q. is highest when respiratory substance is
(a) fat (b) malic acid
(c) glucose (d) protein

23. Electron transport chain is inhibited by
(a) rotenone and amytal
(b) antimycin-A
(c) cyanide (CN^-), azide (N^{3-}) and carbon monoxide (CO)
(d) All of the above

24. Though vertebrates are aerobes, but their ___i___ show anaerobic respiration during ___ii___. During this, ___iii___ of skeletal muscle fibres is broken down to release lactic acid and energy. Lactic acid, if accumulates causes muscle fatigue.
Fill up the blanks in the above paragraph and select the correct option.

	(i)	(ii)	(iii)
(a)	skeletal muscles	heavy exercise	glucose
(b)	skeletal muscles	mild exercise	glycogen
(c)	skeletal muscles	heavy exercise	glycogen
(d)	cardiac muscles	heavy exercise	glycogen

RESPONSE GRID				
10. ⓐⓑⓒⓓ	11. ⓐⓑⓒⓓ	12. ⓐⓑⓒⓓ	13. ⓐⓑⓒⓓ	14. ⓐⓑⓒⓓ
15. ⓐⓑⓒⓓ	16. ⓐⓑⓒⓓ	17. ⓐⓑⓒⓓ	18. ⓐⓑⓒⓓ	19. ⓐⓑⓒⓓ
20. ⓐⓑⓒⓓ	21. ⓐⓑⓒⓓ	22. ⓐⓑⓒⓓ	23. ⓐⓑⓒⓓ	24. ⓐⓑⓒⓓ

Space for Rough Work

25. Identify A and B in the given diagram showing ATP synthesis in oxysomes

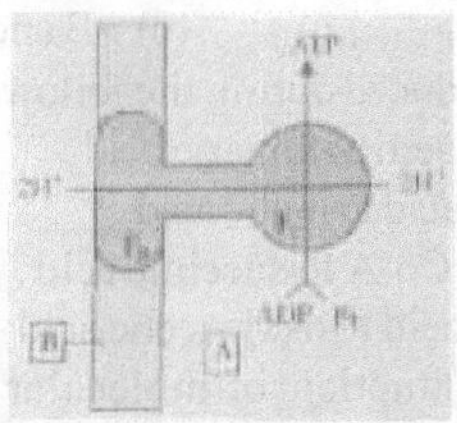

(a) A = Mitochondrial matrix
B = Outer mitochondrial membrane
(b) A = Mitochondrial matrix
B = Inner mitochondrial membrane
(c) A = Cell cytoplasm
B = Inner mitochondrial membrane
(d) A = Cell cytoplasm
B = Outer mitochondrial membrane

26. Potassium cyanide is known to interfere with the formation and use of ATP in cell metabolism.
If the use of potassium cyanide resulted in an accelerated entry of a solute into a cell, it may be reasonably assumed that, under normal circumstances, the solute enters by
(a) active transport (b) osmosis
(c) passive diffusion (d) pinocytosis

27. When yeast growing anaerobically in glucose-rich-environment is exposed to oxygen in sufficient quantity, its rate of glucose utilization decreases. This pheno-menon is known as the Pasteur effect. Adding an uncoupler of oxidative phosphorylation would have what effect on Pasteur effect?
(a) Pasteur effect will be enhanced
(b) Pasteur effect will remain unchanged
(c) Pasteur effect will not take place
(d) Variable effect

28. An inhibitor is added to a cell culture so that succinate accumulates. The enzyme catalysing the formation of which substance has been blocked?
(a) Citrate (b) Oxaloacetate
(c) α -ketoglutarate (d) Fumarate

29. One of the following is common to glycolysis as well as Krebs cycle in eukaryotes :
(a) Substrate level phosphorylation
(b) Photophosphorylation
(c) Localization in mitochondria
(d) Production of $FADH_2$

30. A mutant cell lacking mitochondria will show:
(a) Inability to oxidize carbohydrates
(b) Inability to oxidize fats
(c) Inability to oxidize carbohydrates and fats
(d) Inability to synthesize glucose

31. During the early stages of alcoholic fermentation there is a high rate of growth of yeast. After some time the rate decreases. Which of the following conditions in the culture medium is least likely to have caused this?
(a) Depletion of glucose
(b) Depletion of oxygen
(c) Depletion of mineral salts
(d) Accumulation of waste products

32. Hexose monophosphate shunt does not take place when :
(a) Oxygen is not enough
(b) Glucose is the substrate
(c) NAD is available
(d) NAD is not available due to shortage of oxygen

33. Chemiosmotic theory of ATP synthesis in the chloroplasts and mitochondria is based on
(a) membrane potential (b) accumulation of Na ions
(c) accumulation of K ions (d) proton gradient

34. During the stage in the complete oxidation of glucose are the greatest number of ATP molecules formed from ADP
(a) glycolysis
(b) krebs cycle
(c) conversion of pyruvic acid to acetyl Co-A
(d) electron transport chain

35. How many ATP molecules could maximally be generated from one molecule of glucose, if the complete oxidation of one mole of glucose to CO_2 and H_2O yields 686 kcal and the useful chemical energy available in the high energy phosphate bond of one mole of ATP is 12 kcal ?
(a) Thirty (b) Fifty-seven
(c) One (d) Two

36. All enzymes of TCA cycle are located in the mitochondrial matrix except one which is located in inner mitochondrial membranes in eukaryotes and in cytosol in prokaryotes. This enzyme is
(a) isocitrate dehydrogenase
(b) malate dehydrogenase
(c) succinate dehydrogenase
(d) lactate dehydrogenase.

37. The energy-releasing process in which the substrate is oxidised without an external electron acceptor is called
(a) fermentation (b) photorespiration
(c) aerobic respiration (d) glycolysis

RESPONSE GRID					
	25. ⓐⓑⓒⓓ	26. ⓐⓑⓒⓓ	27. ⓐⓑⓒⓓ	28. ⓐⓑⓒⓓ	29. ⓐⓑⓒⓓ
	30. ⓐⓑⓒⓓ	31. ⓐⓑⓒⓓ	32. ⓐⓑⓒⓓ	33. ⓐⓑⓒⓓ	34. ⓐⓑⓒⓓ
	35. ⓐⓑⓒⓓ	36. ⓐⓑⓒⓓ	37. ⓐⓑⓒⓓ		

Space for Rough Work

38. In TCA cycle the conversion of succinyl Co-A to succinic acid requires
(a) Acetyl Co-A + GTP + iP
(b) Acetyl Co-A + GDP + iP
(c) Co-A + GTP + iP
(d) GDP + iP

39. The diagram shows the reversible conversion of pyruvate to lactate by the enzyme lactate dehydrogenase.

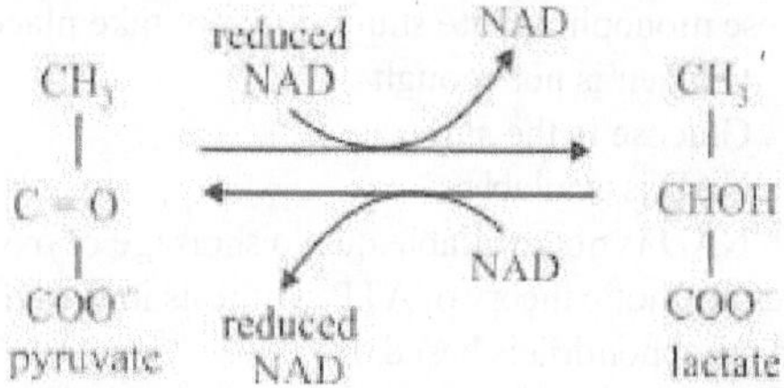

What would be the effect of inhibition of lactate dehydrogenase in a mammalian cell under anaerobic conditions?
(a) A decrease in cell pH, due to the accumulation of lactic acid.
(b) A decrease in glycolysis, due to the lack of NAD.
(c) An increase in ATP production, due to increased amounts of reduced NAD.
(d) An increase in the activity of the Krebs cycle, due to increased amounts of pyruvate.

40. The diagram summarises the pathway of glucose breakdown.

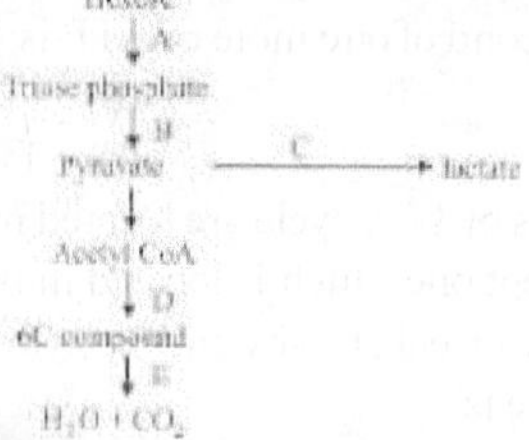

When two steps result in a net increase of ATP?
(a) A and C
(b) A and D
(c) B and D
(d) B and E

41. $FADH_2$ is produced during the following reaction
(a) Succinic acid to fumaric acid
(b) Fumaric acid to malic acid
(c) Succinyl Co-A to succinic acid
(d) Isocitric acid to oxaloacetic acid

42. What is the function of molecular oxygen in cellular respiration?
(a) It causes the breakdown of citric acid.
(b) To combine with glucose to produce carbon dioxide.
(c) To combine with carbon from organic molecules to produce carbon dioxide.
(d) To combine with hydrogen from organic molecules to produce water.

43. Sugars are not as good as fats as a source of energy for cellular respiration, because sugars
(a) produce toxic amino groups when broken down.
(b) contain more hydrogen.
(c) usually bypass glycolysis and the Krebs cycle.
(d) contain fewer hydrogen atoms and electrons.

44. For bacteria to continue growing rapidly when they are shifted from an environment containing O_2 to an anaerobic environment, they must
(a) produce more ATP per mole of glucose during glycolysis.
(b) produce ATP during oxidation of glucose.
(c) increase the rate of glycolysis.
(d) increase the rate of TCA cycle.

45. Protein is used as respiratory substrate only when
(a) Carbohydrates are absent
(b) Fats are absent
(c) Both carbohydrates and fats are exhausted
(d) Fats and carbohydrates are abundant

RESPONSE GRID					
	38. ⓐⓑⓒⓓ	39. ⓐⓑⓒⓓ	40. ⓐⓑⓒⓓ	41. ⓐⓑⓒⓓ	42. ⓐⓑⓒⓓ
	43. ⓐⓑⓒⓓ	44. ⓐⓑⓒⓓ	45. ⓐⓑⓒⓓ		

Space for Rough Work

DAILY PRACTICE PROBLEM DPP CHAPTERWISE 14 - BIOLOGY			
Total Questions	45	Total Marks	180
Attempted		Correct	
Incorrect		Net Score	
Cut-off Score	45	Qualifying Score	60
Success Gap = Net Score – Qualifying Score			
Net Score = (Correct × 4) – (Incorrect × 1)			

DPP - Daily Practice Problems

Chapter-wise Sheets

Date : | Start Time : | End Time :

BIOLOGY

SYLLABUS : Plant Growth and Development

Max. Marks : 180 **Marking Scheme : + 4 for correct & (–1) for incorrect** **Time : 60 min.**

INSTRUCTIONS : This Daily Practice Problem Sheet contains 45 MCQs. For each question only one option is correct. Darken the correct circle/ bubble in the Response Grid provided on each page.

1. Which one is a long day plant?
(a) Tobacco (b) *Glycine max*
(c) *Mirabilis jalapa* (d) Spinach

2. The hormone which controls cell division and cell differentiation is
(a) ABA (b) auxin
(c) gibberellin (d) cytokinin

3. Which one of the following is a long day plant?
(a) Bajra (b) Soyabean
(c) Tobacco (d) Wheat

4. A chemical believed to be involved in flowering is
(a) gibberellin (b) kinetin
(c) florigen (d) IBA

5. Which one of the following acids is a derivative of carotenoids ?
(a) Indole-3 -acetic acid (b) Gibberellic acid
(c) Abscisic acid (d) Indole butyric acid

6. Differentiation of shoot is controlled by
(a) high gibberellin : cytokinin ratio
(b) high auxin : cytokinin ratio
(c) high cytokinin : auxin ratio
(d) high gibberellin : auxin ratio

7. Which one of the following generally acts as an antagonist to gibberellins?
(a) Zeatin (b) Ethylene
(c) ABA (d) IAA

8. To avoid excessive water loss during severe drought stress, the closure of stomata is signalled by the production of
(a) IAA (b) NAA
(c) ABA (d) IBA

RESPONSE GRID					
	1. ⓐⓑⓒⓓ	2. ⓐⓑⓒⓓ	3. ⓐⓑⓒⓓ	4. ⓐⓑⓒⓓ	5. ⓐⓑⓒⓓ
	6. ⓐⓑⓒⓓ	7. ⓐⓑⓒⓓ	8. ⓐⓑⓒⓓ		

Space for Rough Work

9. In short day plants, flowering is induced by
(a) photoperiod less than 12 hours.
(b) photoperiod below a critical length and uninterrupted long night.
(c) long night.
(d) short photoperiod and interrupted long night.

10. Apical dominance in plants is caused by
(a) high concentration of auxins in the terminal bud.
(b) high concentration of gibberellins in the apical bud.
(c) high concentration of auxins in the lateral bud.
(d) absence of auxins and gibberellins in apical bud.

11. Primary precursor of IAA is
(a) phenylalanine (b) tyrosine
(c) tryptophan (d) leucine

12. In a plant growing under dark condition, the leaves turn light coloured, internodes become much elongated and it is termed as
(a) vernalization (b) etiolation
(c) phyllotaxy (d) chlorosis

13. Parthenocarpic fruit can be produced by the application of which of the following auxin?
(a) IBA (Indole butyric acid)
(b) IAA (Indole acetic acid)
(c) 2, 4-D (2-4 dichlorophenoxy acetic acid)
(d) All of the above.

14. The conditions necessary for vernalization are
(a) high temperature and water
(b) low temperature and oxygen
(c) water and carbon dioxide
(d) oxygen and water.

15. Which of the following hormones does not naturally occur in plants?
(a) 2,4–D (2, 4–dichloropheoxy acetic acid)
(b) IAA
(c) ABA
(d) GA

16. Which of the following types of phytohormones resemble the nucleic acids in some structural aspects ?
(a) Auxin (b) Cytokinin
(c) Gibberellin (d) ABA

17. Which one of the following statement regarding auxin is/are correct?
(a) IAA and IBA are natural but NAA, 2, 4-D and 2, 4, 5-T are synthetic auxins.
(b) IAA and NAA are natural but IBA, 2, 4, 5-T and 2, 4-D are synthetic auxins.
(c) NAA and 2, 4, 5-T are natural but IAA, IBA and 2, 4-D are synthetic auxins.
(d) IAA, NAA, IBA, 2, 4-D and 2, 4, 5-T are synthetic auxins.

18. Which one among the following chemicals is used for causing defoliation of forest trees?
(a) Phosphon-D
(b) Malic hydrazide
(c) 2, 4-dichlorophenoxy acetic acid
(d) Amo-1618

19. Which of the following hormone prevents ripening of fruit?
(a) Gibberellin (b) Ethylene
(c) Cytokinin (d) ABA

20. Gibberellin was first extracted from
(a) *Gibberella fujikuroi* (b) *Gelidium*
(c) *Gracelaria* (d) *Aspergillus*

21. A plant completing its life cycle before the onset of dry condition is said to be
(a) short day plant (b) long day plant
(c) drought escaping (d) All of these

22. Phytohormones are
(a) chemicals regulating flowering.
(b) chemicals regulating secondary growth.
(c) hormones regulating growth from seed to adulthood.
(d) regulators synthesized by plants and influencing physiological processes.

RESPONSE GRID					
	9. ⓐⓑⓒⓓ	10. ⓐⓑⓒⓓ	11. ⓐⓑⓒⓓ	12. ⓐⓑⓒⓓ	13. ⓐⓑⓒⓓ
	14. ⓐⓑⓒⓓ	15. ⓐⓑⓒⓓ	16. ⓐⓑⓒⓓ	17. ⓐⓑⓒⓓ	18. ⓐⓑⓒⓓ
	19. ⓐⓑⓒⓓ	20. ⓐⓑⓒⓓ	21. ⓐⓑⓒⓓ	22. ⓐⓑⓒⓓ	

Space for Rough Work

23. Which one of the following is a natural growth inhibitor?
(a) NAA (b) ABA
(c) IAA (d) GA

24. Select the incorrect statement among the following.
(a) Increase in growth per unit time is growth rate.
(b) A sigmoid growth curve is a characteristic of most living organisms in their natural environment.
(c) Rate of growth is constant during geometrical growth.
(d) Exponential phase is also called as log phase.

25. Cabbage is a biennial plant which produces flowers in second year of growth. In an attempt to make it flower in a single year, four potted plants (I, II, III and IV) of cabbage were subjected to different temperatures for several days as given in the table.

Potted plant	Temperature
I	5 °C
II	20 °C
III	30 °C
IV	25 °C

Which potted plant will show flowering?
(a) I (b) II
(c) III (d) IV

26. Etiolated seedlings are produced by germinating seeds and keeping them in total darkness. Under which of the following conditions will plants kept in the dark-begin to synthesize chlorophyll ?
(a) After being given a pulse of blue light
(b) After being given a pulse of red light
(c) After being given a pulse of red light followed by a pulse of far-red light
(d) After being given a pulse of far-red light followed by a pulse of red light

27. Which of the following may function to break dormancy in seeds ?
(a) Penetration of the seed coat
(b) Leaching of inhibitory compounds by water
(c) Exposure to fire
(d) All of the above

28. You have installed an outdoor gas burning grill on your back ratio next to your favorite camellia bush. After the first few chilly nights of using your grill you notice that your camellia, which does not normally lose its leaves, is beginning to do so. Which of the following is the best explanation for what is happening ?
(a) The bush is getting too warm next to your grill.
(b) Ethylene is a by-product of the gas you are burning and is causing senescence in your plant.
(c) Abscisic acid is a by-product of the gas you are burning and is causing senescence in your plant.
(d) The plant is a biennial and is bolting.

29. Cytokinins are known to :
(a) Inhibit cytoplasmic movement
(b) Help in retention of chlorophyll
(c) Influence water movement
(d) Promote abscission layer formation.

30. Which of the following schemes best represents how environmental cues are transduced to changes in a plant ?
(a) Receptors receive environmental cues, a signal transduction pathway is initiated, there is an alteration in the particular genes that are transcribed and translated and a cellular response is generated.
(b) Receptors are triggered, hormones are released, signal transduction pathways are initiated, there is an alteration in expression of genes and a cellular response is generated.
(c) Neither a nor b
(d) Both a and b

31. When a plant is not reproducing most of the cytokinins are produced in its :
(a) Lateral buds (b) Shoot apex
(c) Roots (d) Leaves

RESPONSE GRID					
	23. ⓐⓑⓒⓓ	24. ⓐⓑⓒⓓ	25. ⓐⓑⓒⓓ	26. ⓐⓑⓒⓓ	27. ⓐⓑⓒⓓ
	28. ⓐⓑⓒⓓ	29. ⓐⓑⓒⓓ	30. ⓐⓑⓒⓓ	31. ⓐⓑⓒⓓ	

Space for Rough Work

32. Which of the following light receptors is responsible for absorbing blue and ultraviolet light ?
(a) Phytochrome P_r (b) Photochrome P_{fr}
(c) Cryptochrome (c) Phototropin

33. Phototropic and geotropic movements in plants have been traced to be linked with:
(a) Enzymes (b) Starch
(c) Gibberellins (d) Auxins

34. The natural plant hormones were first isolated from :
(a) Corn germoil and human urine
(b) Cotton fruits, spinach leaves and rice plants
(c) Human urine and rice seedlings
(d) Spinach leaves and fungus Gibberella

35. Auxin regulates cell growth by which of the following mechanisms ?
(a) Altering the elasticity of cell walls
(b) Altering the plasticity of cell walls
(c) Synthesizing new cell walls
(d) Breaking down cell walls in growing cells

36. Which of the following were not observed in studies done on certain dwarf strains of plants ?
(a) Applications of extracts of normal strains promoted growth of dwarf strains.
(b) Dwarf strains grew normally if additional fertilizer was applied
(c) Application of gibberellin A1 promoted growth of dwarf strains
(d) Gibberellin caused little additional growth of normal strains.

37. The closure of lid of pitcher in pitcher plant is :
(a) A paratonic movement
(b) A tropic movement
(c) A turgor movement
(d) An autonomous movement

38. Which of the following instrument can be used to record plant growth by seconds?
(a) Arc auxanometer (b) Arc indicator
(c) Space marker disc (d) Crescograph

39. *Avena* coleoptile test to find out the quantity of growth promoting hormones was discovered by –
(a) F.W. Went (b) L.J. Oudus
(c) K.V. Thimann (d) F. Skoog

40. "Foolish seedling" disease of rice led to the discovery of
(a) ABA (b) 2,4-D
(c) IAA (d) GA

41. Which of the growth substance act as a stimulant during nodule formation in Leguminous plant ?
(a) Ethylene (b) ABA
(c) IAA (d) Morphactin

42. Indole-3 acetic acid called as auxin was first isolated from
(a) Human urine (b) Corn germ oil
(c) *Fusarium* (d) *Rhizopus*

43. Prunning of plants promotes branching, because the axillary buds get sensitized to –
(a) Ethylene (b) Gibberellin
(c) Cytokinin (d) IAA

44. Pineapple can be made to flower in off season by :
(a) Ethylene
(b) Zeatin
(c) Napthalene Acetic Acid (NAA)
(d) Temperature

45. Choose the wrongly matched pair from the following
(a) Auxins – "to grow"
(b) Gibberellins – "*Gibberella fujikuroi*"
(c) Cytokinins – Herring sperm DNA
(d) Abscisic acid – Flowering hormone

RESPONSE GRID					
	32. ⓐⓑⓒⓓ	33. ⓐⓑⓒⓓ	34. ⓐⓑⓒⓓ	35. ⓐⓑⓒⓓ	36. ⓐⓑⓒⓓ
	37. ⓐⓑⓒⓓ	38. ⓐⓑⓒⓓ	39. ⓐⓑⓒⓓ	40. ⓐⓑⓒⓓ	41. ⓐⓑⓒⓓ
	42. ⓐⓑⓒⓓ	43. ⓐⓑⓒⓓ	44. ⓐⓑⓒⓓ	45. ⓐⓑⓒⓓ	

Space for Rough Work

DAILY PRACTICE PROBLEM DPP CHAPTERWISE 15 - BIOLOGY			
Total Questions	45	Total Marks	180
Attempted		Correct	
Incorrect		Net Score	
Cut-off Score	45	Qualifying Score	60
Success Gap = Net Score – Qualifying Score			
Net Score = (Correct × 4) – (Incorrect × 1)			

DPP - Daily Practice Problems

Chapter-wise Sheets

Date : Start Time : End Time :

BIOLOGY CB16

SYLLABUS : Digestion and Absorption

Max. Marks : 180 **Marking Scheme : + 4 for correct & (–1) for incorrect** **Time : 60 min.**

INSTRUCTIONS : This Daily Practice Problem Sheet contains 45 MCQs. For each question only one option is correct. Darken the correct circle/ bubble in the Response Grid provided on each page.

1. Dark purplish gland lying on the left side of abdomen is called :
   (a) liver (b) spleen
   (c) gall bladder (d) appendix
2. Which of the following are water soluble vitamins?
   (a) B, C (b) A, D, K
   (c) A, D, E, K (d) K, B, A, E
3. Jaundice is a disorder of
   (a) Excretory system (b) Skin and eyes
   (c) Digestive system (d) Circulatory system
4. Which one of the following pairs of food components in humans reaches the stomach totally undigested ?
   (a) Starch and cellulose (b) Protein and starch
   (c) Starch and fat (d) Fat and cellulose
5. Which of following teeth are lophodont?
   (a) Incisor and canine (b) Premolar and molar
   (c) Canine and premolar (d) Premolar and incisor
6. Oxyntic cells are located in
   (a) Islets of Langerhans
   (b) Gastric epithelium and secrete pepsin
   (c) Kidneys and secrete renin
   (d) Gastric epithelium and secrete HCl
7. Continued consumption of a diet rich in butter, red meat and eggs for a long period may lead to
   (a) vitamin A toxicity
   (b) kidney stones
   (c) hypercholesterolemia
   (d) urine laden with ketone bodies
8. Which one of the following is a matching pair of a substrate and its particular digestive enzyme?
   (a) Starch — maltase (b) Lactose — rennin
   (c) Maltose — steapsin (d) Casein — chymotrypsin

RESPONSE GRID					
	1. ⓐⓑⓒⓓ	2. ⓐⓑⓒⓓ	3. ⓐⓑⓒⓓ	4. ⓐⓑⓒⓓ	5. ⓐⓑⓒⓓ
	6. ⓐⓑⓒⓓ	7. ⓐⓑⓒⓓ	8. ⓐⓑⓒⓓ		

Space for Rough Work

9. Which one of the following is a matching pair of a vitamin and the deficiency disease related with it ?
(a) Riboflavin — beri beri
(b) Thiamine — xerophthalmia
(c) Niacin — pellagra
(d) Calciferol — scurvy

10. Which one of the following pairs of the cells with their secretion is **correctly** matched?
(a) Oxyntic cells - A secretion with pH between 2.0 and 3.0.
(b) Alpha cells of Islets of Langerhans -Secretion that decreases blood sugar level.
(c) Kupffer cells - A digestive enzyme that hydrolysis nucleic acids.
(d) Sebaceous glands - A digestive enzyme that hydrolysis nucleic acids

11. "Kwashiorkar' and 'Beri-Beri' are
(a) communicable diseases
(b) infectious diseases
(c) deficiency diseases
(d) None of the above

12. First milk produced after child birth is called
(a) sebum (b) cerumen
(c) true milk (d) colostrum

13. Cirrhosis of liver is caused by the chronic intake of
(a) Opium
(b) Alcohol
(c) Tobacco (Chewing)
(d) Cocaine

14. The sphincter of Oddi is present between
(a) Oesophagus and cardiac stomach
(b) Pyloric stomach and duodenum
(c) Hepatic duct and cystic duct
(d) Hepatopancreatic duct and duodenum

15. The structure which prevents the entry of food into the windpipe is
(a) Gullet (b) Glottis
(c) Tonsil (d) Epiglottis

16. Layer of cells that secrete enamel of tooth is
(a) Osteoblast (b) Ameloblast
(c) Odontoblast (d) Dentoblast

17. Cow's milk is slightly yellowish in colour due to the presence of
(a) Carotene
(b) Riboflavin
(c) Xanthophyll
(d) Xanthophyll and Carotene

18. Accessory excretory organs of man are
(a) only skin
(b) only skin and liver
(c) only skin and lungs
(d) skin, lungs, liver and intestine

19. Cholecystokinin is secretion of
(a) Duodenum that causes contraction of gall bladder
(b) Globlet cells of ileum, stimulates secretion of succus entericus
(c) Liver and controls secondary sex characters
(d) Stomach that stimulates pancreas to release juice

20. Which of the following sets represents vestigial organs?
(a) Vermiform appendix, body hair and patella
(b) Wisdom teeth, body hair and atlas vertebra
(c) Ear muscles, cochlea and coccyx
(d) Vermiform appendix, ear muscles and coccyx.

21. Which part of the alimentary canal does not secrete any enzyme?
(a) Mouth (b) Oesophagus
(c) Stomach (d) Duodenum

22. The protein coated, water soluble fat globules are called
(a) Chylomicrons (b) Micelles
(c) Chyle (d) Monoglycerides

23. Where do certain symbiotic micro-organisms normally occur in human body?
(a) Caecum
(b) Oral lining and tongue surface
(c) Vermiform appendix and rectum
(d) Duodenum

RESPONSE GRID					
	9. ⓐⓑⓒⓓ	10. ⓐⓑⓒⓓ	11. ⓐⓑⓒⓓ	12. ⓐⓑⓒⓓ	13. ⓐⓑⓒⓓ
	14. ⓐⓑⓒⓓ	15. ⓐⓑⓒⓓ	16. ⓐⓑⓒⓓ	17. ⓐⓑⓒⓓ	18. ⓐⓑⓒⓓ
	19. ⓐⓑⓒⓓ	20. ⓐⓑⓒⓓ	21. ⓐⓑⓒⓓ	22. ⓐⓑⓒⓓ	23. ⓐⓑⓒⓓ

Space for Rough Work

24. Both the crown and root of a tooth is covered by a layer of bony hard substance. It is called
(a) enamel (b) dentine
(c) bony socket (d) cementum

25. Which one of the following pairs of the kind of cells and their secretion is correctly matched?
(a) Oxyntic cells—a secretion with pH between 2.0 and 3.0
(b) Alpha cells of islets of Langerhans—secretion that decreases blood sugar level
(c) Kupffer cell—a digestive enzyme that hydrolyses nucleic acids
(d) Sebaceous glands—a secretion that evaporates for cooling

26. Which one of the following is the correct matching of the site of action on the given substate, the enzyme acting upon it and the end product?
(a) Small intestine : Proteins $\xrightarrow{\text{Pepsin}}$ Amino acids
(b) Stomach : Fats $\xrightarrow{\text{Lipase}}$ Micelles
(c) Duodenum : Triglycerides $\xrightarrow{\text{Trypsin}}$ Monoglycerides
(d) Small intestine : Starch Disaccharide (maltose)

27. Read the following four statements (i) - (iv) with certain mistakes in two of them.
(i) Fructose is generally absorbed by simple diffusion.
(ii) The digestive wastes, solidified into coherent faeces in the rectum initiate and endocrinal action causing an urge or desire for its removal
(iii) The food mixes thoroughly with the acidic gastric juice of the stomach by the churning movements of its muscular wall and is called the chyme.
(iv) The secretions of the brush border cells of the mucosa alongwith the secretions of the goblet cells constitute the succus entericus.

28. Which of the following dietary deficiencies will affect the functioning of enzymes in the human body most quickly?
(a) Too few carbohydrates
(b) Shortage of fat-soluble vitamins
(c) Shortage of water-soluble vitamins
(d) Insufficient variety of fats

29. Pancreatic cancer is an especially dangerous disease in people because the pancreas is
(a) the organ that produces and stores bile.
(b) the site of synthesis for all of the essential amino acids.
(c) one of the organs through which food must pass on its way to the colon.
(d) an organ in which many different kinds of digestive enzymes are manufacture

30. How does a gastrovascular cavity differ from an alimentary canal? The gastrovascular cavity
(a) stores food but does not digest it.
(b) is usually much larger.
(c) has only one opening.
(d) functions in digestion but not absorption.

31. The largest variety of digestive enzymes function in the
(a) large intestine. (b) oral cavity.
(c) stomach. (d) small intestine.

32. Certain amino acids are essential to the diet of animals because
(a) they prevent overnourishment
(b) they are cofactors and coenzymes that are required for normal physiological function.
(c) an animal cannot directly synthesize them through the transfer of an amino group to an appropriate carbon skeleton.
(d) animals need these substances in order to make the stored fats that are used during hibernation and migration

RESPONSE GRID					
	24. ⓐⓑⓒⓓ	25. ⓐⓑⓒⓓ	26. ⓐⓑⓒⓓ	27. ⓐⓑⓒⓓ	28. ⓐⓑⓒⓓ
	29. ⓐⓑⓒⓓ	30. ⓐⓑⓒⓓ	31. ⓐⓑⓒⓓ	32. ⓐⓑⓒⓓ	

Space for Rough Work

33. Protection of the walls of the stomach against the action of its own digestive juices
(a) results from the presence of an anti-enzyme chemical formed by the gastric glands
(b) results from the nervous reactions of the lining of the stomach.
(c) is controlled by a centre in the medulla of the brain.
(d) results from the neutralizing, buffering, and a coating mucus covering its inner surface

34. Most of the chemical digestion of food in humans is completed in the
(a) small intestine. (b) appendix.
(c) ascending colon (d) stomach

35. Which of the following is characteristic of the large intestine?
(a) It has almost no bacterial populations.
(b) It contains chyme.
(c) It absorbs much of the water remaining in waste materials.
(d) It is the site of most of digestion.

36. The innermost layer of the digestive tract is the
(a) serosa membrane.
(b) mucosa membrane.
(c) submucosa membrane.
(d) lumen.

37. The ______ is primarily a storage chamber within the digestive system, while the ______ reabsorbs water, ions, and generates the faeces.
(a) buccal cavity; midgut
(b) crop; midgut
(c) stomach; hindgut
(d) buccal cavity; hindgut

38. What does an increase in the secretion of insulin produce?
(a) a decrease in glucose metabolism
(b) a decrease in glucose permeability of cells
(c) an increase in blood sugar level
(d) an increase in glucose permeability of cells

39. Which function of the liver results in the production of bile pigments?
(a) breakdown of haemoglobin
(b) deamination of amino acids
(c) detoxification of metabolic poisons
(d) release of stored vitamin A

40. The centre of hunger or centre which regulates the amount of food we eat or out appetite is located in –
(a) Medulla (b) Cerebrum
(c) Hypothalamus (d) Alimentary canal

41. Fatty acid and glycerol are first taken up from alimentary canal by
(a) Villi (b) Blood capillaries
(c) Hepatic portal vein (d) Lymph vessels

42. Mammals may drink water and also get it from
(a) Breakdown of glycogen into glucose
(b) Secretion of saliva
(c) Oxidation of glucose
(d) Conversion of oxyhaemoglobin into haemoglobin

43. When a piece of bread in chewed it tastes sweet because
(a) The sugar contents are drawn out
(b) Saliva converts starch into maltose
(c) It does not taste sweet
(d) The taste buds are stimulated by chewing

44. Anxiety and eating spicy food together in an otherwise normal human, may lead to
(a) Indigestion (b) Jaundice
(c) Diarrhoea (d) Vomiting

45. A young infant may be feeding entirely on mothers milk which is white in colour but the stools which the infant passes out is quite yellowish. What is this yellow colour due to ?
(a) bile pigments passed through bile juice
(b) undigested milk protein casein
(c) pancreatic juice poured into duodenum
(d) intestinal juice

RESPONSE GRID					
	33. ⓐⓑⓒⓓ	34. ⓐⓑⓒⓓ	35. ⓐⓑⓒⓓ	36. ⓐⓑⓒⓓ	37. ⓐⓑⓒⓓ
	38. ⓐⓑⓒⓓ	39. ⓐⓑⓒⓓ	40. ⓐⓑⓒⓓ	41. ⓐⓑⓒⓓ	42. ⓐⓑⓒⓓ
	43. ⓐⓑⓒⓓ	44. ⓐⓑⓒⓓ	45. ⓐⓑⓒⓓ		

Space for Rough Work

DAILY PRACTICE PROBLEM DPP CHAPTERWISE 16 - BIOLOGY			
Total Questions	45	Total Marks	180
Attempted		Correct	
Incorrect		Net Score	
Cut-off Score	45	Qualifying Score	60
Success Gap = Net Score – Qualifying Score			
Net Score = (Correct × 4) – (Incorrect × 1)			

DPP - Daily Practice Problems

Chapter-wise Sheets

Date : | Start Time : | End Time :

BIOLOGY CB17

SYLLABUS : Breathing and Exchange of Gases

Max. Marks : 180 | Marking Scheme : + 4 for correct & (–1) for incorrect | Time : 60 min.

INSTRUCTIONS : This Daily Practice Problem Sheet contains 45 MCQs. For each question only one option is correct. Darken the correct circle/ bubble in the Response Grid provided on each page.

1. Bulk of carbon dioxide (CO_2) released from body tissues into the blood is present as
   (a) bicarbonate in blood plasma and RBCs
   (b) free CO_2 in blood plasma
   (c) 70% carbamino- haemoglobin and 30% as bicarbonate
   (d) carbamino-haemoglobin in RBCs
2. A large proportion of oxygen is left unused in the human blood even after its uptake by the body tissues. This O_2
   (a) acts as a reserve during muscular exercise
   (b) raise the pCO_2 of blood to 75 mm of Hg.
   (c) is enough to keep oxyhaemoglobin saturation at 96%
   (d) helps in releasing more O_2 to the epithelial tissues.
3. Oxygen dissociation curve of haemoglobin is
   (a) Sigmoid (b) Hyperbolic
   (c) Linear (d) Hypobolic
4. Which one of the following organs in the human body is most affected due to shortage of oxygen?
   (a) Intestine (b) Skin (c) Kidney (d) Brain
5. When CO_2 concentration in blood increases, breathing becomes
   (a) shallower and slow
   (b) there is no effect on breathing
   (c) slow and deep
   (d) faster and deeper
6. Although much CO_2 is carried in blood, yet blood does not become acidic, because
   (a) it is absorbed by the leucocytes
   (b) blood buffers play an important role in CO_2 transport.
   (c) it combines with water to form carbonic acid (H_2CO_3) which is neutralized by Na_2CO_3.
   (d) it is continuously diffused through tissues and is not allowed to accumulate.
7. Which one of the following is the correct statement for respiration in humans?
   (a) Workers in grinding and stone-breaking industries may suffer from lung fibrosis.
   (b) About 90% of carbon dioxide (CO_2) is carried by haemoglobin as carbaminohaemoglobin.
   (c) Cigarette smoking may lead to inflammation of bronchi.
   (d) Neural signals from pneumotoxic centre in pons region of brain can increase the duration of inspiration.
8. The largest quantity of air that can be expired after a maximal inspiratory effort is called
   (a) residual volume (b) tidal volume
   (c) vital capacity (d) total lung volume

RESPONSE GRID					
	1. ⓐⓑⓒⓓ	2. ⓐⓑⓒⓓ	3. ⓐⓑⓒⓓ	4. ⓐⓑⓒⓓ	5. ⓐⓑⓒⓓ
	6. ⓐⓑⓒⓓ	7. ⓐⓑⓒⓓ	8. ⓐⓑⓒⓓ		

Space for Rough Work

9. Which one of the following statements is incorrect ?
   (a) The presence of non-respiratory air sacs, increases the efficiency of respiration in birds.
   (b) In insects, circulating body fluids serve to distribute oxygen to tissues.
   (c) The principle of countercurrent flow facilitates efficient respiration in gills of fishes.
   (d) The residual air in lungs slightly decreases the efficiency of respriration in mammals.
10. During oxygen transport the oxyhaemoglobin at the tissue level liberates oxygen to the cells because in tissue
   (a) O_2 concentration is high and CO_2 is low
   (b) O_2 concentration is low and CO_2 is high
   (c) O_2 tension is high and CO_2 tension is low
   (d) O_2 tension is low and CO_2 tension is high
11. Emphysema developes mainly because of
   (a) Allergy or hypersensitisation
   (b) Spasm of the smooth muscles of bronchioles
   (c) Cigarette smoking
   (d) Inflammation of the alveoli
12. During inspiration the diaphragm
   (a) relaxes to become dome-shaped
   (b) contracts and flattens
   (c) expands
   (d) shows no change
13. The volume of air breathed in and out during normal breathing is called
   (a) Vital capacity
   (b) Inspiratory reserve volume
   (c) Explratory reserve volume
   (d) Tidal volume
14. What would happen if human blood becomes acidic (low pH)
   (a) Oxygen carying capacity of haemoglobin increases
   (b) Oxygen carrying capacity of haemoglobin decreases
   (c) RBCs count increases
   (d) RBCs count decreases
15. The respiratory centre, which regulates respiration, is located in
   (a) Cerebral peduncle (b) Vagus nerve
   (c) Pons (d) Medulla oblongata
16. Even when there is no air in it, human trachea does not collapse due to presence of
   (a) Bony rings (b) Turgid pressure
   (c) Chitinous rings (d) Cartilaginous rings
17. The structure which does not contribute to the breathing movements in mammals is
   (a) Larynx (b) Ribs
   (c) Diaphragm (d) Intercostal muscles
18. In emphysema –
   (a) Gas exchange area of lungs is reduced
   (b) Gas exchange area of lungs is increased
   (c) Trachea gets narrowed
   (d) Larynx is permanently closed
19. Which of the following changes occur in diaphragm and intercostal muscles when expiration of air takes place?
   (a) External intercostal muscles relax and diaphargm contracts
   (b) External intercostal muscles contract and diaphargm relaxes
   (c) External intercostal muscles and diaphargm relax
   (d) External intercostal muscles and diaphargm contract
20. The graph below shows how gaseous exchange in an amphibian varies with temperature.

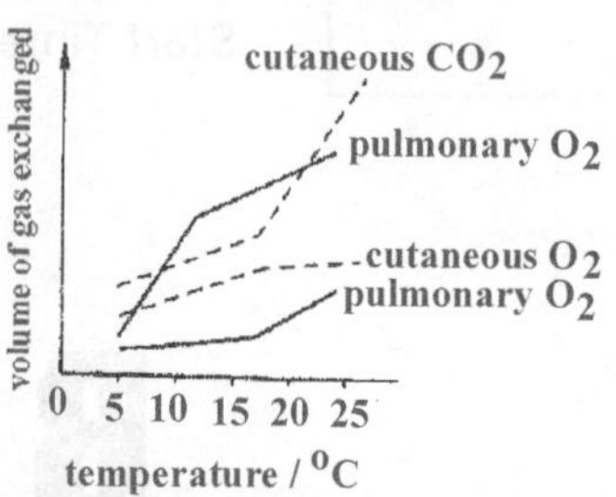

The graph shows that at all temperatures the lungs :
   (a) absorb more oxygen than the skin
   (b) absorb less oxygen than the skin
   (c) release more carbon dioxide than the skin
   (d) release less carbon dioxide than the skin
21. Which statement explains the significance of myoglobin in muscle tissue?
   (a) Myoglobin can associate with oxygen in conditions of high carbon dioxide tension
   (b) Myoglobin can dissociate from oxygen most readily under aerobic conditions
   (c) Myoglobin acts as an oxygen store delaying the onset of anaerobic respiration
   (d) Myoglobin acts as a buffer to hydrogen ions produced during anaerobic respiration
22. When you exhale, the diaphragm
   (a) relaxes and arches.
   (b) relaxes and flattens.
   (c) contracts and arches.
   (d) contracts and flattens.
23. Smoking destroys the cilia in the respiratory passageways. This
   (a) makes it harder to move air in and out of the lungs.
   (b) decreases the surface area for respiration.
   (c) slows blood flow through lung blood vessels.
   (d) makes it harder to keep the lungs clean.

RESPONSE GRID					
	9. ⓐⓑⓒⓓ	10. ⓐⓑⓒⓓ	11. ⓐⓑⓒⓓ	12. ⓐⓑⓒⓓ	13. ⓐⓑⓒⓓ
	14. ⓐⓑⓒⓓ	15. ⓐⓑⓒⓓ	16. ⓐⓑⓒⓓ	17. ⓐⓑⓒⓓ	18. ⓐⓑⓒⓓ
	19. ⓐⓑⓒⓓ	20. ⓐⓑⓒⓓ	21. ⓐⓑⓒⓓ	22. ⓐⓑⓒⓓ	23. ⓐⓑⓒⓓ

Space for Rough Work

24. The graph shows dissociation curves for four respiratory pigments.

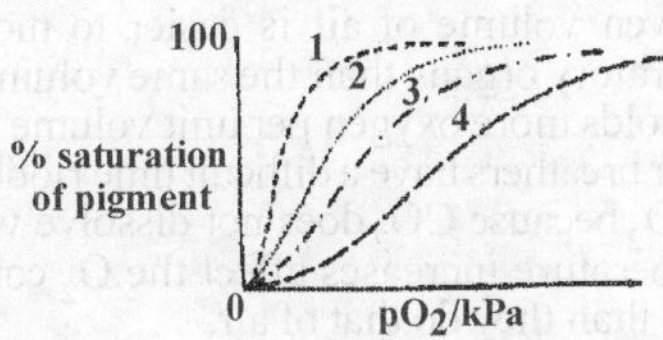

Which combination identifies the four curve?

	adult haemoglobin with high CO_2	adult haemoglobin	myo-globin	foetal haemoglobin
(a)	1	2	3	4
(b)	2	1	4	3
(c)	3	2	1	4
(d)	3	4	2	1

25. Gas flows into the lungs of mammals during inspiration because
(a) The pressure in the lungs falls below atmospheric pressure.
(b) The volume of the lungs decreases.
(c) The pressure in the lungs rises above atmospheric pressure.
(d) The diaphragm moves upward toward the lungs.

26. The Bohr shift describes
(a) The outward movement of Cl^- from the blood cell in exchange for HCO_3^- moving into the cell.
(b) The leftward shift of the entire oxygen equilibrium curve when temperature rises.
(c) The rightward shift of the entire oxygen equilibrium curve when pH rises.
(d) The rightward shift of the entire oxygen equilibrium curve when pH falls.

27. The movement of O_2 and CO_2 between the blood in the tissue capillaries and the cells in tissues depends most directly upon
(a) active transport of O_2 and CO_2.
(b) total atmospheric (barometric) pressure differences across the cell membranes.
(c) diffusion of O_2 and CO_2 down a concentration gradient.
(d) diffusion of O_2 and CO_2 down a partial pressure gradient.

28. The function of the mucus elevator in the mammalian respiratory system is to
(a) move surfactant from the bronchi to the alveoli.
(b) produce negative pressure during inhalation.
(c) stimulate contraction of rib musculature during forcible exhalation.
(d) trap and remove particulate matter that has entered the respiratory system.

29. The oxygen-binding curve for myoglobin is steep and shifted to the left of the same curve for hemoglobin. This indicates that
(a) people with myoglobin are acclimated for life at high altitudes.
(b) myoglobin occurs in high concentration in the human foetus.
(c) myoglobin does not function well at high percent saturation of oxygen.
(d) myoglobin picks up and releases O_2 at lower Po_2 values than does haemoglobin.

30. The graph shows haemoglobin-oxygen dissociation curves at two different partial pressures of carbon dioxide

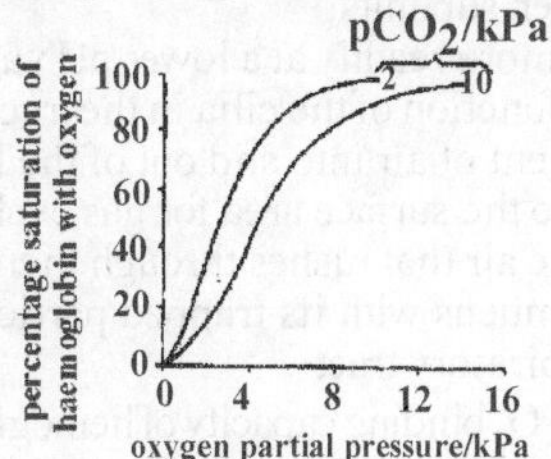

What is the advantage to the organism of the difference between the curve?
(a) CO_2 is readily released into the alveoli
(b) Haemoglobin removes more CO_2 from active tissues
(c) Haemoglobin can release oxygen to myoglobin
(d) Oxygen is readily released into active tissues

31. A person can hold his breath longer if he gradually exhales than if he keeps his lungs fully expanded. This phenomenon is due to the fact that :
(a) concentration of CO_2 in the blood decreases, lessening the excitatory input to the carotid bodies
(b) he is then using less O_2 as his thoracic muscles are not working as hard
(c) stretch receptors in his lungs are then not firing as much thus lessening the excitatory input to the expiratory centres
(d) he is preventing the pressure in his lungs from increasing too much

32. The alveoli contain specialized cells that secrete a phospholipoprotein complex known as pulmonary surfactant. The function of pulmonary surfactant is :
(a) to provide some rigidity to the alveoli to prevent lung collapse
(b) to decrease the surface tension of the alveoli, making it easier to expand them
(c) to facilitate the diffusion of O_2 and CO_2 by providing protein channels through which these gases flow
(d) to provide a sticky surface upon which dust and microbes are trapped and disposed of.

RESPONSE GRID					
	24. ⓐⓑⓒⓓ	25. ⓐⓑⓒⓓ	26. ⓐⓑⓒⓓ	27. ⓐⓑⓒⓓ	28. ⓐⓑⓒⓓ
	29. ⓐⓑⓒⓓ	30. ⓐⓑⓒⓓ	31. ⓐⓑⓒⓓ	32. ⓐⓑⓒⓓ	

Space for Rough Work

33. During inspiration, the air is brought into the lungs by the working of the thoracoabdominal pump whereby the :
(a) diaphragm contracts to lengthen the thoracic cavity
(b) chest muscles pull the ribs up and out to widen the thoracic cavity
(c) abdominal muscles relax
(d) all of the above

34. The binding of an O_2 atom to the first iron atom of a haemoglobin molecule
(a) occurs at a very low partial pressure of oxygen
(b) produces a conformational change that lowers the other subunits' affinity for oxygen
(c) is an example of cooperativity as O_2 readily binds to the other subunits.
(d) occurs more readily at a lower pH value.

35. What is the function of the cilia in the trachea and bronchi?
(a) movement of air into and out of the lungs
(b) increase the surface area for gas exchange
(c) filter the air that rushes through them
(d) sweep mucus with its trapped particles up and out of the respiratory tract

36. A shift in the O_2 binding capacity of hemoglobin as a function of pH is known as
(a) oxygen binding
(b) the Hering-Breuer reflex
(c) the Bohr effect.
(d) respiratory ventilation

37. The most important function of diaphragm is
(a) To divide body cavity into compartments
(b) To protect lungs
(c) To aid in digestion
(d) To aid in ventilation

38. Which is not a structure of the respiratory system ?
(a) the pharynx (b) the bronchus
(c) the larynx (d) the hyoid

39. The alveoli of the lungs do not contain "air" because
(a) we normally do not ventilate our lungs at a high enough rate.
(b) the lungs have too many alveoli to ventilate.
(c) there is "dead space" in the trachea and bronchi.
(d) the trachea and bronchi are too small in volume.

40. Which of the following is not a reason that breathing air is easier than breathing water?
(a) A given volume of air is easier to move across the respiratory organs than the same volume of water.
(b) Air holds more oxygen per unit volume than water.
(c) Water breathers have a difficult time ridding themselves of CO_2 because CO_2 does not dissolve well in water.
(d) Temperature increases affect the O_2 content of water more than they do that of air.

41. Which of the following represents a larger volume of air than is normally found in the resting tidal volume of a human lung ?
(a) Residual volume
(b) Inspiratory reserve volume
(c) Expiratory reserve volume
(d) All of the above

42. The graph shows four dissociation curves.

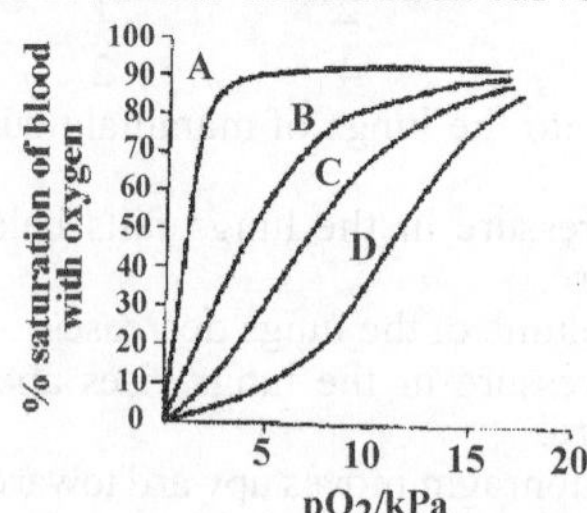

Which curve represents the oxygen dissociation curve for myoglobin?
(a) At A (b) At B (c) At C (d) At D

43. Functional residual capacity can be represented as
(a) TV+ERV (b) ERV+RV
(c) RV+IRV (d) ERV+TV+IRV

44. The process of migration of chloride ions from plasma to RBC and carbonate ions from RBC to plasma is
(a) chloride shift (b) ionic shift
(c) atomic shift (d) Na^+ pump

45. The quantity 1500 ml in the respiratory volumes of a normal human adult refers to
(a) maximum air that can be breathed in and breathed out
(b) residual volume
(c) expiratory reserve volume
(d) total lung capacity

RESPONSE GRID					
	33. ⓐⓑⓒⓓ	34. ⓐⓑⓒⓓ	35. ⓐⓑⓒⓓ	36. ⓐⓑⓒⓓ	37. ⓐⓑⓒⓓ
	38. ⓐⓑⓒⓓ	39. ⓐⓑⓒⓓ	40. ⓐⓑⓒⓓ	41. ⓐⓑⓒⓓ	42. ⓐⓑⓒⓓ
	43. ⓐⓑⓒⓓ	44. ⓐⓑⓒⓓ	45. ⓐⓑⓒⓓ		

Space for Rough Work

DAILY PRACTICE PROBLEM DPP CHAPTERWISE 17 - BIOLOGY			
Total Questions	45	Total Marks	180
Attempted		Correct	
Incorrect		Net Score	
Cut-off Score	50	Qualifying Score	65
Success Gap = Net Score – Qualifying Score			
Net Score = (Correct × 4) – (Incorrect × 1)			

DPP - Daily Practice Problems

Chapter-wise Sheets

Date : | **Start Time :** | **End Time :**

BIOLOGY CB18

SYLLABUS : Body Fluids and Circulation

Max. Marks : 180 **Marking Scheme : + 4 for correct & (–1) for incorrect** **Time : 60 min.**

INSTRUCTIONS : This Daily Practice Problem Sheet contains 45 MCQs. For each question only one option is correct. Darken the correct circle/ bubble in the Response Grid provided on each page.

1. Blood pressure is measured by
(a) sphygmomanometer (b) stethoscope
(c) electrocardiogram (d) phonocardiogram

2. Coronary artery disease (CAD) is often referred to as
(a) Heart failure (b) Cardiac arrest
(c) Atherosclerosis (d) Thrombosis

3. Which one of the following is a **correct** matching pair?
(a) Lubb - Sharp closure of AV valves at the beginning of ventricular systole.
(b) Dup - Sudden opening of semilunar valves at the beginning of ventricular diastole.
(c) Pulsation of the radial artery valves in the blood vessels.
(d) Purkinje fibers - Initiation of the heart beat.

4. Child death may occur in the marriage between
(a) Rh^+ man and Rh^+ woman
(b) Rh^+ man and Rh^- woman
(c) Rh^- man and Rh^- woman
(d) Rh^- man and Rh^+ woman

5. Heart is covered by
(a) Peritoneum (b) Pleural membrane
(c) Pericardium (d) Visceral membrane

6. Which one of the components of ECG in human is correctly interpreted below ?
(a) Complex QRS-One complete Pulse
(b) Peak T - Initiation of total cardiac contraction
(c) Peak P and Peak R together-Systolic and diastolic blood pressures
(d) Peak P- Initiation of left atrial contraction only

7. Pacemaker of heart is
(a) AV node (b) Bundle of His
(c) SA node (d) Purkinje fibres

8. Uricotelism is found in
(a) Frogs and toads
(b) Mammals and birds
(c) Birds, reptiles and insects
(d) Fishes and fresh water protozoans

RESPONSE GRID					
	1. ⓐⓑⓒⓓ	2. ⓐⓑⓒⓓ	3. ⓐⓑⓒⓓ	4. ⓐⓑⓒⓓ	5. ⓐⓑⓒⓓ
	6. ⓐⓑⓒⓓ	7. ⓐⓑⓒⓓ	8. ⓐⓑⓒⓓ		

Space for Rough Work

9. Haemoglobin is having maximum affinity with:
(a) NH_3 (b) O_2
(c) CO_2 (d) CO

10. An adult human with average health has systolic and diastolic pressures as
(a) 80mm Hg and 80 mm Hg
(b) 70 mm Hg and 120 mm Hg
(c) 120mm Hg and 80 mm Hg
(d) 50 mm Hg and 80 mm Hg

11. Blood group AB has
(a) no antigen
(b) no antibody
(c) neither antigen nor antibody
(d) both antigen and antibody

12. With reference to the blood in a normal person, which one of the following statements is correct?
(a) Compared to arteries, veins are less numerous and hold less of the body's blood at any given time.
(b) Blood cells constitute about 70 percent of the total volume of the blood.
(c) White blood cells (WBC) are made by lymph nodes only.
(d) The blood has more platelets than WBC.

13. Thickening of arteries due to cholesterol deposition is
(a) arteriosclerosis (b) rheumatic heart
(c) blood pressure (d) cardiac arrest

14. Hepatic portal system carries
(a) Oxygenated blood from liver to intestine
(b) Deoxygenated blood from liver to intestine
(c) Oxygenated blood from intestine to liver
(d) Deoxygenated blood from intestine to liver

15. 'Bundle of His' is a part of which one of the following organs in humans?
(a) Brain (b) Heart
(c) Kidney (d) Pancreas

16. Chordae tendinae are found in
(a) joints of legs (b) atria of heart
(c) ventricles of brain (d) ventricles of heart

17. The pattern of contraction and relaxation of the heart is referred to as
(a) blood pressure (b) arterial flow
(c) blood flow (d) cardiac cycle

18. Which of the following statements are true ?
(i) The blood transports CO_2 comparatively easily because of its higher solubility.
(ii) Approximately 8-9% of CO_2 is transported being dissolved in the plasma of blood.
(iii) The carbon dioxide produced by the tissues, diffuses passively into the blood stream and passes into red blood corpuscles and react with water to form H_2CO_3.
(iv) The oxyhaemoglobin (HbO_2) of the erythrocytes is basic.
(v) The chloride ions diffuse from plasma into the erythrocytes to maintain ionic balance.
(a) (i), (iii) and (v) are true, (ii) and (iv) are false
(b) (i), (iii) and (v) are false, (ii) and (iv) are true
(c) (i), (ii) and (iv) are true, (iii) and (v) are false
(d) (i), (ii) and (iv) are false, (iii) and (v) are true

19. About 70% of CO_2 is transported as
(a) Carbonic acid (b) Carboxyhaemoglobin
(c) Bicarbonates (d) Carbamino compounds

20. In human heart, which of the following valve remains in contact of the oxygenated blood only?
(a) Tricuspid valve (b) Semi lunar valve
(c) Eustachian valve (d) Mitral valve

21. Myocardial Infarction is caused by
(a) hardening of arteries
(b) lumpy thickness develop in the inner walls of arteries
(c) clot may occur in the lumen of a coronary artery
(d) sudden interruption in blood flow towards a portion of heart

22. The affinity of CO with Hb is more than oxygen by
(a) 2 times (b) 20 times
(c) 200 times (d) 2000 times

23. Life span of human RBCs is
(a) 120 days (b) 20 days
(c) 9 days (d) 90 days

RESPONSE GRID					
	9. ⓐⓑⓒⓓ	10. ⓐⓑⓒⓓ	11. ⓐⓑⓒⓓ	12. ⓐⓑⓒⓓ	13. ⓐⓑⓒⓓ
	14. ⓐⓑⓒⓓ	15. ⓐⓑⓒⓓ	16. ⓐⓑⓒⓓ	17. ⓐⓑⓒⓓ	18. ⓐⓑⓒⓓ
	19. ⓐⓑⓒⓓ	20. ⓐⓑⓒⓓ	21. ⓐⓑⓒⓓ	22. ⓐⓑⓒⓓ	23. ⓐⓑⓒⓓ

Space for Rough Work

24. Which one of the following is a incorrect matching pair?
(a) Lubb-Sharp closure of AV valves at the beginning of atrial systole
(b) Dub-Sudden opening of semilunar valves at the end of ventricular systole
(c) Cardiac Output-Stroke volume and heart beat
(d) Initiation of the heart beat -sino atrial node

25. Erythroblastosis foetalis occurs when
(a) Mother is Rh negative and father is Rh positive
(b) Father is Rh negative and mother is Rh positive
(c) Both are Rh positive
(d) Both are Rh negative

26. Which of the following set of animals has an incomplete double circulation system?
(a) Frog and crocodile (b) Shark and whale
(c) Lizard and pigeon (d) Toad and lizard

27. Find out the incorrect answer from the following?
(a) Veins are typically larger in diameter than arteries
(b) Because of their small size, capillaries contain blood that is moving more quickly than in other parts of the circulatory system
(c) The walls of arteries are elastic, enabling them to stretch and shrink during changes in blood pressure
(d) Veins contain more blood than any other part of the circulatory system

28. Given below are four statements (i-iv) regarding human blood circulatory system
(i) Arteries are thick-walled and have narrow lumen as compared to veins
(ii) Angina is acute chest pain when the blood circulation to the brain is reduced
(iii) Persons with blood group AB can donate blood to any person with any blood group under ABO system
(iv) Calcium ions play a very important role in blood clotting
Which two of the above statements are correct?
(a) (i) and (iv) (b) (i) and (iv)
(c) (ii) and (iii) (d) (iii) and (iv)

29. Haldane effect plays more important role in promoting carbon dioxide transport than that of the Bohr's effect in promoting oxygen transport because
(a) oxyhaemoglobin is a stronger acid which donates hydrogen ion (H^+) which in turn displace carbon dioxide from blood
(b) carbaminohaemoglobin is a stronger acid which splits into hydrogen ion (H^+) and bicarbonate (HCO_3^-)
(c) carbon dioxide reacts with water to form carbonic acid that lowers the pH in tissue
(d) carbon dioxide is less soluble in venous blood than in arterial blood

30. Which of the following factors is known as Christmas factor?
(a) Factor VIII (b) Factor XII
(c) Factor IV (d) Factor IX

31. In veins, values are present to check backward flow to blood flowing at
(a) high pressure (b) atmospheric pressure
(c) low pressure (d) all of these

32. In a cardiac output of 5250 ml per minute, with 75 heartbeats per minute, the stroke volume is
(a) 60 ml (b) 80 ml
(c) 55 ml (d) 70 ml

33. Heart valves function to
(a) keep blood moving forward through the heart.
(b) mix blood thoroughly as it passes through the heart.
(c) control the amount of blood pumped by the heart.
(d) slow blood down as it passes through the heart.

34. The accompanying diagram shows a small part of a normal electrocardiogram. Which region represents a wave of excitation passing through the ventricles?

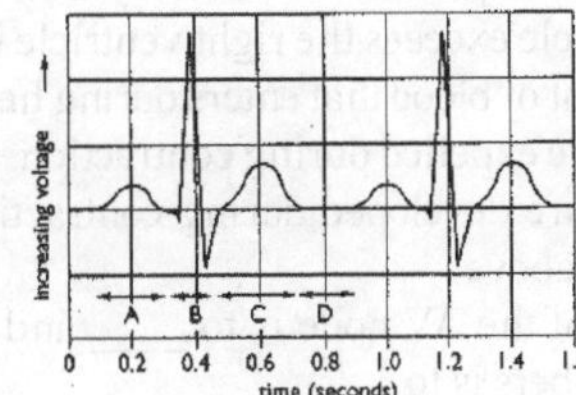

(a) A (b) B
(c) C (d) D

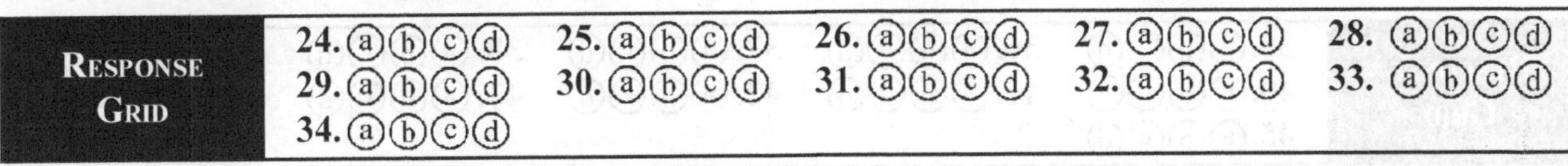

RESPONSE GRID					
	24. ⓐⓑⓒⓓ	25. ⓐⓑⓒⓓ	26. ⓐⓑⓒⓓ	27. ⓐⓑⓒⓓ	28. ⓐⓑⓒⓓ
	29. ⓐⓑⓒⓓ	30. ⓐⓑⓒⓓ	31. ⓐⓑⓒⓓ	32. ⓐⓑⓒⓓ	33. ⓐⓑⓒⓓ
	34. ⓐⓑⓒⓓ				

Space for Rough Work

35. Just after blood leaves the left ventricle of the human heart, it passes through the
(a) pulmonary artery (b) left atrium
(c) aorta (d) superior vena cava

36. In the human body, the highest blood pressure can be recorded in
(a) arteries near the heart
(b) veins returning to the heart
(c) capillaries of the body tissue
(d) the lungs

37. Blood flows rather rapidly as it leaves the human heart. Upon arriving at the capillaries, it slows dramatically. This reduction in speed is due largely to
(a) the loss of pressure as the blood gets farther away from the heart.
(b) changes in pressure that result from water evaporation on the skin.
(c) the resistance offered by the small diameter of the capillaries.
(d) differences in pressure caused by the sugar content of the blood.

38. The atrial walls are ___ than the ventricular wall, and pressure generated in the atrial chambers is ___ than in the ventricles.
(a) thinner, higher (b) thinner, lower
(c) thicker, higher (d) thicker, lower

39. Atheroma consist mainly of
(a) cholesterol and lipid hardened by calcium
(b) cholesterol and lipid hardened by iron
(c) glycogen and protein hardened by calcium
(d) glycogen and protein hardened by iron

40. The left ventricle exceeds the right ventricle in
(a) the amount of blood that enters during heart contraction.
(b) the volume expelled during contraction
(c) the pressure developed during contraction
(d) all of the above

41. The purpose of the AV node is to ___ and the purpose of the Purkinje fibers is to ___ .
(a) create simultaneous atrial and ventricular depolarization; speed up transmission of the cardiac impulse into the ventricle
(b) delay ventricular depolarization relative to atrial depolarization; insulate the cardiac impulse from the general ventricular fibers
(c) delay ventricular depolarization relative to atrial depolarization; transmit the cardiac impulse to very small localized groups of ventricular fibers
(d) delay atrial depolarization relative to ventricular depolarization; transmit the cardiac impulse to very small localized groups of ventricular fibers

42. Persons suffering from high blood pressure should take the following precaution to avoid excessive rise in their blood pressure
(a) sleep as much as possible
(b) avoid standing
(c) increase their weight
(d) avoid emotional disturbances and excitement

43. You are required to draw blood from a patient and to keep the blood in a test tube for analysis of corpuscles and plasma. You are also provided with the following four types of test tubes. Which of them will you not use for the purpose?
(a) Test tube containing calcium bicarbonate
(b) Chilled test-tube
(c) Test-tube containing heparin
(d) Test-tube containing sodium oxalate

44. A drop of each of the following, is placed separately on four slides. Which of them will not coagulate?
(a) Blood serum
(b) Sample from the thoracic duct of lymphatic system
(c) Whole blood from pulmonary vein
(d) Blood plasma

45. Examination of blood of a person suspected of having anemia, shows large, immature, nucleated erythrocytes without haemoglobin. Supplementing his diet with which of the following, is likely to alleviate his symptoms?
(a) Folic acid and cobalamine (b) Riboflavin
(c) Iron compounds (d) Thiamine

RESPONSE GRID					
	35. ⓐⓑⓒⓓ	36. ⓐⓑⓒⓓ	37. ⓐⓑⓒⓓ	38. ⓐⓑⓒⓓ	39. ⓐⓑⓒⓓ
	40. ⓐⓑⓒⓓ	41. ⓐⓑⓒⓓ	42. ⓐⓑⓒⓓ	43. ⓐⓑⓒⓓ	44. ⓐⓑⓒⓓ
	45. ⓐⓑⓒⓓ				

Space for Rough Work

DAILY PRACTICE PROBLEM DPP CHAPTERWISE 18 - BIOLOGY			
Total Questions	45	Total Marks	180
Attempted		Correct	
Incorrect		Net Score	
Cut-off Score	50	Qualifying Score	60
Success Gap = Net Score – Qualifying Score			
Net Score = (Correct × 4) – (Incorrect × 1)			

DPP - Daily Practice Problems

Chapter-wise Sheets

Date : | Start Time : | End Time :

BIOLOGY (CB19)

SYLLABUS : Excretory Products and Their Elimination

Max. Marks : 180 | **Marking Scheme : + 4 for correct & (–1) for incorrect** | **Time : 60 min.**

INSTRUCTIONS : This Daily Practice Problem Sheet contains 45 MCQs. For each question only one option is correct. Darken the correct circle/ bubble in the Response Grid provided on each page.

1. Toxic substances are detoxified in human body in
   (a) kidney (b) lungs (c) liver (d) stomach
2. The function of rennin is
   (a) vasodilation
   (b) reduce blood pressure
   (c) degradation of angiotensinogen
   (d) None of the above
3. Proximal and distal convoluted tubules are parts of
   (a) Seminiferous tubules (b) Nephron
   (c) Oviduct (d) Vas deferens
4. Antidiuretic hormone
   (a) Secretion is determined by plasma osmolarity
   (b) Increases permeability of renal collecting duct cells to water
   (c) Is secreted by nerve cells with their cell bodies in hypothalamus
   (d) All the above
5. Which one of the following is correctly matched pair of the given secretion and its primary role in human physiology?
   (a) Sebum — Sexual attraction
   (b) Sweat — Thermoregulation
   (c) Saliva — Tasting food
   (d) Tears — Excretion of salts
6. A person who is one along hunger strike and is surviving only on water, will have
   (a) less amino acids in his urine
   (b) more glucose in his blood
   (c) less urea in his urine
   (d) more sodium in his urine
7. Uricotelism is found in
   (a) Frogs and toads
   (b) Mammals and birds
   (c) Birds, reptiles and insects
   (d) Fishes and fresh water protozoans
8. If Henle's loop were absent from mammalian nephron which of the following is to be expected?
   (a) The urine will be more concentrated
   (b) The urine will be more dilute
   (c) There will be no urine formation
   (d) There will be hardly any change in the quality and quantity of urine formed
9. The basic functional unit of the human kidney is
   (a) nephron (b) nephridia
   (c) pyramid (d) Henle's loop

RESPONSE GRID					
	1. ⓐⓑⓒⓓ	2. ⓐⓑⓒⓓ	3. ⓐⓑⓒⓓ	4. ⓐⓑⓒⓓ	5. ⓐⓑⓒⓓ
	6. ⓐⓑⓒⓓ	7. ⓐⓑⓒⓓ	8. ⓐⓑⓒⓓ	9. ⓐⓑⓒⓓ	

Space for Rough Work

10. Reabsorption of water in distal parts of kidney tubules/urine formation is controlled by
(a) relaxin (b) calcitonin
(c) oxytocin (d) vasopressin

11. Urea from the blood can be removed by
(a) Uremia (b) Diuresis
(c) Dialysis (d) Micturition

12. In mammals, the urinary bladder opens into
(a) Uterus (b) Urethra
(c) Vestibule (d) Ureter

13. Which of the following components of blood does not enter into the nephron?
(a) urea (b) water
(c) glucose (d) plasma protein

14. Which one of the following correctly explains the function of a specific part of a human nephron ?
(a) Podocytes : create minute spaces (slite pores) for the filtration of blood into the Bowman's capsule.
(b) Henle's loop : most reabsorption of the major substances from the glomerular filtrate.
(c) Distal convoluted tubule : reabsorption of K^+ ions into the surrounding blood capillaries.
(d) Afferent arteriole : carries the blood away from the glomerular towards renal vein.

15. The condition of excess urea in blood is known as
(a) Polyuria (b) Haematuria
(c) Uraemia (d) Diuresis

16. Urine under normal conditions does not contain glucose because
(a) The normal blood sugar is fructose
(b) Glucose of blood is not filtered in the glomerulus
(c) Glucose in glomerular filtrate is reabsorbed in the uriniferous tubules
(d) Glucose in glomerular filtrate is converted into glycogen.

17. Which is not correct with respect to human kidney?
(a) The peripheral region is called cortex and central medulla
(b) Malpighian capsules are present in the cortex region
(c) Blood enters glomerulus through efferent arterioles
(d) The concave part of kidney is called hilus

18. Atrial natriuretic factor (ANF) is released in response to the increase in blood volume and blood pressure. Which of the following is not the function of ANF?
(a) Stimulates aldosterone secretion
(b) Inhibits the release of renin from JGA
(c) Stimulates salt loss in urine
(d) Inhibits sodium reabsorption from collecting duct

19. Reabsorption of chloride ions from glomerular filtrate in kidney tubule occurs by
(a) Active transport (b) Diffusion
(c) Osmosis (d) Brownian movement

20. Metanephric kidneys are found in
(a) Reptiles only (b) Birds only
(c) mammals only (d) All of these

21. In Prawn, excretion is carried out by
(a) Nephrons (b) Malpighian tubules
(c) Flame cells (d) Green glands

22. Consider the following four statements (i - iv) about certain desert animals such as kangaroo rat
(i) They have dark colour and high rate of reproduction and excrete solid urine
(ii) They do not drink water, breathe at a slow rate to conserve water and have their body covered with thick hairs
(iii) They feed on dry seeds and do not require drinking water
(iv) They excrete very concentrated urine and do not use water to regulate body temperature.
Which two of the above statements for such animals are true?
(a) (i) and (ii) (b) (iii) and (iv)
(c) (ii) and (iii) (d) (iii) and (i)

23. The longest loop of Henle is found in
(a) kangaroo rat (b) opposum
(c) rhesus monkey (d) porcupine

24. In peritoneal dialysis
(a) the blood is removed from the body and a natural filter is used
(b) the blood is not removed from the body and a natural filter is used
(c) the blood is not removed from the body and an artificial filter is used
(d) the blood is removed from the body and an artificial filter is used

25. Diabetes insipidus is a condition in which a person is unable to produce sufficient levels of the hormone ADH. The hormone increases the permeability to water of the second (distal) convoluted tubule and collecting duct in the kidney nephrons.
What is produced as a result?
(a) large volumes of concentrated urine
(b) large volumes of dilute urine
(c) small volumes of concentrated urine
(d) small volumes of dilute urine

RESPONSE GRID					
	10. ⓐⓑⓒⓓ	11. ⓐⓑⓒⓓ	12. ⓐⓑⓒⓓ	13. ⓐⓑⓒⓓ	14. ⓐⓑⓒⓓ
	15. ⓐⓑⓒⓓ	16. ⓐⓑⓒⓓ	17. ⓐⓑⓒⓓ	18. ⓐⓑⓒⓓ	19. ⓐⓑⓒⓓ
	20. ⓐⓑⓒⓓ	21. ⓐⓑⓒⓓ	22. ⓐⓑⓒⓓ	23. ⓐⓑⓒⓓ	24. ⓐⓑⓒⓓ
	25. ⓐⓑⓒⓓ				

Space for Rough Work

26. By definition, an ectotherm
(a) is cold-blooded.
(b) is warm-blooded.
(c) obtains most of its heat from its environment.
(d) derives most of its heat from its own metabolism.

27. Which region of the kidney nephron is the main site of amino acid reabsorption?
(a) glomerulus
(b) Bowman's capsule
(c) proximal convoluted tubule
(d) distal convoluted tubule

28. Long term kidney failure can be treated by introducing sterile dialysis fluid into the abdominal cavity. The fluid is drained and replaced regularly using a tube inserted surgically through the abdominal wall.
Why does this method work well?
(a) because osmoregulation and excretion are achieved by diffusion between the blood in the abdominal capillaries and the dialysis fluid.
(b) because osmoregulation and excretion are achieved by the active transport of ions, water and urea between the abdominal capillaries and the dialysis fluid.
(c) because the fluid is in direct contact with the kidneys, the urea and excess ions can pass into it without being filtered by the glomeruli.
(d) because the fluid is in direct contact with the liver and the large intestine, and wastes and excess ions can pass into it from these organs.

29. Which of the following describes the route of urine out of the body after it leaves the kidney?
(a) renal vein-urinary bladder-urethra-ureter
(b) urethra-urinary bladder-ureter
(c) renal vein-ureter-urinary bladder-urethra
(d) ureter-urinary bladder-urethra

30. If the human kidneys filter 150 litres of plasma in a 24 hour period, what is the typical amount of urine produced and eliminated in that time period?
(a) 0.15 litres (b) 1.5 litres
(c) 15 litres (d) 30 litres

31. A number of hormones help to regulate water and solute uptake and release in the nephron. Antidiuretic hormone (ADH) promotes ______ in response to ______.
(a) active transport of Cl^-, increased solute concetration
(b) active transport of Na^+, increased blood pressure
(c) increased permeability of the collecting duct to water, increased blood pressure
(d) decreased permeability of the collecting duct to water, increased solute concetration

32. The filtrate formed by the nephrons in the kidney is not the same as urine. The filtrate is first refined and concentrated by the processes of ______, forming the urine that leaves the body.
(a) filtration and secretion
(b) reabsorption and secretion
(c) reabsorption and excretion
(d) filtration and reabsorption

33. What pathway is taken by water and solutes as they travel through a nephron?
(a) Glomerulus, to Bowman's capsule, to proximal tubule, to loop of Henle, to distal tubule, to collecting ducts
(b) Bowman's capsule, to glomerulus, to distal tubule, to loop of Henle, to proximal tubule, to collecting ducts
(c) Glomerulus, to Bowman's capsule, to distal tubule, to loop of Henle, to proximal tubule, to collecting ducts
(d) Glomerulus, to Bowman's capsule, to proximal tubule, to collecting ducts, to distal tubule, to loop of Henle

34. The kidney's filtration process is nonselective, so
(a) many valuable substances are lost in the urine.
(b) the proportions of substances in urine are the same as in blood.
(c) urine is much less concentrated than blood.
(d) useful substances must be selectively reabsorbed.

35. Na^+ and Cl^- are actively transported out of the tubules to help set up the countercurrent multiplier. Which the following are sites of active Na^+ and Cl^- transport in the nephron?
(a) Proximal tubule, ascending limb of the loop of Henle
(b) Descending limb of the loop of Henle, ascending limb of the loop of Henle
(c) Ascending limb of the loop of Henle, proximal tubule
(d) Collecting duct, descending limb of the loop of Henle

36. Which of the following is not a normal constituent of the glomerular filtrate?
(a) Red blood cells (b) Urea
(c) Sodium ion (d) Glucose

37. Kidneys help in the conservation of useful materials and excretion of wastes and therefore they receive 20% of the hearts output of blood (as much as the heart and brain combined). On a percentage basis which substance is most completely reabsorbed by the kidneys ?
(a) Water (b) Glucose
(c) Urea (d) Sodium

38. The sole mechanism for water reabsorption by the renal tubules is :
(a) active transport
(b) osmosis.
(c) cotransport with sodium ions
(d) cotransport with bicarbonate ions

RESPONSE GRID					
	26. ⓐⓑⓒⓓ	27. ⓐⓑⓒⓓ	28. ⓐⓑⓒⓓ	29. ⓐⓑⓒⓓ	30. ⓐⓑⓒⓓ
	31. ⓐⓑⓒⓓ	32. ⓐⓑⓒⓓ	33. ⓐⓑⓒⓓ	34. ⓐⓑⓒⓓ	35. ⓐⓑⓒⓓ
	36. ⓐⓑⓒⓓ	37. ⓐⓑⓒⓓ	38. ⓐⓑⓒⓓ		

Space for Rough Work

39. Figure shown human urinary system with structures labelled A to D. Select option which correctly identifies them and gives their characteristics and /or functions.

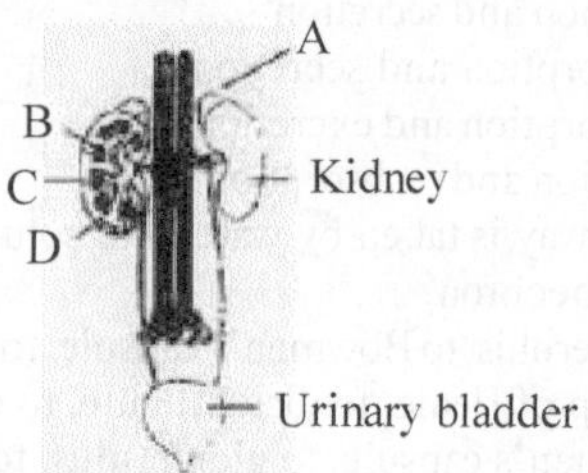

(a) B-pelvis-broad funnel shaped space inner to hilum, directly connected to loops of Henle.
(b) C-Medulla-inner zone of kidney and contains complex nephrons.
(c) D - Cortex - outer part of kidney and do not contain any part of nephrons
(d) A-Adrenal gland - located at the anterior part of kidney. Secrete Catecholamines which stimulate glycogen breakdown.

40. Refer the following diagram and identify the parts of a kidney indicated

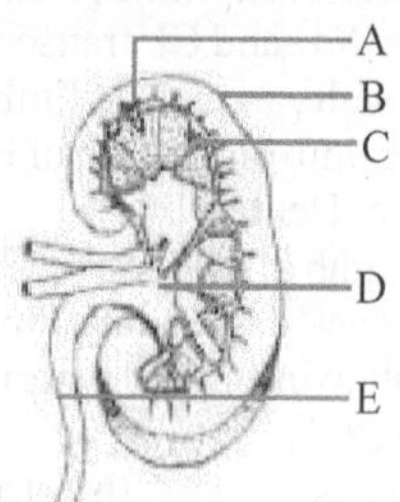

(a) A = cortex, B = nephron, C = pelvis, D= medulla, E = ureter
(b) A = cortex, B = medulla, C = nephron, D = pelvis, E = ureter
(c) A = nephron, B = cortex, C = medulla, D = ureter, E = pelvis
(d) A = nephron, B = cortex, C = medulla, D = pelvis, E = ureter

41. Select the option which shows correct matching of animal with excretory organs and excretory product

	Animal	Excretory organs	Excretory product
(a)	Housefly	Renal tubules	Uric acid
(b)	*Labeo* (Rohu)	Nephridial tubes	Ammonia
(c)	Salamander	Kidney	Urea
(d)	Peacock	Kidney	Urea

42. The appearance of albumin in the urine is most likely due to
(a) Increase in the blood pressure
(b) Decrease in the blood osmotic corpuscles
(c) Damage to the Malpighian corpuscles
(d) Damage to the proximal convoluted tubules

43. Kidney crystals are solid clusters of
(a) Calcium nitrate and uric acid
(b) Phosphate and uric acid
(c) Calcium carbonate and uric acid
(d) Calcium metabisulphite and uric acid

44. The liquid which is collected in the cavity of Bowman's capsule is
(a) Concentrated urine
(b) Blood plasma minus blood proteins
(c) Glycogen and water
(d) Sulphates and water

45. A person who is on a long hunger strike and is surviving only on water, will have –
(a) less amino acids in his urine
(b) more glucose in his blood
(c) less urea in his urine
(d) more sodium in his urine

RESPONSE GRID	39. ⓐⓑⓒⓓ	40. ⓐⓑⓒⓓ	41. ⓐⓑⓒⓓ	42. ⓐⓑⓒⓓ	43. ⓐⓑⓒⓓ
	44. ⓐⓑⓒⓓ	45. ⓐⓑⓒⓓ			

Space for Rough Work

DAILY PRACTICE PROBLEM DPP CHAPTERWISE 19 - BIOLOGY			
Total Questions	45	Total Marks	180
Attempted		Correct	
Incorrect		Net Score	
Cut-off Score	45	Qualifying Score	60
Success Gap = Net Score – Qualifying Score			
Net Score = (Correct × 4) – (Incorrect × 1)			

DPP - Daily Practice Problems

Chapter-wise Sheets

Date : | Start Time : | End Time :

BIOLOGY

CB20

SYLLABUS : Locomotion and Movement

Max. Marks : 180 **Marking Scheme : + 4 for correct & (–1) for incorrect** **Time : 60 min.**

INSTRUCTIONS : This Daily Practice Problem Sheet contains 45 MCQs. For each question only one option is correct. Darken the correct circle/ bubble in the Response Grid provided on each page.

1. Which ion is essential for muscle contraction?
   (a) Na^+ (b) K^+
   (c) Ca^{2+} (d) Cl^-
2. Elbow joint is an example of:
   (a) hinge joint (b) gliding joint
   (c) ball and socket joint (d) pivot joint
3. Two of the body parts which do not appear in MRI may be
   (a) molar teeth and eye lens
   (b) scapula and canines
   (c) ligaments and ribs
   (d) tendons and premolars
4. Which of the following is made up of a single bone in mammal?
   (a) Dentary (b) Hyoid
   (c) Upper jaw (d) All of these
5. Intercoastal muscles are found attached with
   (a) diaphragm (b) ribs
   (c) pleura (d) lungs
6. Ball and socket joint is found between
   (a) ribs and vertebral
   (b) femur and tibio-fibula
   (c) humerus and olecranon fossa
   (d) humerus and pectoral girdle
7. Which of the following is the contractile protein of a muscle?
   (a) Myosin (b) Tropomyosin
   (c) Actin (d) Tubulin

RESPONSE GRID					
	1. ⓐⓑⓒⓓ	2. ⓐⓑⓒⓓ	3. ⓐⓑⓒⓓ	4. ⓐⓑⓒⓓ	5. ⓐⓑⓒⓓ
	6. ⓐⓑⓒⓓ	7. ⓐⓑⓒⓓ			

Space for Rough Work

8. Myofibrils are made up of
(a) Myosin and actin
(b) Myosin and troponin
(c) Actin and tropomyosin
(d) All the above components

9. Which one of the following is the correct matching of three items and their grouping category?

	Items		Group
(a)	Ilium, ischium, pubis	-	coxal bones of pelvic girdle
(b)	Actin, myosin, rhodopsin.	-	muscle proteins
(c)	Cytosine, uracil, thiamine	-	pyrimidines
(d)	Malleus, incus, cochlea	-	ear ossicles

10. The number of floating ribs, in the human body, is
(a) 6 pairs (b) 5 pairs
(c) 3 pairs (d) 2 pairs

11. Select the **correct** statement regarding the specific disorder of muscular or skeletal system :
(a) *Myasthenia gravis* - Autoimmune disorder which inhibits sliding of myosin filaments.
(b) *Gout* - inflammation of joints due to extra deposition of calcium.
(c) *Muscular dystrophy* - age related shortening of muscles.
(d) *Osteoporosis* - decrease in bone mass and higher chances of fractures with advancing age.

12. The sensation of fatigue in the muscles after prolonged strenuous physical work, is caused by
(a) a decrease in the supply of oxygen
(b) minor wear and tear of muscle fibres
(c) the depletion of glucose
(d) the accumulation of lactic acid

13. Synovial fluid is found in
(a) cranial cavity (b) spinal cavity
(c) immovable joints (d) freely movable joints

14. Humerus differs from the femur in having:
(a) Sigmoid notch (b) Trochanter
(c) Deltoid ridge (d) None of these

15. The most abundant mineral in human body is
(a) Magnesium (b) Sodium
(c) Calcium (d) Potassium

16. Ankle joint is
(a) Pivot Joint (b) Ball and socket joint
(c) Hinge joint (d) Gliding joint

17. The major function of the intervertebral disc is to
(a) Absorb shock
(b) String the vertebrae together
(c) Prevent injuries
(d) Prevent hyperextension

18. Which one of the following pairs of chemical substances is correctly categorized?
(a) Calcitonin and thymosin - Thyroid hormones
(b) Pepsin and prolactin - Two digestive enzymes secreted in stomach
(c) Troponin and myosin - Complex proteins in striated muscles
(d) Secretin and rhodopsin - Polypeptide hormones

19. The functional unit of contractile system of a striated muscle is
(a) Sarcomere (b) Z-band
(c) Sarcosome (d) Myofibril

20. Joint between bones of human skull is
(a) Hinge joint (b) Synovial joint
(c) Cartilaginous joint (d) Fibrous joint

RESPONSE GRID					
	8. ⓐⓑⓒⓓ	9. ⓐⓑⓒⓓ	10. ⓐⓑⓒⓓ	11. ⓐⓑⓒⓓ	12. ⓐⓑⓒⓓ
	13. ⓐⓑⓒⓓ	14. ⓐⓑⓒⓓ	15. ⓐⓑⓒⓓ	16. ⓐⓑⓒⓓ	17. ⓐⓑⓒⓓ
	18. ⓐⓑⓒⓓ	19. ⓐⓑⓒⓓ	20. ⓐⓑⓒⓓ		

Space for Rough Work

21. Which one of the following is the *correct description* of a certain part of the normal human skeleton ?
(a) Parietal bone and the temporal bone of the skull are joined by fibrous joint
(b) First vertebra is axis which articulates with the occipital condyles
(c) The 9th and 10th pairs of ribs are called the floating ribs
(d) Glenoid cavity is a depression to which the thigh bone articulates.

22. Which of the following is an autoimmune disorder ?
(a) Myasthenia gravis (b) Osteoporosis
(c) Muscular dystrophy (d) Gout

23. The joint in our neck which allows us to rotate our head left to right is
(a) pivot joint (b) hinge joint
(c) saddle joint (d) ellipsoid joint

24. Read the following statements carefully and select the correct ones.
(i) Cardiac fibres are branched with one or more nuclei
(ii) Smooth muscles are unbranched and cylindrical
(iii) Skeletal muscles can be branched or unbranched
(iv) Smooth muscles are non-striated
(a) only (iv) (b) (ii) and (iii)
(c) (iii) and (iv) (d) only (iii)

25. A cricket player is fast chasing a ball in the field. Which one of the following groups of bones is directly contributing in this movement?
(a) Femur, malleus, tibia, metatarsals
(b) Pelvis, ulna, patella, tarsals
(c) Sternum, femur, tibia, fibula
(d) Tarsals, femur, metatarsals, tibia

26. Which of the following statement is incorrect w.r.t. bone?
(a) If born is kept in HCl it becomes soft
(b) Bone is made up of 60&70% organic matter and 30&40% inorganic matter
(c) If bone is heated then the organic part disappears and inorganic part is retained
(d) Hydroxyapatite salts and fluorapatite salts are found in matrix

27. One of the following is a location of most abundant cartilage in the human body.
(a) Tracheal rings and costal cartilages
(b) Intevertebral disc and public symphysis
(c) Pinna and tip of nose
(d) Pectoral girdle and pelvic girdle

28. Upon preventing acetylcholine from diffusing across a neuromuscular junction, which of the following will *not* result?
(a) No action potential will be produced in the affected muscle fibre's plasma membrane.
(b) The endoplasmic reticulum releases calcium ions (Ca^{2+}) into the cytoplasm.
(c) Myosin will not bind to actin in the affected muscle fibre.
(d) The affected muscle fiber will fail to contract.

29. Muscle A and muscle B are the same size, but muscle A is capable of much finer control than muscle B. Which one of the following is likely to be true of muscle A?
(a) It contains fewer motor units than muscle B.
(b) It has larger sarcomeres than muscle B.
(c) It is controlled by more neurons than muscle B.
(d) It is controlled by fewer neurons than muscle B.

30. Much discussion of muscle is related to 'striated' muscle, but 'smooth' muscles are also important for ________
(a) protecting and nourishing striated muscle cells.
(b) conveying action potentials from nerve endings to the deepest parts of striated muscle.
(c) involuntary activities, such as movement of food in the gut and controlling blood pressure.
(d) sheathing the striated muscles so that they do not damage each other as they slide past one another.

RESPONSE GRID	21. ⓐⓑⓒⓓ	22. ⓐⓑⓒⓓ	23. ⓐⓑⓒⓓ	24. ⓐⓑⓒⓓ	25. ⓐⓑⓒⓓ
	26. ⓐⓑⓒⓓ	27. ⓐⓑⓒⓓ	28. ⓐⓑⓒⓓ	29. ⓐⓑⓒⓓ	30. ⓐⓑⓒⓓ

Space for Rough Work

31. What is a hydrostatic skeleton largely composed of?
(a) Bone (b) Fluid
(c) Cartilage (d) Connective tissue

32 The only movable bone in the skull is:
(a) Mandible (b) Maxilla
(c) Ethmoid (d) None

33. The smallest irregular bone in man is:
(a) Patella (b) Stapes
(c) Nasal (d) Palatine

34. ATP provides the energy for muscle contraction by allowing for
(a) an action potential formation in the muscle cell.
(b) cross-bridge detachment of myosin from action.
(c) cross-bridge attachment of myosin to actin
(d) release of calcium by sarcoplasmic reticulum.

35. The primary difference between an endoskeleton and an exoskeleton has to do with
(a) the presence of both circular and longitudinal muscles.
(b) whether or not the skeleton is on the inside of the body.
(c) the presence or absence of joints.
(d) the amount of fluid in the body.

36. Deposition of uric acid crystals within the synovial joint causes:
(a) osteoarthritis (b) rheumatoid arthritis
(c) gout (d) paralysis

37. Total no. of muscles in our body is
(a) 256 muscles (b) 639 muscles
(c) 400 muscles (d) 421 muscles

38. The muscle band that remains unchanged during contraction and relaxation of the skeletal muscle is
(a) I (b) H
(c) A (d) Z line

39. What is not true about human skull ?
(a) It is dicondylic
(b) It includes 6 ear ossicles
(c) It includes 14 facial bones
(d) Hyoid is not included in skull bones

40. Which is part of pectoral girdle?
(a) Glenoid cavity (b) Sternum
(c) Ilium (d) Acetabulum

41. Number of bones in hind limb of human is-
(a) 21 (b) 24
(c) 30 (d) 14

42. Haversian system is diagnostic feature of-
(a) Avian bones (b) Reptilian bones
(c) Mammalian bones (d) Bones of all animal

43. In children the bones are more flexible and brittle because their bones have
(a) large quantity of salts and little organic substances
(b) large quantity of organic substances and little salts
(c) well developed haversian system
(d) large number of osteoblasts

44. Pneumatic bones are found in
(a) House lizard (b) Pigeon
(c) Flying fish (d) Frog's tadpole

45. Actin protein occurs in two forms
(a) Polymeric F- actin and monomeric G- actin
(b) Monomeric F- actin and polymeric G-actin
(c) The tail and a head
(d) F-actin and G- actin, but both globular

RESPONSE GRID					
	31. ⓐⓑⓒⓓ	32. ⓐⓑⓒⓓ	33. ⓐⓑⓒⓓ	34. ⓐⓑⓒⓓ	35. ⓐⓑⓒⓓ
	36. ⓐⓑⓒⓓ	37. ⓐⓑⓒⓓ	38. ⓐⓑⓒⓓ	39. ⓐⓑⓒⓓ	40. ⓐⓑⓒⓓ
	41. ⓐⓑⓒⓓ	42. ⓐⓑⓒⓓ	43. ⓐⓑⓒⓓ	44. ⓐⓑⓒⓓ	45. ⓐⓑⓒⓓ

Space for Rough Work

DAILY PRACTICE PROBLEM DPP CHAPTERWISE 20 - BIOLOGY			
Total Questions	45	Total Marks	180
Attempted		Correct	
Incorrect		Net Score	
Cut-off Score	45	Qualifying Score	60
Success Gap = Net Score – Qualifying Score			
Net Score = (Correct × 4) – (Incorrect × 1)			

DPP - Daily Practice Problems

Chapter-wise Sheets

Date : **Start Time :** **End Time :**

BIOLOGY CB21

SYLLABUS : Neural Control and Co-ordination

Max. Marks : 180 **Marking Scheme : + 4 for correct & (–1) for incorrect** **Time : 60 min.**

INSTRUCTIONS : This Daily Practice Problem Sheet contains 45 MCQs. For each question only one option is correct. Darken the correct circle/ bubble in the Response Grid provided on each page.

1. Afferent nerve fibres carry impulses from
(a) effector organs to CNS (b) receptors to CNS
(c) CNS to receptors (d) CNS to muscles
2. The nerve centres which control the body temperature and the urge for eating are contained in:
(a) hypothalamus (b) pons
(c) cerebellum (d) thalamus
3. Rods and cones of eyes are modified
(a) multipolar neuron (b) unipolar neuron
(c) bipolar neuron (d) None of these
4. The nerve transmitter, produced at the synapse and neuromuscular junction, is
(a) GTP (b) ATP
(c) acetylcholine (d) phosphokinase
5. Alzheimer disease in humans is associated with the deficiency of:
(a) glutamic acid
(b) acetylcholine
(c) gamma aminobutyric acid (GABA)
(d) dopamine
6. Which is the example of conditioned reflex ?
(a) Eyes closed when anything enter into it.
(b) Hand took up when piercing with needle.
(c) Salivation in a hungry dog in response to ringing of a bell.
(d) Digestion food goes forward in alimentary canal.
7. The black pigment in the eye which reduces the internal reflection is located in
(a) retina (b) iris
(c) cornea (d) sclerotic
8. Which one of the following does not act as a neurotransmitter ?
(a) Epinephrine (b) Norepinephrine
(c) Cortisone (d) Acetylcholine

RESPONSE GRID	
	1. ⓐⓑⓒⓓ 2. ⓐⓑⓒⓓ 3. ⓐⓑⓒⓓ 4. ⓐⓑⓒⓓ 5. ⓐⓑⓒⓓ
	6. ⓐⓑⓒⓓ 7. ⓐⓑⓒⓓ 8. ⓐⓑⓒⓓ

Space for Rough Work

9. During the propagation of a nerve impulse, the action potential results from the movement of:
(a) K^+ ions from extracellular fluid to intracellular fluid.
(b) Na^+ ions from intracellular fluid to extracellular fluid.
(c) K^+ ions from intracellular fluid to extracellular fluid.
(d) Na^+ ions from extracellular fluid to intracellular fluid.

10. The cochlea of ear contains
(a) perilymph
(b) aqueous humour
(c) perilymph and endolymph
(d) only endolymph

11. Eustachian tube connects
(a) External ear with middle ear
(b) External ear with internal ear
(c) Middle ear with pharynx
(d) Internal pharyax

12. Mark the incorrect statement
(a) The ear ossicle attached to tympanic membrane is malleus
(b) Opsin (of Rhodopsin) develops from vitamin A
(c) The pressure on ear drum is equalized by eustachian tube
(d) Otolith organ consists of saccule and utricle

13. Frequency of sound is discriminated by
(a) The intensity of movement of basilar fibres of cochlea
(b) The type of fluid - perilymph or endolymph
(c) The site at the cochlear coil
(d) All of these

14. Which one of the following is the correct difference between rod cells and cone cells of our retina?

		Rod Cells	Cone Cells
(a)	Distribution	More concentrated in centre of retina	Evenly distributed all over retina
(b)	Visual activity	High	Low
(c)	Visual pigment contained	Iodopsin	Rhodopsin
(d)	Over all function	Vision in poor light	Colour vision and detailed vision in bright light

15. Saltatory conduction of nerve impulse takes place through :
(a) Myelinated fibre
(b) Non-myelinated fibre
(c) Grey fibres
(d) None of these

16. In which part of the brain corpora quadrigemina is located?
(a) Diencephalon (b) Mesencephalon
(c) Prosencephalon (d) Rhombencephalon

17. A person entering an empty room suddenly finds a snake right in front on opening the door. Which one of the following is likely to happen in his neuro-hormonal control system ?
(a) Sympathetic nervous system is activated releasing epinephrine and norepinephrine from adrenal medulla.
(b) Neurotransmitters diffuse rapidly across the cleft and transmit a nerve impulse.
(c) Hypothalamus activates the parasympathetic division of brain.
(d) Sympathetic nervous system is activated releasing epinephrine and norepinephrine from adrenal cortex.

18. The blind spot is the region where
(a) Image is formed
(b) Cones are numerous
(c) The optic nerve leaves out
(d) Image is formed during the dark

19. In myopia or short sightedness
(a) Image is formed slightly in front of retina because eye ball is longer
(b) Eye ball is normal but image is formed over blind spot
(c) Eye ball is normal but images is formed slightly behind the retina due to faulty lens
(d) Curvature of cornea becomes irregular

20. Sequence of meninges from inner to outside is
(a) Duramater – Archnoid – Piamater
(b) Duramater – Piamater – Arachnoid
(c) Arachnoid – Duramater - Piamater
(d) Piamater- Arachnoid - Duramater

21. Among which one of the following groups of chemicals, all are neurotransmitters?
(a) Glycine, dopamine, melatonin
(b) Somatostatin, serotonin, acetylcholine
(c) Noradrenaline, somatostatin, threonine
(d) Acetylcholine, noradrenaline, dopamine

RESPONSE GRID					
	9. ⓐⓑⓒⓓ	10. ⓐⓑⓒⓓ	11. ⓐⓑⓒⓓ	12. ⓐⓑⓒⓓ	13. ⓐⓑⓒⓓ
	14. ⓐⓑⓒⓓ	15. ⓐⓑⓒⓓ	16. ⓐⓑⓒⓓ	17. ⓐⓑⓒⓓ	18. ⓐⓑⓒⓓ
	19. ⓐⓑⓒⓓ	20. ⓐⓑⓒⓓ	21. ⓐⓑⓒⓓ		

Space for Rough Work

22. The enzyme required for the conduction of nerve impulse across synapse is:
(a) peroxidase (b) choline acetylase
(c) ascorbic acid oxidase (d) succinic dehydrogenase

23. The purplish red pigment rhodopsin contained in the rods type of photoreceptor cells of the human eye, is a derivative of
(a) Vitamin A (b) Vitamin B_1
(c) Vitamin C (d) Vitamin D

24. During stress condition which of the following nerves start working?
(a) Sympathetic nerves
(b) Parasympathetic nerves
(c) Autonomic nerves
(d) Cranial nerves

25. The gelatinous membrane covering the sensorys hair cells of the ear is known as
(a) Reissner's membrane
(b) tectorial membrane
(c) basilar membrane
(d) neuro-sensory membrane

26. Nerve cells do not divide because they do not have
(a) nucleus (b) centrosome
(c) Golgi body (d) mitochondria

27. Reflex arcs, which involve no processing in the brain, are beneficial because they permit
(a) instantaneous signaling to motor neurons.
(b) avoidance of dangerous stimuli, even without a previously "learned" response.
(c) immediate response to dangerous stimuli.
(d) All of the above

28. Drugs that alter the permeability of the plasma membrane of a neuron have which of the following effects on nerve impulse conduction?
(a) They prevent movement of sodium and potassium ions into and out of the cell, thereby stopping the transmission of the impulse.
(b) They primarily affect conduction along the axon.
(c) They prevent the release of neurotransmitters at synapses.
(d) All of the above

29. The local depolarization of a receptor-cell membrane is called as
(a) inhibitory postsynaptic potential (or IPSP).
(b) action potential.
(c) resting potential.
(d) threshold potential.

30. Action potentials travel along a neuron because ________.
(a) the neuron cytoskeleton conducts electricity as long as an ion gradient is maintained by the sodium-potassium pump
(b) of cytoplasmic streaming within the neuron
(c) they are pulled along by positive-negative attraction
(d) depolarization of the membrane at one point causes an increase of permeability to sodium at the next point

31. Which one of the following structures constitutes the mammalian forebrain?
(a) Cerebrum and cerebellum
(b) Olfactory bulb
(c) Cerebellum and medulla
(d) Thalamus, hypothalamus, and cerebrum

32. If you electrically stimulate a resting neuron in the middle of its axon and cause it to fire an action potential, which of the following will happen ?
(a) The action potential will propagate toward the axon terminal.
(b) The action potential will propagate toward the cell body.
(c) The action potential will propagate toward both the axon terminal and the cell body.
(d) The action potential will not propagate at all.

33. The resting membrane potential for neuron A is –70mV, while the resting potential for neuron B is –50mV. The threshold voltage for the production of an action potential is –35mV for both neurons. Which of the following statements is *false* ?
(a) Neuron A must depolarize by 35mV to reach the threshold voltage.
(b) Neuron B must hyperpolarize by 15mV to reach the threshold voltage.
(c) The inside of both neurons is negatively charged with respect to the outside.
(d) A single EPSP received by neuron A would cause it to depolarize slightly.

RESPONSE GRID					
	22. ⓐⓑⓒⓓ	23. ⓐⓑⓒⓓ	24. ⓐⓑⓒⓓ	25. ⓐⓑⓒⓓ	26. ⓐⓑⓒⓓ
	27. ⓐⓑⓒⓓ	28. ⓐⓑⓒⓓ	29. ⓐⓑⓒⓓ	30. ⓐⓑⓒⓓ	31. ⓐⓑⓒⓓ
	32. ⓐⓑⓒⓓ	33. ⓐⓑⓒⓓ			

Space for Rough Work

34. You are sitting in biology class daydreaming. Your intrinsic heartbeat is controlled by .....
(a) the spinal cord.
(b) the cerebrum.
(c) the medulla.
(d) levels of adrenaline in the blood.

35. The threshold of a neuron is
(a) The amount of inhibitory neurotransmitter required to inhibit an action potential.
(b) The membrane voltage at which an axon potential will be suppressed.
(c) the amount of excitatory neurotransmitter required to elicit an action potential.
(d) The membrane voltage at which the membrane potential develops into an action potential.

36. Which of the following statements about gap juctions or electrical synapses is *false*?
(a) Connexons form molecular tunnels between two cells.
(b) Electrical synapses cannot be inhibitory
(c) Electrical synapses do not allow for temporal summation.
(d) Electrical transmission is very slow and is bidirectional.

37. All but one of the following chemicals are neurotransmitters that function in the human brain. Select the *exception.*
(a) Dopamine (b) Glycine
(c) Atropine (d) Glutamic acid

38. Which of the following cell stop dividing after birth ?
(a) Epithelium (b) Neuron
(c) Glial cells (d) Liver

39. Arbor vitae is composed of
(a) Neuroglea cells (b) Grey matter
(c) White matter (d) All of these

40. Twilight vision is also called
(a) Scotopic vision and is the function of rods
(b) Scotopic vision and is the function of cones
(c) Photopic vision and is the function of rods
(d) Photopic vision and is the function of cones

41. The white matter of the CNS is always
(a) deep to the grey matter
(b) unmyelinated
(c) arranged into tracts
(d) composed of sensory fibers only

42. The pneumotaxic centre in the body is
(a) Heart (b) Lung
(c) Medulla (d) Liver

43. Identify the organ/innervation mismatch
(a) Glossopharyngeal nerve-tongue
(b) Optic nerve-Eye
(c) Facial nerve-Olfactory epithelium
(d) Cochlear nerve-Spiral organ

44. Our ear can hear the frequency of sound waves
(a) 20 to 20,000 cycles/sec
(b) 1000 to 2000 cycles/sec
(c) 5000 to 7000 cycles/sec
(d) 5000 to 10,000 cycles/sec

45. Fenestra ovalis is the opening of –
(a) Cranium (b) Tympanum
(c) Tympanic cavity (d) Brain

RESPONSE GRID					
	34. ⓐⓑⓒⓓ	35. ⓐⓑⓒⓓ	36. ⓐⓑⓒⓓ	37. ⓐⓑⓒⓓ	38. ⓐⓑⓒⓓ
	39. ⓐⓑⓒⓓ	40. ⓐⓑⓒⓓ	41. ⓐⓑⓒⓓ	42. ⓐⓑⓒⓓ	43. ⓐⓑⓒⓓ
	44. ⓐⓑⓒⓓ	45. ⓐⓑⓒⓓ			

Space for Rough Work

DAILY PRACTICE PROBLEM DPP CHAPTERWISE 21 - BIOLOGY			
Total Questions	45	Total Marks	180
Attempted		Correct	
Incorrect		Net Score	
Cut-off Score	45	Qualifying Score	60
Success Gap = Net Score – Qualifying Score			
Net Score = (Correct × 4) – (Incorrect × 1)			

DPP - Daily Practice Problems

Chapter-wise Sheets

Date : | Start Time : | End Time :

BIOLOGY (CB22)

SYLLABUS : Chemical Co-ordination and Integration

Max. Marks : 180 **Marking Scheme : + 4 for correct & (–1) for incorrect** **Time : 60 min.**

INSTRUCTIONS : This Daily Practice Problem Sheet contains 45 MCQs. For each question only one option is correct. Darken the correct circle/ bubble in the Response Grid provided on each page.

1. Which hormone possesses anti-insulin effect?
(a) Cortisol (b) Calcitonin
(c) Oxytocin (d) Aldosterone
2. Which of the following is both exocrine and endocrine gland ?
(a) Liver (b) Pancreas
(c) Thyroid (d) Adrenal
3. Chemically hormones are
(a) biogenic amines only
(b) proteins, steroids and biogenic amines
(c) proteins only
(d) steroids only
4. The blood calcium level is lowered by the deficiency of
(a) parathormone
(b) thyroxine
(c) both calcitonin and parathormone
(d) calcitonin
5. The technique used for estimation of minute amounts of hormones and drugs is called
(a) electrophoresis
(b) electroencephalogram
(c) fractionation
(d) radioimmunoassay
6. Testosterone is produced by
(a) sertoli cells (b) leydig's cells
(c) oxyntic cells (d) pituitary gland
7. Which one of the following pairs correctly matches a hormone with a disease resulting from its deficiency?
(a) Luteinizing - Failure of ovulation
(b) Insulin - Diabetes insipidus
(c) Thyroxine - Tetany
(d) Parathyroid - Diabetes mellitus
8. Which one of the following is not a second messenger in hormone action ?
(a) Calcium (b) Sodium
(c) cAMP (d) cGMP

RESPONSE GRID					
	1. ⓐⓑⓒⓓ	2. ⓐⓑⓒⓓ	3. ⓐⓑⓒⓓ	4. ⓐⓑⓒⓓ	5. ⓐⓑⓒⓓ
	6. ⓐⓑⓒⓓ	7. ⓐⓑⓒⓓ	8. ⓐⓑⓒⓓ		

Space for Rough Work

9. Which of the following statements regarding glucagon is false?
(a) It is secreted by α-cells islets of Langerhans.
(b) It acts antagonistically to insulin.
(c) It decreases blood sugar level.
(d) The gland responsible for its secretion is heterocrine gland.

10. Which one of the following statements is correct?
(a) Neurons regulate endocrine activity, but not *vice versa.*
(b) Endocrine glands regulate neural activity and nervous system regulates endocrine glands.
(c) Neither hormones control neural activity nor the neurons control endocrine activity.
(d) Endocrine glands regulate neural activity but not *vice versa.*

11. Match the source gland with respective hormone as well as the function correctly.

	Source gland	Hormone	Function
(a)	Anterior pituitary	Oxytocin	Contraction of uterus muscles during child birth
(b)	Posterior pituitary	Vasopressin	Stimulates reabsorption of water in the distal tubules in the nephron
(c)	Corpus luteum	Estrogen	Supports pregnancy
(d)	Thyroid	Thyroxine	Regulates blood calcium level

12. Which of the following is incorrect?
(a) Iodine is needed for thyroxine formation.
(b) Calcium regulates the excitibility of nerve fibres.
(c) Potassium plays an important role in the regulation of acid base balance in cell.
(d) Phosphorus helps to maintain the osmotic pressure of the body fluids.

13. Sertoli cells are found in
(a) ovaries and secrete progesterone
(b) adrenal cortex and secrete adrenaline
(c) seminiferous tubules and provide nutrition to germ cells
(d) pancreas and secrete cholecystokinin

14. The phase of menstrual cycle in humans that lasts for 7-8 days, is
(a) follicular phase (b) ovulatory phase
(c) luteal phase (d) menstruation

15. Blood glucose level in man is regulated by:
(a) insulin
(b) adrenaline
(c) glucagon and insulin
(d) All of the above

16. Which of the following glands grows to the maximum size at puberty and then diminishes gradually?
(a) Thymus (b) Pituitary
(c) Thyroid (d) Adrenal

17. Hypoglycemic hormone is
(a) Insulin (b) Glucagon
(c) Thyroxine (d) Calcitonin

18. Which of the following diseases is caused by the under secretion of cortisol?
(a) Anaemia
(b) Addison's disease
(c) Hyperglycemia
(d) Mental illness or retardation

19. Glycogen is converted to glucose by
(a) Insulin
(b) Glucagon
(c) Galactose
(d) Both glucagons and insulin

20. A decrease in the level of oestrogen and progesterone causes
(a) Growth and dilation of myometrium
(b) Growth of endometrium
(c) Constriction of uterine blood vessels leading to sloughing of endometrium or uterine epithelium
(d) Release of ovum from the ovary.

21. Which of the following is a minerelocorticoid?
(a) Calciferol (b) Progesterone
(c) Adrenalin (d) Aldosterone

22. Which hormone interacts with membrane bound receptor and does not normally enter the target cell ?
(a) Follicle stimulating hormone
(b) Estrogen
(c) Thyroxin
(d) Cortisol

RESPONSE GRID				
9. ⓐⓑⓒⓓ	10. ⓐⓑⓒⓓ	11. ⓐⓑⓒⓓ	12. ⓐⓑⓒⓓ	13. ⓐⓑⓒⓓ
14. ⓐⓑⓒⓓ	15. ⓐⓑⓒⓓ	16. ⓐⓑⓒⓓ	17. ⓐⓑⓒⓓ	18. ⓐⓑⓒⓓ
19. ⓐⓑⓒⓓ	20. ⓐⓑⓒⓓ	21. ⓐⓑⓒⓓ	22. ⓐⓑⓒⓓ	

Space for Rough Work

23. The 24 hour (diurnal) rhythm of our body such as the sleep-wake cycle is regulated by the hormone
(a) Adrenaline (b) Melatonin
(c) Calcitonin (d) Prolactin

24. Steroid hormones transmit their information by
(a) stimulating the receptors present on cell membrance
(b) entering into the cell and modifying cellular contents
(c) entering into the cell and modifying nuclear organization
(d) the help of an intracellular second messenger

25. Increase in bleeding time and delay in blood coagulation is due to the deficiency of which hormone?
(a) Adrenaline (b) Noradrenaline
(c) Parathormone (d) Thyroxine

26. Estrogen and testosterone are steroid hormones, and most likely bind to
(a) membrane ion channels
(b) enzyme-linked membrane reseptors
(c) G-protein coupled membrane receptors
(d) cytoplasmic receptors

27. The only unicellular exocrine glands in our body are
(a) Sweat glands
(b) Mucus secreting goblet cells
(c) Mammary glands
(d) Sebaceous glands

28. Steroid-based hormones are able to act inside the cell. This is possible because
(a) there are no receptors for hormones on the cell surface.
(b) hormones must interact with the nucleus to have an effect.
(c) proteins carry them into the cell.
(d) steroid-based hormones are hydrophobic molecules that can pass through the cell membrane.

29. Every time you eat a cookie or candy bar, your blood sugar increases. This triggers an increase in the hormone
(a) thyroxine. (b) epinephrine.
(c) glucagon. (d) insulin.

30. Target cells
(a) react specifically to a chemical messenger.
(b) have receptors for chemical messengers
(c) secrete hormones
(d) Both a and b

31. Researchers have found increased levels of hormones from the _____ in the blood of students preparing for final exams. These hormones are produced in response to stress.
(a) thyroid gland (b) pineal gland
(c) posterior pituitary (d) adrenal glands

32. Hormones generally cause a response in a cell by
(a) interacting directly with the cell's DNA.
(b) binding with a receptor and stimulating protein production.
(c) changing the polarity of the cell membrane and causing a cascade of events within the cell.
(d) halting all other cellular activity except the required response

33. Hormones are produced at a particular centralized site and transported throughout the organism by means of
(a) a series of synapses
(b) an integrated neural pathway system.
(c) cellular communication
(d) an internal transport system

34. Injections of a hormone are sometimes given to strengthen contractions of the uterus during childbirth. What hormone might this be?
(a) adrenocorticotropic hormone (ACTH)
(b) thyroxine
(c) oxytocin (d) insulin

35. The graph below illustrates the changes in blood sugar concentration after a healthy man has drunk a glucose solution.

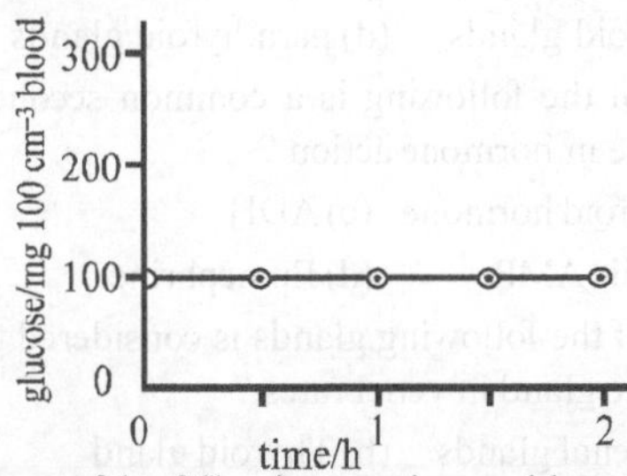

Which one of the following graphs would apply to a diabetic man in similar circumstances ?

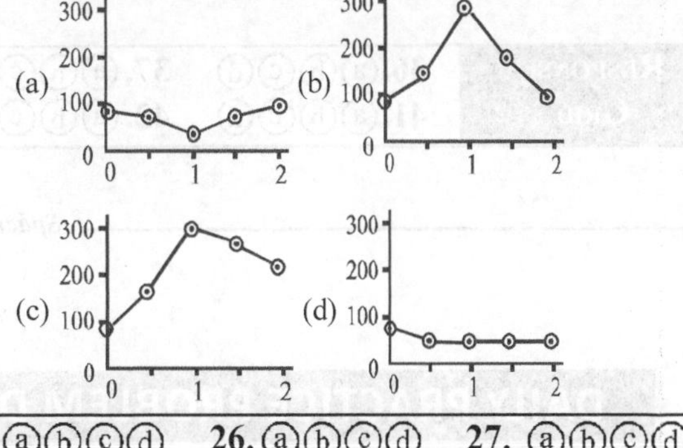

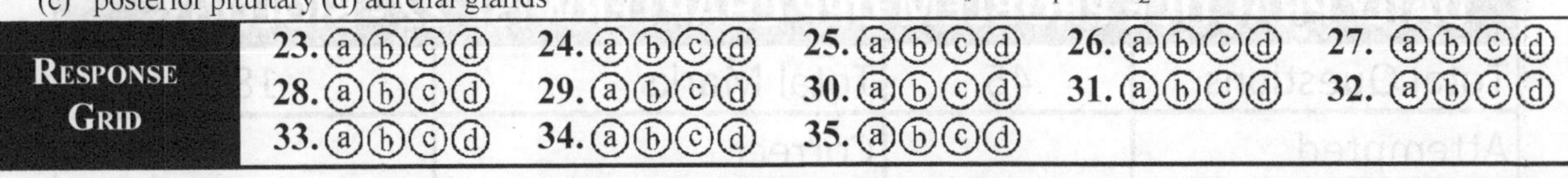

RESPONSE GRID					
	23. ⓐⓑⓒⓓ	24. ⓐⓑⓒⓓ	25. ⓐⓑⓒⓓ	26. ⓐⓑⓒⓓ	27. ⓐⓑⓒⓓ
	28. ⓐⓑⓒⓓ	29. ⓐⓑⓒⓓ	30. ⓐⓑⓒⓓ	31. ⓐⓑⓒⓓ	32. ⓐⓑⓒⓓ
	33. ⓐⓑⓒⓓ	34. ⓐⓑⓒⓓ	35. ⓐⓑⓒⓓ		

Space for Rough Work

36. A paracrine hormone is
(a) a local hormone that acts on the cell that releases it.
(b) always acting on a wide variety of target tissues.
(c) a local hormone produced at one site but active at a different site in the body.
(d) none of the above

37. "Upregulation" of hormone receptors refers to
(a) increase in hormone receptor numbers with low hormone levels.
(b) increase in hormone receptor numbers with high neurotransmitter levels
(c) increase in hormone levels produced by increase in hormone receptor numbers
(d) decrease in hormone levels produced by decrease in hormone receptor numbers

38. In an experiment, researchers removed the ______ of young mice, and as a result, these mice were able to accept organ transplants without rejection.
(a) pineal glands (b) thymus glands
(c) thyroid glands (d) parathyroid glands

39. Which of the following is a common second messenger substance in hormone action ?
(a) Thyroid hormone (b) ADH
(c) Cyclic AMP (d) Epinephrine

40. Which of the following glands is considered the "master" endocrine gland in vertebrates ?
(a) Adrenal glands (b) Thyroid gland
(c) Hypothalamus (d) Pituitary gland

41. Through negative feedback, a hormone may shut off the secretion of an anterior pituitary hormone by :
(a) stimulating the release of a (hypothalamic) releasing hormone
(b) inhibiting the release of a (hypothalamic) inhibiting hormone
(c) inhibiting the release of a (hypothalamic) releasing hormone
(d) all of the preceding.

42. ANF is a peptide hormone and is secrected from
(a) Gastrointestinal tract (b) Kidney
(c) Post. Pituitary (d) None of these

43. Steroid hormones –
(a) have only cell surface receptors
(b) are lipophobic
(c) act through altering the activity of proteins in the target cell
(d) are produced by only adrenal cortex.

44. The hormones that initiate ejection of milk, stimulates milk production and growth of ovarian follicles are respectively known as
(a) PRL, OT and LH (b) OT, PRL and FSH
(c) LH, PRL and FSH (d) PRH, OT and LH

45. A pregnant female deliver a baby who suffers from stunted growth, mental retardation/low intelligence quotient and abnormal skin. This is the result of :
(a) Low secretion of growth hormone
(b) Cancer of the thyroid gland
(c) Over secretion of pars distalis
(d) Deficiency of iodine in diet

RESPONSE GRID					
	36. ⓐⓑⓒⓓ	37. ⓐⓑⓒⓓ	38. ⓐⓑⓒⓓ	39. ⓐⓑⓒⓓ	40. ⓐⓑⓒⓓ
	41. ⓐⓑⓒⓓ	42. ⓐⓑⓒⓓ	43. ⓐⓑⓒⓓ	44. ⓐⓑⓒⓓ	45. ⓐⓑⓒⓓ

Space for Rough Work

DAILY PRACTICE PROBLEM DPP CHAPTERWISE 22 - BIOLOGY			
Total Questions	45	Total Marks	180
Attempted		Correct	
Incorrect		Net Score	
Cut-off Score	40	Qualifying Score	50
Success Gap = Net Score – Qualifying Score			
Net Score = (Correct × 4) – (Incorrect × 1)			

DPP - Daily Practice Problems

Chapter-wise Sheets

Date : | Start Time : | End Time :

BIOLOGY

SYLLABUS : Reproduction in Organism

Max. Marks : 180 **Marking Scheme : + 4 for correct & (–1) for incorrect** **Time : 60 min.**

INSTRUCTIONS : This Daily Practice Problem Sheet contains 45 MCQ's. For each question only one option is correct. Darken the correct circle/ bubble in the Response Grid provided on each page.

1. Which one of the following is not a method of vegetative propagation?
(a) Budding (b) Layering
(c) Sowing (d) Tissue culture

2. A population of genetically identical individuals, obtained from asexual reproduction is
(a) Callus (b) Clone
(c) Deme (d) Aggregate

3. Natural parthenogenesis occurs in:
(a) Protozoans (b) Earthworm
(c) All insects (d) Honeybee

4. Retention of larval characters even after sexual maturity is called
(a) Parthenogenesis (b) Ontogenesis
(c) Paedogenesis (d) Neoteny

5. Asexual reproduction is an effective strategy in stable environments because
(a) gametogenesis is most efficient under these conditions.
(b) the offspring, genetically identical to their parents, are preadapted to the environment.
(c) asexual parthenogenesis produces a large amount of genetic diversity.
(d) animal cells tend to be more totipotent under stable conditions.

6. If you compared the genetic makeup of an animal produced by parthenogenesis with that of its mother, which of the following would you expect?
(a) About 100 percent genetic similarity
(b) About 50 percent genetic similarity
(c) No genetic similarity
(d) Parthenogenetic animals have no mother

RESPONSE GRID	1. ⓐⓑⓒⓓ	2. ⓐⓑⓒⓓ	3. ⓐⓑⓒⓓ	4. ⓐⓑⓒⓓ	5. ⓐⓑⓒⓓ
	6. ⓐⓑⓒⓓ				

Space for Rough Work

7. Which of the following statements about animals that utilize external fertilization is *false*?
(a) They are divided equally between terrestrial and aquatic species.
(b) Many produce large numbers of gametes to ensure successful reproduction.
(c) The behaviours associated with mating are often highly synchronized.
(d) The probability of any one egg being fertilized and developing into an adult can be low.

8. Which of the following statements about animal reproduction is *false* ?
(a) Species that reproduce sexually cannot also reproduce asexually.
(b) Viviparity, but not ovoviviparity, is common in mammals.
(c) Male insects can remove spermatophores deposited in a female by other males.
(d) Oogenesis and spermatogenesis both occur in simultaneous hermaphrodites.

9. Which of the following animals qualifies as a sexually reproducing, oviparous species ?
(a) Human (b) Chicken
(c) Kangaroo (d) Sea star

10. Benefits of asexual reproduction include all of the following except
(a) it often allows for the production of many more offspring at the same time
(b) it is advantageous in changing environments in which population variety is the key to successful propagation of a species
(c) it is easier in certain environments to have offspring without searching for a mate
(d) allowing the conservation of resources otherwise allocated to finding mates and performing ritualized courtship.

11. All the 'eyes' of a potato tuber are taken out and it is sown in the ground normally. New plants will
(a) Not emerge (b) Emerge normally
(c) Be weaker (d) Be healthier

12. Basal half of an onion bulb is removed and upper half is sown in the ground. New plant will
(a) Emerge normally (b) Not emerge
(c) Be without leaves (d) Be without flowers

13. A small portion of cane-sugar stem between the two successive nodes is cut off and sown in the soil normally. New plants will
(a) Be formed normally (b) Not be formed
(c) Be without juice (d) Without nodes

14. A cutting of rose plant is thoroughly waxed and planted in the soil normally, It will form
(a) New rose plant
(b) A dead piece of rose stem after some time
(c) A rose plant of improved variety
(d) None of these

15. When an ovary develops into a fruit, without fertilization, it is called
(a) apospory (b) apogamy
(c) parthenocarpy (d) porogamy

16. Asexual reproduction is the best strategy for plants
(a) that are well adapted to their stable environment.
(b) as winter approaches
(c) when new genes must be introduced
(d) that have underground stems.

17. Bamboo reproduces by
(a) rhizomes (b) tubers
(c) corms (d) stolons

18. Grafting is an example of asexual reproduction. Which of the following choices is an example of asexual reproduction involving nonvegetative parts of a plant ?
(a) Apomixis (b) Production of corms
(c) Production of bulbs (d) Production of rhizomes

19. What is necessary for successful grafting to occur ?
(a) Each section must be able to form roots.
(b) The grafted section must be able to form seeds.
(c) Fusion of the two vascular tissues must occur.
(d) Fusion of the two cambial tissues must occur.

20. Banana fruits are seedless, because
(a) auxins are sprayed for rapid development of fruit.
(b) vegetative propagation of plant.
(c) they are triploid plants.
(d) fruits are artificially ripened.

RESPONSE GRID					
	7. ⓐⓑⓒⓓ	8. ⓐⓑⓒⓓ	9. ⓐⓑⓒⓓ	10. ⓐⓑⓒⓓ	11. ⓐⓑⓒⓓ
	12. ⓐⓑⓒⓓ	13. ⓐⓑⓒⓓ	14. ⓐⓑⓒⓓ	15. ⓐⓑⓒⓓ	16. ⓐⓑⓒⓓ
	17. ⓐⓑⓒⓓ	18. ⓐⓑⓒⓓ	19. ⓐⓑⓒⓓ	20. ⓐⓑⓒⓓ	

Space for Rough Work

21. Consider the following statements and choose the correct option.
(i) The genetic constitution of a plant is unaffected in vegetative propagation.
(ii) Rhizome in ginger serves as an organ of vegetative reproduction.
(iii) Totipotency of cells enables us to micropropagate plants.
(a) Statements (i) and (ii) alone are true
(a) Statements (ii) and (iii) alone are true
(c) Statement (ii) alone is true
(d) All the three statements (i), (ii) and (iii) are true

22. Plants identical to mother plants can be obtained from
(a) seeds (b) stem cutting
(c) Both (a) and (b) (d) None of these

23. Ploidy of ovary, anther, egg, pollen, male gamete and zygote are respectively-
(a) 2n, 2n, n, 2n, n, 2n (b) 2n, 2n, n, n, n, 2n
(c) 2n, n, n, n, n, n (d) 2n, 2n, n, 2n, 2n, 2n

24. Offsprings of oviparous animals are at greater risk as compared to offsprings of viviparous animals because-
(a) Proper embryonic care and protection is lesser
(b) Embryo is not developed
(c) Progenies are with more variation
(d) Progenies are larger

25. The parameter(s) of senescence or old age is/are-
(a) End of the reproductive phase
(b) Concomitant change in body (like slowing metabolism)
(c) Failure of metabolism decreases
(d) Both (a) and (b)

26. The terms homothallic and monoecious are used to denote
(a) bisexual condition (b) unisexual condition
(c) staminate flowers (d) pistillate flowers

27. During regeneration, modification of an organ to other organ is known as
(a) Morphogenesis (b) Epimorphosis
(c) Morphallaxis (d) Accretionary growth

28. Cells become variable in morphology and function in different regions of the embryo. The process is
(a) differentiation (b) metamorphosis
(c) organisation (d) rearrangement

29. Earthworms, sponges, tapeworms are
(a) bisexual animals (b) unisexual animals
(c) hermaphrodites (d) Both (a) and (c)

30. The site of origin of the new plantlets in potato, dahlia, ginger and banana is-
(a) Floral buds present on stem
(b) Internodes of modified stem
(c) Nodes of modified stem
(d) Adventitious buds present on root

31. Among the following which one is not a method of vegetative propagation?
(a) Budding (b) Layering
(c) Sowing (d) Tissue culture

32. Vegetative propagation in mint occurs by:
(a) offset (b) rhizome
(c) sucker (d) runner

33. What is common between vegetative reproduction and apomixis?
(a) Both are applicable to only dicot plants
(b) Both bypass the flowering phase
(c) Both occur round the year
(d) Both produces progeny identical to the parent

34. Individuals of a clone-
(a) Are genetically similar but morphologically different
(b) Are morphologically similar but genetically different
(c) Are morphologically and genetically similar
(d) Are genetically and phenotypically different

35. Some organisms are capable of asexual or sexual reproduction. Under favourable conditions, reproduction proceeds asexually. When conditions become more stressful reproduction switchess to a sexual mode. Why?
(a) Sexual reproduction is simple and more rapid allowing larger numbers of offspring to be produced
(b) Sexual reproduction requires two separate individuals, who can mutually provide nutrient support during stress
(c) Sexual reproduction produces individuals with new combinations of recombined chromosomes increasing diversity
(d) Asexual reproduction requires more energy

RESPONSE GRID					
	21. ⓐⓑⓒⓓ	22. ⓐⓑⓒⓓ	23. ⓐⓑⓒⓓ	24. ⓐⓑⓒⓓ	25. ⓐⓑⓒⓓ
	26. ⓐⓑⓒⓓ	27. ⓐⓑⓒⓓ	28. ⓐⓑⓒⓓ	29. ⓐⓑⓒⓓ	30. ⓐⓑⓒⓓ
	31. ⓐⓑⓒⓓ	32. ⓐⓑⓒⓓ	33. ⓐⓑⓒⓓ	34. ⓐⓑⓒⓓ	35. ⓐⓑⓒⓓ

Space for Rough Work

36. Apomixis in plant means development of a plant
(a) from root cuttings
(b) without fusion of gametes
(c) from fusion of gametes
(d) from cuttings of stem

37. Which of the following is not vegetative propagule ?
(a) Rhizome and sucker
(b) Tuber and offset
(c) Bulbil (*e.g.* in *Agave*), leaf buds and bulb
(d) Antherozoid

38. Which of the following is false about external fertilization?
(a) Organisms showing external fertilization exhibit great synchrony between the sexes and release gametes.
(b) Gametes are produced in large number in water to enhance the chances of fertilization.
(c) A large number of gametes are wasted.
(d) A major advantage is that the offspring are protected from predators and there is a great chance of their survival upto adulthood.

39. Modified stem present in *Gladiolus* is:
(a) bulb (b) rhizome
(c) corm (d) bulbil

40. Which of the following are seasonal breeders?
(a) Frogs (b) Birds
(c) Lizards (d) All of these

41. Select the wrong statement:
(a) Anisogametes differ either in structure, function or behaviour.
(b) In oomycetes female gamete is smaller and motile, while male gamete is larger and non-motile.
(c) *Chalmydomonas* exhibits both isogamy and anisogamy and *Fucus* shows oogamy.
(d) Isogametes are similar in structure, function and behaviour.

42. Monoecious plant of *Chara* shows occurrence of :
(a) stamen and carpel of the same plant
(b) upper antheridium and lower oogonium on the same plant
(c) upper oogonium and lower antheridium on the same plant
(d) antheridiophore and archegoniophore on the same plant

43. Which of the following statement(s) is/are false about internal fertilization?
(i) Male gametes are motile.
(ii) Male gametes are non-motile.
(iii) Male gametes are produced in large number.
(iv) Male gametes are produced in small number.
(v) There is a significant reduction in the number of eggs produced.
(a) (i), (iii) and (v) (b) (iii) and (iv)
(c) (ii) and (iv) (d) Only (v)

44. Syngamy can occur outside the body of the organism in
(a) Fungi (b) Mosses
(c) Algae (d) Ferns

45. Select the correct sequence from the following.
(i) Juvenile phase → Senescent phase → Reproductive phase
(ii) Juvenile phase → Reproductive phase → Senescent phase
(iii) Reproductive phase → Juvenile phase → Senescent phase
(iv) Vegetative phase → Reproductive phase → Senescent phase
(a) (i) and (ii) (b) (i) and (iv)
(c) (iii) and (iv) (d) (ii) and (iv)

RESPONSE GRID					
	36. ⓐⓑⓒⓓ	37. ⓐⓑⓒⓓ	38. ⓐⓑⓒⓓ	39. ⓐⓑⓒⓓ	40. ⓐⓑⓒⓓ
	41. ⓐⓑⓒⓓ	42. ⓐⓑⓒⓓ	43. ⓐⓑⓒⓓ	44. ⓐⓑⓒⓓ	45. ⓐⓑⓒⓓ

Space for Rough Work

DAILY PRACTICE PROBLEM DPP CHAPTERWISE 23 - BIOLOGY			
Total Questions	45	Total Marks	180
Attempted		Correct	
Incorrect		Net Score	
Cut-off Score	55	Qualifying Score	65
Success Gap = Net Score – Qualifying Score			
Net Score = (Correct × 4) – (Incorrect × 1)			

DPP - Daily Practice Problems

Chapter-wise Sheets

Date : | Start Time : | End Time :

BIOLOGY CB24

SYLLABUS : Sexual Reproduction in Flowering Plants

Max. Marks : 180 | **Marking Scheme :** + 4 for correct & (–1) for incorrect | **Time : 60 min.**

INSTRUCTIONS : This Daily Practice Problem Sheet contains 45 MCQs. For each question only one option is correct. Darken the correct circle/ bubble in the Response Grid provided on each page.

1. Entry of pollen tube through micropyle is
   (a) Chalazogamy (b) Mesogamy
   (c) Porogamy (d) Pseudogamy
2. Dry indehiscent single-seeded fruit formed from bicarpellary syncarpous inferior ovary is
   (a) Caryopsis (b) Cypsela
   (c) Berry (d) Cremocarp
3. One of the most resistant biological material is
   (a) lignin (b) hemicellulose
   (c) lignocellulose (d) sporopollenin
4. When funiculum, chalaza, and micropyle lie in one straight line, then ovule is called –
   (a) Amphitropous (b) Orthotropous
   (c) Campylotropous (d) Anatropous
5. Which one of the following is a fruit ?
   (a) Ginger (b) Sweet potato
   (c) Radish (d) Lady's finger
6. Female gametophyte of angiosperms is represented by
   (a) Ovule
   (b) Megaspore mother cell
   (c) Embryo sac
   (d) Nucellus
7. Which of the following pair has haploid structures?
   (a) Nucellus and antipodal cells
   (b) Antipodal cells and egg cell
   (c) Antipodal cells and megaspore mother cell
   (d) Nucellus and primary endosperm nucleus
8. Polyembryony commonly occurs in
   (a) citrus (b) turmeric
   (c) tomato (d) potato

RESPONSE GRID					
	1. ⓐⓑⓒⓓ	2. ⓐⓑⓒⓓ	3. ⓐⓑⓒⓓ	4. ⓐⓑⓒⓓ	5. ⓐⓑⓒⓓ
	6. ⓐⓑⓒⓓ	7. ⓐⓑⓒⓓ	8. ⓐⓑⓒⓓ		

Space for Rough Work

9. Chasmogamy refers to the condition where –
   (a) Flowers remains closed
   (b) Flowers are absent
   (c) Flowers open
   (d) Flowers gamopetalous
10. In oogamy, fertilization involves
   (a) a small non-motile female gamete and a large motile male gamete
   (b) a large non-motile female gamete and a small motile male gamete
   (c) a large non-motile female gamete and a small nonmotile male gamete
   (d) a large motile female gamete and a small non-motile male gamete
11. Cotyledons and testa respectively are edible parts in
   (a) walnut and tamarind
   (b) french bean and coconut
   (c) cashew nut and litchi
   (d) groundnut and pomegranate
12. A *Polygonum* type of embryo sac is
   (a) 7-celled and 8-nucleate
   (b) 8-celled and 7-nucleate
   (c) 7-celled and 7-nucleate
   (d) 8-celled and 8-nucleate
13. In dicot embryo the radicle is formed by
   (a) epibasal tier of embryo
   (b) hypobasal tier of embryo
   (c) hypophysis of suspensor
   (d) terminal cell of suspensor
14. A drupe develops in
   (a) mango (b) wheat
   (c) pea (d) tomato
15. When pollen grains are not transferred from anthers to stigma in a flower, due to the barrier, it is called
   (a) herkogamy (b) heterogamy
   (c) cleistogamy (d) dichogamy
16. The parenchyma tissue which forms the bulk of ovule where the sporogenous tissue is produced is –
   (a) Megaspore mother cell
   (b) Nucellus
   (c) Ovule
   (d) Embryo sac
17. Unisexuality of flowers prevents
   (a) autogamy, but not geitonogamy
   (b) both geitonogamy and xenogamy
   (c) geitonogamy, but not xenogamy
   (d) autogamy and geitonogamy
18. Which one of the following represents an ovule, where the embryo sac becomes horse-shoe shaped and the funiculus and micropyle are close to each other?
   (a) Amphitropous (b) Circinotropous
   (c) Atropous (d) Anotropous
19. If an angiospermic male plant is diploid and female plant tetraploid, the ploidy level of endosperm will be
   (a) haploid (b) triploid
   (c) tetraploid (d) pentaploid
20. These plants flower and fruit only once in their life time and die after fruiting. These are
   (a) monocarpic plants
   (b) polycarpic plants
   (c) vegetative plants
   (d) reproductive plants
21. Select the mismatched pair.
   (a) Microsporangium — Pollen sac
   (b) Megasporangium — Nucellus
   (c) Pollen grain — Male gamete
   (d) Embryo sac — Female gametophyte

RESPONSE GRID					
	9. ⓐⓑⓒⓓ	10. ⓐⓑⓒⓓ	11. ⓐⓑⓒⓓ	12. ⓐⓑⓒⓓ	13. ⓐⓑⓒⓓ
	14. ⓐⓑⓒⓓ	15. ⓐⓑⓒⓓ	16. ⓐⓑⓒⓓ	17. ⓐⓑⓒⓓ	18. ⓐⓑⓒⓓ
	19. ⓐⓑⓒⓓ	20. ⓐⓑⓒⓓ	21. ⓐⓑⓒⓓ		

Space for Rough Work

22. A typical angiospermous ovule is attached to the placenta by means of a stalk called X. Body of the ovule fuses with X in the region called Y. Identify X and Y.

	X	Y
(a)	Funicle	Hilum
(b)	Hilum	Funicle
(c)	Funicle	Micropyle
(d)	Hilum	Chalaza

23. Which of the following options is correct?
(a) Transfer of pollen grains from the anther to the stigma of the same flower – Autogamy.
(b) Transfer of pollen grains from the anther of one flower to the stigma of another flower of same plant – Geitonogamy.
(c) Transfer of pollen grains from the anther to the stigma of a genetically different plant – Xenogamy.
(d) All of these

24. Which of the following is not a water pollinated plant?
(a) *Zostera* (b) *Vallisneria*
(c) *Hydrilla* (d) *Cannabis*

25. Identify the parts labelled A, B, C and D in the given figure and select the correct option.

	A	B	C	D
(a)	Synergids	Antipodals	Egg	Filiform apparatus
(b)	Antipodals	Synergids	Egg	Filiform apparatus
(c)	Antipodals	Synergids	Filiform appartus	Egg
(d)	Polar nuclei	Antipodals	Filiform appartus	Egg

26. Growth of pollen tube towards embryo sac is
(a) chemotropic (b) thigmotaxis
(c) geotropic (d) none of these

27. The part of gynoecium that determines the complatible nature of pollen is
(a) stigma (b) style
(c) ovary (d) synergids

28. The innermost layer of anther is tapetum whose function is
(a) dehiscence (b) mechanical
(c) nutrition (d) protection

29. The female gametophyte of a typical dicot at the time of fertilization is
(a) 8-celled (b) 7-celled
(d) 6-celled (d) 5-celled

30. One of the most resistant biological material present in the exine of pollen grain is
(a) pectocellulose (b) sporopollenin
(c) suberin (d) cellulose

31. What is the function of germ pore?
(a) Emergence of radicle
(b) Absorption of water for seed germination
(c) Initiation of pollen tube
(d) All of these

32. How many meiotic divisions are required for the formation of 100 functional megaspores?
(a) 100 (b) 50
(c) 25 (d) 26

33. Study of pollen grains is called
(a) micrology (b) anthology
(c) palynology (d) pomology

RESPONSE GRID					
	22. ⓐⓑⓒⓓ	23. ⓐⓑⓒⓓ	24. ⓐⓑⓒⓓ	25. ⓐⓑⓒⓓ	26. ⓐⓑⓒⓓ
	27. ⓐⓑⓒⓓ	28. ⓐⓑⓒⓓ	29. ⓐⓑⓒⓓ	30. ⓐⓑⓒⓓ	31. ⓐⓑⓒⓓ
	32. ⓐⓑⓒⓓ	33. ⓐⓑⓒⓓ			

Space for Rough Work

34. Which of these is a condition that makes flowers invariably autogamous?
(a) Dioecy
(b) Self incompatibilty
(c) Cleistogamy
(d) Xenogamy

35. An embryo may sometimes develop from any cell of embryo sac other than egg. It is termed as
(a) apospory (b) apogamy
(c) parthenogenesis (d) parthenocarpy

36. Endosperm is completely consumed by the developing embryo in
(a) pea and groundnut
(b) maize and castor
(c) castor and groundnut
(d) maize and pea

37. The portion of embryonal axis between plumule (future shoot) and cotyledons is called
(a) hypocotyl (b) epicotyl
(c) coleohize (d) coleoptile

38. Stigma of a flower is removed before pollination. The flower will
(a) Form fruit normally
(b) Not form fruit
(c) Form deformed fruit
(d) Form fruit smaller than normal size

39. In a pea flower, all petals are removed before pollination. The flower will
(a) Form fruit normally
(b) Not form fruit
(c) Form smaller pod
(d) Form deformed pod

40. A drop of glue is placed on the stigma of a flower before pollination. The flower will
(a) Not form fruit
(b) Form normal fruit
(c) Form sticky fruit
(d) Form fruit filled with glue

41. An apomictic seed contains an embryo that is
(a) produced when two sperm fertilize one egg.
(b) developed from one egg alone.
(c) the result of parental self-fertilization
(d) genetically identical to its parent.

42. The embryo is carefully taken out of pea seed and sown in the soil and watered normally. New plant will
(a) Be healthier
(b) Be weaker
(c) Not be formed
(d) Be formed normally

43. Angiospermic plant has chromosome number of 24. The number of chromosomes in pollens will be
(a) 6 (b) 12
(c) 24 (d) 48

44. What is the fate of the seven cells of the embryo sac ?
(a) All but one disintegrate upon fertilization.
(b) Two become fertilized; the others disintegrate.
(c) Two become fertilized; the others fuse to form endosperm.
(d) All are involved in nuclear fusion events.

45. A close relation between flower and pollinating agent is best exhibited by :
(a) *Cocos* (b) *Salvia*
(c) *Yucca* (d) *Avena*

RESPONSE GRID					
	34. ⓐⓑⓒⓓ	35. ⓐⓑⓒⓓ	36. ⓐⓑⓒⓓ	37. ⓐⓑⓒⓓ	38. ⓐⓑⓒⓓ
	39. ⓐⓑⓒⓓ	40. ⓐⓑⓒⓓ	41. ⓐⓑⓒⓓ	42. ⓐⓑⓒⓓ	43. ⓐⓑⓒⓓ
	44. ⓐⓑⓒⓓ	45. ⓐⓑⓒⓓ			

Space for Rough Work

DAILY PRACTICE PROBLEM DPP CHAPTERWISE 24 - BIOLOGY			
Total Questions	45	Total Marks	180
Attempted		Correct	
Incorrect		Net Score	
Cut-off Score	40	Qualifying Score	55
Success Gap = Net Score – Qualifying Score			
Net Score = (Correct × 4) – (Incorrect × 1)			

DPP - Daily Practice Problems

Chapter-wise Sheets

Date : | Start Time : | End Time :

BIOLOGY

SYLLABUS : Human Reproduction

Max. Marks : 180 **Marking Scheme : + 4 for correct & (–1) for incorrect** **Time : 60 min.**

INSTRUCTIONS : This Daily Practice Problem Sheet contains 45 MCQs. For each question only one option is correct. Darken the correct circle/ bubble in the Response Grid provided on each page.

1. Breast feeding suspends pregnancy due to
   (a) post pregnancy lower levels of FSH and LH
   (b) post pregnancy higher levels of FSH and LH which put negative check on ovulation
   (c) inhibiting the release of LH by prolactin and thus countering the effects of LH on the ovarian follicles
   (d) increasing the release of inhibin by prolactin and thus countering the effects of FSH on the ovarian follicles
2. The human embryo, with 8 to 16 blastomere is called
   (a) Morula (b) Blastula
   (c) Gastrula (d) Foetus
3. A person which shows the secondary sexual characters of both male and female is called –
   (a) Intersex (b) Hermaphrodite
   (c) Bisexual (d) Gynandromorph
4. Which one of the following is the correct matching of the events occurring during menstrual cycle?
   (a) Proliferative phase: Rapid regeneration of myometrium and maturation of Graffian follicle.
   (b) Development of corpus luteum : Secretory phase and increased secretion of progesterone.
   (c) Menstruation: Breakdown of myometrium and ovum not fertilised.
   (d) Ovulation: LH and FSH attain peak level and sharp fall in the secretion of progesterone.
5. The second maturation division of the mammalian ovum occurs:
   (a) in the graafian follicle following the first maturation division
   (b) shortly after ovulation before the ovum makes entry into the Fallopian tube
   (c) until after the ovum has been penetrated by a sperm
   (d) until the nucleus of the sperm has fused with that of the ovum

RESPONSE GRID	1. ⓐⓑⓒⓓ	2. ⓐⓑⓒⓓ	3. ⓐⓑⓒⓓ	4. ⓐⓑⓒⓓ	5. ⓐⓑⓒⓓ

Space for Rough Work

6. Foetal ejection reflex in human female is induced by
(a) release oxytocin from pituitary
(b) fully developed foetus and placenta
(c) differentiation of mammary glands
(d) pressure exerted by amniotic fluid

7. Which one of the following statements about human sperm is correct?
(a) Acrosome has a conical pointed structure used for piercing and penetrating the egg, resulting in fertilisation
(b) The sperm lysine in the acrosome dissolve the egg envelope facilitating fertilisation
(c) Acrosome serves as a sensory structure leading the sperm towards the ovum
(d) Acrosome serves no particular function

8. Sertoli cells are found in
(a) ovaries and secrete progesterone
(b) adrenal cortex and secrete adrenaline
(c) seminiferous tubules and provide nutrition to germ cells
(d) pancreas and secrete cholecystokinin

9. The phase of menstrual cycle in humans that lasts for 7-8 days, is
(a) follicular phase (b) ovulatory phase
(c) luteal phase (d) menstruation

10. Cessation of menstrual cycle in women is called
(a) menopause (b) lactation
(c) ovulation (d) parturition

11. In human female the blastocyst
(a) Forms placenta even before implantation
(b) Gets implanted into uterus 3 days after ovulation
(c) Gets nutrition from uterine endometrial secretion only after implantation
(d) Gets implanted in endometrium by the trophoblast cells

12. Bulbourethral gland is present in
(a) Males and is another name for Uterus masculina
(b) Females and is another name for Bertholin's gland
(c) Males and is another name for Cowper's gland
(d) None of these

13. Which one of the following are rich in fructose, calcium and some enzymes?
(a) Male accessory glands
(b) Liver
(c) Pancreas
(d) Salivary glands

14. Umbilical cord has
(a) Two arteries carrying blood to placenta and one vein returning blood to foetus
(b) One artery carrying blood to placenta and two veins returning blood to foetus
(c) Two arteries bringing blood to foetus and one vein carrying blood to placenta
(d) One artery bringing blood to foetus and two veins carrying blood to placenta

15. Which of the following induces parturition ?
(a) Vasopressin
(b) Oxytocin
(c) Growth hormone
(d) Thyroid stimulating hormone

16. What happens during fertilisation in humans after many sperms reach close to the ovum?
(a) Cells of corona radiata trap all the sperms except one
(b) Only two sperms nearest the ovum penetrate zona pellucida
(c) Secretions of acrosome helps one sperm enter cytoplasm of ovum through zona pellucida
(d) All sperms except the one nearest to the ovum lose their tails

17. The part of fallopian tube closest to the ovary is
(a) isthmus (b) infundibulum
(c) cervix (d) ampulla

18. The chemical substance found in the surface layer of cytoplasm of spermatozoa is:
(a) fertilizin (b) agglutinin
(c) antifertilizin (d) hyaluronidase

RESPONSE GRID					
	6. ⓐⓑⓒⓓ	7. ⓐⓑⓒⓓ	8. ⓐⓑⓒⓓ	9. ⓐⓑⓒⓓ	10. ⓐⓑⓒⓓ
	11. ⓐⓑⓒⓓ	12. ⓐⓑⓒⓓ	13. ⓐⓑⓒⓓ	14. ⓐⓑⓒⓓ	15. ⓐⓑⓒⓓ
	16. ⓐⓑⓒⓓ	17. ⓐⓑⓒⓓ	18. ⓐⓑⓒⓓ		

Space for Rough Work

19. Which of the following diseases is caused by the under secretion of cortisol?
(a) Anaemia
(b) Addison's disease
(c) Hyperglycemia
(d) Mental illness or retardation

20. Reabsorption of chloride ions from glomerular filtrate in kidney tubule occurs by
(a) Active transport (b) Diffusion
(c) Osmosis (d) Brownian movement

21. If for some reason, the vasa efferentia in the human reproductive system get blocked, the gametes will not be transported from
(a) testes to epididymis
(b) epididymis to vas deferens
(c) ovary to uterus
(d) vagina to uterus

22. The nutritive cells found in seminiferous tubules are
(a) Leydig's cells (b) atretic follicular cells
(c) Sertoli cells (d) chromaffin cells.

23. Seminal plasma in humans is rich in
(a) fructose and calcium but has no enzymes
(b) glucose and certain enzymes but has no calcium
(c) fructose and certain enzymes but poor in calcium
(d) fructose, calcium and certain enzymes

24. The function of the secretion of prostate gland is to
(a) inhibit sperm activity
(b) attract sperms
(c) stimulate sperm activity
(d) none of these

25. The head of the epididymis at the head of the testis is called
(d) cauda epididymis
(b) vas deferens
(c) caput epididymis
(d) gubernaculum

26. Which part of ovary in mammals acts as an endocrine gland after ovulation?
(a) Stroma
(b) Germinal epithelium
(c) Vitelline membrane
(d) Graafian follicle

27. The female external genitalia include
(i) Ovary (ii) Mammary gland
(iii) Mons pubis (iv) Clitoris
(a) (i) and (ii) (b) (ii) and (iii)
(c) (iii), (iv) and (v) (d) (ii), (iii) and (v)

28. 2n = 16 is in a primary spermatocyte which is in metaphase of first meiotic division. What shall be the total number of chromatids in each of the secondary spermatocyte?
(a) 16 (b) 24
(c) 32 (d) 8

29. In humans, at the end of the first meiotic division, the male germ cells differentiate into the
(a) spermatids
(b) spermatogionia
(c) primary spermatocytes
(d) secondary spermatocytes

30. The sperms undergo physiological maturation, acquiring increased motility and fertilizing capacity in
(a) seminiferous tubules
(b) vasa efferentia
(c) epididymis
(d) vagina

31. At what stage of life is oogenesis initiated in a human female?
(a) At puberty
(b) During menarch
(c) During menopause
(d) During embryonic development

32. The middle piece of the sperm contains
(a) proteins (b) mitochondria
(c) centriole (d) nucleus

RESPONSE GRID					
	19. ⓐⓑⓒⓓ	20. ⓐⓑⓒⓓ	21. ⓐⓑⓒⓓ	22. ⓐⓑⓒⓓ	23. ⓐⓑⓒⓓ
	24. ⓐⓑⓒⓓ	25. ⓐⓑⓒⓓ	26. ⓐⓑⓒⓓ	27. ⓐⓑⓒⓓ	28. ⓐⓑⓒⓓ
	29. ⓐⓑⓒⓓ	30. ⓐⓑⓒⓓ	31. ⓐⓑⓒⓓ	32. ⓐⓑⓒⓓ	

Space for Rough Work

33. Corpus luteum is developed from
(a) Graafian follicle (b) nephrostome
(c) oocyte (d) none of these

34. Spermatogenesis is induced by
(a) FSH (b) ICSH
(C) STH (d) ATH

35. Layers of an ovum from outside to inside is
(a) corona radiata, zona pellucida and vitelline membrane
(b) zona pellucida, corona radiata and vitelline membrane
(c) vitelline membrane, zona pellucida and corona radiata
(d) zona pellucida, vitelline membrane and corona radiata

36. Name the hormone that has no role in menstruation.
(a) LH (b) FSH
(c) Estradiol (d) TSH

37. The time for optimum chances of conception in a woman is ________ starting from the day of menstruation.
(a) 1st day (b) 4th day
(c) 14th day (d) 26th day

38. Repair of endometrium is undertaken by
(a) LH (b) FSH
(c) estrogen (d) prolactin

39. The sex of the foetus will be decided at
(a) fertilization by male gamete
(b) implantation
(c) fertilization by female gamete
(d) the start of cleavage

40. Match Column-I with Column-II and select the correct option from the codes given below.

	Column-I		Column-II
A.	Cleavage	(i)	Fertilization
B.	Morula	(ii)	Mitotic divisions
C.	Polyspermy	(iii)	Endometric
D.	Implantation	(iv)	Little mulberry

(a) A-(ii), B-(iv), C-(i), D-(iii)
(b) A-(i), B-(iv), C-(ii), D-(iii)
(c) A-(iv), B-(ii), C-(i), D-(iii)
(d) A-(ii), B-(iv), C-(iii), D-(i)

41. Which part of the sperm plays an important role in penetrating the egg membrane?
(a) Allosome (b) Tail
(c) Autosome (d) Acrosome

42. Which of the following hormones is not a secretory product of human placenta?
(a) Human chorionic gonadotropin
(b) Prolactin
(c) Estrogen
(d) Progesterone

43. After birth, colostrum is released from mammary glands which is rich in
(a) fat and low in proteins
(b) proteins and low in fat
(c) proteins, antibodies and low in fat
(d) proteins, fat and low in antibodies

44. The correct sequence in the process of development of human embryo is
(a) fertilization—zygote—cleavage—morula—blastula—gastrula
(b) fertilization—cleavage—morula—zygote—blastula—gastrula
(c) fertilization—zygote—blastula—morula—cleavage—gastrula
(d) cleavage—zygote—fertilization—morula—blastula—gastrula

45. The given figure refers to which type of reproduction in yeast?

(a) Binary fission (b) Budding
(c) Layering (d) Fusion

RESPONSE GRID				
33. ⓐⓑⓒⓓ	34. ⓐⓑⓒⓓ	35. ⓐⓑⓒⓓ	36. ⓐⓑⓒⓓ	37. ⓐⓑⓒⓓ
38. ⓐⓑⓒⓓ	39. ⓐⓑⓒⓓ	40. ⓐⓑⓒⓓ	41. ⓐⓑⓒⓓ	42. ⓐⓑⓒⓓ
43. ⓐⓑⓒⓓ	44. ⓐⓑⓒⓓ	45. ⓐⓑⓒⓓ		

Space for Rough Work

DAILY PRACTICE PROBLEM DPP CHAPTERWISE 25 - BIOLOGY			
Total Questions	45	Total Marks	180
Attempted		Correct	
Incorrect		Net Score	
Cut-off Score	55	Qualifying Score	60
Success Gap = Net Score – Qualifying Score			
Net Score = (Correct × 4) – (Incorrect × 1)			

DPP - Daily Practice Problems

Chapter-wise Sheets

Date : **Start Time :** **End Time :**

BIOLOGY CB26

SYLLABUS : Reproductive Health

Max. Marks : 180 **Marking Scheme : + 4 for correct & (–1) for incorrect** **Time : 60 min.**

INSTRUCTIONS : This Daily Practice Problem Sheet contains 45 MCQs. For each question only one option is correct. Darken the correct circle/ bubble in the Response Grid provided on each page.

1. Consider the statements given below regarding contraception and answer as directed thereafter:
   (i) Medical Termination of Pregnancy (MTP) during first trimester is generally safe
   (ii) Generally chances of conception are nil until mother breast-feeds the infant upto two years
   (iii) Intrauterine devices like copper-T are effective contraceptives
   (iv) Contraception pills may be taken upto one week after coitus to prevent conception
   Which two of the above statements are correct?
   (a) ii and iii (b) iii and iv
   (c) i and iii (d) i and ii
2. Amniocentesis is a technique
   (a) by which the essential amino acids in the body can be estimated
   (b) by which any chromosomal anomalies in the foetus can be detected
   (c) in which the sex of the foetus can be reversed
   (d) that can be used for correcting genetic disorders of the foetus
3. Tying up or removing a small part of fallopian duct is called
   (a) Vasectomy (b) Ductus arteriosus
   (c) Archidectomy (d) Tubectomy
4. Copper-T is a device that prevents
   (a) implantation of blastocyst
   (b) ovulation
   (c) fertilization
   (d) egg maturation
5. The diaphragm, cervical cap and vaults are
   (a) Disposable contraceptive devices
   (b) Reusable contraceptives
   (c) IUDs
   (d) Implants
6. The transfer of zygote or early embryo (up to 8 blastomere) into fallopian tube is
   (a) IVF and ET (b) ZIFT
   (c) GIFT (d) IUT

RESPONSE GRID	1. ⓐⓑⓒⓓ	2. ⓐⓑⓒⓓ	3. ⓐⓑⓒⓓ	4. ⓐⓑⓒⓓ	5. ⓐⓑⓒⓓ
	6. ⓐⓑⓒⓓ				

Space for Rough Work

7. The technique called gamete intrafallopian transfer (GIFT) is recommended for those females:
   (a) who cannot produce an ovum
   (b) who cannot retain the foetus inside uterus.
   (c) whose cervical canal is too narrow to allow passage for the sperms
   (d) who cannot provide suitable environment for fertilisation
8. The copper ions of IUDs
   (a) Suppress the motility and fertilization capacity of sperms
   (b) Make the uterus unsuitable for implantation
   (c) Increase phagocytosis of sperms
   (d) Make cervix hostile to sperms
9. Condoms are barriers that cover–
   (a) Penis in male and ovary in female.
   (b) Penis in male and cervix and vagina in female.
   (c) Scrotum in male and cervix and vagina in female.
   (d) Cervix in male and vagina in female.
10. Birth control pills check ovulation in female by inhibiting the secretion of –
   (a) follicle stimulating hormone
   (b) luteinizing hormone
   (c) Both (a) and (b)
   (d) None of the above
11. Identical twins are produced when
   (a) One fertilized egg divides and two blastomeres separate
   (b) One sperm fertilizes two eggs
   (c) One egg is fertilized with two sperms
   (d) Two eggs are fertilized
12. Progestasert and LNG-20 are
   (a) Implants (b) Copper releasing IUDs
   (c) Non-medicated IUDs (d) Hormone releasing IUDs
13. Read the following 4-statements and mark the option that has both correct statements
   (i) MTP was legalized in 1971
   (ii) Inability to conceive or produce children even after 2 years of unprotected sexual cohabitation is called infertility
   (iii) Surgical method of contraception prevents gamete formation
   (iv) MTPs are relatively safe up to 12 weeks of pregnancy
   (a) (i) and (ii) (b) (ii) and (iii)
   (c) (iii) and (iv) (d) (i) and (iii)
14. Medical Termination of Pregnancy (MTP) is considered safe up to how many weeks of pregnancy?
   (a) Eight weeks (b) Twelve weeks
   (c) Eighteen weeks (d) Six weeks
15. Ultrasound of how much frequency is beamed into human body for sonography ?
   (a) 15 – 30 MHZ (b) 1 – 15 MHZ
   (b) 45 – 70 MHZ (d) 30 – 45 MHZ.
16. What is the figure given below showing in particular ?

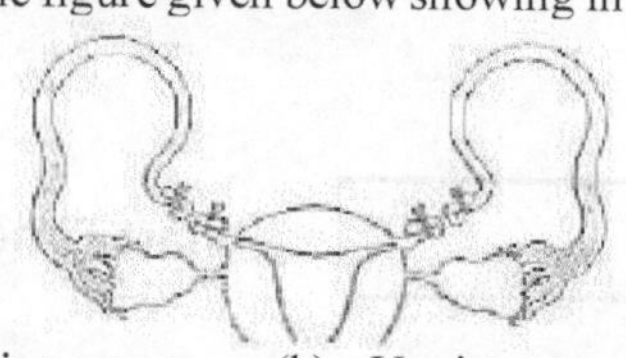

   (a) Ovarian cancer (b) Uterine cancer
   (c) Tubectomy (d) Vasectomy
17. *In vitro* fertilisation is a technique that involves transfer of which one of the following into the fallopian tube ?
   (a) Zygote only
   (b) Embryo only, upto 8 cell stage
   (c) Either zygote or early embryo upto 8 cell stage
   (d) Embryo of 32 cell stage
18. MTP is practised mainly to –
   (a) get rid of unwanted female child legally .
   (b) get rid of unwanted pregnancies due to failure of contraception.
   (c) both (a) and (b).
   (d) decrease population size.
19. In amniocentesis, the fluid is taken from –
   (a) foetal blood
   (b) mother's blood
   (c) body fluid of mother
   (d) fluid surrounding foetus
20. Which of the following is not *true* about the birth control pill ?
   (a) The pill works by preventing ovulation.
   (b) The pill works by preventing implantation.
   (c) The ovarian cycle is suspended by the birth control pill.
   (d) The birth control pill contains low doses of estrogen and progesterone.
21. Which of the following birth control measures can be considered as the safest?
   (a) The rhythm method
   (b) The use of physical barriers
   (c) Temination of unwanted pregnancy
   (d) Sterilization techniques

RESPONSE GRID					
	7. ⓐⓑⓒⓓ	8. ⓐⓑⓒⓓ	9. ⓐⓑⓒⓓ	10. ⓐⓑⓒⓓ	11. ⓐⓑⓒⓓ
	12. ⓐⓑⓒⓓ	13. ⓐⓑⓒⓓ	14. ⓐⓑⓒⓓ	15. ⓐⓑⓒⓓ	16. ⓐⓑⓒⓓ
	17. ⓐⓑⓒⓓ	18. ⓐⓑⓒⓓ	19. ⓐⓑⓒⓓ	20. ⓐⓑⓒⓓ	21. ⓐⓑⓒⓓ

Space for Rough Work

22. Using which contraceptive also provides protection from contracting STDs and AIDS ?
(a) Diaphragms (b) Spermicidal foams
(c) Condoms (d) Lactational amenorrhoea

23. Copper-T is a device that prevents
(a) implantation of blastocyst
(b) ovulation
(c) fertilization
(d) egg maturation

24. Cu ions released from copper - releasing Intra Uterine Devices (IUDs):
(a) make uterus unsuitable for implantation
(b) increase phagocytosis of sperms
(c) suppress sperm motility
(d) prevent ovulation

25. Which one of the following is the most widely accepted method of contraception in India, as at present ?
(a) Cervical caps
(b) Tubectomy
(c) Diaphragms
(d) IUDs. (Intra uterine devices)

26. The test-tube Baby Programme employs which one of the following techniques?
(a) Intra cytoplasmic sperm injection (ICSI)
(b) Intra uterine insemination (IUI)
(c) Gamete intra fallopian transfer (GIET)
(d) Zygote intra fallopian transfer (ZIFT)

27. Oral contraceptive pills are composed of
(a) progestrogens only
(b) progestrogen and estrogen combination
(c) progestrogen-testosterone combination
(d) both (a) or (b)

28. Which of the following cannot be detected in a developing foetus by amniocentesis?
(a) Sex of the foetus
(b) Down syndrome
(c) Jaundice
(d) Klinefelter syndrome

29. One of the legal methods of birth control is :
(a) by abstaining from coitus from day 10 to 17 of the menstrual cycle
(b) by having coitus at the time of day break
(c) by a premature ejaculation during coitus
(d) abortion by taking an appropriate medicine

30. One of the following is **not** a method of contraception – which one?
(a) Tubectomy
(b) Condoms
(c) Pills of a combination of oxytocin and vasopressin
(d) Lippes loop

31. Which one of the following statements is correct regarding Sexually Transmitted Diseases (STD)?
(a) The chances of a 5 year boy contacting a STD are very little
(b) A person may contact syphilis by sharing milk with one already suffering from the disease
(c) Haemophilia is one of the STD
(d) Genital herpes and sickle-cell anaemia are both STD

32. The stage transferred into the uterus after induced fertilization of ova in the laboratory is:
(a) Zygote
(b) Embryo at 4 blastomere stage
(c) Embryo at 2 blastomere stage
(d) Morula

33. Reproductive health in society can be improved by –
(i) Introduction of sex education in schools.
(ii) Increased medical assistance.
(iii) Awareness about contraception and STDs.
(iv) Equal opportunities to male and female child.
(v) Ban on aminocentesis.
(vi) Encouraging myths and misconceptions.
(a) All of these (b) (i), (ii), (iv) and (vi)
(c) (i), (ii), (iii), (iv) and (v) (d) (ii) and (v)

34. What is true for " Lactational amenorrhoea"?
(i) It means absence of menstruation.
(ii) Ovulation does not occur during the lactational period.
(iii) Chances of failure of contraception are almost nil upto six months following parturition.
(iv) Side effects are almost nil.
(v) Contraceptive efficiency reduces after the period of intense lactation.
(vi) It is natural method of contraception.
(vii) It increases phagocytosis of sperms
(a) (ii), (iii), (iv), (v) and (vi)
(b) All except (vii)
(c) (ii), (iii), (iv) and (v)
(d) All of these

35. IUDs-
(i) Increase phagocytosis of sperms.
(ii) Release Cu ions that suppress sperm motility and fertilising capacity of sperms.
(iii) Make the uterus unsuitable for implantation.
(iv) Make the cervix hostile to sperms.
(v) Prevent semen from entering the female reproductive tract.
(a) (i), (ii), (iii), (iv) and (v) (b) (i), (ii), (iii) and (iv)
(c) (i) and (ii) (d) (ii), (iii) and (iv)

RESPONSE GRID					
	22. ⓐⓑⓒⓓ	23. ⓐⓑⓒⓓ	24. ⓐⓑⓒⓓ	25. ⓐⓑⓒⓓ	26. ⓐⓑⓒⓓ
	27. ⓐⓑⓒⓓ	28. ⓐⓑⓒⓓ	29. ⓐⓑⓒⓓ	30. ⓐⓑⓒⓓ	31. ⓐⓑⓒⓓ
	32. ⓐⓑⓒⓓ	33. ⓐⓑⓒⓓ	34. ⓐⓑⓒⓓ	35. ⓐⓑⓒⓓ	

Space for Rough Work

36. Match Column -I with Column - II.

	Column I		Column II
A.	Non-medicated IUDs	(i)	Lippes loop
B.	Hormone releasing IUDs	(ii)	Multiload (iii)7
C.	Copper releasing IUDs	(iii)	CuT
		(iv)	Cu7
		(v)	LNG-(ii)0
		(vi)	Progestasert

(a) A-(i); B-(ii, vi); C-(iii, iv, v)
(b) A-(i); B-(v, vi); C-(ii, iii, iv)
(c) A-(ii); B-(v, vi); C-(i, iii, iv)
(d) A-(ii); B-(i, vi); C-(iii, iv, v)

37. Progesterone pill helps in preventing pregnancy by not allowing –
(a) ova formation (b) fertilization
(c) implantation (d) None of these

38. The permissible use of the technique amniocentesis is for
(a) detecting sex of the unborn foetus
(b) artificial insemination
(c) transfer of embryo into the uterus of the surrogate mother
(d) detecting any genetic abnormality

39. The family planning programmes in India were initiated in
(a) 1951 (b) 1961
(c) 1971 (d) 1981

40. RCH stands for
(a) Routine Check-up of Health
(b) Reproduction Cum Hygiene
(c) Reversible Contraceptive Hazards
(d) Reproductive and Child Health Care

41. 'Saheli' a new oral contraceptive developed by
(a) All Indian Institute of Medical Science
(b) Central Drug Research Institute
(c) Health Care Pvt. Ltd.
(d) Bharat Immunologicals & Biologicals corp. Ltd.

42. Artificial insemination mean
(a) transfer of sperms of husband to a test tube containing ova.
(b) artificial introduction of sperms of a healthy donor into the vagina.
(c) introduction of sperms of a healthy donor directly into the ovary.
(d) transfer of sperms of a healthy donor to a test tube containing ova.

43. What is true for an ideal contraceptive ?
(i) It should be user-friendly.
(ii) It should be easily available.
(iii) It should be ineffective and reversible with least side effects.
(iv) It should be effective and reversible with least side effects.
(v) It should interfere with the sexual act of the user.
(a) All of these (b) (i), (ii) and (iii)
(c) (i), (ii) and (iv) (d) (i),(ii), (iv) and (v)

44. Choose the correct statements –
(i) According to the WHO, reproductive health is total well-being in the physical, social, emotional, behavioural aspects of reproduction.
(ii) According to the WHO, reproductive health is total well being in the physical, social and emotional aspects of reproduction.
(iii) A reproductively healthy society has people with physically and functionally normal reproductive organs.
(iv) Reproductively healthy societies have abnormal sex-related emotional and behavioural interactions.
(a) (i), (ii) and (iii) (b) (ii) and (iv)
(c) (i) and (iii) (d) (i) only

45. Study the given reasons on the basis of which pregnancy can be terminated. Identify the correct reasons.
(i) To get rid of unwanted pregnancies.
(ii) To prevent the fatality or harmfulness to the mother or to foetus or both due to the continuation of pregnancy.
(iii) Termination of pregnancy is safe in each and every case.
(iv) If the foetus is male.
(v) It plays an important role in decreasing the population.
(a) (i), (ii), only (b) (ii), (iii) only
(c) (iii), (iv), (v) only (d) All of these

RESPONSE GRID					
	36. ⓐⓑⓒⓓ	37. ⓐⓑⓒⓓ	38. ⓐⓑⓒⓓ	39. ⓐⓑⓒⓓ	40. ⓐⓑⓒⓓ
	41. ⓐⓑⓒⓓ	42. ⓐⓑⓒⓓ	43. ⓐⓑⓒⓓ	44. ⓐⓑⓒⓓ	45. ⓐⓑⓒⓓ

Space for Rough Work

DAILY PRACTICE PROBLEM DPP CHAPTERWISE 26 - BIOLOGY			
Total Questions	45	Total Marks	180
Attempted		Correct	
Incorrect		Net Score	
Cut-off Score	48	Qualifying Score	55
Success Gap = Net Score – Qualifying Score			
Net Score = (Correct × 4) – (Incorrect × 1)			

DPP - Daily Practice Problems

Chapter-wise Sheets

Date : | **Start Time :** | **End Time :**

BIOLOGY CB27

SYLLABUS : Principles of inheritance and variation

Max. Marks : 180 **Marking Scheme :** + 4 for correct & (–1) for incorrect **Time : 60 min.**

INSTRUCTIONS : This Daily Practice Problem Sheet contains 45 MCQs. For each question only one option is correct. Darken the correct circle/ bubble in the Response Grid provided on each page.

1. Cross between AaBB and aaBB will form
(a) 1 AaBB : 1aaBB (b) All AaBB
(c) 3 AaBB : 1 aaBB (d) 1 AaBB : 3 aaBB

2. Red (RR) *Antirrhinum* is crossed with white (WW) one. Offspring RW are pink. This is an example of
(a) dominant -recessive
(b) incomplete dominance
(c) hybrid
(d) supplementary genes

3. The dihybrid ratio in F_2 - generation is
(a) 1 : 1 : 1 : 1 (b) 2 : 1 : 2 : 1
(c) 3 :1 (d) 9 : 3 : 3 : 1

4. Genetic map is one that
(a) shows the distribution of various species in a region
(b) establishes sites of the genes on a chromosome
(c) establishes the various stages in gene evolution
(d) show the stages during the cell division

5. A gene pair hides the effect of another gene. The phenomenon is
(a) epistasis (b) dominance
(c) mutation (d) None of the above

6. Independent assortment of genes does not take place when
(a) genes are located on homologous chromosomes
(b) genes are linked and located on same chromosomes
(c) genes are located on non-homologous chromosomes
(d) All the above

7. Extranuclear inheritance occurs in
(a) peroxisome and ribosome
(b) chloroplast and mitochondria
(c) mitochondria and ribosome
(d) chloroplast and lysosome

8. Test cross involves
(a) crossing between two F_1 hybrids
(b) crossing the F_1 hybrid with a double recessive genotype
(c) crossing between two genotypes with dominant trait
(d) crossing between two genotypes with recessive trait

9. When one gene controls two or more different characters simultaneously, the phenomenon is called
(a) apomixis (b) pleiotropy
(c) polyploidy (d) polyteny

RESPONSE GRID					
	1. ⓐⓑⓒⓓ	2. ⓐⓑⓒⓓ	3. ⓐⓑⓒⓓ	4. ⓐⓑⓒⓓ	5. ⓐⓑⓒⓓ
	6. ⓐⓑⓒⓓ	7. ⓐⓑⓒⓓ	8. ⓐⓑⓒⓓ	9. ⓐⓑⓒⓓ	

Space for Rough Work

10. ABO blood group system is due to
(a) multifactor inheritance
(b) incomplete dominance
(c) multiple allelism
(d) epistasis

11. In humans, male XXY and female XXXX occur due to
(a) aneuploidy
(b) euploidy
(c) Nutosomal syndrome
(d) none of these

12. Haemophilia is more common in males because it is a
(a) Recessive character carried by Y-chromosome
(b) Dominant character carried by Y-chromosome
(c) Dominant trait carried by X-chromosome
(d) Recessive trait carried by X-chromosome

13. The most striking example of point mutation is found in a disease called
(a) thalassemia (b) night blindness
(c) down's syndrome (d) sickle cell anaemia

14. In Down's syndrome of a male child, the sex complement is
(a) XO (b) 45 + XY
(c) 45+XX (d) XXY

15. Barr body in mammals represents
(a) all the heterochromatin in female cells
(b) Y-chromosomes in somatic cells of male
(c) all heterochromatin in male and female cells
(d) one of the two X-chromosomes in somatic cells of females

16. A person with the sex chromosomes XXY suffers from
(a) Down's syndrome
(b) Klinefelter's syndrome
(c) Turner's syndrome
(d) Gynandromorphism

17. *Drosophila* flies with XXY genotype are females, but human beings with such genotype are abnormal males. It shows that
(a) Y-chromosome is essential for sex determination in *Drosophila.*
(b) Y-chromosome is female determinant in *Drosophila.*
(c) Y-chromosome is male determination in human beings.
(d) Y-chromosome has no role in sex determination either in *Drosophila* or in human beings.

18. Lack of independent assortment of two genes A and B in fruit fly *Drosophila* is due to
(a) repulsion (b) recombination
(c) linkage (d) crossing over

19. Select the **incorrect** statement from the following:
(a) Galactosemia is an inborn error of metabolism
(b) Small population size results in random genetic drift in a population
(c) Baldness is a sex -limited trait
(d) Linkage is an exception to the principle of independent assortment in heredity

20. The "Cri-du-Chat" syndrome is caused by change in chromosome structure involving
(a) deletion (b) duplication
(c) inversion (d) translocation

21. Biometric genetics deals with :
(a) the biochemical explanations of various genetical phenomena
(b) the effect of environment on genetic set up organisms
(c) the genetical radiations on the living organisms
(d) the inheritance of quantitative traits

22. Which one of the following conditions correctly describes the manner of determining the sex in the given example?
(a) Homozygous sex chromosomes (ZZ) determines female sex in birds.
(b) XO type of sex chromosomes determines male sex in grasshopper.
(c) XO condition in human as found in Turner syndrome, determines female sex.
(d) Homozygous sex chromosomes (XX) produces male in *Drosophila.*

23. Select the correct statement from the ones given below with respect to dihybrid cross.
(a) Tightly linked genes on the same chromosome show higher recombinations
(b) Genes far apart on the same chromosome show very few recombinations
(c) Genes loosely linked on the same chromosome show similar recombinations as the tightly linked ones
(d) Tightly linked genes on the same chromosome show very few recombinations

24. Chromosome complement with 2n – 1 is called
(a) Monosomy (b) Trisomy
(c) Nullisomy (d) Tetrasomy

RESPONSE GRID				
10. ⓐⓑⓒⓓ	11. ⓐⓑⓒⓓ	12. ⓐⓑⓒⓓ	13. ⓐⓑⓒⓓ	14. ⓐⓑⓒⓓ
15. ⓐⓑⓒⓓ	16. ⓐⓑⓒⓓ	17. ⓐⓑⓒⓓ	18. ⓐⓑⓒⓓ	19. ⓐⓑⓒⓓ
20. ⓐⓑⓒⓓ	21. ⓐⓑⓒⓓ	22. ⓐⓑⓒⓓ	23. ⓐⓑⓒⓓ	24. ⓐⓑⓒⓓ

Space for Rough Work

25. In a cross between AABB × aabb, the ratio of F_2 genotypes between AABB, AaBB, Aabb and aabb would be
(a) 9 : 3 : 3 : 1 (b) 2 : 1 : 1 : 2
(c) 1 : 2 : 2 : 1 (d) 7 : 5 : 3 : 1

26. The basis of karyotaxonomy is
(a) Number of nucleoli
(b) Sedimentation rate of ribosomes
(c) Chromosome banding
(d) Chromosome number

27. Study the pedigree chart given below:

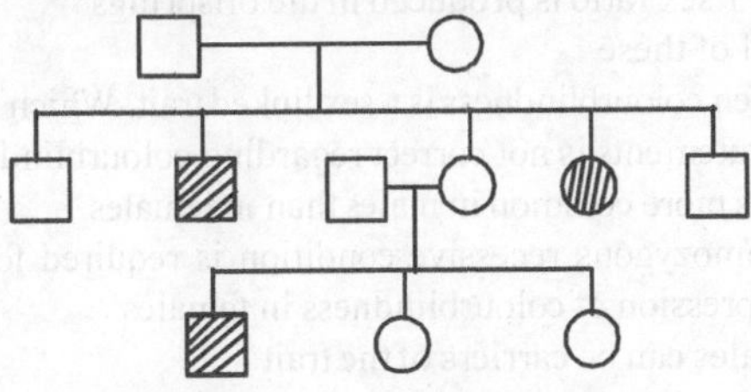

What does it show?
(a) Inheritance of a recessive sex-linked disease like haemophilia
(b) Inheritance of a sex-linked inborn error of metabolism like phenylketonuria
(c) Inheritance of a condition like phenylketonuria as an autosomal recessive trait
(d) The pedigree chart is wrong as this is not possible

28. Which one is a hereditary disease ?
(a) Cataract (b) Leprosy
(c) Blindness (d) Phenylketonuria

29. Diploid chromosome number in humans is
(a) 46 (b) 44
(c) 48 (d) 42

30. Given below is a pedigree chart showing the inheritance of a certain sex-linked trait in humans

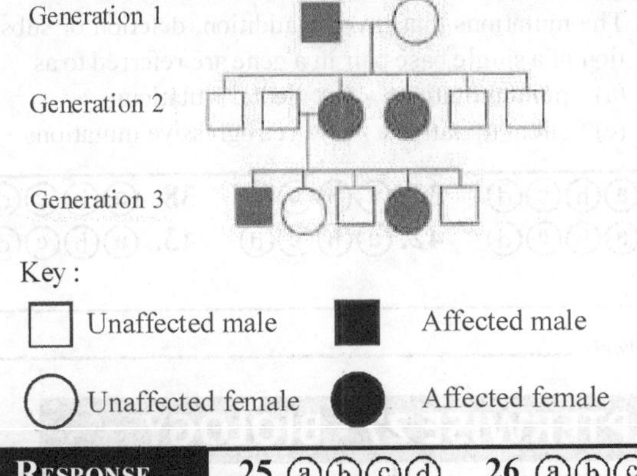

The trait traced in the above pedigree chart is
(a) dominant X - linked
(b) recessive X-linked
(c) dominant Y - linked
(d) recessive Y-linked

31. In maize, coloured endosperm (C) is dominant over colourless (c); and full endosperm (R) is dominant over shrunken (r). When a dihybrid of F_1 generation was test crossed, it produced your phenotypes in the following percentage:
Coloured full - 48% Coloured shrunken - 5%
Colourless full - 7% Colourless shrunken - 40%
From this data, what will be the distance between two non-allelic genes?
(a) 48 units (b) 5 units
(c) 7 units (d) 12 units

32. Inheritance of which of the following traits is shown in the given cross?

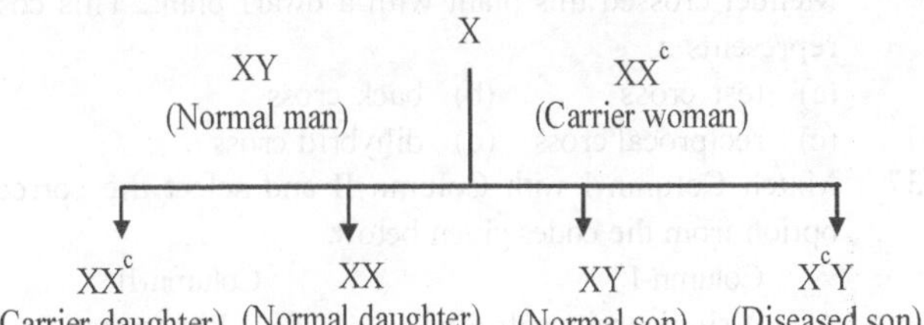

(a) X-linked dominant trait
(b) X-linked recessive trait
(c) Autosomal dominant trait
(d) Autosomal recessive trait

33. More than two alternate forms of a gene present on the same locus are called (i). They are produced due to repeated (ii) of the same gene but in different directions. Their well known example is (iii).

	(i)	(ii)	(iii)
(a)	Epistatic genes	Crossing over	polydactyly
(b)	Multiple alleles	mutations	human blood groups
(c)	Supplementary	mutations	hypertrichosis
(d)	Linked genes	Crossing over	aljaptonuria

RESPONSE GRID					
	25. ⓐⓑⓒⓓ	26. ⓐⓑⓒⓓ	27. ⓐⓑⓒⓓ	28. ⓐⓑⓒⓓ	29. ⓐⓑⓒⓓ
	30. ⓐⓑⓒⓓ	31. ⓐⓑⓒⓓ	32. ⓐⓑⓒⓓ	33. ⓐⓑⓒⓓ	

Space for Rough Work

34. In fruit files, long wing is dominant to vestigial wing. When heterozygous long-winged files were crossed with vestigial-winged files, 192 offsprings were produced. Of these, 101 had long wings and 91 had vestigial wings.
If an exact Mendelian ratio had been obtained, then the number of each phenotype would have been

	Long-winged	Vestigial-winged
(a)	64	128
(b)	96	96
(c)	128	64
(d)	192	0

35. Phenotypic and genotypic ratio is similar in case of
(a) complete dominance
(b) incomplete dominance
(c) over dominance
(d) epistasis

36. To determine the genotype of a tall plant of F_2 generation, Mendel crossed this plant with a dwarf plant. This coss represents a
(a) test cross (b) back cross
(c) reciprocal cross (d) dihybrid cross

37. Match Column-I with Column-II and select the correct option from the codes given below.

	Column-I	Column-II
A.	A single trait controlled by three or more than three alleles	(i) Pleiotropy
B.	A single trait controlled by three or more than three genes	(ii) Multiple alleles
C.	A single gene exhibits multiple phenotypic expression	(iii) Polygenic inheritance

38. Chromosomal theory of inheritanc was given by
(a) Morgan et al (b) Sutton and Boveri
(c) Hugo de Vries (d) Gregor J. Mendel

39. What is true about the crossing over between linked genes?
(a) No crossing over at all
(b) High percentage of crossing over
(c) Hardly any crossing over
(d) None of these

40. Which of the following is incorrect regarding ZW-ZZ type of sex determination?
(a) It occurs in birds and some reptiles
(b) Females are homogametic and males are heterogametic
(c) 1 : 1 sex ratio is produced in the offsprings
(d) All of these

41. Red green colourblindness is a sex linked trait. Which of the given statements is not correct regarding colourblindness?
(a) It is more common in males than in females
(b) Homozygous recessive condition is required for the expression of colourblindness in females
(c) Males can be carriers of the trait
(d) Colourblind women always have colourblind father and always produce colourblind son.

42. At a particular locus, frequency of allele A is 0.6 and that of allele a is 0.4. What would be the frequency of heterozygotes in a random mating population at equilibrium?
(a) 0.36 (b) 0.16
(c) 0.24 (d) 0.48

43. The distance between the genes is measured by
(a) Angstrom (b) map unit
(c) Dobson unit (d) millimetre

44. Which of the following trait is controlled by dominant autosomal genes?
(a) Polydactyly
(b) Huntigton's chorea
(c) PTC (phenylthiocarbomide) tasting
(d) All of these

45. The mutations that involve addition, deletion or substitution of a single base pair in a gene are referred to as
(a) point mutations (b) lethal mutations
(c) silent mutations (d) retrogressive mutations

RESPONSE GRID					
	34. ⓐⓑⓒⓓ	35. ⓐⓑⓒⓓ	36. ⓐⓑⓒⓓ	37. ⓐⓑⓒⓓ	38. ⓐⓑⓒⓓ
	39. ⓐⓑⓒⓓ	40. ⓐⓑⓒⓓ	41. ⓐⓑⓒⓓ	42. ⓐⓑⓒⓓ	43. ⓐⓑⓒⓓ
	44. ⓐⓑⓒⓓ	45. ⓐⓑⓒⓓ			

Space for Rough Work

DAILY PRACTICE PROBLEM DPP CHAPTERWISE 27 - BIOLOGY			
Total Questions	45	Total Marks	180
Attempted		Correct	
Incorrect		Net Score	
Cut-off Score	45	Qualifying Score	60
Success Gap = Net Score – Qualifying Score			
Net Score = (Correct × 4) – (Incorrect × 1)			

DPP - Daily Practice Problems

Chapter-wise Sheets

Date : ______ **Start Time :** ______ **End Time :** ______

BIOLOGY

CB28

SYLLABUS : Molecular Basis of Inheritance

Max. Marks : 180 **Marking Scheme : + 4 for correct & (–1) for incorrect** **Time : 60 min.**

INSTRUCTIONS : This Daily Practice Problem Sheet contains 45 MCQs. For each question only one option is correct. Darken the correct circle/ bubble in the Response Grid provided on each page.

1. Which form of RNA has a structure resembling clover leaf?
   (a) rRNA (b) hn RNA
   (c) m RNA (d) t RNA
2. Other than DNA polymerase, which is/ are the enzyme(s) involved in DNA synthesis?
   (a) Topoisomerase (b) Helicase
   (c) RNA primase (d) All of these
3. The enzyme used for joining two DNA fragments is called:
   (a) ligase (b) restriction endonuclease
   (c) DNA polymerase (d) gyrase
4. Nucleotide arrangement in DNA can be seen by
   (a) X-ray crystallography
   (b) electron microscope
   (c) ultracentrifuge
   (d) light microscope
5. The okazaki fragments in DNA chain growth
   (a) polymerize in the 3' - to - 5' direction and forms replication fork
   (b) prove semi-conservative nature of DNA replication
   (c) polymerize in the 5' - to - 3' direction and explain 3' - to - 5' DNA replication
   (d) result in transcription
6. Reverse transcriptase is
   (a) RNA dependent RNA polymerase
   (b) DNA dependent RNA polymerase
   (c) DNA dependent DNA polymerase
   (d) RNA dependent DNA polymerase
7. Telomerase is an enzyme which is a
   (a) simple protein
   (b) RNA
   (c) ribonucleoprotein
   (d) repetitive DNA
8. DNA template sequence of CTGATAGC is transcribed over mRNA as
   (a) GUCTUTCG (b) GACUAUCG
   (c) GAUTATUG (d) UACTATCU

RESPONSE GRID					
	1. ⓐⓑⓒⓓ	2. ⓐⓑⓒⓓ	3. ⓐⓑⓒⓓ	4. ⓐⓑⓒⓓ	5. ⓐⓑⓒⓓ
	6. ⓐⓑⓒⓓ	7. ⓐⓑⓒⓓ	8. ⓐⓑⓒⓓ		

Space for Rough Work

9. During replication of a bacterial chromosome DNA synthesis starts from a replication origin site and
(a) RNA primers are involved
(b) is facilitated by telomerase
(c) moves in one direction of the site
(d) moves in bi-directional way

10. The enzyme, which helps to cut one strand of DNA duplex to release tension of coiling of two strands is
(a) DNA ligase
(b) DNA polymerase I
(c) topoisomerase
(d) helicase or unwindases

11. Histones are
(a) glycoproteins (b) mucoproteins
(c) basic proteins (d) acid proteins

12. One gene-one enzyme relationship was established for the first time in
(a) *Salmonella typhimurium*
(b) *Escherichia coli*
(c) *Diplococcus pneumoniae*
(d) *Neurospora crassa.*

13. The transforming principle of *Pneumococcus* as found out by Avery, MacLeod and McCarty was
(a) mRNA (b) DNA
(c) protein (d) polysaccharide

14. How many pair of nucleotides are present in one helix of B-DNA?
(a) 10 (b) 12
(c) 5 (d) 6

15. Read the following four statements (i - iv).
(i) In transcription, adenosine pairs with uracil.
(ii) Regulation of *lac* operon by repressor is referred to as positive regulation.
(iii) The human genome has approximately 50,000 genes.
(iv) Haemophilia is a sex-linked recessive disease.
How many of the above statements are correct?
(a) Two (b) Three
(c) Four (d) One

16. In negative operon
(a) co-repressor binds with repressor
(b) co-repressor does not bind with repressor
(c) co-repressor binds with inducer
(d) cAMP have negative effect on *lac* operon

17. Transfer of genetic material from one bacterium to another in the transduction process is through
(a) conjugation
(b) bacteriophages
(c) another bacterium
(d) physical contact between donor and recipient strain

18. Telomere and eukaryotic chromosome possesses short segments of
(a) guanine rich repeats.
(b) thymine rich repeats.
(c) cytosine rich repeats.
(d) adenine rich repeats.

19. Refer the given figure of nucleosome and select the option that correctly identifies the parts A, B and C.

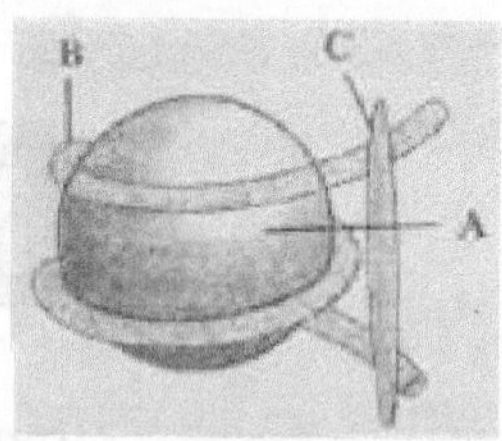

	A	B	C
(a)	DNA	Histone octamer	H_1 histone
(b)	Histone octamer	H_1 histone	DNA
(c)	Histone	DNA	H_1 histone
(d)	DNA octamer	H_1 histone	Histone octamer

20. The process of transformation is not affected by which of the following enzymes?
A. DNase B. RNase
C. Peptidase D. Lipase
(a) A, B (b) A, B, C, D
(c) B, C, D (d) A, B, C

21. Select the incorrect statement regarding DNA replication
(a) Leading strand is formed is 5' $\rightarrow$ 3' direction
(b) Okazaki fragments are formed in 5' $\rightarrow$ 3' direction
(c) DNA polymerase catalyses polymerisation in 5' $\rightarrow$ 3' direction
(d) DNA polymerase catalyses polymerisation in 3' $\rightarrow$ 5' direction

RESPONSE GRID				
9. ⓐⓑⓒⓓ	10. ⓐⓑⓒⓓ	11. ⓐⓑⓒⓓ	12. ⓐⓑⓒⓓ	13. ⓐⓑⓒⓓ
14. ⓐⓑⓒⓓ	15. ⓐⓑⓒⓓ	16. ⓐⓑⓒⓓ	17. ⓐⓑⓒⓓ	18. ⓐⓑⓒⓓ
19. ⓐⓑⓒⓓ	20. ⓐⓑⓒⓓ	21. ⓐⓑⓒⓓ		

Space for Rough Work

22. Identify the labels A, B, C and D in the given structure of tRNA and select the correct option.

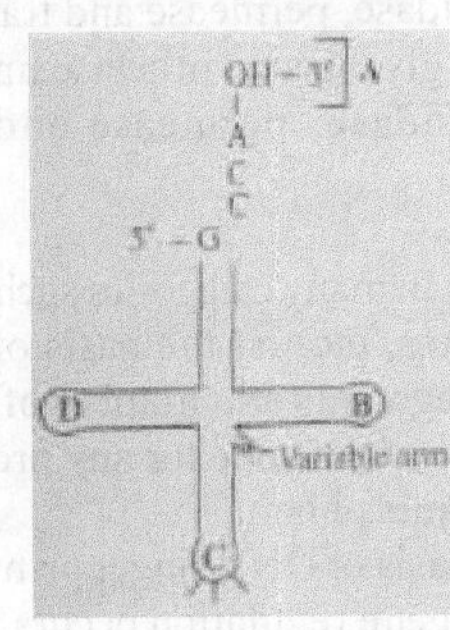

	A	B	C	D
(a)	Anticodon loop	T C loop	AA binding stie	DHU loop
b	AA binding site	T C loop	Anticodon loop	DHU loop
c	AA binding site	DHU loop	Anticodon loop	T C loop
d	AA binding site	DHU loop	T C loop	Anticodon loop

23. In one polynucleotide strand of a DNA molecule the ratio of A + T/G + C is 0.3. What is the A + G/T + C ratio of the entire DNA molecule?
(a) 0.3 (b) 0.6
(c) 1.2 (d) 1

24. Chemically, RNA is __i__ reactive and __ii__ stable as compared to DNA.
(a) (i) equally, (ii) equally
(b) (i) less, (ii) more
(c) (i) more, (ii) less
(d) (i) more, (ii) equally

25. Given diagram represents the components of a transcription unit. Select the correct answer regarding it.

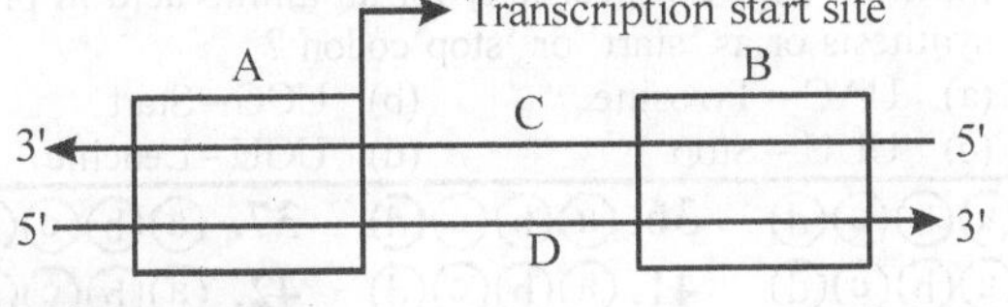

	A	B	C	D
(a)	Teminator	Promoter	Template strand	Coding strand
(b)	Promoter	Terminator	Coding strand	Template strand
(c)	Promoter	Terminator	Template strand	Coding strand
(d)	Terminator	Promoter	Coding strand	Template strand

26. In transcription in eukaryotes, heterogenous nuclear RNA (hnRNA) is transcribed by
(a) RNA polymerase I
(b) RNA polymerase II
(c) RNA polymerase III
(d) all of these

27. What would happen if in a gene encoding a polypeptide of 50 amino acids, 25th codon (UAU) is mutated to UAA?
(a) A polypeptide of 24 amino acids will be formed.
(b) Two polypeptides of 24 and 25 amino acids will be formed
(c) A polypeptide of 49 amino acids will be formed
(d) A polypeptide of 25 amino acids will be formed

28. Regulation of gene expression occurs at the level of
(a) transcription (b) processing/splicing
(c) translation (d) all of these

29. Find out the wrong sentence about heterochromatin.
(a) It is densely packed
(b) It stains dark
(c) It is transcriptionally active
(d) It is late replicating

30. Other than DNA polymerase, which are the enzymes involved in DNA synthesis?
(a) topoisomerase (b) helicase
(c) RNA primase (d) all of these

31. What does "*lac*" refer to in what we call the *lac* operon?
(a) The number 1,00,000 (b) Lactose
(c) Lactase (d) *Lac* insect

32. The error rate of changing an incorrect base with another incorrect base during proofreading is
(a) 1 in 10 bases (b) 1 in 100 bases
(c) 1 in 1,000 bases (d) 1 in 10,000 bases

RESPONSE GRID					
	22. ⓐⓑⓒⓓ	23. ⓐⓑⓒⓓ	24. ⓐⓑⓒⓓ	25. ⓐⓑⓒⓓ	26. ⓐⓑⓒⓓ
	27. ⓐⓑⓒⓓ	28. ⓐⓑⓒⓓ	29. ⓐⓑⓒⓓ	30. ⓐⓑⓒⓓ	31. ⓐⓑⓒⓓ
	32. ⓐⓑⓒⓓ				

Space for Rough Work

33. DNA fingerprinting refers to
(a) molecular analysis of profiles of DNA samples
(b) analysis of DNA samples using imprinting devices
(c) techniques used for molecular analysis of different specimens of DNA
(d) techniques used for identification of fingerprints of individuals

34. In negative operon
(a) co-repressor binds with repressor
(b) co-repressor does not bind with repressor
(c) co-repressor binds with inducer
(d) cAMP have negative effect on *lac* operon

35. Genes that are involved in turning on or off the transcription of a set of structural genes are called
(a) Operator genes (b) Redundant genes
(c) Regulator genes (d) Polymorphic genes

36. If the gene encoding the *lac* repressor is mutated so that it can no longer bind the operator, will transcription of that operon occur?
(a) Yes, but only when lactose is present.
(b) No, because RNA polymerase is need to transcribe the genes.
(c) Yes, because the operator will not be bound by repressor and RNA polymerase can transcribe the *lac* operon.
(d) No, because cAMP levels are low when the repressor is nonfunctional.

37. Transcriptional regulation in prokaryotes can occur by
(a) a repressor binding an operator and preventing transcription.
(b) an activator binding upstream from a promoter and positively affecting transcription.
(c) different promoter sequences binding RNA polymerase more tightly, resulting in more effective transcriptional initiation.
(d) All of the above

38. Process used for amplication or multiplication of DNA for finger printing is
(a) polymerse chain reaction
(b) nesslerisation
(c) southern blotting
(d) northern blotting

39. Lactose operon produces enzymes
(a) β-galactosidase, permease and glycogen synthetase.
(b) β-galactosidase, permease and transacetylase.
(c) Permease, glycogen synthetase and transacetylase.
(d) β-galactosidase, permease and phosphoglucose isomerase.

40. Satellite DNA
(a) is classified in many categories such as micro-satellites, minisatellites, etc. on the basis of base composition length of segments and number of repetitive units.
(b) normally does not code for any protein.
(c) shows polymorphism.
(d) forms the basis of DNA finger printing.

41. In prokaryotes, gene regulation occurs at the level of
(a) transcription (b) translation
(c) post-transcription (d) post-translation

42. The regulation of tryptophan synthesis in *E. coli* is an example of affecting gene expression through
(a) translational control.
(b) transcriptional control.
(c) homeotic gene control.
(d) breaking down mRNA molecules.

43. Transcription in prokaryotic cell is :
(a) initiated at a promoter using one of three RNA polymerases (RNA polymerase II).
(b) initiated at a start codon with the help of initiation factors and the small subunit of the ribosome.
(c) initiated at a promoter and uses only one strand of DNA, the template strand, to synthesize a complementary RNA strand.
(d) is terminated at stop codons.

44. DNA replication is an ______ process and ______ energy.
(a) exergonic; does not require
(b) endothermic; does require
(c) endergonic; does require
(d) endothermic; does not require

45. Which one of the following triplet codes, is correctly matched with its specificity for an amino acid in protein synthesis or as 'start' or 'stop' codon ?
(a) UAC – Tyrosine (b) UCG – Start
(c) UUU – Stop (d) UGU – Leucine

RESPONSE GRID					
	33. ⓐⓑⓒⓓ	34. ⓐⓑⓒⓓ	35. ⓐⓑⓒⓓ	36. ⓐⓑⓒⓓ	37. ⓐⓑⓒⓓ
	38. ⓐⓑⓒⓓ	39. ⓐⓑⓒⓓ	40. ⓐⓑⓒⓓ	41. ⓐⓑⓒⓓ	42. ⓐⓑⓒⓓ
	43. ⓐⓑⓒⓓ	44. ⓐⓑⓒⓓ	45. ⓐⓑⓒⓓ		

Space for Rough Work

DAILY PRACTICE PROBLEM DPP CHAPTERWISE 28 - BIOLOGY			
Total Questions	45	Total Marks	180
Attempted		Correct	
Incorrect		Net Score	
Cut-off Score	55	Qualifying Score	65
Success Gap = Net Score – Qualifying Score			
Net Score = (Correct × 4) – (Incorrect × 1)			

DPP - Daily Practice Problems

Chapter-wise Sheets

Date : | **Start Time :** | **End Time :**

BIOLOGY CB29

SYLLABUS : Evolution

Max. Marks : 180 **Marking Scheme : + 4 for correct & (–1) for incorrect** **Time : 60 min.**

INSTRUCTIONS : This Daily Practice Problem Sheet contains 45 MCQs. For each question only one option is correct. Darken the correct circle/ bubble in the Response Grid provided on each page.

1. Genetic drift operates only in
   (a) smaller populations
   (b) larger populations
   (c) mendelian populations
   (d) island populations
2. *Archaeopteryx* is connecting link between
   (a) Reptiles and birds
   (b) Reptiles and mammals
   (c) Fishes and reptiles
   (d) Chordates and nonchordates
3. Which of the following is true?
   (a) Wings of birds and insects are homologous organs.
   (b) Human hands and bird's wings are analogous organs.
   (c) Human hands and bat's wings are analogous organs.
   (d) Flipper of seal and wing of birds are homologous organs.
4. Which one of the following phenomena supports Darwin's concept of natural selection in organic evolution?
   (a) Development of transgenic animals.
   (b) Production of 'Dolly', the sheep by cloning.
   (c) Prevalence of pesticide resistance insects.
   (d) Development of organs from 'stem cells' for organ transplantation.
5. What is the correct arrangement of periods of palaeozoic era in ascending order in geological time scale?
   (a) Cambrian $\rightarrow$ Devonian $\rightarrow$ Ordovician $\rightarrow$ Silurian $\rightarrow$ Carboniferous $\rightarrow$ Permian
   (b) Cambrian $\rightarrow$ Ordovician $\rightarrow$ Silurian $\rightarrow$ Devonian $\rightarrow$ Carboniferous $\rightarrow$ Permian
   (c) Cambrian $\rightarrow$ Ordovician $\rightarrow$ Devonian $\rightarrow$ Silurian $\rightarrow$ Carboniferous $\rightarrow$ Permian
   (d) Silurian $\rightarrow$ Devonian $\rightarrow$ Cambrian $\rightarrow$ Ordovician $\rightarrow$ Permian $\rightarrow$ Carboniferous
6. The brain capacity of *Homo erectus* was about
   (a) 650 cc (b) 900 cc
   (c) 1500 cc (d) 1400 cc

RESPONSE GRID	
	1. ⓐⓑⓒⓓ 2. ⓐⓑⓒⓓ 3. ⓐⓑⓒⓓ 4. ⓐⓑⓒⓓ 5. ⓐⓑⓒⓓ
	6. ⓐⓑⓒⓓ

Space for Rough Work

7. Following is the digrammatic representation of the operation of natural selection on different traits. Which of the following options correctly identifies all the three graphs A, B and C.

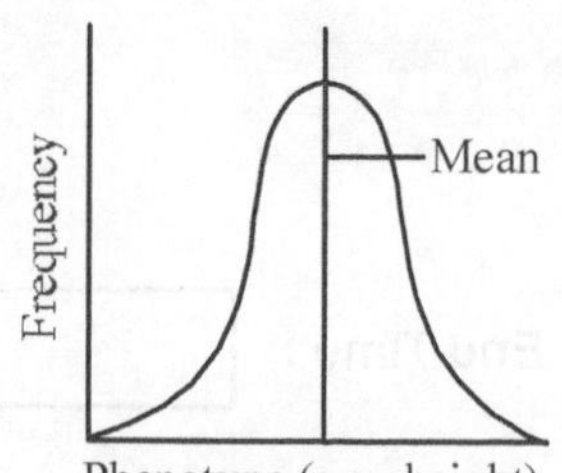

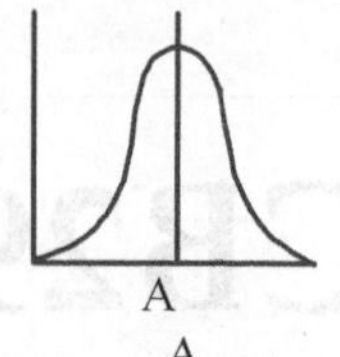

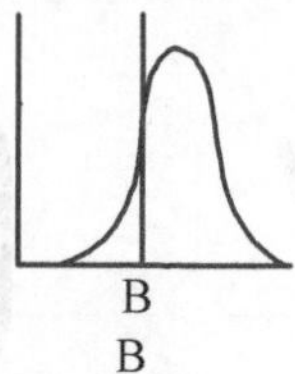

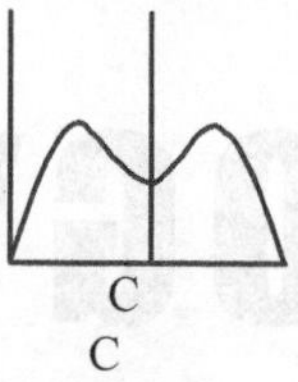

	A	B	C
(a)	Directional	Stabilizing	Disruptive
(b)	Stabilizing	Directional	Disruptive
(c)	Disruptive	Stabilizing	Directional
(d)	Directional	Disruptive	Stabilizing

8. First life form on earth was a
(a) cyanobacterium (b) chemoheterotroph
(c) autotroph (d) photoautotroph

9. The cranial capacity was largest among the
(a) Peking man (b) Java ape man
(c) African man (d) Neanderthal man

10. The Hardy-Weinberg principle cannot operate if
(a) the population is very large
(b) frequent mutations occur in the population
(c) the population has no chance of interaction with other populations
(d) free interbreeding occurs among all members of the population

11. The different forms of interbreeding species that live in different geographical regions are called
(a) sibling species (b) sympatric species
(c) allopatric species (d) polytypic species

12. At a particular locus, frequency of allele A is 0.6 and that of allele a is 0.4. What would be the frequency of heterozygotes in a random mating population at equilibrium?
(a) 0.36 (b) 0.16
(c) 0.24 (d) 0.48

13. On the primitive earth, polymers such as proteins and nucleic acids in aqueous suspension formed the spherical aggregates. These are called
(a) primitosomes (b) liposomes
(c) primitogens (d) coacervates

14. Which of the following evidences does not favour the Lamarckian concept of inheritance of acquired characters?
(a) Lack of pigment in cave-dwelling animals
(b) Melanization in peppered moth
(c) Absence of limbs in snakes
(d) Presence of webbed toes in aquatic birds.

15. Genetic drift is change of
(a) gene frequency in same generation
(b) appearance of recessive genes
(c) gene frequency from one generation to next
(d) None of these

16. The Miller-Urey abiotic synthesis experiment (and other subsequent, similar experiments) showed that...
(a) simple organic molecules can form spontaneously under conditions like those thought to prevail early in the earth's history.
(b) the earliest life forms introduced large amounts of oxygen to the atmosphere.
(c) life can be created in a test tube.
(d) long chains of DNA can form under abiotic conditions.

17. In evolutionary terms, an organism's fitness is measured by its
(a) health
(b) contribution to the gene pool of the next generation
(c) mutation rate
(d) genetic variability

18. Frequency of a character or a mutant gene in a population is expected to increase if the phenotype or its gene is.
(a) dominant (b) favourably selected
(c) recessive (d) sex linked

RESPONSE GRID

7. ⓐⓑⓒⓓ 8. ⓐⓑⓒⓓ 9. ⓐⓑⓒⓓ 10. ⓐⓑⓒⓓ 11. ⓐⓑⓒⓓ
12. ⓐⓑⓒⓓ 13. ⓐⓑⓒⓓ 14. ⓐⓑⓒⓓ 15. ⓐⓑⓒⓓ 16. ⓐⓑⓒⓓ
17. ⓐⓑⓒⓓ 18. ⓐⓑⓒⓓ

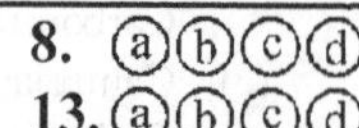

Space for Rough Work

19. In a population of bears, which would be considered the fittest?
   (a) The biggest bear
   (b) The bear having the largest number of mutations
   (c) The bear that blends in with its environment the best
   (d) The bear that leaves the most descendants
20. Presence of temporary gill pouches in embryos of snakes. birds and mammals indicates that
   (a) these embryos need the pouches for breathing
   (b) common ancestor of these animals had gill pouches
   (c) lungs evolved from gills
   (d) fluid medium in which these embryos develop has abundant
21. Which of the following statements is correct?
   (a) Present apes are ancestors of man anatomically
   (b) Proconsul was perhaps the common ancestor of apes and man
   (c) Proconsul was the ancestor of man and not of apes
   (d) None of these
22. What is common to whale, seal and shark?
   (a) Thick subcutaneous fat
   (b) Convergent evolution
   (c) Homoiothermy
   (d) Seasonal migration
23. One of the important consequences of geographical isolation is:
   (a) preventing speciation
   (b) speciation through reproductive isolation
   (c) random creation of new species
   (d) no change in the isolated fauna
24. Which one of the following in birds, indicates their reptilian ancestry?
   (a) Scales on their hind limbs
   (b) Eggs with a calcareous shell
   (c) Four-chambered heart
   (d) Both (a) and (b)
25. Jurassic period of the mesozoic era was characterised by:
   (a) Radiation of reptiles and origin of mammal-like reptiles
   (b) Dinosaurs become extinct and angiosperms appeared
   (c) Flowering plants and first dinosaurs appeared
   (d) Gymnosperms were dominant plants and first birds appeared
26. Which one of the following scientists name is correctly matched with the theory put forth by him?
   (a) Weismann - Theory of continuity of Germplasm
   (b) Pasteur - Inheritance of acquired characters
   (c) De Vries - Natural selection
   (d) Mendel - Theory of Pangenesis
27. The eye of octopus and eye of cat show different patterns of structure, yet they perform similar function. This is an example of:
   (a) Homologous organs that have evolved due to divergent evolution.
   (b) Analogous organs that have evolved due to convergent evolution.
   (c) Analogous organs that have evolved due to divergent evolution.
   (d) Homologous organs that have evolved due to convergent evolution.
28. Variation in gene frequencies within populations can occur by chance rather than by natural section. The is referred to as:
   (a) genetic drift
   (b) random mating
   (c) genetic load
   (d) genetic flow
29. The finch species of Galapagos Islands are grouped according to their food sources. Which of the following is not a finch food?
   (a) Seeds (b) Carrion
   (c) Insects (d) Tree buds
30. Evolution is
   (a) progressive development of a race
   (b) history and development of a race alongwith variations
   (c) history of a race
   (d) development of a race
31. Homologous organs are
   (a) Wings of Insects and Bat
   (b) Gills of Fish and lungs of Rabbit
   (c) Pectoral fins of Fish and fore limbs of Horse
   (d) Wings of Grasshopper and Crow

RESPONSE GRID					
	19. ⓐⓑⓒⓓ	20. ⓐⓑⓒⓓ	21. ⓐⓑⓒⓓ	22. ⓐⓑⓒⓓ	23. ⓐⓑⓒⓓ
	24. ⓐⓑⓒⓓ	25. ⓐⓑⓒⓓ	26. ⓐⓑⓒⓓ	27. ⓐⓑⓒⓓ	28. ⓐⓑⓒⓓ
	29. ⓐⓑⓒⓓ	30. ⓐⓑⓒⓓ	31. ⓐⓑⓒⓓ		

Space for Rough Work

32. Theory of inheritance of acquired characters was given by
(a) Wallace (b) Lamarck
(c) Darwin (d) De Vries

33. The kind of evolution in which two species of different geneology come to resemble one another closely, is termed as
(a) progressive evolution
(b) convergent evolution
(c) parallel evolution
(d) retrogressive evolution

34. The presence of gill slits in the embryos of all vertebrates, supports the theory of
(a) biogenesis
(b) recapitulation
(c) metamorphosis
(d) organic evolution

35. The change of the light-coloured variety of peppered moth *(Biston betularia)* to its darker variety *(Biston carbonaria)* is due to
(a) mutation (b) regeneration
(c) genetic isolation (d) temporal isolation

36. Which of the following primate is the closest relative of humans?
(a) Rhesus monkey (b) Orangutan
(c) Gorilla (d) Gibbon

37. Which of the following is the relatively most accurate method for dating of fossils?
(a) Radio-carbon method
(b) Potassium-argon method
(c) Electron-spin resonance method
(d) Uranium-lead method

38. Adaptive radiation refers to
(a) evolution of different species from a common ancestor
(b) migration of members of a species to different geographical areas
(c) power of adaptation in an individual to a variety of environments
(d) adaptations due to geographical isolation.

39. What was the most significant trend in evolution of modern man (*Homo sapiens*) from his ancestors ?
(a) Upright posture
(b) Shortening of jaws
(c) Binocular vision
(d) Increasing brain capacity

40. According to Oparin, which one of the following was ***not*** present in the primitive atmosphere of the earth?
(a) Methane (b) Oxygen
(c) Hydrogen (d) Water vapour

41. There is no life on moon due to the absence of
(a) O_2 (b) water
(c) light (d) temperature

42. Which one of the following statements is correct?
(a) Cro-Magnon man's fossil has been found in Ethiopia
(b) *Homo erectus* is the ancestor of man
(c) Neanderthal man is the direct ancestor of *Homo sapiens*
(d) *Australopithecus* is the real ancestor of modern man

43. Atmosphere of earth just before the origin of life consisted of
(a) water vapours, CH_4, NH_3 and oxygen.
(b) CO_2, NH_3, and CH_2
(c) CH_4, NH_3, H_2 and water vapours.
(d) CH_4, O_3, O_2 and water vapours.

44. The extinct human ancestor who ate only fruits and hunted with stone weapons was
(a) *Ramapithecus* (b) *Australopithecus*
(c) *Dryopithecus* (d) *Homo habilis*

45. Tasmanian Wolf is a marsupial while Wolf is a placental mammal. This shows
(a) convergent evolution
(b) divergent evolution
(c) inheritance of acquired characters
(d) None of these

RESPONSE GRID					
	32. ⓐⓑⓒⓓ	33. ⓐⓑⓒⓓ	34. ⓐⓑⓒⓓ	35. ⓐⓑⓒⓓ	36. ⓐⓑⓒⓓ
	37. ⓐⓑⓒⓓ	38. ⓐⓑⓒⓓ	39. ⓐⓑⓒⓓ	40. ⓐⓑⓒⓓ	41. ⓐⓑⓒⓓ
	42. ⓐⓑⓒⓓ	43. ⓐⓑⓒⓓ	44. ⓐⓑⓒⓓ	45. ⓐⓑⓒⓓ	

Space for Rough Work

DAILY PRACTICE PROBLEM DPP CHAPTERWISE 29 - BIOLOGY			
Total Questions	45	Total Marks	180
Attempted		Correct	
Incorrect		Net Score	
Cut-off Score	45	Qualifying Score	60
Success Gap = Net Score – Qualifying Score			
Net Score = (Correct × 4) – (Incorrect × 1)			

DPP - Daily Practice Problems

Chapter-wise Sheets

Date : | Start Time : | End Time :

BIOLOGY CB30

SYLLABUS : Human Health and Disease

Max. Marks : 180 **Marking Scheme : + 4 for correct & (–1) for incorrect** **Time : 60 min.**

INSTRUCTIONS : This Daily Practice Problem Sheet contains 45 MCQs. For each question only one option is correct. Darken the correct circle/ bubble in the Response Grid provided on each page.

1. A cell coded protein formed in response to infection with most animal viruses is
   (a) Antigen (b) Antibody
   (c) Interferon (d) Histone
2. A sexually transmitted disease caused by bacteria is
   (a) Leprosy (b) AIDS
   (c) Syphilis (d) Pertussis
3. What is true about T-lymphocytes in mammals?
   (a) These are produced in thyroid
   (b) There are of three main types — cytotoxic T-cells, helper T-cells and suppressor T-cells
   (c) These originate in lymphoid tissues
   (d) They scavenge damaged cells and cellular debris
4. Obstacle to large scale transplantation of organs is
   (a) insufficiency of organ donors
   (b) immunological rejection of foreign bodies
   (c) religious or ethnic considerations
   (d) lack of effective surgical techniques
5. The use of *Cannabis* products (bhang, ganja, charas, marijuana or hashish) causes
   (a) depression of brain activity and feeling of calmness
   (b) alters thoughts, perceptions and feelings
   (c) suppresses brain function and relieves pain
   (d) stimulates the nervous system and increases alertness and activity.
6. Which one of the following is an Indian medicinal plant ?
   (a) *Saccharum officinarum* (b) *Rauwolfia serpentina*
   (c) *Oryza sativa* (d) *Solanum melongena*
7. In which one of the following pairs of diseases both are caused by viruses ?
   (a) Tetanus and typhoid
   (b) Whooping cough and sleeping sickness
   (c) Syphilis and AIDS
   (d) Measles and Rabies
8. Hypersensitivity to an allergen is associated with
   (a) aberrant functioning of the immune mechanism
   (b) increase in ambient temperature
   (c) age of the individual
   (d) food habits

RESPONSE GRID					
	1. ⓐⓑⓒⓓ	2. ⓐⓑⓒⓓ	3. ⓐⓑⓒⓓ	4. ⓐⓑⓒⓓ	5. ⓐⓑⓒⓓ
	6. ⓐⓑⓒⓓ	7. ⓐⓑⓒⓓ	8. ⓐⓑⓒⓓ		

Space for Rough Work

9. Carcinoma refers to
(a) benign tumours of the connective tissue
(b) malignant tumours of the connective tissue
(c) malignant tumours of the skin or mucous membrane
(d) malignant tumours of the colon

10. Human immuno deficiency virus (HIV) has a protein coat and a genetic material which is
(a) single stranded DNA
(b) double stranded DNA
(c) single stranded RNA
(d) double stranded RNA

11. Short-lived immunity acquired from mother to foetus across placenta or through mother's milk to the infant is categorised as
(a) innate non-specific immunity
(b) active immunity
(c) passive immunity
(d) cellular immunity

12. Which one is a correctly match sexually transmitted disease with its pathogen?
(a) Syphilis – *Leishmania donovani*
(b) AIDS – *Bacillus anthracis*
(c) Urethritis – *Entamoeba gingivalis*
(d) Gonorrhoea – *Neisseria gonorrhoeae*

13. To which type of barriers under innate immunity, do the saliva in the mouth and the tears from the eyes, belong?
(a) Cytokine barriers
(b) Cellular barriers
(c) Physiological barriers
(d) Physical barriers

14. Widal test is used for the diagnosis of
(a) Malaria (b) Pneumonia
(c) Tuberculosis (d) Typhoid

15. Which one of the following depresses brain activity and produced feelings of calmness, relaxation and drowsiness?
(a) Morphine (b) Valium
(c) Amphetamines (d) Hashish

16. Antibodies in our body are complex
(a) steroids (b) prostaglandins
(c) glycoproteins (d) lipoproteins

17. Select the correct statement with respect to diseases and immunisation?
(a) If due to some reason B-and T-lymphocytes are damaged, the body will not produce antibodies against a pathogen
(b) Injection of dead / inactivated pathogens causes passive immunity
(c) Certain protozoans have been used to mass produce hepatitis B vaccine.
(d) Injection of snake antivenom against snake bite is an example of active immunisation

18. Increased asthmatic attacks in certain seasons are related to
(a) eating fruits preserved in tin containers
(b) inhalation of seasonal pollen
(c) low temperature
(d) hot and humid environment.

19. A person likely to develop tetanus is immunised by administering
(a) preformed antibodies
(b) wide spectrum antibiotics
(c) weakened germs
(d) dead germs

20. Which one of the following statements is correct with respect to AIDS?
(a) The HIV can be transmitted through eating food together with an infected person
(b) Drug addicts are least susceptible to HIV infection .
(c) AIDS patients are being fully cured with proper care and nutrition
(d) The causative HIV retrovirus enters helper T-lymphocytes thus reducing their numbers

21. Damage to thymus in a child may lead to –
(a) a reduction in haemoglobin content of blood
(b) a reduction in stem cell production
(c) loss of antibody mediated immunity
(d) loss of cell mediated immunity

RESPONSE GRID					
	9. ⓐⓑⓒⓓ	10. ⓐⓑⓒⓓ	11. ⓐⓑⓒⓓ	12. ⓐⓑⓒⓓ	13. ⓐⓑⓒⓓ
	14. ⓐⓑⓒⓓ	15. ⓐⓑⓒⓓ	16. ⓐⓑⓒⓓ	17. ⓐⓑⓒⓓ	18. ⓐⓑⓒⓓ
	19. ⓐⓑⓒⓓ	20. ⓐⓑⓒⓓ	21. ⓐⓑⓒⓓ		

Space for Rough Work

22. Thromboplastin is secreted by
(a) platelets (b) lymphocytes
(c) helper T-cells (d) mast cells

23. Which one of the following techniques is safest for the detection of cancers?
(a) Magnetic resonance imaging (MRI)
(b) Radiography (X-ray)
(c) Computed tomography (CT)
(d) Histopathological studies

24. Common cold differs from pneumonia in, that
(a) Pneumonia is caused by a virus while the common cold is caused by the bacterium *Haemophilus influenzae*
(b) Pneumonia pathogen infects alveoli whereas the common cold affects nose and respiratory passage but not the lungs
(c) Pneumonia is a communicable disease whereas the common cold is a nutritional deficiency disease
(d) Pneumonia can be prevented by a live attenuated bacterial vaccine whereas the common cold has no effective vaccine

25. Which one of the following statements is correct with respect to immunity?
(a) Preformed antibodies need to be injected to treat the bite by a viper snake.
(b) The antibodies against small pox pathogen are produced by T – lymphocytes.
(c) Antibodies are protein molecules, each of which has four light chains.
(d) Rejection of a kidney graft is the function of B-lymphocytes.

26. Which of the following pairs contains an infections and a non-infectious disease respectively?
(a) Typhoid and AIDS
(b) AIDS and Cancer
(c) Pneumonia and malaria
(d) Cancer and malaria

27. The comon cold is caused by
(a) Rhino viruses
(b) *Streptococcus pneumoniae*
(c) *Salmonella typhimurium*
(d) *Plasmodium vivax*

28. Elephantiasis, a chronic inflammation that results in gross deformities is caused by
(a) *Ascaris* (b) *E.coli*
(c) *Wuchereria* (d) *Trichophyton*

29. Match Column-I with Column-II and select the correct answer from codes given below.

	Column-I		Column-II
A.	Sporozoites	(I)	Infectious form
B.	Filariasis	(II)	Aedes mosquitoes
C.	Typhoid	(III)	Wuchereria
D.	Chikungunya	(IV)	Widal test

(a) A-(IV), B-(II), C-(I), D-(III)
(b) A-(III), B-(IV), C-(II), D-(I)
(c) A-(II), B-(III), C-(I), D-(IV)
(d) A-(I), B-(III), C-(IV), D-(II)

30. Hepatitis B is transmitted through
(a) sneezing (b) female Anopheles
(c) coughing (d) blood transfusions

31. Which of the following is the bacterial disease in humans?
(a) Dysentery (b) Malaria
(c) Plague (d) Both (a) and (c)

32. The first line of defence in the immune system is provident by
(a) skin and mucous membrane
(b) inflammatory response
(c) the complement system
(d) none of these

33. The most abundant antiody produced against allergens
(a) lgE (b) lgA
(c) lgG (d) lgM

Response Grid					
	22. ⓐⓑⓒⓓ	23. ⓐⓑⓒⓓ	24. ⓐⓑⓒⓓ	25. ⓐⓑⓒⓓ	26. ⓐⓑⓒⓓ
	27. ⓐⓑⓒⓓ	28. ⓐⓑⓒⓓ	29. ⓐⓑⓒⓓ	30. ⓐⓑⓒⓓ	31. ⓐⓑⓒⓓ
	32. ⓐⓑⓒⓓ	33. ⓐⓑⓒⓓ			

Space for Rough Work

34. An auto-immune disease is
(a) SCID
(b) rheumatoid arthritis
(c) myasthenia gravis
(d) both (b) and (c)

35. The primary lymphoid organs are
(a) spleen and thymus
(b) bone marrow and thymus
(c) bone marrow and lymph node
(d) thymus and MALT

36. Vaccine against polio viruses is an example of
(a) auto-immunization
(b) passive immunization
(c) active immunization
(d) simple immunization

37. Which of the following is celebrated as 'World AIDS Day'?
(a) 31st March (b) 1st March
(c) 1st December (d) 31st December

38. Which of the following approaches are used for the treatment of cancer?
(a) Gene therapy and immunotherapy
(b) Surgery
(c) Radiotherapy and Chemotherapy
(d) All of these

39. Which of these is a member of the group of chemicals whose chemical structure is given below?

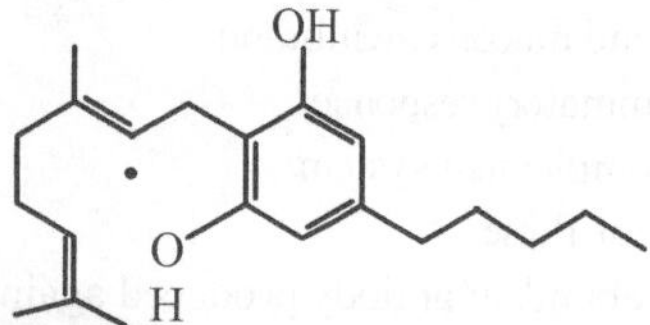

(a) Marijuana (b) Hashish
(c) Ganja (d) All of these

40. Which drug is being excessively taken by some sports persons nowadays?
(a) Opioids
(b) Barbiturates
(c) Cannabinoids
(d) Lysergic acid diethyl amides (LSD)

41. Take the odd one out
(a) Rabies, Influenza, AIDS
(b) Amoebiasis, Giardiasis, Trypanosomiasis
(c) Taeniasis, Ascariasis, Elephantiasis
(d) Cancer, Tuberculosis, Tetanus

42. Small proteins produced by vertebrate cells in response to viral infections inhibiting viral multiplication are known as
(a) Lipoproteins (b) Immuglobulins
(c) Interferons (d) Antitoxins

43. Which one of the following acts as a physiological barrier to the entry of micro-organisms in human body ?
(a) Epithelium of urogenital tract
(b) Tears
(c) Monocytes
(d) Skin

44. The treatment of snake-bite by antivenine is an example of
(a) Artificially acquired active immunity
(b) Artificially acquired passive immunity
(c) Naturally acquired passive immunity
(d) Specific natural immunity

45. The drug useful to increase cardiovascular effects in human beings is
(a) Cocaine (b) Barbiturate
(c) Benzodiazepine (d) Insulin

RESPONSE GRID					
	34. ⓐⓑⓒⓓ	35. ⓐⓑⓒⓓ	36. ⓐⓑⓒⓓ	37. ⓐⓑⓒⓓ	38. ⓐⓑⓒⓓ
	39. ⓐⓑⓒⓓ	40. ⓐⓑⓒⓓ	41. ⓐⓑⓒⓓ	42. ⓐⓑⓒⓓ	43. ⓐⓑⓒⓓ
	44. ⓐⓑⓒⓓ	45. ⓐⓑⓒⓓ			

Space for Rough Work

DAILY PRACTICE PROBLEM DPP CHAPTERWISE 30 - BIOLOGY			
Total Questions	45	Total Marks	180
Attempted		Correct	
Incorrect		Net Score	
Cut-off Score	48	Qualifying Score	55
Success Gap = Net Score – Qualifying Score			
Net Score = (Correct × 4) – (Incorrect × 1)			

DPP - Daily Practice Problems

Chapter-wise Sheets

Date : | **Start Time :** | **End Time :**

BIOLOGY (CB31)

SYLLABUS : Strategies for Enchancement in Food Production

Max. Marks : 180 | **Marking Scheme : + 4 for correct & (–1) for incorrect** | **Time : 60 min.**

INSTRUCTIONS : This Daily Practice Problem Sheet contains 45 MCQs. For each question only one option is correct. Darken the correct circle/ bubble in the Response Grid provided on each page.

1. Parthenocarpic tomato fruits can be produced by
   (a) treating the plants with low concentrations of gibberellic acid and auxins
   (b) raising the plants from vernalized seeds
   (c) treating the plants with phenylmercuric acetate
   (d) removing androecium of flowers before pollen grains are released
2. The technique of obtaining large number of plantlets by tissue culture method is called
   (a) Plantlet culture
   (b) Organ culture
   (c) Micropropagation
   (d) Macropropagation
3. Breeding of crops with high levels of minerals, vitamins and proteins is called
   (a) Somatic hybridisation
   (b) Biofortification
   (c) Biomagnification
   (d) Micropropagation
4. Jaya and Ratna developed for green revolution in India are the varieties of
   (a) maize (b) rice
   (c) wheat (d) bajra
5. Which one of the following pesticides is banned now a-days?
   (a) DDT (b) Eldrin
   (c) Aldrin (d) Toxaphene
6. Which one of the following is an exotic Indian fish ?
   (a) *Catla catla*
   (b) *Heteropneustes fossilis*
   (c) *Cyprinus carpio*
   (d) *Labeo rohita*

RESPONSE GRID	1. ⓐⓑⓒⓓ	2. ⓐⓑⓒⓓ	3. ⓐⓑⓒⓓ	4. ⓐⓑⓒⓓ	5. ⓐⓑⓒⓓ
	6. ⓐⓑⓒⓓ				

Space for Rough Work

7. Consider the following statements concerning food chains :
   (i) Removal of 80% tigers from an area resulted in greatly increased growth of vegetation
   (ii) Removal of most of the carnivores resulted in an increased population of deers
   (iii) The length of food chains is generally limited to 3-4 trophic levels due to energy loss
   (iv) The length of food chains may vary from 2 to 8 trophic levels

   Which two of the above statements are correct?
   (a) (ii) and (iii) (b) (iii) and (iv)
   (c) (i) and (iv) (d) (i) and (ii)
8. Which one is a neem product used as insect repellent?
   (a) Azadirachtin (b) Rotenone
   (c) Parathione (d) Endrin
9. Bull semen for the purpose of artificial insemination is stored in
   (a) Ice (b) Liquid oxygen
   (c) Liquid nitrogen (d) Liquid CO_2
10. The 'Mule' is the result of
    (a) Inbreeding depression
    (b) Out breeding
    (c) Cross breeding
    (d) Inter-specific hybridization
11. In live stock breeding experiments the following stage is transferred to surrogate mothers
    (a) Unfertilized eggs
    (b) Fertilized eggs
    (c) 8 to 32 celled embryo
    (d) Frozen semen
12. The animal husbandry deals with the care, breeding and management of
    (a) Domesticated animals
    (b) Fishes
    (c) Honey bees and silk worms
    (d) All of these
13. Which of the following has been recently used for increasing productivity of super milk cows?
    (a) Artificial insemination by a pedigree bull only
    (b) Superovulation of a high production cow only
    (c) Embryo transplantation only
    (d) A combination of superovulation, artificial insemination and embryo transplantation into a 'carrier cow' (surrogate mother)
14. Which one of the following statements is correct in relation to honey bees?
    (a) *Apis indica* is the largest wild honey bee in India
    (b) Honey is predominantly sucrose and arabinose
    (c) Beewax is a waste product of honey bees
    (d) Communication in honey bees was discovered by Karl Von Frisch
15. 'Inland fishery' refers to
    (a) Culturing fish in fresh water
    (b) Trapping and capturing fishes from sea coast
    (c) Deep sea fishing
    (d) Extraction of oil from fishes
16. Which of the following combinations is generally recommended for composite fish farming in India?
    (a) *Catla*, *Labeo*, *Cirrhinus*
    (b) *Catla*, *Cyprinus*, *Clarias*
    (c) *Clarias*, *Channa*, *Cyprinus*
    (d) *Cirrhinus*, *Cyprinus*, *Channa*
17. Hinny is a hybrid of male
    (a) Horse and female donkey
    (b) Donkey and female horse
    (c) Goat and female lamb
    (d) Sheep and female goat.
18. When cross is made between two species of the same genus, then the cross is known as
    (a) intraspecific hybridization
    (b) interspecific hybridization
    (c) intergeneric hybridization
    (d) intervarietal hybridization
19. Which of the following is a viral disease of poultry birds?
    (a) Anthrax (b) Ranikhet
    (c) Coccidiosis (d) None of these

RESPONSE GRID					
	7. ⓐⓑⓒⓓ	8. ⓐⓑⓒⓓ	9. ⓐⓑⓒⓓ	10. ⓐⓑⓒⓓ	11. ⓐⓑⓒⓓ
	12. ⓐⓑⓒⓓ	13. ⓐⓑⓒⓓ	14. ⓐⓑⓒⓓ	15. ⓐⓑⓒⓓ	16. ⓐⓑⓒⓓ
	17. ⓐⓑⓒⓓ	18. ⓐⓑⓒⓓ	19. ⓐⓑⓒⓓ		

Space for Rough Work

20. Which one of the following processes of breeding increases homozygosity?
(a) Inbreeding
(b) Out breeding
(c) Cross breeding
(d) Inter-specific breeding
21. The most popular breed of fowl in India is
(a) White leg horn (b) Aseel
(c) Plymouth (d) Langshan
22. Murrah, Mehsana, Jaffarbadi are breeds of
(a) Buffalo (b) Cow
(c) Cattle (d) Horse
23. Consumption of which one of the following foods can prevent the kind of blindness associated with vitamin 'A' deficiency ?
(a) '*Flavr Savr*' tomato
(b) Canolla
(c) Golden rice
(d) Bt-Brinjal
24. Golden rice is a transgenic crop of the future with the following improved trait :
(a) insect resistance
(b) high lysine (essential amino acid) content
(c) high protein content
(d) high vitamin-A content
25. Use of transgenic plants as biological factories for the production of special chemicals is called–
(a) Molecular farming (b) Molecular genetics
(c) Molecular mapping (d) Dry farming
26. Main objective of production/use of herbicide resistant GM crops is to
(a) eliminate weeds from the field without the use of manual labour
(b) eliminate weeds from the field without the use of herbicides
(c) encourage eco-friendly herbicides
(d) reduce herbicide accumulation in food articles for health safety
27. In order to obtain virus- free plants through tissue culture, the best method is
(a) protoplast culture (b) embryo rescue
(c) anther culture (d) meristem culture
28. A collection of plants and seeds having diverse alleles of all the genes of a crop is called
(a) herbarium (b) germplasm
(c) gene library (d) genome
29. Which one of the following is a wrong matching?
(a) Somatic hybridization - Fusion of two diverse cells
(b) Vector DNA - Site for t-RNA synthesis
(c) Micropropagation - *in vitro* production of plants in large numbers
(d) Callus - Unorganised mass of cell produced in tissue culture
30. In tissue culture roots can be induced by
(a) no cytokinin and only auxins.
(b) higher concentration cytokinin and lower concentration auxins.
(c) lower concentration of cytokinin and higher concentration of auxins.
(d) only cytokinin and no auxins.
31. Three crops that contribute maximum to global food grain production are
(a) Wheat, rice and maize
(b) Wheat, rice and barley
(c) Wheat, maize and sorghum
(d) Rice, maize and sorghum
32. Himgiri developed by hybridisation and selection for disease resistance against rust pathogens is a variety of
(a) chilli (b) maize
(c) sugarcane (d) wheat
33. Bee dances are meant for
(a) Courtship (b) Communication
(c) Recreation (d) Instinct
34. Artificial insemination involves
(a) super ovulation (b) semen collection
(c) egg collection (d) embryo collection

RESPONSE GRID					
	20. ⓐⓑⓒⓓ	21. ⓐⓑⓒⓓ	22. ⓐⓑⓒⓓ	23. ⓐⓑⓒⓓ	24. ⓐⓑⓒⓓ
	25. ⓐⓑⓒⓓ	26. ⓐⓑⓒⓓ	27. ⓐⓑⓒⓓ	28. ⓐⓑⓒⓓ	29. ⓐⓑⓒⓓ
	30. ⓐⓑⓒⓓ	31. ⓐⓑⓒⓓ	32. ⓐⓑⓒⓓ	33. ⓐⓑⓒⓓ	34. ⓐⓑⓒⓓ

Space for Rough Work

35. The term "breed" refers to
(a) a group of animals not related by descent but similar in most characters
(b) a group of animals related by descent and similar in most characters
(c) a group of animals related by descent but have almost different characteristics
(d) a group of animals neither related by descent nor have similar characteristics.

36. What strategy would you suggest if a person wants to evolve a pureline in an animal?
(a) Cross-breeding (b) Inbreeding
(c) Out-breeding (d) Artificial insemination

37. Select the incorrect statement from the following.
(a) Apiculture provides additional income generating industry to the farmers.
(b) Bee keeping is labour intensive process.
(c) Bee venom is used to cure certain diseases like gout and arthritis.
(d) Honey is used as laxative, antiseptic and sedative.

38. Germplasm collection is the collection of
(a) germ cells
(b) semens
(c) plants/seeds with all the diverse alleles for all genes
(d) egg cells

39. Which of the following diseases is caused by virus?
(a) Tobacco mosaic (b) Late blight of potato
(c) Turnip mosaic (d) Both (a) and (c)

40. Single cell protein can be obtained from
(a) bacteria (b) algae
(c) fungi (d) all of these

41. Meristem culture is used
(a) to produce disease free plants
(b) in germplasm conservation
(c) in rapid clonal multiplication
(d) all of these

42. A plant cell without cell wall is called
(a) proplast (b) protoplast
(c) nucleoplasm (d) explant

43. Refer the given figures and answer the questions.

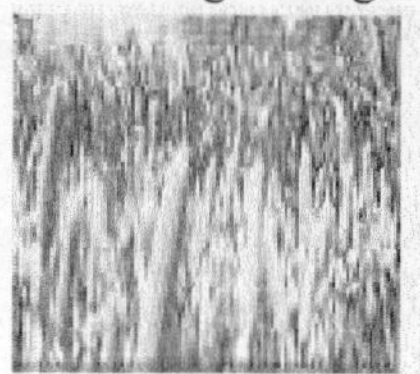  

Which of the following statements is correct regarding the above figures?
(i) These are all Indian hybrid crops of low yielding varieties.
(ii) These are all Indian hybrid crops of high yielding varieties.
(iii) The production of the above crops led to dramatic increase in food production.
(iv) These crops are produced as a result of various plants breeding technique.
(a) (i), (ii), and (iii) (b) (ii), (iii) and (iv)
(c) (iii) and (iv) only (d) (i) and (iii) only

44. Match Column - I with Column - II

	Column-I		Column-II
A.	Hybrid variety	(I).	X-ray
B.	Mutation	(II).	Allopolyploidy
C.	Pure line	(III).	F_1 generation
D.	*Triticale*	(IV).	Selection in self pollinated crops
		(V).	Genetic engineering

(a) A-(III); B-(II); C-(I); D-(IV)
(b) A-(III); B-(I); C-(IV); D-(II)
(c) A-(I); B-(III); C-(IV); D-(II)
(d) A-(II); B-(I); C-(III); D-(IV)

45. Farmers in a particular region were concerned that pre-mature yellowing of leaves of a pulse crop might cause decrease in the yield. Which treatment could be most beneficial to obtain maximum seed yield ?
(a) Treatment of the plants with cytokinins along with a small dose of nitrogenous fertilizer
(b) Removal of all yellow leaves and spraying the remaining green leaves with 2, 4, 5- trichlorophenoxy acetic acid
(c) Application of iron and magnesium to promote synthesis of chlorophyll
(d) Frequent irrigation of the crop

RESPONSE GRID					
	35. ⓐⓑⓒⓓ	36. ⓐⓑⓒⓓ	37. ⓐⓑⓒⓓ	38. ⓐⓑⓒⓓ	39. ⓐⓑⓒⓓ
	40. ⓐⓑⓒⓓ	41. ⓐⓑⓒⓓ	42. ⓐⓑⓒⓓ	43. ⓐⓑⓒⓓ	44. ⓐⓑⓒⓓ
	45. ⓐⓑⓒⓓ				

Space for Rough Work

DAILY PRACTICE PROBLEM DPP CHAPTERWISE 31 - BIOLOGY			
Total Questions	45	Total Marks	180
Attempted		Correct	
Incorrect		Net Score	
Cut-off Score	55	Qualifying Score	60
Success Gap = Net Score – Qualifying Score			
Net Score = (Correct × 4) – (Incorrect × 1)			

DPP - Daily Practice Problems

Chapter-wise Sheets

Date : | **Start Time :** | **End Time :**

BIOLOGY CB32

SYLLABUS : Microbes in human welfare

Max. Marks : 180 | **Marking Scheme : + 4 for correct & (–1) for incorrect** | **Time : 60 min.**

INSTRUCTIONS : This Daily Practice Problem Sheet contains 45 MCQs. For each question only one option is correct. Darken the correct circle/ bubble in the Response Grid provided on each page.

1. Select the correct statement from the following.
(a) Biogas is produced by the activity of aerobic bacteria on animal waste
(b) *Methanobacterium* is an aerobic bacterium found in rumen of cattle
(c) Biogas, commonly called gobar gas, is pure methane
(d) Activated sludge-sediment in settlement tanks of sewage treatment plant is a rich source of aerobic bacteria

2. Which one thing is not true about antibiotics?
(a) The term "antibiotic" was coined by Selman Waksman in 1942
(b) First antibiotic was discovered by Alexander Flemming
(c) Each antibiotic is effective only against one particular kind of germ
(d) Some persons can be allergic to a particular antibiotic

3. *Monascus purpureus* is a yeast used commercially in the production of :
(a) ethanol
(b) streptokinase for removing clots from the blood vessels.
(c) citric acid
(d) blood cholesterol lowering statins

4. A common biocontrol agent for the control of plant diseases is
(a) Baculovirus (b) *Bacillus thuringiensis*
(c) *Glomus* (d) *Trichoderma*

5. Continuous addition of sugars in 'fed batch' fermentation is done to:
(a) produce methane (b) obtain antibiotics
(c) purify enzymes (d) degrade sewage

RESPONSE GRID	1. ⓐⓑⓒⓓ	2. ⓐⓑⓒⓓ	3. ⓐⓑⓒⓓ	4. ⓐⓑⓒⓓ	5. ⓐⓑⓒⓓ

Space for Rough Work

6. Streptomycin is obtained from
(a) *Streptomyces griseus*
(b) S. *aureofaciens*
(c) S. *venezuelae*
(d) S. *ramosus*

7. For biogas production besides dung an extensive use of which weed is recommended in our country–
(a) *Mangifera indica*
(b) *Hydrilla*
(c) *Eicchornia crassipes*
(d) *Solanum*

8. Chloramphenicol and erythromycin (broad spectrum antibiotics) are produced by
(a) *Streptomyces* (b) *Nitrobacter*
(c) *Rhizobium* (d) *Penicillium*

9. Which one of the microorganism is used for production of citric acid in industries ?
(a) *Lactobacillus bulgaricus*
(b) *Penicillium citrinum*
(c) *Aspergillus niger*
(d) *Rhizopus nigricans*

10. A genetically engineered bacteria used for clearing oil spills is :
(a) *Escherichia coli*
(b) *Bacillus subtilis*
(c) *Agrobacterium tumifaciens*
(d) *Pseudomonas putida*

11. Human insulin is being commercially produced from a transgenic species of
(a) *Escherichia* (b) *Mycobacterium*
(c) *Rhizobium* (d) *Saccharomyces*

12. Which one of the following is not used in organic farming?
(a) *Glomus* (b) Earthworm
(c) *Oscillatoria* (d) Snail

13. What is mode of bacterial resistance against antibiotics ?
(a) Development of thick mucilaginous layer
(b) Alteration of cell membrane
(c) Mutation in bacteria
(d) All the above

14. Which one of the following is a wrong matching of a microbe and its industrial product, while the remaining three are correct ?
(a) Yeast - statins
(b) *Acetobacter aceti* - acetic acid
(c) *Clostridium butylicum* - lactic acid
(d) *Aspergillus niger* - citric acid

15. During anaerobic digestion of organic waste, such as in producing biogas, which one of the following is left undegraded ?
(a) Lipids (b) Lignin
(c) Hemi-cellulose (d) Cellulose

16. Rennin used in cheese industry is –
(a) Antibiotic (b) Enzyme
(c) Alkaloid (d) Inhibitor

17. Vitamin B_{12} is formed during fermentation of
(a) *Ashloya gossipii*
(b) *Rhizopus stalonifer*
(c) *Propionibacteria*
(d) *Saccharomyces cerevisiae*

18. The term "antibiotic" was coined by –
(a) Edward Jenner (b) Louis Pasteur
(c) Selman Waksman (d) Alexander Flemming

19. Which one of the following micro-organisms is used for production of citric acid in industries?
(a) *Penicillium citrinum*
(b) *Aspergillus niger*
(c) *Rhizopus nigricans*
(d) *Lactobacillus bulgaris*

20. Streptokinase which is used as a 'clot buster' obtained from
(a) *Streptococcus* (b) *Staphylococcus*
(c) *Lactobacillus* (d) *Saccharomyces*

RESPONSE GRID					
	6. ⓐⓑⓒⓓ	7. ⓐⓑⓒⓓ	8. ⓐⓑⓒⓓ	9. ⓐⓑⓒⓓ	10. ⓐⓑⓒⓓ
	11. ⓐⓑⓒⓓ	12. ⓐⓑⓒⓓ	13. ⓐⓑⓒⓓ	14. ⓐⓑⓒⓓ	15. ⓐⓑⓒⓓ
	16. ⓐⓑⓒⓓ	17. ⓐⓑⓒⓓ	18. ⓐⓑⓒⓓ	19. ⓐⓑⓒⓓ	20. ⓐⓑⓒⓓ

Space for Rough Work

21. Baculoviruses are excellent candidates for
(a) species-specific narrow spectrum pesticidal applications.
(b) species-specific broad spectrum pesticidal applications.
(c) species-specific narrow spectrum insecticidal applications.
(d) species-specific broad spectrum insecticidal applications.

22. Farmers have reported over 50% higher yields of rice by using the biofertilizer
(a) *Azolla pinnata*
(b) *Cyanobacteria*
(c) *Legume-Rhizobium* symbiosis
(d) Mycorrhiza

23. Microbes are present in
(a) soil (b) thermal vents
(c) polluted water (d) all of these

24. Which of the following microbes is a proteinacious infectious agent?
(a) Fungi (b) Prions
(c) Bacteria (d) Protozoa

25. Probiotics are
(a) cancer inducing microbes
(b) new kind of food allergens
(c) live microbial food supplement
(d) safe antibiotics

26. *Saccharomyces cerevisiae* is a yeast commercially used in
(a) citric acid
(b) ethanol
(c) baking
(d) streptokinase for removing clots from blood vessels

27. The masses of bacteria held together by slime and fungal filaments to form mesh like structures are called as
(a) primary sludge (b) flocs
(c) activated sludge (d) anaerobic sludge

28. The purpose of biological treatment of waste water is to
(a) reduce BOD
(b) increase BOD
(c) reduce sedimentation
(d) increase sedimentation

29. These bacteria grow anaerobically on cellulosic material, produce large amount of methane along with CO_2 and H_2, and are collectively called as methanogen. Examples of such bacteria are
(a) *Methanobacterium*
(b) *Methanobrevibacter*
(c) *Methanococcus*
(d) All of these

30. Biogas is produced by
(a) aerobic breakdown of biomass
(b) anaerobic breakdown of biomass
(c) with the help of methanogenic bacteria
(d) both (b) and (c)

31. Match Column-I with Column-II and select the correct answer from the codes given below.

	Column-I		Column-II
A.	*Trichoderma*	(I)	Nitrification
B.	*Streptomyces*	(II)	Biocontrol agent
C.	*Nitrosomonas*	(III)	Lactic acid
D.	*Lactobacillus*	(IV)	Source of antibiotic

(a) A-(II), B-(III), C-(IV), D-(I)
(b) A-(II), B-(IV), C-(I), D-(III)
(c) A-(III), B-(I), C-(II), D-(IV)
(d) A-(IV), B-(II), C-(I), D-(III)

32. Organic farming does not include
(a) green manures (b) chemical fertilizers
(c) farmyard manures (d) compost

33. The symbiotic association between fungi and roots of higher plants is referred to as
(a) lichen (b) Mycorrhiza
(c) biofertilizer (d) biocontrol agent

34. Which of the following options includes biofertilizers?
(a) Cowdung manure and farmyard waste
(b) A quick growing crop ploughed back into the field
(c) Nostoc, Oscillatoria
(d) All of these

RESPONSE GRID					
	21. ⓐⓑⓒⓓ	22. ⓐⓑⓒⓓ	23. ⓐⓑⓒⓓ	24. ⓐⓑⓒⓓ	25. ⓐⓑⓒⓓ
	26. ⓐⓑⓒⓓ	27. ⓐⓑⓒⓓ	28. ⓐⓑⓒⓓ	29. ⓐⓑⓒⓓ	30. ⓐⓑⓒⓓ
	31. ⓐⓑⓒⓓ	32. ⓐⓑⓒⓓ	33. ⓐⓑⓒⓓ	34. ⓐⓑⓒⓓ	

Space for Rough Work

35. Yeast *Saccharomyces cervisiae* is used in the industrial production of
(a) butanol (b) citric acid
(c) tetracycline (d) ethanol

36. In cheese manufacture, the micro-organisms are used for
(a) the souring of milk only
(b) the ripening only
(c) development of resistance to spoilage
(d) Both (a) and (b)

37. Brewer's yeast lack
(a) diastase and amylase (b) amylase only
(c) diastase only (d) maltose

38. Baggasse is related to the manufacture of
(a) cinchonidine (b) cellulose materials
(c) resin (d) cane sugar

39. Which of the following bacteria is used for the production of butanol and acetone from starch ?
(a) *Lactobacillus bulgaricus*
(b) *Clostridium acetobutylicum*
(c) *Streptococcum thermophilus*
(d) Both (a) and (c)

40. Which bacterium helps in the production of 'Swiss cheese'?
(a) *Propionibacterium sharmanii*
(b) *Trichoderma polysporum*
(c) *Saccharomyces cerevisiae*
(d) *Aspergillus niger*

41. Statins, a bioactive molecule, inhibiting the enzyme responsible for synthesis of
(a) carbohydrate (b) protein
(c) vitamins (d) cholestrol

42. Gallic acid is obtained from
(a) *Pseudomonas species*
(b) *Penicillium purpurogenum*
(c) *Aspergillus niger*
(d) *Streptomyces species*

43. The diagram below shows a typical biogas plant. With few structure labelled as A, B and C. Identify A, B and C.

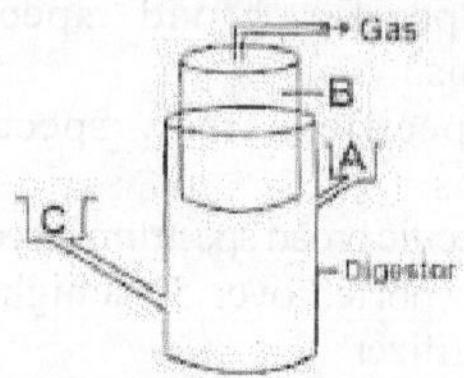

(a) A – Sludge, B – Methane, Oxygen, C – Dung, water
(b) A – Sludge, B – Methane, Carbon dioxide, C– Dung, water
(c) A – Sludge, B – Ethylin, Carbon dioxide, C – Dung, water
(d) A – Sludge, B – Methane, Carbon dioxide, C – Sewage

44. Match column-I with column-II and choose the correct option

	Column-I		Column-II
A.	Statins	I.	Yeast
B.	Ethanol	II.	Blood-cholesterol lowering agent
C.	Dung	III.	Insect-resistant plant
D.	Bt-cotton	IV.	Biogas

(a) A-II; B-I; C-IV; D-III
(b) A-III; B-IV; C-I; D-II
(c) A-I; B-II; C-III; D-IV
(d) A-IV; B- II; C-I; D-III

45. Which one of the following statement regarding BOD is true?
(a) The greater the BOD of waste water, more is its polluting potential.
(b) The greater the BOD of waste water, less is its polluting potential.
(c) The lesser the BOD of waste water, more is its polluting potential.
(d) The lesser the BOD of waste water, less is its polluting potential.

RESPONSE GRID					
	35. ⓐⓑⓒⓓ	36. ⓐⓑⓒⓓ	37. ⓐⓑⓒⓓ	38. ⓐⓑⓒⓓ	39. ⓐⓑⓒⓓ
	40. ⓐⓑⓒⓓ	41. ⓐⓑⓒⓓ	42. ⓐⓑⓒⓓ	43. ⓐⓑⓒⓓ	44. ⓐⓑⓒⓓ
	45. ⓐⓑⓒⓓ				

Space for Rough Work

DAILY PRACTICE PROBLEM DPP CHAPTERWISE 32 - BIOLOGY			
Total Questions	45	Total Marks	180
Attempted		Correct	
Incorrect		Net Score	
Cut-off Score	45	Qualifying Score	60
Success Gap = Net Score – Qualifying Score			
Net Score = (Correct × 4) – (Incorrect × 1)			

DPP - Daily Practice Problems

Chapter-wise Sheets

Date : | **Start Time :** | **End Time :**

BIOLOGY CB33

SYLLABUS : Biotechnology: Principles and Processes

Max. Marks : 180 **Marking Scheme : + 4 for correct & (–1) for incorrect** **Time : 60 min.**

INSTRUCTIONS : This Daily Practice Problem Sheet contains 45 MCQ's. For each question only one option is correct. Darken the correct circle/ bubble in the Response Grid provided on each page.

1. The linking of antibiotic resistance gene with the plasmid vector became possible with
   (a) DNA ligase
   (b) Endonucleases
   (c) DNA polymerase
   (d) Exonucleases
2. DNA or RNA segment tagged with a radioactive molecule is called
   (a) Vector (b) Probe
   (c) Clone (d) Plasmid
3. Restriction endonucleases are enzymes which
   (a) make cuts at specific positions within the DNA molecule
   (b) recognize a specific nucleotide sequence for binding of DNA ligase
   (c) restrict the action of the enzyme DNA polymerase
   (d) remove nucleotides from the ends of the DNA molecule
4. Agarose extracted from sea weeds finds use in :
   (a) Spectrophotometry
   (b) Tissue culture
   (c) PCR
   (d) Gel electrophoresis
5. PCR and Restriction Fragment Length Polymorphism are the methods for :
   (a) Study of enzymes
   (b) Genetic transformation
   (c) DNA sequencing
   (d) Genetic Fingerprinting
6. 'Cloning' is meant for/to
   (a) production of HGH gene in *E. coli*
   (b) preserve the genotype of organism
   (c) replace the original gene
   (d) All of the above
7. Which one of the following is used as vector for cloning genes into higher organisms?
   (a) Baculovirus
   (b) *Salmonella typhimurium*
   (c) *Rhizopus nigricans*
   (d) Retrovirus

RESPONSE GRID					
	1. ⓐⓑⓒⓓ	2. ⓐⓑⓒⓓ	3. ⓐⓑⓒⓓ	4. ⓐⓑⓒⓓ	5. ⓐⓑⓒⓓ
	6. ⓐⓑⓒⓓ	7. ⓐⓑⓒⓓ			

Space for Rough Work

8. The figure below is the diagrammatic representation of the *E.Coli* vector pBR 322. Which one of the given options correctly identifies its certain component (s) ?

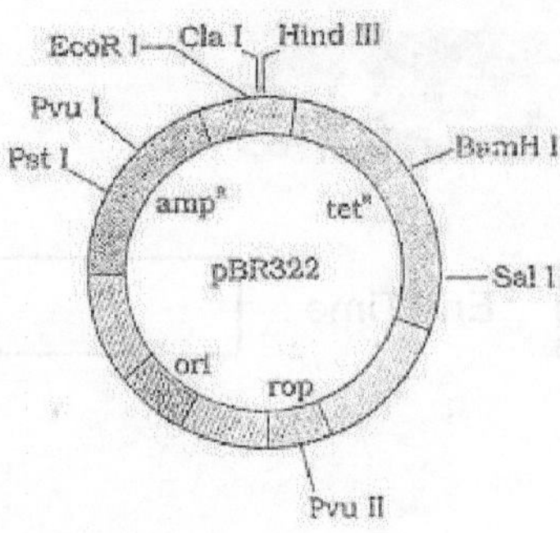

(a) ori - original restriction enzyme
(b) rop-reduced osmotic pressure
(c) Hind III, EcoRI - selectable markers
(d) amp^R, tet^R - antibiotic resistance genes

9. Electroporation procedure involves
(a) fast passage of food through sieve pores in phloem elements with the help of electric stimulation.
(b) opening of stomatal pores during night by artificial light.
(c) making transient pores in the cell membrane to introduce gene constructs.
(d) purification of saline water with the help of a membrane system.

10. What is the first step in the Southern blot technique?
(a) Denaturation of DNA on the gel for hybridization with specific probe.
(b) Production of a group of genetically identical cells.
(c) Digestion of DNA by restriction enzyme.
(d) Denaturation of DNA from a nucleated cell such as the one from the scene of crime.

11. The polymerase chain reaction (PCR) technology was discovered by
(a) Karry Mullis (b) Saiki *et al*
(c) Craig Venter (d) Maxam and Gilbert

12. For transformation, micro-particles coated with DNA to be bombarded with gene gun are made up of :
(a) Silver or Platinum
(b) Platinum or Zinc
(c) Silicon or Platinum
(d) Gold or Tungsten

13. Plasmid used to construct the first recombinant DNA was isolated from which bacterium species?
(a) *Escherichia coli*
(b) *Salmonella typhimurium*
(c) *Agrobacterium tumefaciens*
(d) *Thermus aquaticus*

14. Genetic engineering is possible because
(a) we can cut DNA at specific sites by restriction endonucleases
(b) restriction endonucleases purified from virus can be used in bacteria
(c) the phenomenon of transduction in bacteria is well understood
(d) we can seen DNA by electron microscope

15. Gel electrophoresis is a
(a) technique of separation of charged molecules under the influence of magnetic field
(b) technique of incorporation of DNA molecules into the cell through transient pores made due to electrical impulses
(c) technique of separation of DNA fragments through the pores of agarose gel under the influence of electric field
(d) technique of separation and purification of gene products.

16. In recombinant DNA technology, the term vector refers to
(a) the enzyme that cuts DNA into restriction fragments
(b) the sticky end of a DNA fragment
(c) a plasmid used to transfer DNA into a living cell
(d) a DNA fragment which carries only ori gene

17. In agarose gel electrophoresis
(a) DNA migrates towards the negative electrode
(b) supercoiled plamids migrate slower than their nicked counterparts
(c) larger molecules migrate faster than smaller molecules
(d) ethidium bromide can be used to visualize the DNA

18. Which of the following is based upon the principle of antigen-antibody interaction?
(a) PCR (b) ELISA
(c) R DNA technology (d) RNA

19. Two microbes found to be very useful in genetic engineering are
(a) *Vibrio cholerae* and a tailed bacteriophage
(b) *Diplococcus* sp. and *Pseudomonas* sp.
(c) Crown gall bacterium and *Caenorhabditis elegans*
(d) *Escherichia coli* and *Agrobacterium tumefaciens*

RESPONSE GRID					
	8. ⓐⓑⓒⓓ	9. ⓐⓑⓒⓓ	10. ⓐⓑⓒⓓ	11. ⓐⓑⓒⓓ	12. ⓐⓑⓒⓓ
	13. ⓐⓑⓒⓓ	14. ⓐⓑⓒⓓ	15. ⓐⓑⓒⓓ	16. ⓐⓑⓒⓓ	17. ⓐⓑⓒⓓ
	18. ⓐⓑⓒⓓ	19. ⓐⓑⓒⓓ			

Space for Rough Work

20. The figure below shows three steps (A, B, C) of Polymerase Chain Reaction (PCR). Select the option giving correct identification together with what it represents?

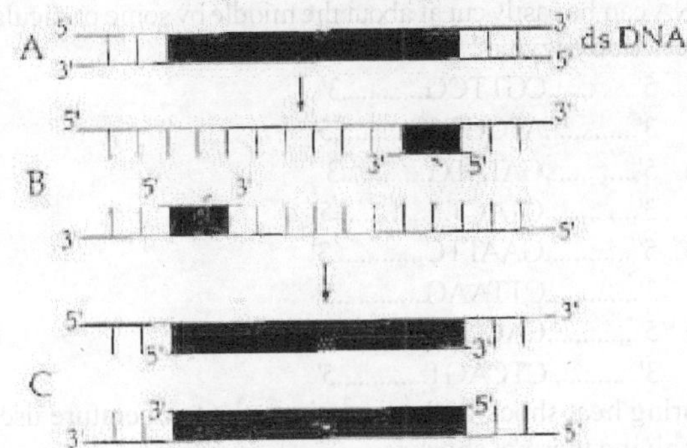

(a) B - Denaturation at a temperature of about 98°C separating the two DNA strands.
(b) A - Denaturation at a temperature of about 50°C.
(c) C - Extension in the presence of heat stable DNA polymerase.
(d) A - Annealing with two sets of primers.

21. Biolistics (gene-gun) is suitable for
(a) DNA finger printing
(b) Disarming pathogen vectors
(c) Transformation of plant cells
(d) Constructing recombinant DNA by joining with vectors

22. Baculoviruses are excellent candidates for
(a) species-specific narrow spectrum pesticidal applications.
(b) species-specific broad spectrum pesticidal applications.
(c) species-specific narrow spectrum insecticidal applications.
(d) species-specific broad spectrum insecticidal applications.

23. Which of the following technique is used for the separation of DNA fragments ?
(a) Gel electrophoresis (b) Chromatography
(c) Transformation (d) Transduction

24. Plasmids are suitable vectors for gene cloning because
(a) these are small circular DNA molecules which can integrate with host chromosomal DNA.
(b) these are small circular DNA molecules with their own replication origin site.
(c) these can shuttle between prokaryotic and eukaryotic cells.
(d) these often carry antibiotic resistance genes.

25. The term "competent" refers to
(a) increasing the competition between cells
(b) making cells impermeable for DNA
(c) increasing the efficiency with which DNA enters the bacterium through pores in its cell wall
(d) making cells permeable for divalent cations

26. The correct sequence of different steps of polymerase chain reaction is
(a) Annealing → Denaturation → Extension
(b) Denaturation → Extension → Annealing
(c) Denaturation → Annealing → Extension
(d) Extension → Denaturation → Annealing

27. Eukaryotic genes do not function properly when cloned into a bacterial cell, because
(a) of high pH present in bacterial cells
(b) of inability to excise introns and destruction by bacterial restriction enzymes
(c) of inappropriate insertion of genes
(d) both (a) and (b).

28. Which structure involved in genetic engineering?
(a) Plastid (b) Plasmid
(c) Codon (d) None of these

29. *Ti*-plasmid used in genetic engineering has been modified by
(a) adding tumour forming genes.
(b) deleting tumour forming genes.
(c) adding genes for endonucleases.
(d) deleting genes for endonucleases.

30. Which of the following technique is used for the detection of RNA fragments ?
(a) Northern blotting (b) Chromatography
(c) Transformation (d) Transduction

31. Which of these is not correctly matched ?
(a) Gene gun—biolistic gun
(b) Plasmids—extrachromosomal DNA
(c) DNA ligase—Biological scissors
(d) Bacteriophages—viruses.

32. Polyethylene glycol method is used for
(a) biodiesel production
(b) seedless fruit production
(c) energy production from sewage
(d) gene transfer without a vector

RESPONSE GRID					
	20. ⓐⓑⓒⓓ	21. ⓐⓑⓒⓓ	22. ⓐⓑⓒⓓ	23. ⓐⓑⓒⓓ	24. ⓐⓑⓒⓓ
	25. ⓐⓑⓒⓓ	26. ⓐⓑⓒⓓ	27. ⓐⓑⓒⓓ	28. ⓐⓑⓒⓓ	29. ⓐⓑⓒⓓ
	30. ⓐⓑⓒⓓ	31. ⓐⓑⓒⓓ	32. ⓐⓑⓒⓓ		

Space for Rough Work

33. Which of the following is a molecular scissors?
(a) *EcoRI* (b) *Hind*III
(c) *Bam* H II (d) All of these

34. Rennin used in cheese industry is
(a) antibiotic (b) alkaloid
(c) enzyme (d) inhibitor

35. The primary reason why the same basic techniques can be used to analyze the DNA from species as diverse as bacteria and humans is that
(a) all cells are identical.
(b) every organism has the same amount of DNA.
(c) the DNA sequences of all organisms are the same.
(d) DNA has a consistent structure in all organisms.

36. Which of the following is a plasmid?
(a) pBR 322 (b) *Bam H* I
(c) *Sal* I (d) *Eco R*I

37. There is a restriction endonuclease called *EcoR*I. What does *.co* part in it stand for ?
(a) Colon (b) Coelom
(c) Coenzyme (d) *coli*

38. Restriction endonuclease - *Hind* II always cuts DNA molecules at a particular point by recognizing a specific sequence of
(a) six base pairs. (b) five base pairs.
(c) four base pairs. (d) seven base pairs.

39. Which of the following enzyme is used in case of fungus to cause release of DNA along with other macromolecules ?
(a) Lysozyme (b) Cellulase
(c) Chitinase (d) Amylase

40. Match Column I with Column II and identify the correct option.

Column - I	Column - II
A. Primers	I. PCR
B. Separation and purification of products	II. C_2H_5OH
C. Precipitation of DNA	III. Uptake of foreign DNA by bacterium
D. Transformation	IV. Down stream processing

(a) A-IV; B-I; C-II; D-III
(b) A-II; B-I; C-IV; D-III
(c) A-IV; B-I; C-III; D-II
(d) A-I; B-IV; C-II; D-III

41. Which one of the following palindromic base sequences in DNA can be easily cut at about the middle by some particular restriction enzyme?
(a) 5'............CGTTCG............3'
3'............ATGGTA............5'
(b) 5'............GATATG............3'
3'............CTACTA............5'
(c) 5'............GAATTC............3'
3'............CTTAAG............5'
(d) 5'............CACGTA............3'
3'............CTCAGT............5'

42. During heat shock to the bacterium, the temperature used for giving thermal shock is
(a) 52°C (b) 100°C
(c) liquid nitrogen (d) 42°C

43. Stirred-tank bioreactors have been designed for
(a) addition of preservatives to the product.
(b) purification of the product.
(c) ensuring anaerobic conditions in the culture vessel.
(d) availability of oxygen throughout the process.

44. After completion of biosynthetic stage, the product has to be subjected through a series of processes before it is ready to marketing as a finished product. This series of processes is called
(a) upstream processing (b) downstream processing
(c) elution (d) insertional inactivation

45. Which of the following is not necessary to execute a polymerase chain reaction successfully?
(a) All four DNA bases
(b) Short DNA base primers
(c) DNA polymerase
(d) DNA library

RESPONSE GRID					
	33. ⓐⓑⓒⓓ	34. ⓐⓑⓒⓓ	35. ⓐⓑⓒⓓ	36. ⓐⓑⓒⓓ	37. ⓐⓑⓒⓓ
	38. ⓐⓑⓒⓓ	39. ⓐⓑⓒⓓ	40. ⓐⓑⓒⓓ	41. ⓐⓑⓒⓓ	42. ⓐⓑⓒⓓ
	43. ⓐⓑⓒⓓ	44. ⓐⓑⓒⓓ	45. ⓐⓑⓒⓓ		

Space for Rough Work

DAILY PRACTICE PROBLEM DPP CHAPTERWISE 33 - BIOLOGY			
Total Questions	45	Total Marks	180
Attempted		Correct	
Incorrect		Net Score	
Cut-off Score	48	Qualifying Score	60
Success Gap = Net Score – Qualifying Score			
Net Score = (Correct × 4) – (Incorrect × 1)			

DPP - Daily Practice Problems

Chapter-wise Sheets

Date : | Start Time : | End Time :

BIOLOGY (CB34)

SYLLABUS : Biotechnology and its Applications

Max. Marks : 180 **Marking Scheme : + 4 for correct & (–1) for incorrect** **Time : 60 min.**

INSTRUCTIONS : This Daily Practice Problem Sheet contains 45 MCQ's. For each question only one option is correct. Darken the correct circle/ bubble in the Response Grid provided on each page.

1. Silencing of mRNA has been used in producing transgenic plants resistant to:
   (a) bollworms (b) nematodes
   (c) white rusts (d) bacterial blights
2. Important objective of biotechnology in agriculture section is –
   (a) To produce pest resistant varieties of plants
   (b) To increase the nitrogen content
   (c) To decrease the seed number
   (d) To increase the plant weight
3. Tobacco plants resistant to a nematode have been developed by the introduction of DNA that produced (in the host cells)
   (a) both sense and anti-sense RNA
   (b) a particular hormone
   (c) an antifeedant
   (d) a toxic protein
4. Which one of the following techniques made it possible to genetically engineered living organism ?
   (a) Recombinant DNA techniques
   (b) X-ray diffraction
   (c) Heavier isotope labelling
   (d) Hybridization
5. Consider the following four statements
   (i) The first transgenic buffalo, Rosie produced milk which was human alpha-lactal albumin enriched.
   (ii) Restriction enzymes are used in isolation of DNA from other macro-molecules.
   (iii) Downstream processing is one of the steps of R-DNA technology.
   (iv) Disarmed pathogen vectors are also used in transfer of R-DNA into the host.

   Which are the two statements incorrect?
   (a) Statement (ii) and (iii)
   (b) Statement (iii) and (iv)
   (c) Statement (i) and (iii)
   (d) Statement (i) and (ii)

RESPONSE GRID	1. ⓐⓑⓒⓓ	2. ⓐⓑⓒⓓ	3. ⓐⓑⓒⓓ	4. ⓐⓑⓒⓓ	5. ⓐⓑⓒⓓ

Space for Rough Work

6. cDNA probes are copied from the messenger RNA molecules with the help of–
   (a) Restriction enzymes
   (b) Reverse transcriptase
   (c) DNA polymerase
   (d) Adenosine deaminase
7. In history of biology, human genome project led to the development of:
   (a) biotechnology (b) biomonitoring
   (c) bioinformatics (d) biosystematics
8. Dolly sheep was obtained by –
   (a) Cloning the udder cell (somatic cell) fused with enucleated oocyte
   (b) Cloning of gametes
   (c) Tissue culture
   (d) None of the above
9. Which of the following forms the basis of DNA Finger printing?
   (a) The relative proportions of purines and pyrimidines in DNA.
   (b) Satellite DNA occurring as highly repeated short DNA segments.
   (c) The relative difference in the DNA occurrence in blood, skin and saliva.
   (d) The relative amount of DNA in the ridges and grooves of the fingerprints.
10. The genetically-modified (GM) brinjal in India has been developed for:
   (a) insect-resistance
   (b) enhancing shelf life
   (c) enhancing mineral content
   (d) drought-resistance
11. Which of the following statement(s) is/are true ?
   (i) Biowar is the use of biological weapons against humans and /or their crops and animals.
   (ii) Bioethics is the unauthorized use of bioresources and traditional knowledge related to bioresources for commercial benefits.
   (iii) Biopatent is exploitation of bioresources of other nations without proper authorisation.
   (a) (ii) only (b) (i) only
   (c) (i) and (ii) only (d) (i) and (iii) only
12. Which of the following statements is/are correct with regard to the disadvantages of GM crops?
   (a) GM crops can affect human health by causing allergic reactions.
   (b) Transgenes in commercial crops can endanger native species e.g., the Bt toxin gene expressed in pollen might endanger pollinators like honeybees.
   (c) Production of GM crops causes damage to the natural environment and is always costly.
   (d) All of these
13. Which of the following genes were introduced in cotton to protect it from cotton bollworms?
   (a) Cry Ac and Cry ab
   (b) Bt Ac and Bt Ab
   (c) Cry IAc and Cry II Ab
   (d) Nif genes
14. Hirudin is
   (a) a protein produced by *Hordeum vulgare*, which is rich in lysine
   (b) a toxic molecuel isolated from *Gossypium hirsutum*, which reduces human fertility
   (c) a protein produced from transgenic *Brassica napus* which prevents blood clotting
   (d) an antibiotic produced by a genetically engineered bacterium *Escherichia coli*.
15. Early detection of a disease is possible by
   (a) PCR
   (b) Gene therapy
   (c) recombination DNA technology ad ELISA
   (d) both (a) and (c)
16. Gene therapy can be referred to as
   (a) pre-clinical testing for inherited diseases in new-borns\
   (b) treatment of diseases caused by genetic defect
   (c) genetic engineering using rDNA technology
   (d) cancer treatment using in vitro cultured stem cells
17. Rules of conduct that may be used to regulate our activities in relation to the biological world is called
   (a) bioethics (b) biowar
   (c) biopatent (d) biopiracy
18. *Flavr Savr* is the transgenic variety of
   (a) cotton (b) rice
   (c) tomato (d) potato
19. Biopiracy is related to
   (a) Traditional knowledge
   (b) Biomolecules and regarding bioresources genes isolated from bioresources
   (c) Bioresources
   (d) All of the above

RESPONSE GRID					
	6. ⓐⓑⓒⓓ	7. ⓐⓑⓒⓓ	8. ⓐⓑⓒⓓ	9. ⓐⓑⓒⓓ	10. ⓐⓑⓒⓓ
	11. ⓐⓑⓒⓓ	12. ⓐⓑⓒⓓ	13. ⓐⓑⓒⓓ	14. ⓐⓑⓒⓓ	15. ⓐⓑⓒⓓ
	16. ⓐⓑⓒⓓ	17. ⓐⓑⓒⓓ	18. ⓐⓑⓒⓓ	19. ⓐⓑⓒⓓ	

Space for Rough Work

20. Genetically engineered bacteria have been used in commercial production of
(a) thyroxine (b) human insulin
(c) testosterone (d) penicillin

21. Golden rice is a promising transgenic crop, when released for cultivation, it will help in
(a) producing petrol like fuel from rice.
(b) alleviation of vitamin A.
(c) pest resistance.
(d) herbicide tolerance.

22. Chloramphenicol and erythromycin (broad spectrum antibiotics) are produced by
(a) *Streptomyces* (b) *Nitrobacter*
(c) *Rhizobium* (d) *Penicillium*

23. The transgenic animals are those which have
(a) foreign DNA in some cells.
(b) foreign DNA in all of their cells.
(c) foreign RNA in all of their cells.
(d) Both (a) and (c).

24. The transgenic plants are the plants having
(a) no gene.
(b) genes in transposition.
(c) genes with no function to perform.
(d) genes of an other organism.

25. Recombinant DNA technology can be used to produce quantities of biologically active form of which one of the following products in *E.coli*?
(a) Luteinizing hormone (b) Ecdyson
(c) Rifamycin (d) Interferon

26. Which of the following combinations of risk are associated with genetically modified food ?
I. Toxicity
II. Allergic reaction
III. Antibiotic resistance in micro-organisms present in alimentary canal. –
(a) I and II (b) I, II and III
(c) I and III (d) II and III

27. Cheese and Yogurt are products obtained by
(a) distillation (b) pasteurization
(c) fermentation (d) dehydration

28. Main objective of producing herbicide resistant GM crops is
(a) encourage ecofriendly herbicides.
(b) reduce herbicide accumulation in food articles for health safety.
(c) eliminate weeds from fields without the use of manual labour.
(d) eliminate weeds from the fields without the use of herbicides.

29. Which one of the following is the correctly matched pair of a product and the microorganism responsible for it ?
(a) Ethyl alcohol- Yeast
(b) Acetic acid- *Lactobacillus*
(c) Cheese - *Nitrobacter*
(d) Curd - *Azotobacter*

30. Humulin is
(a) A form of chitin
(b) A powerful antibiotic
(c) A new digestive enzyme
(d) Human insulin

31. A genetically engineered micro-organism used successfully in bioremediation of oil spills is a species of
(a) *Pseudomonas* (b) *Trichoderma*
(c) *Xanthomonas* (d) *Bacillus*

32. Prenatal screening in humans, AIDS virus testing in humans, and genetic engineering for protection against insect attack in plants all
(a) are types of gene therapy.
(b) are examples of the application of r-DNA technology.
(c) require use of the same restriction enzyme.
(d) are carried out using gene guns.

33. Genetically engineered bovine growth hormone (BGH), which is highly effective for improving overall growth and milk production in cattle, remains a hotly debated issue because
(a) BGH is clearly hazardous to human health.
(b) BGH is an environmental hazard.
(c) BGH could drive traditional family farmers out of business.
(d) scientists remain unconvince d that BGH really works.

34. DNA vaccines are
(a) mixture of hormones (b) recombinant vaccines
(c) synthetic vaccines (d) pure DNA or RNA

35. Toxic component of *Bacillus thuringiensis* is protein. Commercial preparation of *Bacillus thuringiensis* consists of mixture of spores, cry protein and inert carrier. *Bt* stands for *Bacillus thuringiensis* for in popular crop of *Bt* cotton. *Tricoderma* is used in biocontrol of fungal pathogen because it has capacity to secrete the enzyme
(a) DNAase (b) RNAase
(c) Chitinase (d) Cry protein

RESPONSE GRID					
	20. ⓐⓑⓒⓓ	21. ⓐⓑⓒⓓ	22. ⓐⓑⓒⓓ	23. ⓐⓑⓒⓓ	24. ⓐⓑⓒⓓ
	25. ⓐⓑⓒⓓ	26. ⓐⓑⓒⓓ	27. ⓐⓑⓒⓓ	28. ⓐⓑⓒⓓ	29. ⓐⓑⓒⓓ
	30. ⓐⓑⓒⓓ	31. ⓐⓑⓒⓓ	32. ⓐⓑⓒⓓ	33. ⓐⓑⓒⓓ	34. ⓐⓑⓒⓓ
	35. ⓐⓑⓒⓓ				

Space for Rough Work

36. GEAC stands for
(a) Genome Engineering Action Committee
(b) Ground Environment Action Committee
(c) Genetic Engineering Approval Committee
(d) Genetic and Environment Approval committee

37. A transgenic food crop'which may help in solving the problem of night blindness in developing countries is
(a) *Flavr Savr* tomatoes (b) Starlink maize
(c) *Bt* Soybean (d) Golden rice

38. Some of the characteristics of Bt cotton are:
(a) long fibre and resistance to aphids
(b) medium yield, long fibre and resistance to beetle pests
(c) high yield and production of toxic protein crystals which kill dipteran pests

39. It is sometimes necessary to genetically engineered mammalian cells to produce proteins because they
(a) can produce larger quantities of protein than bacteria.
(b) can read eukaryotic genes and bacteria cannot.
(c) can add sugars to make glycoproteins and bacteria cannot.
(d) are easier to grow than bacteria.

40. Which one of the following is commonly used in transfer of foreign DNA into crop plants?
(a) *Meloidogyne incognita*
(b) *Agrobacterium tumefaciens*
(c) *Penicillium expansum*
(d) *Trichoderma harzianum*

41. α-I antitrypsin is
(a) an antacid
(b) an enzyme
(c) used to treat arthritis
(d) used to treat emphysema

42. Which one of the following genes is defective in patients suffering from severe combined immuno-deficiency syndrome (SCID)?
(a) RNAase
(b) ADA
(c) Ribonucleotide reductase
(d) DNAase

43. Which one of the following option is incorrect?
(a) The majority of baculoviruses used as biological control agents are in the genus Nucleopolyhedrovirus.
(b) Nucleopolyhedrovirus are excellent model for broad-spectrum insecticidal applications.
(c) Nucleopolyhedrovirus have no negative impacts on plants, mammals, birds, fish or even on non-target insects.
(d) This is especially desirable when beneficial insects are being conserved to aid in an overall IPM programme.

44. Select the correct set of the names for A, B, C and D.

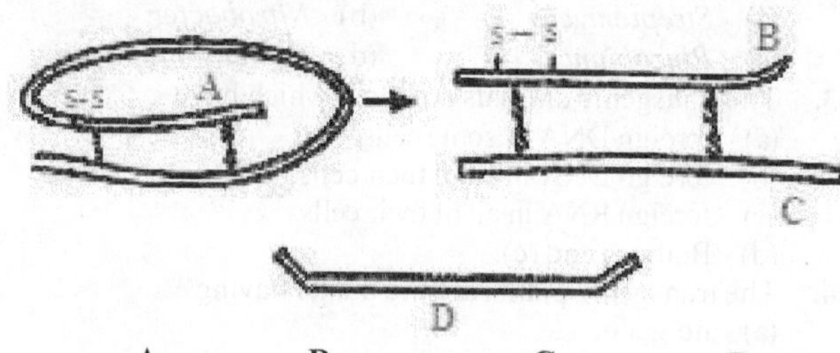

	A	B	C	D
(a)	Proinsulin	B-peptide	A-peptide	Insulin
(b)	Proinsulin	A-peptide	B-Peptide	Free C-Peptide
(c)	Proinsulin	A-peptide	B-peptide	Insulin
(d)	Proinsulin	B-Peptide	A-peptide	Free C-Peptide

45. Match the Column I with Column II and choose the correct option.

	Column I		Column II
A.	Golden rice	I.	Increased shelf life
B.	Flavr Savr tomato	II.	HGH
C.	Mouse	III.	Vitamin A
D.	Transgenic pig	IV.	Organ transplantation

(a) A-III; B-I; C-II; D-IV
(b) A-II; B-I; C-III; D-IV
(c) A-II; B-III; C-I; D-IV
(d) A-IV; B-I; C-II; D-III

RESPONSE GRID					
	36. ⓐⓑⓒⓓ	37. ⓐⓑⓒⓓ	38. ⓐⓑⓒⓓ	39. ⓐⓑⓒⓓ	40. ⓐⓑⓒⓓ
	41. ⓐⓑⓒⓓ	42. ⓐ(	ⓒⓓ	44. ⓐⓑⓒⓓ	45. ⓐⓑⓒⓓ

DAILY PRACTICE PROBLEM DPP CHAPTERWISE 34 - BIOLOGY			
Total Questions	45	Total Marks	180
Attempted		Correct	
Incorrect		Net Score	
Cut-off Score	40	Qualifying Score	55
Success Gap = Net Score – Qualifying Score			
Net Score = (Correct × 4) – (Incorrect × 1)			

DPP - Daily Practice Problems

Chapter-wise Sheets

Date : Start Time : End Time :

BIOLOGY

SYLLABUS : Organisms and Populations

Max. Marks : 180 **Marking Scheme : + 4 for correct & (–1) for incorrect** **Time : 60 min.**

INSTRUCTIONS : This Daily Practice Problem Sheet contains 45 MCQ's. For each question only one option is correct. Darken the correct circle/ bubble in the Response Grid provided on each page.

1. Which part of the world has a high density of organisms?
   (a) Grasslands
   (b) Savannahs
   (c) Deciduous forests
   (d) Tropical rain forests
2. Competition for light, nutrients and space is most severe between
   (a) closely related organism growing in different habitats
   (b) closely related organisms growing in the same habitat
   (c) distantly related organisms growing in the same habitat
   (d) distantly related organisms growing in different habitats
3. A mutually beneficial association necessary for survival of both partners is
   (a) mutualism/symbiosis (b) commensalism
   (c) amensalism (d) both (a) and (b)
4. Praying mantis is a good example of
   (a) warning colouration
   (b) social insects
   (c) camouflage
   (d) mullerian mimicry
5. In increasing order of organizational complexity, which one of the following is the correct sequence?
   (a) Population, species, community, ecosystem
   (b) Population, variety, species, ecosystem
   (c) Population, ecosystem, species, community
   (d) Species, variety, ecosystem, community
6. Keystone species in an ecosystem are those
   (a) present in maximum number
   (b) that are most frequent
   (c) attaining a large biomass
   (d) contributing to ecosystem properties
7. Niche overlap indicates
   (a) two different parasites on the same host
   (b) sharing of one or more resources between the two species
   (c) mutualism between two species
   (d) active cooperation between two species

RESPONSE GRID					
	1. ⓐⓑⓒⓓ	2. ⓐⓑⓒⓓ	3. ⓐⓑⓒⓓ	4. ⓐⓑⓒⓓ	5. ⓐⓑⓒⓓ
	6. ⓐⓑⓒⓓ	7. ⓐⓑⓒⓓ			

Space for Rough Work

8. Animals that can tolerate a narrow range of salinity are
(a) stenohaline (b) euryhaline
(c) anadromous (d) catadromous

9. Human population growth in India
(a) tends to follow a sigmoid curve as in case of many other animal species
(b) tends to reach a zero population growth as in case of some animal species
(c) can be reduced by permitting natural calamities and enforcing birth control measures
(d) can be regulated by following the national programme of family planning

10. Consider the following four conditions (i) - (iv) and select the correct pair of them as adaptation to environment in desert lizards.
The conditions :
(i) Burrowing in soil to escape high temperature
(ii) Losing heat rapidly from the body during high temperature
(iii) Bask in sun when temperature is low
(iv) Insulating body due to thick fatty dermis
(a) (iii), (iv) (b) (i)
(c) (iii), (ii), (iv) (d) (i), (ii)

11. A force acting against achievement of highest possible level of population growth is
(a) Carrying capacity
(b) Environment resistance
(c) Population pressure
(d) Saturation level

12. Natality is the characteristic of a population which means
(a) the total number of individuals present per unit area at a given time
(b) the increase in number of individuals in a population under given environmental conditions
(c) loss of individuals due to death in a population under given environmental conditions
(d) the movement of individuals into and out of population

13. The number of births per thousand people in the population is expressed as :
(a) Reproduction rate (b) Conception rate
(c) Crude birth rate (d) Growth rate

14. What type of human population is represented by the following age pyramid?

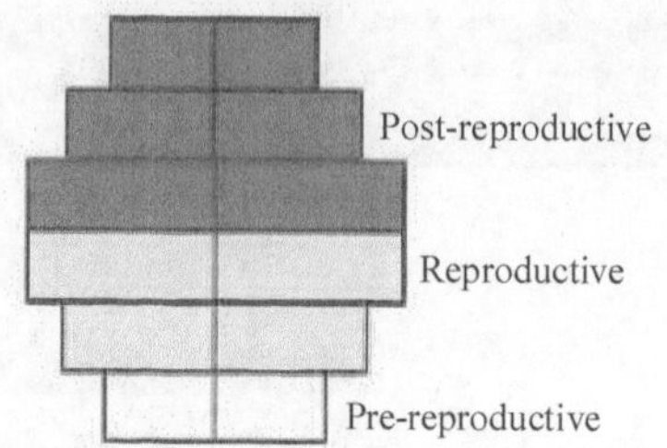

(a) Vanishing population
(b) Stable population
(c) Declining population
(d) Expanding population

15. The logistic population growth is expressed by the equation:
(a) $dt/dN = Nr \frac{K \quad N}{K}$
(b) $dN/dt = rN \frac{K \quad N}{K}$
(c) $dN/dt = rN$
(d) $dN/dt = rN \frac{N \quad K}{N}$

16. What is the best pH of soil for cultivation of plants ?
(a) 3.4 – 5.4 (b) 6.5 – 7.5
(c) 4.5 – 8.5 (d) 5.6 – 6.5

17. The maintenance of internal favourable conditions, by a self-regulated mechanism, inspite of the fact that there are changes in the environment, is known as
(a) entropy (b) enthalpy
(c) steady state (d) homeostasis

18. Two different species cannot live for long duration in the same niche or habitat. This law is
(a) Allen's law
(b) Gause's hypothesis
(c) Dollo's rule
(d) Weisman's theory

19. One of the important consequences of geographical isolation is
(a) preventing speciation
(b) speciation through reproductive isolation
(c) random creation of new species
(d) no change in the isolated fauna.

RESPONSE GRID					
	8. ⓐⓑⓒⓓ	9. ⓐⓑⓒⓓ	10. ⓐⓑⓒⓓ	11. ⓐⓑⓒⓓ	12. ⓐⓑⓒⓓ
	13. ⓐⓑⓒⓓ	14. ⓐⓑⓒⓓ	15. ⓐⓑⓒⓓ	16. ⓐⓑⓒⓓ	17. ⓐⓑⓒⓓ
	18. ⓐⓑⓒⓓ	19. ⓐⓑⓒⓓ			

Space for Rough Work

20. Which one of the following areas in India, is a hotspot of biodiversity?
(a) Eastern Ghats (b) Gangetic Plain
(c) Sunderbans (d) Western Ghats

21. A taxon facing an extremely high risk of extinction in wild in the immediate future is called :
(a) critical endangered
(b) endangered
(c) vulnerable
(d) extinct in wild

22. The species that invade a bare area are called
(a) keystone species (b) extinct species
(c) pioneer species (d) rare species

23. Several plant and animal species present together at a place constitute a
(a) genus (b) population
(c) biome (d) community

24. Deserts, rainforests, tundra, etc. are example of
(a) community (b) biome
(c) ecosystem (d) population

25. Life is sustainable with water only because
(a) it makes 90% of the protoplasm
(b) translocation of nutrients inside the body occurs with the help of water
(c) water loss in form of sweating and transpiration helps to maintain body temperature
(d) all of these

26. Which of the following algae are found in deepest ocean waters?
(a) Red algae (b) Yellow algae
(c) Green algae (d) Brown algae

27. When organisms change their location to escape from harsh environment, it is called as
(a) hibernation (b) vernalization
(c) migration (d) aestivation

28. Species interaction with negative influence on both is referred to as
(a) amensalism (b) mutualism
(c) commensalism (d) competition

29. Total number of individual of a species per unit area and per unit time is called
(a) population size
(b) population density
(c) demography
(d) population dyamics.

30. In a life table, the number of individuals alive at the begining of the 1st year to 2nd year interval is 800. During this interval, 200 individuals die. The death rate for this interval is
(a) 0.25 (b) 200
(c) 800 (d) 0.2

31. Which one of the following is the most productive ecosystem?
(a) Temperate forest (b) Grassland
(c) Desert (d) Tropical rain forest

32. Pedology is the study of –
(a) Locomotion of animals
(b) Rocks
(c) Soil
(d) Reproduction

33. Sequence of humification and mineralisation is-
(a) Dead organic matter → Litter → Duff → Humus → Minerals
(b) Humus → Minerals → Litter → Duff
(c) Minerals → Humus → Litter → Duff
(d) Dead organic matter → Duff → Litter → Minerals → Humus

34. Soil is composed of –
(a) Mineral + Water + Air
(b) Mineral + Organic matter + Water
(c) Mineral + Organic matter + Air + Water
(d) Organic matter + Water

35. Who employed the term ecology to study plants ?
(a) Haeckel (b) Odum
(c) Warming (d) Dudgeon

36. Biotic factors are –
(a) Chemical factors of soil which effect life
(b) Physical factors of soil which effect life
(c) All living organisms which influence other organisms
(d) Atmospheric factors which influence life

RESPONSE GRID					
	20. ⓐⓑⓒⓓ	21. ⓐⓑⓒⓓ	22. ⓐⓑⓒⓓ	23. ⓐⓑⓒⓓ	24. ⓐⓑⓒⓓ
	25. ⓐⓑⓒⓓ	26. ⓐⓑⓒⓓ	27. ⓐⓑⓒⓓ	28. ⓐⓑⓒⓓ	29. ⓐⓑⓒⓓ
	30. ⓐⓑⓒⓓ	31. ⓐⓑⓒⓓ	32. ⓐⓑⓒⓓ	33. ⓐⓑⓒⓓ	34. ⓐⓑⓒⓓ
	35. ⓐⓑⓒⓓ	36. ⓐⓑⓒⓓ			

Space for Rough Work

37. The age of pyramid with broad base indicates
(a) High percentage of young individuals
(b) Low percentage of young individuals
(c) High percentage of old individuals
(d) Low percentage of old individuals

38. Permafrost is the most distinctive feature of
(a) Taiga (b) Temperate grasslands
(c) Tundra (d) Pacific northwest

39. Which statement does not reflect a reason why grasses predominate in savannas ?
(a) There are few native herbivores in this biome.
(b) Grasses grow well in fine, sandy soils.
(c) The above ground portions of grass plants die during dry seasons.
(d) The deep roots of the grasses can survive many months of drought.

40. Desert regions are characterized by ___ centimeters of rainfall per year.
(a) less than 5 (b) less than 15
(c) less than 25 (d) over 50

41. To avoid summer - related problems such as heat and dessication fish undergoes
(a) hibernation (b) diapause
(c) aestivation (d) none of these

42. Seals have a thick layer of fat (blubber) below their skin that acts as an
(a) thermostat (b) capacitor
(c) resistor (d) insulator

43. Match the following

	Population		**Example**
A.	Predation	I.	*Cuscuta* and hedge plants
B.	Commensalism	II.	*Balanus* and *Chathamalus*
C.	Parasitism	III.	Cactus and moth
D.	Competition	IV.	Orchid and mango

(a) A-III; B-IV; C-I; D-II
(b) A-IV; B-III; C-II; D-I
(c) A-I; B-III; C-II; D-IV
(d) A-III; B-IV; C-II; D-I

44. The density of a population in a given habitat during a given period, fluctuates due to changes in four basic processes On this basis choose the correct option to fill up A and B boxes in the given diagram.

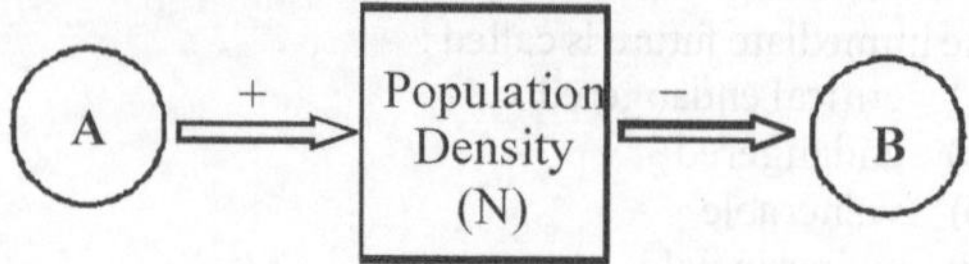

(a) A = Natality + Immigration, B = Mortality + Emigration
(b) A = Natality + Mortality, B = Immigration + Emigration
(c) A = Birth rate + Death rate, B = Mortality + Emigration
(d) A = Natality + Emigration, B = Mortality + Immigration

45. In laboratory experiments, two species of the protist *Paramecium* were grown alone and in the presence of the other species. The following graphs show growth of species 1 (left) and species 2 (right), both along and when in mixed culture.

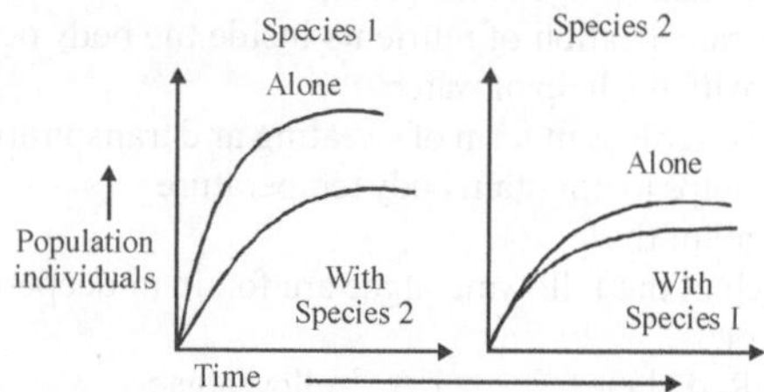

Interpretation of these graphs shows that
(a) competitive exclusion occurred in these experiments.
(b) both species are affected by interspecific competition but species 1 is less affected.
(c) both species are affected by interspecific competition but species 2 is less affected.
(d) both species are affected equally by interspecific competition.

RESPONSE GRID					
	37. ⓐⓑⓒⓓ	38. ⓐⓑⓒⓓ	39. ⓐⓑⓒⓓ	40. ⓐⓑⓒⓓ	41. ⓐⓑⓒⓓ
	42. ⓐⓑⓒⓓ	43. ⓐⓑⓒⓓ	44. ⓐⓑⓒⓓ	45. ⓐⓑⓒⓓ	

Space for Rough Work

DAILY PRACTICE PROBLEM DPP CHAPTERWISE 35 - BIOLOGY			
Total Questions	45	Total Marks	180
Attempted		Correct	
Incorrect		Net Score	
Cut-off Score	40	Qualifying Score	55
Success Gap = Net Score – Qualifying Score			
Net Score = (Correct × 4) – (Incorrect × 1)			

DPP - Daily Practice Problems

Chapter-wise Sheets

Date : | Start Time : | End Time :

BIOLOGY CB36

SYLLABUS : Ecosystem

Max. Marks : 180 **Marking Scheme : + 4 for correct & (–1) for incorrect** **Time : 60 min.**

INSTRUCTIONS : This Daily Practice Problem Sheet contains 45 MCQs. For each question only one option is correct. Darken the correct circle/ bubble in the Response Grid provided on each page.

1. Deep black soil is productive due to high proportion of
   (a) sand and zinc (b) gravel and calcium
   (c) clay and humus (d) silt and earthworm
2. Which one of the following statements is correct for secondary succession ?
   (a) It begins on a bare rock
   (b) It occurs on a deforested site
   (c) It follows primary succession
   (d) It is similar to primary succession except that it has a relatively fast pace
3. The zone of atmosphere that lies near the ground is
   (a) troposphere (b) stratosphere
   (c) homosphere (d) ionosphere
4. Food chain in which micro-organisms breakdown the food formed by primary producers is
   (a) parasitic food chain (b) detritus food chain
   (c) consumer food chain (d) predator food chain
5. Which of the following pairs is a sedimentary type of biogeochemical cycle?
   (a) Oxygen and nitrogen
   (b) Phosphorus and sulphur
   (c) Phosphorus and nitrogen
   (d) Phosphorus and carbon dioxide
6. Mass of living matter at a trophic level in an area at any time is called
   (a) standing crop (b) detritus
   (c) humus (d) standing state
7. If we completely remove the decomposers from an ecosystem, its functioning will be adversely affected because
   (a) energy flow will be blocked
   (b) herbivores will not receive solar energy
   (c) mineral movement will be blocked
   (d) rate of decomposition will be very high
8. Humus is essential for plant growth because:
   (a) it is rich in nutrients and increases the water holding capacity of soil
   (b) it increases aeration of soil
   (c) it increases porocity of soil
   (d) All of the above

RESPONSE GRID				
1. ⓐⓑⓒⓓ	2. ⓐⓑⓒⓓ	3. ⓐⓑⓒⓓ	4. ⓐⓑⓒⓓ	5. ⓐⓑⓒⓓ
6. ⓐⓑⓒⓓ	7. ⓐⓑⓒⓓ	8. ⓐⓑⓒⓓ		

Space for Rough Work

9. The rate at which light energy is converted to the chemical energy of organic molecules in the ecosystem is
(a) net primary productivity
(b) gross primary productivity
(c) net secondary productivity
(d) gross secondary productivity

10. An ecosystem which can be easily damaged but can recover after some time if damaging effect stops will be having
(a) low stability and high resilience
(b) high stability and low resilience
(c) low stability and low resilience
(d) high stability and high resilience

11. Both hydrarch and xerarch successions lead to:
(a) medium water conditions
(b) xeric conditions
(c) highly dry conditions
(d) excessive wet conditions

12. Which of the following ecosystems has highest rate of gross primary production?
(a) Grasslands
(b) Mangroves
(c) Coral reefs
(d) Equatorial rain forest

13. Largest amount of fresh water is found in
(a) lakes and streams
(b) underground
(c) polar ice caps and glaciers
(d) rivers

14. Which one of the following is not a function of an ecosystem?
(a) Energy flow (b) Decomposition
(c) Productivity (d) Stratification

15. Ecosystem is
(a) always open
(b) always closed
(c) both open and closed depending upon community
(d) both open and closed depending upon biomass

16. Transition zone between two vegetations is
(a) ecotone (b) ecotype
(c) ecocline (d) ecosystem

17. Which one of the following statement is correct?
(a) Warm and moist environment favour decomposition whereas low temperature and anaerobiosis inhibit decomposition
(b) Warm and moist environment inhibit decomposition whereas low temperature and anaerobiosis favour decomposition
(c) Warm and anaerobiosis favour decomposition as well as low temperature favours decomposition
(d) Warm and low temperature inhibit decomposition whereas anaerobiosis favours decomposition

18. _______ is the rate of production of organic matter by consumers.
(a) Primary productivity
(b) Secondary productivity
(c) Net primary productivity
(d) Gross primary productivity

19. Which of the following is not a characteristic of humus?
(a) It is rich in organic matter such as lignin and cellu-lose.
(b) It is collodial in nature and serves as a reservoir of nutrients.
(c) It is highly resistant to microbial action and under-goes slow decomposition.
(d) It is further degraded by the process of humification.

20. Percentage of photosynthetically active radiation (PAR) in the incident solar radiation is
(a) 1 - 5% (b) 2 - 10%
(c) less than 50 % (d) approx. 100%

21. Which one of the following animals may occupy more than one trophic levels in the same ecosystem at the same time?
(a) Sparrow (b) Lion
(c) Goat (d) Frog

22. In lithosere, foliose lichens make the conditions favourable for the growth of
(a) crustose lichens (b) mosses
(c) annual grasses (d) perennial grasses

23. Which one of the following is not one of the three aspects studied in biogeochemical cycling?
(a) The nature and size of natural reservoir
(b) The rate of movement between reservoirs
(c) How different biogeochemical cycles interact
(d) How new species create their own biogeochemical cycles

RESPONSE GRID					
	9. ⓐⓑⓒⓓ	10. ⓐⓑⓒⓓ	11. ⓐⓑⓒⓓ	12. ⓐⓑⓒⓓ	13. ⓐⓑⓒⓓ
	14. ⓐⓑⓒⓓ	15. ⓐⓑⓒⓓ	16. ⓐⓑⓒⓓ	17. ⓐⓑⓒⓓ	18. ⓐⓑⓒⓓ
	19. ⓐⓑⓒⓓ	20. ⓐⓑⓒⓓ	21. ⓐⓑⓒⓓ	22. ⓐⓑⓒⓓ	23. ⓐⓑⓒⓓ

Space for Rough Work

24. The largest ecosystem of the world is-
(a) Forests (b) Grasslands
(c) Great lakes (d) Oceans

25. Which of the following is a man made artificial ecosystem?
(a) Grassland ecosystem
(b) Agro ecosystem
(c) Ecosystem of artificial lakes and dams
(d) Forest ecosystem

26. What is the annual net primary productivity of whole biosphere?
(a) 170 billion tons (dry weight) of organic matter
(b) 165 billion tons (dry weight) of organic matter
(c) 160 billion tons (dry weight) of organic matter
(d) 155 billion tons (dry weight) of organic matter

27. Select the formula for ecological efficiency
(a) $\frac{\text{Gross primary productivity} \times 100}{\text{Incident total solar radiation}}$
(b) $\frac{\text{Food primary assimilated} \times 100}{\text{Food energy ingested}}$
(c) $\frac{\text{Net primary productivity} \times 100}{\text{Gross primary productivity}}$
(d) $\frac{\text{Energy in biomass production at a trophic level} \times 100}{\text{Energy in biomass production at previous trophic level}}$

28. Vegetation of a geographic region with low rainfall, high temperature, loose and sandy soil is of the type called:
(a) grassland (b) scrub forest
(c) xerophytic (d) evergreen tropical forest

29. Which one is nature's cleaner ?
(a) Consumers
(b) Producers
(c) Decomposers and scavengers
(d) Symbionts

30. Which one of the following statement is correct?
(a) Decomposition rate is slower if detritus is rich in lignin and chitin, and quicker, if detritus is rich in nitrogen and water - soluble substances like sugars.
(b) Decomposition rate is slower if detritus is rich in nitrogen and water - soluble substances like sugars, and quicker, if detritus is rich in lignin and chitin.
(c) Decomposition rate is slower if detritus is rich in cellulose, and quicker, if detritus is rich in phosphorus.
(d) Decomposition rate is quicker if detritus is rich in lignin, and quicker, if detritus is rich in sulphur.

31. What are the two most important climatic factors that regulate decomposition through their effects on soil microbes ?
(a) Temperature and rainfall
(b) Temperature and soil - moisture
(c) Temperature and humidity
(d) Temperature and pressure

32. Which of the following process helps in nutrient conservation?
(a) Mineralisation (b) Immobilisation
(c) Leaching (d) Nitrification

33. Bacteria and fungi in a forest ecosystem are generally
(a) Producers
(b) Decomposers
(c) Primary consumers
(d) Secondary consumers

34. Which one of the following is not a gaseous biogeochemical cycle in ecosystem ?
(a) Sulphur cycle (b) Phosphorus cycle
(c) Nitrogen cycle (d) Carbon cycle

35. In grass-deer-tiger food chain, grass biomass is one tonne. The tiger biomass shall be
(a) 100 kg (b) 10 kg
(c) 200 kg (d) 1 kg

36. The early settlers on a barren area (rock) are
(a) Ferns (b) Mosses
(c) Lichens (d) Diatoms

37. A community which starts succession in a habitat is
(a) Pioneer community (b) Seral community
(c) Biotic community (d) Ecosere

38. Last stabilised community in a plant succession is known as
(a) Seral community (b) Pioneer community
(c) Ecosere (d) Climax community

39. Identification and enumeration of plant and animal species of an ecosystem gives its
(a) productivity (b) stratification
(c) species composition (d) all of these

RESPONSE GRID					
	24. ⓐⓑⓒⓓ	25. ⓐⓑⓒⓓ	26. ⓐⓑⓒⓓ	27. ⓐⓑⓒⓓ	28. ⓐⓑⓒⓓ
	29. ⓐⓑⓒⓓ	30. ⓐⓑⓒⓓ	31. ⓐⓑⓒⓓ	32. ⓐⓑⓒⓓ	33. ⓐⓑⓒⓓ
	34. ⓐⓑⓒⓓ	35. ⓐⓑⓒⓓ	36. ⓐⓑⓒⓓ	37. ⓐⓑⓒⓓ	38. ⓐⓑⓒⓓ
	39. ⓐⓑⓒⓓ				

Space for Rough Work

40. Which of the following factors influence communities ?
(i) Climate
(ii) Species interaction
(iii) Feeding relationships among organisms
(iv) Succession
(a) (i) and (iii) (b) (i) and (iv)
(c) (i), (ii) and (iii) (d) All of these

41. Grasslands can support greater grazing rates by herbivores than forests because
(a) net production of grassland is greater.
(b) more of the grassland is above the ground.
(c) grasslands receive more sunlight.
(d) grasslands produce less woody plant tissue.

42. Given below is an imaginary pyramid of numbers. What could be one of the possibilities about certain organisms at some of the different levels ?

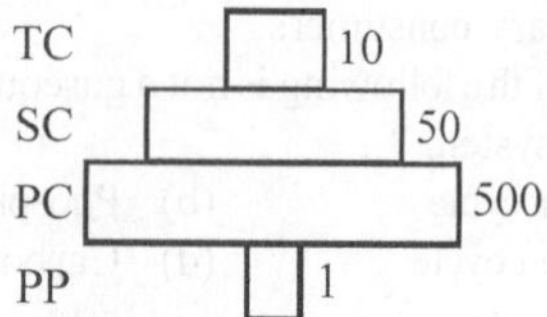

(a) Level PC is "insects" and level SC is "small insectivorous birds".
(b) Level PP is "phytoplanktons" in sea and "whale" is on top level TC.
(c) Level PP is "peepal trees" and the level SC is "sheep".
(d) Level PC is "rats" and level SC is "cats".

43. The given diagram shows (I, II, III, and IV) the flow of materials between different trophic levels. Which arrow is incorrect?

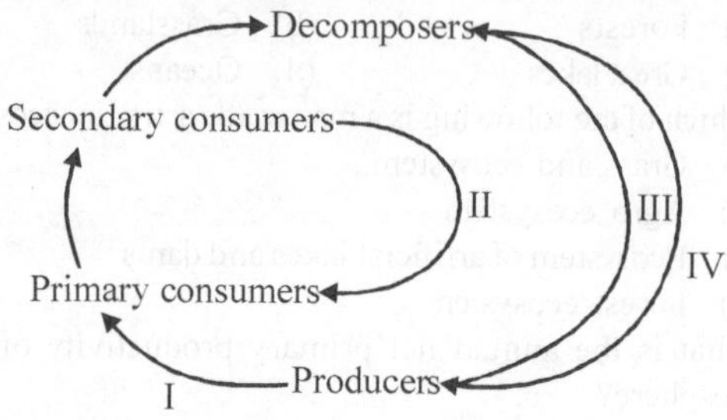

(a) I (b) II
(c) III (d) IV

44. Match the following and choose the correct option.

	Column - I		Column - II
A.	Presence of 3-4 storey of plants grown in a forest	I.	Blue-green algae
B.	A biome having grasses with scattered trees	II.	Stratification
C.	Man made ecosystem	III.	Savannah
D.	Pioneer in hydrosere	IV.	Dam

(a) A – III; B – II; C – IV; D – I
(b) A – III; B – I; C – IV; D – II
(c) A – I; B – III; C – II; D – IV
(d) A – II; B – III; C – IV; D – I

45. The correct order of the process of decomposition is
(a) Catabolism → Fragmentation → Leaching → Humification → Mineralization
(b) Catabolism → Fragmentation → Humification → Leaching → Mineralization
(c) Fragmentation → Humification → Catabolism → Leaching → Mineralization
(d) Fragmentation → Leaching → Catabolism → Humification → Mineralization

RESPONSE GRID	40. ⓐⓑⓒⓓ	41. ⓐⓑⓒⓓ	42. ⓐⓑⓒⓓ	43. ⓐⓑⓒⓓ	44. ⓐⓑⓒⓓ
	45. ⓐⓑⓒⓓ				

Space for Rough Work

DAILY PRACTICE PROBLEM DPP CHAPTERWISE 36 - BIOLOGY			
Total Questions	45	Total Marks	180
Attempted		Correct	
Incorrect		Net Score	
Cut-off Score	50	Qualifying Score	70
Success Gap = Net Score – Qualifying Score			
Net Score = (Correct × 4) – (Incorrect × 1)			

DPP - Daily Practice Problems

Chapter-wise Sheets

Date : | **Start Time :** | **End Time :**

BIOLOGY CB37

SYLLABUS : Biodiversity and its conservation

Max. Marks : 180 **Marking Scheme : + 4 for correct & (–1) for incorrect** **Time : 60 min.**

INSTRUCTIONS : This Daily Practice Problem Sheet contains 45 MCQ's. For each question only one option is correct. Darken the correct circle/ bubble in the Response Grid provided on each page.

1. Which one of the following is not included under *in-situ* conservation ?
   (a) Botanical garden
   (b) Biosphere reserve
   (c) National park
   (d) Sanctuary
2. Biodiversity Act of India was passed by the parliament in the year
   (a) 1992 (b) 1996
   (c) 2000 (d) 2002
3. Biodiversity of a geographical region represents
   (a) endangered species found in the region.
   (b) the diversity in the organisms living in the region.
   (c) genetic diversity present in the dominant species of the region.
   (d) species endemic to the region.
4. Which animal has become extinct from India?
   (a) Snow Leopard (b) Hippopotamus
   (c) Wolf (d) Cheetah
5. Which of the following is considered a hot-spot of biodiversity in India ?
   (a) Indo-Gangetic Plain (b) Eastern Ghats
   (c) Aravalli Hills (d) Western Ghats
6. Which group of vertebrates comprises the highest number of endangered species ?
   (a) Birds (b) Mammals
   (c) Fishes (d) Reptiles
7. One of endangered species of Indian medicinal plants is that of
   (a) *Ocimum* (b) Garlic
   (c) *Nepenthes* (d) *Podophyllum*
8. Which one of the following is the correct matched pair of an endangered animal and National Park ?
   (a) Rhinoceros - Kaziranga National Park
   (b) Wild Ass - Dudhwa National Park
   (c) Great Indian - Keoladeo National Park Bustard
   (d) Lion - Corbett National Park

RESPONSE GRID					
	1. ⓐⓑⓒⓓ	2. ⓐⓑⓒⓓ	3. ⓐⓑⓒⓓ	4. ⓐⓑⓒⓓ	5. ⓐⓑⓒⓓ
	6. ⓐⓑⓒⓓ	7. ⓐⓑⓒⓓ	8. ⓐⓑⓒⓓ		

Space for Rough Work

9. The area where wild populations, traditional life styles and genetic resources are protected is :
(a) Core Zone
(b) Buffer Zone
(c) Biosphere reserve
(d) Manipulation Zone

10. Which one of the following species of plant is considered as the world's most problematic aquatic weed?
(a) *Lantana*
(b) *Eichhornia*
(c) *Parthenium* (carrot grass)
(d) Brown algae

11. Which one of the following is not observed in biodiversity hotspots ?
(a) Endemism
(b) Accelerated species loss
(c) Lesser inter-specific competition
(d) Species richness

12. Which one of the following shows maximum genetic diversity in India ?
(a) Groundnut (b) Rice
(c) Maize (d) Mango

13. Study the four statements (i–iv) given below and select the two correct ones out of them:
(i) A lion eating a deer and a sparrow feeding on grain are ecologically similar in being consumers
(ii) Predator star fish *Pisaster* helps in maintaining species diversity of some invertebrates
(iii) Predators ultimately lead to the extinction of prey species
(iv) Production of chemicals such as nicotine, strychnine by the plants are metabolic disorders
The two correct statements are:
(a) (ii) and (iii) (b) (iii) and (iv)
(c) (i) and (iv) (d) (i) and (ii)

14. Reason of diversity in living beings is due to
(a) mutation
(b) long term evolutionary change
(c) gradual change
(d) short term evolutionary change

15. Which of the following is a reason for the greater biological diversity of tropical regions?
(a) Tropical latitudes have remained almost undisturbed for millions of years.
(b) Tropical environments are less seasonal, relatively more constant and predictable.
(c) More solar energy is available in the tropics resulting inhigh productivity.
(d) All of these.

16. An exotic species that is introduced to a new area, spreads rapidly and eliminates native species is called
(a) immigrant species
(b) invasive species
(c) *Eicchornia crasspies*
(d) all of these

17. Characters of stable community are
(a) it should not show too much variations in year-to-year productivity.
(b) it must be resistant to occasional natural or man-made disturbances.
(d) it should be resistant to invasions by alien species.
(d) All of these

18. What is the decreasing order of numbr of animals species as far as India is concerned?
(a) Mammals, Aves, Reptiles, Amphibians
(b) Aves, Reptiles, Mammals, Amphibians
(c) Mammals, Reptiles, Amphibians, Aves
(d) Reptiles, Amphibians, Mammals, Aves

19. Organization responsible for maintaning Red Data Book is
(a) IUCN (b) WWF
(c) CITES (d) IBWL

RESPONSE GRID				
9. ⓐⓑⓒⓓ	10. ⓐⓑⓒⓓ	11. ⓐⓑⓒⓓ	12. ⓐⓑⓒⓓ	13. ⓐⓑⓒⓓ
14. ⓐⓑⓒⓓ	15. ⓐⓑⓒⓓ	16. ⓐⓑⓒⓓ	17. ⓐⓑⓒⓓ	18. ⓐⓑⓒⓓ
19. ⓐⓑⓒⓓ				

Space for Rough Work

20. Which option correctly describes are equations for curves A and B, in the given graph of species - area relationhip?

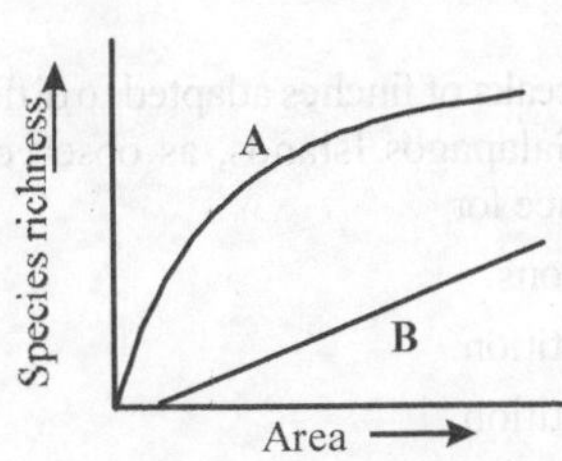

	A	B
(a)	$S = CA^Z$	$\text{Log } S = \text{Log } C + Z \text{ Log} A$
(b)	$\text{Log } S = \text{Log } C + Z \text{ Log} A$	$S = CA^Z$
(c)	$\text{Log } C = \text{Log } S + Z \log A$	$S = CA^Z$
(d)	$S = CA^Z$	$\text{Log } C = \text{Log } S + Z \text{ Log} A$

21. Genetic variations affect the productionof the drug reserpine in the medicinal plant *Rauwolfia vomitoria* growing in different Himalayan ranges. What kind of diversity does it indicate?
(a) Species diversity
(b) Genetic diversity
(c) Ecological diversity
(d) None of these

22. Keystone species deserve protection because these
(a) are capable of surviving in hash environmental conditions
(b) indicate the presence of certain minerals in the soil
(c) have become rare due to overexploitation
(d) play an important role in supporting other species.

23. In a National Park, protection is provided to
(a) flora and fauna (b) entire ecosystem
(c) fauna only (d) flora only

24. Which of thse organisms are protected by people of 'Bishnoi' community of Rajasthan?
(a) *Prosopis cineraria* (b) Black buck
(c) Bhojpatra (d) Both (a) and (b).

25. Which o the following statements regarding biodiversity hot spots are incorrect?
(i) High endemism
(ii) High levels of species richness.
(iii) Total number is 34 in the world.
(iv) Five of these occur in India.
(v) High alien species invasions.
(vi) Cover less than 2% of the earth's land area; but if properly conserved, they can reduce extinctions by about 30%.
(a) (i), (ii), (iii) (b) (i), (ii), (iii), (iv)
(c) (iv), (v), (vi) (d) (i), (ii), (iv)

26. Ex situ conservation is used for the conservation of
(a) all plants
(b) all animals
(c) threatened animals and plants
(d) both (a) and (b).

27. _______ National Park was the first National Park of India.
(a) Jim Corbett (b) Nanda Devi
(c) Kaziranga (d) Jaldapara

28. First 'Earth Summit' for 'Convention on Biological Diversity' (CBD) was held at
(a) Johannesberg (2002), S. Africa
(b) Rio de Janeiro (1992' Brazil
(c) Dehradun (1992), India
(d) New York (2000), U.S.A.

29. *Antilopa carvicapra*/Black Buck is –
(a) Vulnerable (b) Endangered
(c) Critically endangered (d) Extinct in the wild

30. Alpha diversity is present –
(a) Within community (b) Between community
(c) Ranges of communities (d) All of the above

31. Biodiversity is determined by –
(a) Number of individuals in an area
(b) Species richness
(c) Evenness
(d) Both (b) and (c)

RESPONSE GRID					
	20. ⓐⓑⓒⓓ	21. ⓐⓑⓒⓓ	22. ⓐⓑⓒⓓ	23. ⓐⓑⓒⓓ	24. ⓐⓑⓒⓓ
	25. ⓐⓑⓒⓓ	26. ⓐⓑⓒⓓ	27. ⓐⓑⓒⓓ	28. ⓐⓑⓒⓓ	29. ⓐⓑⓒⓓ
	30. ⓐⓑⓒⓓ	31. ⓐⓑⓒⓓ			

Space for Rough Work

32. Which of the following species is endangered ?
(a) *Panthera leo perrica* (b) *Pythan molurus*
(c) *Chelonia mydas* (d) *Pallas siberian crane*

33. Decrease in species diversity in tropical countries is mainly due to –
(a) Urbanisation (b) Pollution
(c) Deforestation (d) Soil erosion

34. Which of the following animal has become almost extinct in India ?
(a) Wolf (b) Rhinoceros
(c) Hippopotamus (d) Cheetah

35. Those species whose populations have been seriously depleted and whose ultimate security is not assured are known as –
(a) Threatened species (b) Endangered species
(c) Vulnerable species (d) Rare species

36. Which of the following is threatened by illegal trade in wildlife products?
(a) Nile perch (b) Elephants
(c) Key deer (d) Galapagos tortoises

37. Humans have altered _____ of Earth's land surface.
(a) very little (b) about 25%
(c) about 50% (d) about 80%

38. Which endangered animal is the source of world's finest, lightest, warmest and most expensive wool-the shahtoosh ?
(a) Nilgai (b) Cheetal
(c) Kashmiri Goat (d) Chiru

39. Which of the following is the most serious threat to biodiversity?
(a) Competition from exotic species
(b) Commercial harvesting
(c) Habitat destruction
(d) Overexploitation

40. Diversity in the type of breaks of finches adapted to different feeding habits on the Galapagos Islands, as observed by Darwin, provides evidence for
(a) intraspecific variations.
(b) intraspecific competition.
(c) interspecific competition.
(d) origin of species by natural selection.

41. Which of the following is not a step in understanding biodiversity ?
(a) Naming the species
(b) Looking at other related species
(c) Assessing the species geographic range
(d) Quantifying the species genome

42. Diversity of habitat over the total landscape is called
(a) β diversity (b) γ diveristy (gamma)
(c) landscape diversity (d) ecosystem diversity

43. The term hot spot for high diversity ecological regions was coined by –
(a) IUCN (b) Myers
(c) Odum (d) Kormandy

44. Following arrangement is correct from the point of view of decreasing biodiversity in angiosperms (N), fungi (F), pteridophytes (P) and algae(A)
(a) N > F > P > A (b) N > F > A > P
(c) F > N > P > A (d) F > N > A > P

45. The number of plant species recorded from India, is
(a) 30,000 (b) 45,000
(c) 70,000 (d) 90,000

RESPONSE GRID					
	32. ⓐⓑⓒⓓ	33. ⓐⓑⓒⓓ	34. ⓐⓑⓒⓓ	35. ⓐⓑⓒⓓ	36. ⓐⓑⓒⓓ
	37. ⓐⓑⓒⓓ	38. ⓐⓑⓒⓓ	39. ⓐⓑⓒⓓ	40. ⓐⓑⓒⓓ	41. ⓐⓑⓒⓓ
	42. ⓐⓑⓒⓓ	43. ⓐⓑⓒⓓ	44. ⓐⓑⓒⓓ	45. ⓐⓑⓒⓓ	

Space for Rough Work

DAILY PRACTICE PROBLEM DPP CHAPTERWISE 37 - BIOLOGY			
Total Questions	45	Total Marks	180
Attempted		Correct	
Incorrect		Net Score	
Cut-off Score	50	Qualifying Score	60
Success Gap = Net Score – Qualifying Score			
Net Score = (Correct × 4) – (Incorrect × 1)			

DPP - Daily Practice Problems

Chapter-wise Sheets

Date : | Start Time : | End Time :

BIOLOGY (CB38)

SYLLABUS : Environmental issues

Max. Marks : 180 **Marking Scheme : + 4 for correct & (–1) for incorrect** **Time : 60 min.**

INSTRUCTIONS : This Daily Practice Problem Sheet contains 45 MCQs. For each question only one option is correct. Darken the correct circle/ bubble in the Response Grid provided on each page.

1. Acid rains are produced by
   (a) excess emissions of NO_2 and SO_2 from burning fossil fuels
   (b) excess production of NH_3 by industry and coal gas
   (c) excess release of carbon monoxide by incomplete combustion
   (d) excess formation of CO_2 by combustion and animal respiration
2. DDT residues are rapidly passed through food chain causing biomagnification because DDT is
   (a) moderately toxic
   (b) non-toxic to aquatic animals
   (c) water soluble
   (d) lipo soluble
3. Which of these is a pollution related to occupational health hazard disease ?
   (a) Pneumoconiosis (b) Asthma
   (c) Flurosis (d) Silicosis
4. Secondary sewage treatment is mainly a
   (a) physical process (b) mechanical process
   (c) chemical process (d) biological process
5. Biochemical Oxygen Demand *(BOD)* in a river water
   (a) has no relationship with concentration of oxygen in the water.
   (b) gives a measure of *Salmonella* in the water.
   (c) increases when sewage gets mixed with river water.
   (d) remains unchanged when algal bloom occurs.
6. Today, concentration of green house gases is high because of
   (a) Use of refrigerator
   (b) Increased combustion of oil and coal
   (c) Deforestation
   (d) All the above
7. The electrostatic precipitator is used for removing particulate matter from
   (a) Exhaust of the thermal power plant
   (b) Exhaust from the automobiles
   (c) Industrial effluents
   (d) Kitchen waste

RESPONSE GRID	1. ⓐⓑⓒⓓ	2. ⓐⓑⓒⓓ	3. ⓐⓑⓒⓓ	4. ⓐⓑⓒⓓ	5. ⓐⓑⓒⓓ
	6. ⓐⓑⓒⓓ	7. ⓐⓑⓒⓓ			

Space for Rough Work

8. Photochemical smog formed in congested metropolitan cities mainly consists of
   (a) Ozone, peroxyacetyl nitrate and NOX
   (b) Smoke, peroxyacetyl nitrate and SO_2
   (c) Hydrocarbons, SO_2 and CO_2
   (d) Hydrocarbon, ozone and SOX
9. Ozone hole means
   (a) hole in the ozone layer in stratosphere.
   (b) decrease in the thickness of ozone in the stratosphere.
   (c) decrease in concentration of ozone in trophosphere.
   (d) increase in the concentration of ozone in trophosphere.
10. A lake near a village suffered heavy mortality of fishes within a few days. Consider the following reasons for this.
   (i) Lots of urea and phosphate fertilizer were used in the crops in the vicinity
   (ii) The area was sprayed with DDT by an aircraft
   (iii) The lake water turned green and stinky
   (iv) Phytoplankton populations in the lake declined initially thereby greatly reducing photosynthesis.
   (a) (i) and (iii) (b) (i), (ii) and (iv)
   (c) (iii) and (iv) (d) (ii), (iii) and (iv)
11. Wildlife is destroyed most when
   (a) there is lack of proper care
   (b) mass scale hunting for foreign trade
   (c) its natural habitat is destroyed
   (d) natural calamity occurs
12. The 'Earth Summit' held at Rio de Janerio in 1992 resulted into
   (a) Compilation of Red list
   (b) Establishment of Biosphere Reserves
   (c) Convention on Biodiversity
   (d) Development of Hot Spots of Biodiversity
13. Which one of the following expanded forms of the following acronyms is correct?
   (a) IPCC= International Panel for Climate Change
   (b) UNEP= United Nations Environmental Policy
   (c) EPA= Environmental Pollution Agency
   (d) IUCN= International Union for Conservation of Nature and Natural Resources
14. Photochemical smog pollution does not contain
   (a) Nitrogen dioxide
   (b) Carbon dioxide
   (c) PAN (peroxy acetyl nitrate)
   (d) Ozone
15. When huge amount of sewage is dumped into a river, its B.O.D. will
   (a) increase (b) decrease
   (c) sharply decrease (d) remain unchanged
16. Which of the following statements is correct?
   (a) Primary pollutants are more harmful than secondary pollutants.
   (b) Primary pollutants and secondary pollutants are equally harmful.
   (c) Secondary pollutants are more harmful than primary pollutants.
   (d) DDT is a secondary pollutant
17. Which one of the following statements is incorrect regarding Bhopal gas tragedy?
   (a) Methyl isocyanate gas leakage took place
   (b) Thousands of human beings died
   (c) Radioactive fall out engulfed Bhopal
   (d) It took place in the night of December 2/3, 1984
18. Lichens are the indicators of
   (a) water pollution
   (b) air pollution
   (c) soil pollution
   (d) all of these.
19. Noise pollution may cause nervousness and irritability by stimulating the secretion of
   (a) thyroid hormone
   (b) adernaline hormone
   (c) parathyroid hormone
   (d) none of these
20. Atmospheric pollutant is –
   (a) 0.0034% CO_2 (b) 0.034% CO_2
   (c) 0.34% CO_2 (d) 3.4% CO_2

RESPONSE GRID				
8. ⓐⓑⓒⓓ	9. ⓐⓑⓒⓓ	10. ⓐⓑⓒⓓ	11. ⓐⓑⓒⓓ	12. ⓐⓑⓒⓓ
13. ⓐⓑⓒⓓ	14. ⓐⓑⓒⓓ	15. ⓐⓑⓒⓓ	16. ⓐⓑⓒⓓ	17. ⓐⓑⓒⓓ
18. ⓐⓑⓒⓓ	19. ⓐⓑⓒⓓ	20. ⓐⓑⓒⓓ		

Space for Rough Work

21. Given pie-diagram represents the relative contribution of various GHGz to total global warming. Select the correct statement(s) regarding A, B and C.

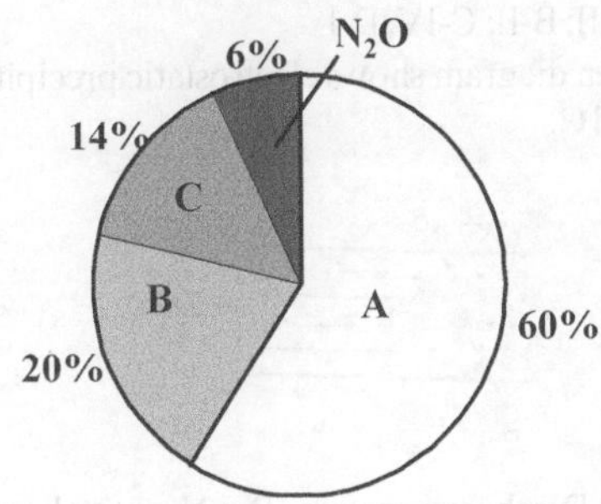

(a) A is the gas which is produced during the combustion of fossil fuels.
(b) B are the chemicals which are used as coolants in refrigerators
(c) C is the gas which is the major constituent of biogas
(d) All of these

22. Which of the following can cause DNA damage and mutations in humans?
(a) Absorption of UV-A and UV-B
(b) Absorption of UV-B
(c) Absorption of UV-A
(d) Absorption of UV-A and UV-C

23. Which of the following statements is not correct regarding jhum cultivation?
(a) It is also called as shifting cultivation and has resulted in deforestation
(b) It helps in increasing crop yield to a considerable extent.
(c) A time-gap of several years is required for the recovery of the land after cultivation.
(d) It involves cutting down of trees of the forest, burning of the plant remains and then using the land for farming.

24. Which of the following statement is incorrect ?
(a) Carbon monoxide is the major environment pollutant
(b) All pollutants are not waste
(c) Nitrogen and magnesium can pollute water
(d) Lichens are affected by sulphur

25. Noise pollution is created if noise is in excess to –
(a) 70-75 dB (b) 50-60 dB
(c) 80-99 dB (d) 40-65 dB

26. Which of the following does not cause pollution ?
(a) Thermal power project (b) Steel power plant
(c) Automobiles (d) Hydro electric project

27. Eutrophication causes reduction in
(a) dissolved CO_2 (b) dissolved O_2
(c) dissolved nutrients (d) none of the above

28. It is said, the Taj Mahal may be destroyed due to
(a) Flood in Yamuna river
(b) Decomposition of marble as a result of high temperature
(c) Air pollutants released from oil refinery of Mathura
(d) All of the above

29. The scrubber is used mainly to remove the following gas/es from the exhaust after spraying water/ lime
(a) CO_2 (b) SO_2
(c) O_2 & N_2 (d) CO & CO_2

30. Volcano is ______ source of pollution.
(a) artificial (b) natural
(c) Both (a) and (b) (d) man-made

31. Montreal Protocol was called to check emission of
(a) e-wastes (b) UV radiation
(c) CFCs (d) All of these

32. Which of the following is normally not an atmosphere pollutant ?
(a) CO_2 (b) SO_2
(c) Hydrogen (d) CO

33. Which of the following is not properly matched
(a) Formaldehyde - carcinogenic
(b) Sulphur dioxide - respiratory problems
(c) Nitrogen oxide - brown air
(d) Mean annual temperature - 25° C

34. Cleaning Environment with biological options such as microbes and plants is called
(a) Bioremediation (b) Biotechnology
(c) Biowarfare (d) Incineration

35. The 'Jhum cultivation' in India is practiced in
(a) North- East states (b) South–West coast
(c) South- East coast (d) Gangetic plain

36. Why is the concentration of ozone is less over the north and south poles ?
(a) CFCs accumulate only in area where the air is cold
(b) CFC use is highest in these areas
(c) CFCs stick to frozen water vapour and are able to act catalysts
(d) UV rays are stronger in the atmosphere

RESPONSE GRID					
	21. ⓐⓑⓒⓓ	22. ⓐⓑⓒⓓ	23. ⓐⓑⓒⓓ	24. ⓐⓑⓒⓓ	25. ⓐⓑⓒⓓ
	26. ⓐⓑⓒⓓ	27. ⓐⓑⓒⓓ	28. ⓐⓑⓒⓓ	29. ⓐⓑⓒⓓ	30. ⓐⓑⓒⓓ
	31. ⓐⓑⓒⓓ	32. ⓐⓑⓒⓓ	33. ⓐⓑⓒⓓ	34. ⓐⓑⓒⓓ	35. ⓐⓑⓒⓓ
	36. ⓐⓑⓒⓓ				

Space for Rough Work

37. The waste water from Industries may contain toxic heavy metals having density of more than
(a) 5 g / cm^3 (b) 10 g /cm^3
(c) 2 g /cm^3 (d) 15 g /cm^3

38. DDT causes egg shell thinning in birds because it inhibits
(a) calmodulin (b) calcium ATPase
(c) magnesium ATPase (d) carbonic anhydrase

39. FOAM (Friends of the Arcata Marsh) is a group of citizens responsible for the integrated process of
(a) reducing eutrophication
(b) sewage and water treatment
(c) radio- active waste treatment
(d) minimizing global warming

40. Without Green house effect the average temperature of earth surface would have been
(a) 18 °C (b) 8°C
(c) – 8°C (d) –18°C

41. Read the following statements.
A lake near a village suffered heavy mortality of fishes within a few days. Consider the following reasons for this;
(i) Lots of urea and phosphate fertilizer were used in the crops in the vicinity.
(ii) The area was sprayed with DDT by an aircraft.
(iii) The lake water turned green and stinky.
(iv) Phytoplankton populations in the lake declined intially thereby greatly reducing photosynthesis.
Which two of the above were the main causes of fish mortality in the lake ?
(a) (i) and (iii) (b) (i) and (ii)
(c) (ii) and (iii) (d) (iii) and (iv)

42. Match column-I with column-II and select the correct option.

	Column-I		Column-II
A.	Catalytic converter	I.	Particulate matter
B.	Electrostatic precipitator	II.	Carbon monoxide and nitrogen oxides
C.	Earmuffs	III.	High noise level
D.	Land fills	IV.	Solid wastes

(a) A-I; B-II; C-III; D-IV
(b) A-II; B-I; C-III; D-IV
(c) A-IV; B-III; C-II; D-I
(d) A-III; B-II; C-IV; D-I

43. The given diagram shows electrostatic precipitator. Identify A, B and C.

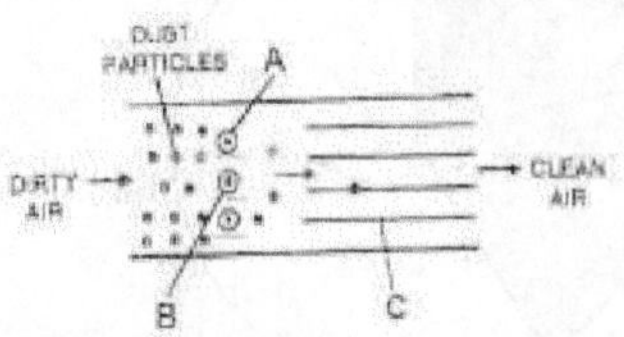

(a) A - Discharge corona, B - Negatively charged wire, C - Collection plate grounded
(b) A - Discharge corona, B - Positively charged wire, C - Collection plate grounded
(c) A - Discharge corona, B - Negatively charged wire, C - Collection plate burnt
(d) A - Uncharge corona, B - Positively charged wire, C - Collection plate never grounded

44. Which of the following statements is/are correct about 'Eco San' toilets?
(a) It is a sustainable system for handling human excreta or faecal matter by using dry 'composting toilets'.
(b) These are very useful for the rural areas where sewer systems are not possible.
(c) These toilets are hygienic, efficient, practical & most effective for the disposal of human waste.
(d) All of the above

45. Algal blooms impart a distinct colour to water due to
(a) their pigments.
(b) excretion of coloured substances.
(c) formation of coloured chemicals in water facilitated by physiological degradation of algae.
(d) absorption of light by algal cell wall.

RESPONSE GRID					
	37. ⓐⓑⓒⓓ	38. ⓐⓑⓒⓓ	39. ⓐⓑⓒⓓ	40. ⓐⓑⓒⓓ	41. ⓐⓑⓒⓓ
	42. ⓐⓑⓒⓓ	43. ⓐⓑⓒⓓ	44. ⓐⓑⓒⓓ	45. ⓐⓑⓒⓓ	

Space for Rough Work

DAILY PRACTICE PROBLEM DPP CHAPTERWISE 38 - BIOLOGY			
Total Questions	45	Total Marks	180
Attempted		Correct	
Incorrect		Net Score	
Cut-off Score	40	Qualifying Score	60
Success Gap = Net Score – Qualifying Score			
Net Score = (Correct × 4) – (Incorrect × 1)			

HINTS & SOLUTIONS

DPP /CB01

1. **(b)** There are several factors and processes which differentiate living beings with non-living beings like reproduction, respiration, growth, etc. But among them reproduction is the only characteristic which differentiates without any exception, living being from non-living beings.
2. **(a)** *Panthera tigris* is the scientific name of tiger.
3. **(a)** 4. **(b)** 5. **(c)**
6. (a) The common characteristics between brinjal and wheat can be observed maximum at the level of their division.
7. **(c)** Cytotaxonomy is the study of chromosome number, size and behaviour.
8. **(b)** The original scientific names were taken from Latin and Greek languages. New names are now derived either from Latin language or are latinised. This is because Latin language is dead and therefore, it will not change in the form or spellings with the passage of time.
9. **(d)**
10. **(b)** The number of similar characters of categories decreases from lowest rank to highest rank in a taxonomic hierarchy. Out of the given categories i.e., family, class, genus and species, the class being the highest category possesses the least similar characters.
11. **(c)** Hierarchy of categories is the classification of organisms in a definite sequence of categories (taxonomic categories) in a descending order starting from kingdom and reaching upto species or an ascending order from species to kingdom. The number of similar characters of categories decreases from lowest rank (species) to highest rank (kingdom). The taxonomic hierarchy includes seven obligate categories-kingdom, division or phylum, class, order, family, genus and species. Some subcategories like tribe, subfamily, subspecies are also used whenever required.

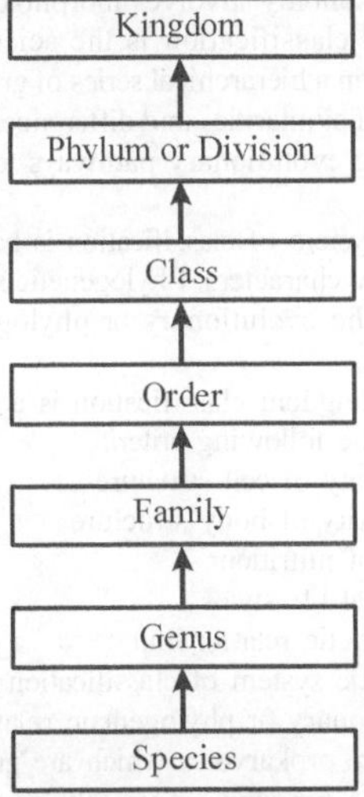

Fig: Taxonomical hierarchy showing arrangement in ascending order

12. **(c)** Teleology is the unscientific explanation of phenomena, while study of vestigeal organs is known as dysteleology.
13. **(d)** International code for Zoological Nomenclature is responsible for setting rules for naming of animals and resolution of nomenclature problems.
14. **(b)** When external temperature is warm, the superficial blood vessels dilate to lose the body heat.
15. **(c)** The biological species concept was formulated by Theodosium Dobzhansky and Ernst Mayr. According to this concept "A species is a reproductive community of populations whose members can interbreed with each other but not with the members of other species.
16. **(b)** Carolus Von Linnaeus in his book Genera Plantanum made use of the artificial system of classification. He distinguished between the natural and artificial systems. He used the binomial nomenclature system and classified organisms into genus and species.
17. **(a)** Species is the lowest taxonomic category. Class is a category made of one or more related orders possessing similar correlated characters, Family is composed of one to many related genera. Division comprises of several related classes.
18. **(d)** Taxon is a grouping of organisms of any level in hierarchy of classification based on some common characteristic.
19. **(c)** Species is the lowest or basic taxonomic category comprising of one or more natural population of individuals that interbreed freely.
20. **(b)** Carolus Linnaeus system of classification is artificial.
21. **(a)** Carolus Von Linnaeus wrote the book systema Naturae.
22. **(d)** Benthem and Hooker system was published well before there were internationally accepted rules for botanical nomenclature. In this system, terminology for families was not so as it is now.
23. **(c)** Individuals of the same species can interbreed. No two individuals share the same ecological niche.
24. **(d)** Biosystematics is the study of diversity of organism and all their comparative and evolutionary relationships.
25. **(b)** Trinomial nomenclature of gorilla is *Gorilla gorilla gorilla.*
26. **(a)** Isotype refers to the duplicate of holotype. When the original material is missing, the new nomenclatural type is termed as neotype. When there is no holotype, any of two or more specimens cited by the author is named as syntype.
27. **(c)** Carnivora and Insectivora are two orders of mammals. Mastigophora is a class of Protozoa. Any plant-eating animal can be termed as a herbivore.
28. **(c)** Many related families can be grouped under an order.
29. **(d)** NBRI is a research institute situated in Lucknow that is engaged in the field of taxonomy and Modern biology.
30. **(c)** Holotype- a specimen on which the original description of the species is based.
Syntype- any of two or more specimens cited by author when there is no holotype.
Neotype- a specimen designated when the original types are known to have been destroyed.
Paratype- a specimen other than the holotype available with the author.
Isotype- a duplicate of the type.
31. **(a)** Chemotaxonomy is the method of biological classifi cation based on similarities in the structure of certain compounds among the organisms being classified.
32. **(b)** Tautonym is the name in which both genus and species epithet are identical
33. **(c)** Lectotype is a specimen later selected to serve as the single type specimen for species originally described from a set of syntypes.

34. (d) Neotype is a type specimen that is selected subsequent to the description of a species to replace a pre-existing type that has been cost or destroyed.
35. (a) Dried specimens are poisoned not by DDT but by others like formaldehyds.
36. (a) Reproduction is a characteristic feature of all living organisms.
37. (b) Animals in lower hierarchy possess more characters in common.
38. (c) Phylum, class and family are classificatory units.
39. (c) Homonym is a name for a taxon that is identical in spelling to another such name, that belongs to a different taxon.
40. (a) Autonyms are automatically created names as regulated by the International code of Nomenclature for algae, fungi, and plants.
41. (d) A binomial name in which the specific name repeats the generic.
42. (b) The most important process for formation of new species is reproductive isolation as this prevents interbreeding in between members of a population, thus giving rise to new species.
43. (c) Two or more species which are reproductively isolated from each other but are morphologically quite similar are known as sibling species. Sympatric species are developed due to reproductive isolation and occur in over lapping or same area of geographical distribution. Parapatric species are developed in adjacent geographical areas meeting in very narrow regions of overlap.
44. (b) Members of same species form population and members of different species form community.
45. (d) Exobiology is the study of possible extra-terrestrial life.

DPP /CB02

1. (b) 2. (d)
3. (c) Cyanobacteria, chlorophyll a, PS I and II.
4. (d)
5. (c) Vegetative phase-animal-like and reproductive phase- plant-like.
6. (c) Desmids i.e golden algae belongs to group chrysophytes. These are found in freshwater as well as marine environment.
7. (a)
8. (c) T_2 phage consists of linear ds DNA.
9. (b) Cell wall is absent in Mycoplasma.
10. (d) Non-motile spores, saprophytic protists.
11. (c) Eubacteria can be differentiated from archaebacteria and bacteria on the basis of cell wall. Eubacteria possess true cell wall, made up of NAM and NAG i.e. N-acetyl muramic acid and N-acetyl glucoseamine. Cell wall of archaebacteria is made up of N-acetyl talosaminuric acid.
12. (d) Pathogen of cholera is *Vibrio cholerae* bacterium. Cholera is transmitted by contaminated water. Typhoid or enteric fever spreads through contaminated water in which bacterium *Salmonella typhi* is present. Citrus canker and crown gall are bacterial disease of plants caused by *Xanthomonas citri* and *Agrobacterium tumefaciens* respectively.
13. (a) Cyanobacteria or blue-green algae are Gram + photosynthetic prokaryotes which perform oxygenic photosynthesis. Photosynthetic pigments include chlorophyll *a*, carotenoids and phycobilins. Cyanobacteria are classified under kingdom Monera. Cyanobacterial cell structure is typically prokaryotic – one envelope organisation with peptidogolycan wall, naked DNA, 70S ribosomes and absence of membrane bound structure like endoplasmic reticulum, mitochondria, golgi bodies, plastids, lysosomes, sap vacuoles. The outer part of the protoplast, called chromoplasm, contains a number of photosynthetic thylakoids.
14. (d) Saprophytic bacteria are free living bacteria which obtain their food from organic remains, plant and animal origin. Aerobic breakdown of organic compounds is known as decay. In nature saprophytic bacteria alongwith saprotrophic fungi are the decomposers of organic remains.
15. (a) Diatoms are very important photosynthesizers. About half of all the organic matter synthesized in the world is believed to be produced by them. Diatoms are probably the most numerous of all plants like protists. Because of this abundance, they are one of the most important primary producers of the sea. There are about 5,500 species of diatoms, mainly marine. The diatoms constitute an important phytoplankton component of the oceans.
16. (c) Fungi is a large kingdom of over 100,000 species. They are achlorophyllous, heterotrophic, spore-forming, non-vascular, eukaryotic organisms which often contain chitin or fungal cellulose in their walls and possess glycogen as food reserve. They are cosmopolitan in occurrence being present in air, water, soil, over and inside animals and plants.
17. (d) Virus is an obligate parasite and is inert outside the host cell. An inert virus outside host is called virion.
18. (d) Monera is the kingdom of all prokaryotes and includes bacteria, blue green algae (cyanobacteria) and archae- bacteria- a group of ancient bacteria kingdom. Protista includes slime unicellular and colonial eukaryotes. The important members are diatoms, dinoflagellates, euglenoids, alone moulds and protozoans. Fungi the kingdom of multicellular or multinucleate heterophyllous and spore producing eukaryotic organisms like *Rhizopus* mildews, mushroom etc. Kingdom plantae includes all coloured multicellular photosynthetic organisms (plants).
19. (c) Viruses that get integrated with the bacterial host genome are called Lysogenic. Lysozymes are present in the saliva and are antibacterial agents. Lipolytic enzymes are the enzymes which catalyse breakdown (lysis) of fats (Lipids).
20. (a) Phenetic classification is purely based on appearances. Phylogenetic classifiaction is based on ancestral lineage. Karyotaxonomy is based on DNA characteristics. Morphotaxonomy involves morphological characters.
21. (d) Biological classification is the scientific arrangement of organisms in a hierarchical series of groups and subgroups on the basis of similarities and differences in their traits. It helps in building evolutionary pathways and in identifying new organisms.
22. (c) Artificial system of classification is based on comparison of on or a few characters. Phylogenetic system of classification indicates the evolutionary or phylogenetic relationship of organisms.
23. (b) The five kingdom classification is a mode of classification based on the following criteria.
   - Complexity of cell structure
   - Complexity of body structure
   - Modes of nutrition
   - Ecological life styles
   - Phylogenetic relationship
24. (c) Phylogenetic system of classification is a system indicating the evolutionary or phylogenetic relationship of organisms.
25. (a) Bacteria are prokaryotes which are grouped under Monera. Protista is a kingdom of unicellular eukaryotes. Fungi is a kingdom of multicellular spore-producing eukaryotes. Plantae are photosynthetic eukaryotes.
26. (a) Retroviruses have RNA as the genetic material and hence they exhibit reverse transcription whereby DNA is synthesized on RNA template. They have reverse transcriptase as the enzyme.

27. (b) Four criteria are : (i) complexity of cell, (ii) complexity of organism, (iii) mode of nutrition and (iv) major ecological role
28. (b) First phylogenetic classification was given by Eichler but that was partially phylogenetic and first true phylogenetic classification was given by Engler and Prantl.
29. (a) *Taenia*, commonly known as tapewom is not a protist, it belongs to phylum Platyhelminthes (Kingdom Animalia)
30. (c)
31. (b) Physiological characters
32. (c) During the origin of life, the first organisms evolved were chemoheterotrophs. They obtained the organic material from outside which they utilized in energy production and synthesis of their own organic material.
33. (d) Bacteria posses various forms and shapes and are of 4 different types - coccus (round), bacillus (rod shaped), vibrio (comma shaped) and spirillum (spiral like cork screw).
34. (c) Slime moulds in the division of myxomycota have spores that develop into flagellated gametes.
35. (b) While working at the Rockefeller Institute, Brown reported isolation of a PPLO from human arthritic joint tissue in 1938. In discussing the significance of this observation, Brown reported successful treatment of arthritic patients in 1949 with a new antibiotic called aureomycin.
36. (c)
37. (c) *Saccharomyces cervisiae* is a yeast used in making bread (Baker's yeast) and commercial production of ethanol. *Paramecium* & *Plasmodium* are of animal kingdom while *Pencillium* is a fungi. Lichen is composite organism formed from the symbiotic association of an algae and a fungus. *Nostoc* & *Anabaena* are examples of kingdom monera.
38. (a) 39. (a) 40. (b) 41. (b) 42. (a)
43. (a)
44. (b) The correct labeling in the figures of bacterial cell and *Nostoc* are - A- cell wall, B - cell membranes, C - DNA, D- heterocysts, E - mucilaginous sheath.
45. (a) A - Cocci (spherical), B - Bacilli (rod shaped), C - Spirilla (spiral), D - Vibrio (comma shaped). These are all the shapes of the bacteria.

DPP /CB03

1. (b) Reserve food in red algae (Rhodophyceae) is floridean starch. In green algae (Chlorophyceae), reserve food is starch. In brown algae Phaeophyceae, reserve food is laminarin and mannitol. In Cyanophyceae, reserve food is cyanophycean starch.
2. (a) Peat moss is another name for *Sphagnum.* It is also known as famine food in China.
3. (b) Well developed vascular system is present in the members of pteridophytes but absent in mosses as the plant body is sporophyte which is distinguished into true root, stem and leaves.
4. (c) The leaves appear green because of the pigment chlorophyll which does not absorb green light rather reflects it back.
5. (b) In chlorophyceae, sexual reproduction takes place by all the three processes :
Isogamy – fusion of morphologically and physiologically similar gametes. Anisogamy – morphologically similar but physiologically dissimilar gametes. Oogamy – fusing gametes are dissimilar in all respect.
6. (d) *Azolla* has been used as a green manure crop in Vietnam and China for centuries. The ability of *Azolla* to shade out weeds in wetland rice and taro has been noted by farmers and researchers since the early 20th century.
7. (a) *Rafflesia* or Corpse flower is a total root parasite. It obtains the total nourishment from the root of the host plant. Its flowers only come out of the host plant and they have a diameter of 1 m and weight around 10 kg.
8. (a)
9. (b) *Laminaria* (kelp) and *Fucus* (rock weed) are marine algae. They are the members of class- phaeophyceae (brown) algae.
10. (b) Algae *Spirulina* is one of the richest sources of protein, containing 40–50 percent crude protein on dry weight basis which under favourable condition may reach upto 70 percent. So, the people recovering from long illness are advised to take it in the diet.
11. (b) The membrane surrounding a lysosome allows the digestive enzymes to work at the 4.5 pH they require. They are created by the addition of hydrolytic enzymes to early endosomes from the Golgi apparatus.
12. (d) 13. (a) 14. (c)
15. (a) Statement (i) and (ii) are correct.
Riccia is liverwort in which simplest sporophyte consists of capsule only while *Polytrichum* is moss in which sporophyte consists of foot, seta & capsule. *Volvox* is a fresh water green colonial alga. Reproduction is both sexual and asexual. Sexual reproduction is oogamous. Slime moulds are consumer decomposer protists. They possess characters of plants (cellulosic cell wall), animals (phagotrophic nutrition) and fungi (spores).
16. (d)
17. (d)
(a) Gametophyte of bryophytes bears protonemal & leafy stage.
(b) In gymnosperm female gametophyte is not free living.
(c) They are present in *Marchantia* or which is a bryophyte.
(d) Origin of seed habit started in pteridophyte *Selaginaella.*
18. (d) 19. (b) 20. (a) 21. (a) 22. (d)
23. (a) 24. (b) 25. (d) 26. (b)
27. (a) Prothallus of pteridophytes is free-living, mostly photosynthetic small but multicellular and inconspicuous.
28. (d)
- Psilopsida – *Psilotum* – rootless
- Sphenopsida – *Equisetum* – Strobilus
- Lycopsida – *Lycopodium* – Homosporous
- Lycopsida – *Selaginella* – Heterosporous

29. (a) Angiosperms are highly evolved and well adapted land plants. They have both vessels and tracheids in xylem for better conduction of water. Roots are modified into tap roots, adventitious roots, pneumatophores etc. to suit the desired climate.
Sex organs are highly developed, sporophylls are organised into flowers and the flowers are highly coloured or modified to attract pollinators ar different times and places. Insect pollination is more prevalent because it is more efficient and leads to less wastage of pollen grains as compared to wind pollination. So the flowers are made attractive to attract a variety of insects. Seed are more protected as they are enclosed inside a fruit.
All these adaptations have made angiosperms more adaptative in diverse habitats.
30. (d) Tracheophyta are those plants which possess conducting or vascular tissue, xylem and phloem. Xylem transports water and minerals while phloem conducts organic food. Tracheophytes include pteridophytes, gymnosperms and angiosperms.
31. (c) Nearly 50% of total carbon dioxide fixation of photosynthesis of world is carried out by algae.
Photosynthesis by algae releases oxygen in the immediate aquatic environment. It is essential for respiration of aquatic life. Algae are primary producers of food in large bodies of fresh, brackish and sea water.

32. (d) In bryophytes, the zygote develops into a multicellular, undifferentiated structure called embryo. The embryo, within venter of archegonium, by further segmentation and differentiation finally develops into a full fledged sporophyte called sporogonium.
33. (d)
34. (d) The gymnosperms are comparatively more ancient than the angiosperms in evolutionary terms. Most of the gymnosperms have now become extinct and the group is presently represented by only 900 living species. The living gymnosperms are widely distributed in the cold climates where snow is the source of water. *Cycas* and *Ginkgo* are referred to as living fossils.
35. (d) In gymnosperms, pteridophytes and angiosperms, the sporophytic phase is dominant and the gametophytic phase is dependent on sporophyte.
36. (b) In diplontic life cycle, the dominant free living phase is the diploid (2n). Sporophyte is photosynthetic. The gametophytic phase is represented by single to few-celled haploid gametophyte e.g., all seed-bearing plants i.e., gymnosperms and angiosperms.
37. (d) Algae and fungi are grouped under thallophyta. Bryophytes are non vascular embryophytes having an independent gametophyte and parasitic sporophyte.
38. (a) Angiosperms have adapted themeslves to all kinds of habitat-terrestial, aquatic, tropical, deciduous and alpine. Self pollination is seen in very few angiosperms. Production of large number of seeds ensure that at least some will germinate. Not all plants have been domesticated by man.
39. (d) Blue-green algae which grow in hot water springs (at about 70°C) are *Oscillatoria terebriformis*, *Oscillatoria brevis* etc. These blue-green algae are also called thermophytes. The temperature tolerance of these algae is due to the presence of homopolar bonds in their proteins.
40. (b) *Cladophora* is an autotrophic and filamentous algae which occurs in fresh as well as in marine water.
41. (a) The pyrenoids are small spherical protein bodies surrounded by starch deposition. They are found singly or in numbers embedded in the chloroplast of many algae and bryophytes.
42. (c) Phycocyanin pigment is found in blue-green algae (myxophycean algae). The reserve food material is in the form of cyanophycean starch.
43. (c) *Cephaleuros virescens* grows as a parasite on the leaves of tea and causes red rust of tea. It is the common intercellular parasitic algae of chlorophyceae.
44. (b)
45. (b)

DPP /CB04

1. (b) The terms Porifera was given by Grant, the phylum includes animals with pores in their body. Its classification based on skeleton or spicules.
2. (c) Chordates are featured by the presence of the following characters : Notochord (cephalic or head region), Pharyngeal gill silts, Postanal tail.
3. (b) Lampreys and hagfishes are unusual, jawless fish that comprise the order Cyclostomata, so named because of the circular shape of the mouth.
4. (d)
5. (a) Aschelminthes is bilaterally symmetrical and triploblastic. These are mostly aquatic, free living or parasitic. Their body is three layered which is ectoderm, mesoderm and endoderm.
6. (c)
7. (c) Characteristic of phylum-Annelida is pseudocoelomate. Pseudocoelomate is any invertebrate animal whose body cavity is a pseudocoel, a cavity between the gut and the outer body wall derived from a persistent blastocoel, rather than a true coelom. Pseudocoelomate animals include the Rotifera and Nematoda.
8. (c) Mammals have 12 pairs of cranial nerves.
9. (b)
10. (d) Poison glands of snake are modified parotid salivary gland.
11. (c)
12. (a) All sponges are marine and have collared cells without any exception. Sponges are animals of the phylum Porifera. They are multicellular organisms which have bodies full of pores and channels allowing water to circulate through them, consisting of jelly-like substance sandwiched between two thin layers of cells.
13. (b) 14. (b) 15. (c) 16. (c) 17. (a)
18. (c)
19. (d) Annelids are first to evolve true metameric segmentation. They are present in embryo as well as in adults. They are present both in ectoderm as well as in mesoderm and are repetitive.
20. (c) Presence of coelom posses a problem for circulation of oxygen and nutrients.
21. (c) Birds belong to class Aves. Bats belong to class Mammalia. Both are warm blooded, endothermic and maintain a constant body temperature. Non-vascular air sacs connected to lungs to supplement respiration are present only in birds not bats.
22. (d) In the molluscs mantle is a loose fold of skin, not concerned with locomotion.
23. (a) In birds syrinx is the sound box.
24. (c) *Tubifex* is blood worm, it is an annelid.
25. (c) In all chordates notochord is present at some stage of embryonic development. Hemichordates do not possess notochord; instead of this they possess stomochord, ectodermal in origin.
26. (d) Birds have internal fertilization.
27. (b) *Adamsia* – Sea anemone (coelenterate)
28. (a) In whales, *Balaenoptera* external ear pinna is absent.
29. (a) A body cavity can mean any internal space, or a series of spaces present inside body, whereas coelom or true body cavity generally refers to a large fluid-filled space (cavity) lying between the outer body wall and the inner digestive tube. In acoelomates, no body cavity or coelom is present. Embryonic mesoderm remains as a solid layer, space between endoderm (gut wall) and ectoderm (body wall) is filled with mesenchyme and muscle fibres. In pseudocoelomates, body space is a pseudocoelom or false coelom. In coelomates or eucoelomates, body space is a true coelom, enclosed by mesoderm on both sides. Aschelminthes are pseudocoelomates. Molluscs and insects are coelomates while flatworms are acoelomates.
30. (b) Mollusca is normally oviparous and its development may be direct or indirect. If development is indirect then it occurs through trochophore or veliger larva. Echinoderms show bilateral symmetry in larvae but pentamerous radial symmetry adult. Fertilization is external. In arthopods, body is divided into head, thorax and abdomen. Respiratory organs consists of book gills, book lungs and tracheae. Notochord is present in chordates at some stage of their lives. Urinary and anal openings, may or may not be separate.
31. (b) In reptiles, fertilization is internal and development is direct.

32. (d) The body cavity of coelomates develops from the mesoderm and contains a peritoneum. The acoelomates lack a body cavity.
33. (b) Because they are not structurally robust, heavy wave action would destroy large, upright sponges.
34. (c) The Ctenophora have evolved a complete gut with a mouth and two anal pores.
35. (c) There are no eyes but photoreceptors do occur on prostomium and dorsal epidermis. They have been, of course, called "ocelli of earthworm".
36. (b) The phylum Rotifera are pseudocoelomates and have a pseudocoel.
37. (a) The segmentation of the annelids allows for more complex coordinated movement.
38. (c) Similar body halves could be obtained with either radial or biradial symmetry. Spherical symmetry has no main body axis along which to cut, and bilateral symmetry produces mirror-image halves.
39. (d) Bilateral animals tend to move through the environment. cephalization is important in the control and coordination of this locomotion.
40. (b) Olfactory chemoreceptors means olfactoreceptors.
41. (d) Corals and sea anemones are members of the cnidarian class Anthozoa.
42. (d) Nematocysts, the stinging cells found in cnidocytes, are the only item in this list that the cnidarians possess exclusively.
43. (d) Flatworms are bilaterally symmetrical and triploblastic animals.
44. (d) A garden snail possesses all of these structures.
45. (a) Medusa and polyp are two types of body shapes of Cnidaria, which includes jellyfish, sea anemones, corals and hydrozoans. Medusa has an umbrella shape. Polyp is umbrella in shape also but inverted. Both medusae and polyps have gastrovascular cavities, tentacles and a mouth. Spongin is a protein found in the inside layer of a sponge. it helps in its structural support.

DPP /CB05

1. (d) The fruit is a mature or ripened ovary. When a fruit develops exclusively from the ovary, it is said to be true fruit. When in addition to the ovary, some other floral part also participates in the formation of fruits, then it is known as false fruit. Apple, pear, cashewnut, mulberry etc. are all false fruits.
2. (a)
3. (a) Single cotyledon of embryo in cereal grain is represented by scutellum. Coleoptile represents the covering of stem. Coleorrhiza represents the covering of root.
4. (d) Desert plants have well developed root system so that they can absorb water from the deeper layers of soil. They have sunken stomata and reduced leaves which reduce the rate of water loss through transpiration.
5. (a)
6. (d) Insectivorous plants grown in nitrogen deficient soil. Therefore, these plants capture insects and have the ability to digest them (their protein). Since proteins are made up of amino acids, having nitrogen in their structure (amino group), these plants overcome the deficiency of nitrogen which is essential for their growth.
7. (b) 8. (b) 9. (d) 10. (d) 11. (d)
12. (c) 13. (a)
14. (c) In *Cocos nucifera* (coconut) milky endosperm is found in which many nuclei, vitamins and growth hormone e.g., cytokinins, auxin and induced cytokinin is found.
15. (a) 16. (d) 17. (c) 18. (a)
19. (a) Epigynous flower ⇒ $\overline{G}$ e.g. Cucumber
Perigynous flower ⇒ G – e.g. Rose and plum
Hypogynous flower ⇒ $\underline{G}$ e.g. Brinjal
20. (c)
21. (b) Caryopsis is a small, indehiscent, one seeded fruit developing from a monocarpellary ovary in which the pericarp is fused with the seed coat. The seed completely fills the chamber, e.g., wheat, maize.
22. (c) Polyarch condition
23. (a)
24. (c) Opuntia has phylloclade for food synthesis.
25. (a) Sub-aerial stem
26. (c) Inferior ovary
27. (c) Drupe is the fruit type in mango & coconut.
28. (c) When stamens are attached to the perianth, they are known as epiphyllous, e.g., Asparagus, lily.
29. (b) In perigynous condition of a flower, the gynoecium is situated in the centre and other floral parts are located on the rim of the thalamus almost at the same level. Ovary is said to be half-inferior, e.g., *Rosa* (Flask-shaped thalamus), *Prunus* (Cup-shaped thalamus).
30. (a) The given floral diagram is of family Solanaceae (potato family). Its flower is bisexual and actinomorphic, abracteate or bracteate, pentamerous, cyclic. Calyx 5, gamosepalous, persistent. Corolla 5, gamopetalous, often plicate in bud. Androecium 5, polyandrous and epipetalous. Gynoecium bicarpellary and syncarpous. Ovary superior, placed obliquely, placentation axile with swollen-placenta. Fruit is berry or capsule.
31. (c) Liliaceae (Lily family) and Poaceae (= Gramineae, gross family) are the two monocot families.
32. (b) In *Nepenthes*, the pitchers are meant for catching and digesting insects. The lamina is modified into pitcher. The leaf apex gives rise to a coloured lid for attracting the insects.
33. (a) In Dahlia, roots do not originate from radicles and are therefore, adventitious. These roots are fleshy having no definite shape, i.e. tuberous in nature. The tuberous roots occur in group or fascicle and are also called fasciculated. Roots of radish, carrot and beet that originate from radicle are the examples of modified tap root.
34. (d) The main function of root-cap is to protect the growing apex from soil particles. Plant growing in water (hydrophytes) or on another plant (epiphytes) or in another plant (parasites) are devoid of root-cap.
35. (d) The coralloid root of *Cycas* and pneuamatophores of mangroves (like *Rhizophora*) become negatively geotropic i.e., come above the soil surface, due to bacterial infection and for aeration, respectively.
36. (b) *Santalum album* (Sandal wood plant) is a small tree, but at the young stage remains as a parasite on the roots of other plants.
37. (b) For storage, tap roots are modified into four ways i. e., napiform, fusiform, conical and tuberous. In the latter form there is no definite shape, as found in *Mirabilis*. A point to note that tuberous root may develop either from tap root or from adventitious root.
38. (c)
39. (b) In many monocots, the stem is represented by underground modifications. However, the flowers are developed on a axis called scape or pseudostem. Such type of development is found in onion, aroids, banana etc.
40. (a) Sweet potato represents the adventitious modified root of *Ipomoea* plant.
41. (c) Epiphyllous bud is a type of adventitious bud, i.e, not originating from stem apex or axil of a leaf. Usually it develops from margin (or leaf surface) of leaf as in *Bryophytlum, Kalanchoe etc*. It serves the function of vegetative propagation.

42. (d) Tuber is a modified stem. A stem can not be developed on root. In potato plant, tubers develop on a special branch of the stem called stolon.

43. (b) 44. (b)

45. (b) Like phylloclade, cladode is also a modification of stem. But here the branch or axillary bud is only modified into a flat, tree like structure with only one internode.

DPP /CB06

1. (b)
2. (d) Function of companion cell is to load sugar and amino acids into sieve elements. These cells use transmembrane proteins to take up by active transport.
3. (c) Phellem, phellogen and phelloderm are collectively called periderm.
4. (d) Submerged hydrophytes do not have stomata (a protective mechanism in aquatic plants against water logging of internal cells and tissues).
5. (d) Tracheids are chief water conducting elements of xylem in gymnosperms. They are devoid of protoplasm and hence dead. The wall constituting the tracheids is hard, thick and lignified. These are elongated cells with tapering ends.
6. (d)
7. (b) Vascular cambium is produced by two types of meristems, fascicular and interfascicular cambium.
8. (d) Vessels are elongated, multicellular water conducting channels with wide lumen formed by end to end fusion of a large number of vessel elements. Tracheids are elongated dead cells with tapering ends having lignified walls with large or wide lumen. Their main function is conduction of water and minerals from root to leaf.
9. (c) The concept of quiescent centre was proposed by Clowes in 1961. On the basis of autoradiographic studies of DNA synthesis in the root tip of *Zea*, he found a reservoir of cells having low DNA, RNA and protein concentration. They may or may not divide. It is resistant to damages.
10. (d) 11. (b)
12. (b) The common bottle cork is the product of phellogen. Phellogen produces cork or phellem on the outer side. It consists of dead and compactly arranged rectangular cells that possess suberised cells walls. The cork cells contain tannins. Hence, they appear brown or dark brown in colour. The cork cells of some plants are filled with air *e.g., Quercus suber* (Cork Oak or Bottle Cork).
13. (b)
14. (d) In submerged hydrophytes, whole plant body remains under water. *e.g. Ceratophyllum* and *Utricularia*. In these plants, stomata is absent and gaseous exchange takes place through general body surface.
15. (d) In collenchyma cell walls show localized thickenings due to presence of approximately 45% pectin, 35% hemicellulose and 20% cellulose.
16. (c) Being a xerophytic plant, *Nerium* bears multiple epidermis to cut the rate of transpiration.
17. (a)
18. (a) Secondary wall situated near the plasma membrane after the formation of primary wall.
19. (d) 20. (a) 21. (d) 22. (d)
23. (a) Based on position of protoxylem in relation to metaxylem, the xylem may be exach/centripetal, endarch/centrifugal, mesarch and centerarch. In endarch condition, protoxylem lies on the inner side of metaxylem e.g., dicot and monocot stems.
24. (a) A monocot stem lacks secondary growth. The vascular bundles are oval or rounded in outline. They contain both phloem and xylem. Phloem lies towards the outside and the xylem on the inner side. Cambium is absent as the whole procambium is consumed in the formation of vascular tissues. Xylem is in the form of the letter Y. It is endarch, i.e., protoxylem lies towards the centre of the stem. Xylem is made up of vessels, tracheids, xylem parenchyma and a few xylem fibres. Metaxylem generally consists of two large oval of rounded vessels lying at the upper two angles of xylem. Protoxylem cavity is present at the end of protoxylem vessels.
25. (d) Heart wood (duramen) is the central wood of an old stem. It is dark coloured. Living cells are absent. Heart wood is the part of secondary xylem. The tracheary elements are plugged by tyloses. Tracheary elements have deposition of tannins, resins, gums, etc. Heart wood is heavier. It is more durable due to its little susceptibility to the attack of pathogens and insects. Heart wood is mechanical in function. The outer or peripheral portion of the trunk is lighter in colour and soft which performs the functions of conduction of water and minerals and it is known as sap wood or alburnum.
26. (d) Intercalated between mature tissues, it helps in regeneration.
27. (d) Spring wood or earlywood
28. (a) Shoot apical meristem
29. (c) Companion cells control activities of sieve tube.
30. (c)
31. (d) Internal (physiological) and external (environmental) factors.
32. (d)
33. (d) Meristematic zone
34. (b) Thickening of wall is due to deposition of cellulose, hemicellulose and pectin.
35. (b) Endodermis with casparian strips is found in dicot stem.
36. (d)
37. (d) The electron microscope revealed al the structural details of the cell.
38. (c)
39. (d) Important functions of veins are : (i) Conduction of water through xylem; (ii) Providing channels for translocation of organic nutrients; (iii) Conduction of minerals; (iv) By their large number, the veins and veinlets provide skeletal support to the lamina so that it can remain stretched for its optimum functioning; (v) Veins and veinlets reduce the effect of wilting.
40. (b) *Cuscuta* is a total parasite so it absorbs the prepared food from the phloem of the host, hence haustoria reaches upto phloem of host.
41. (c) Tylose is a balloon-like outgrowth of paratracheal parenchyma into a pit in the wall of a vessel or tracheid and a xylem parenchyma cell lying next to it. This xylem parenchyma occurs at the edge of annual ring around the vessels.
42. (c) Epidermis is generally uniseriate. i.e., compsoed of single layer of epidermal cells.
    In some cases, it is multilayered also. e.g., *Ficus, Nerium, Peperomia*, etc.
43. (a) In the cortex of monocot root, we observe a few layers of cell below the epidermis which have thick walls due to the deposition of a chemical suberin. These layers of suberized cells are called exodermis.
    There are also number alternating zones of xylem present in root. The number of xylem patches far exceeds to ten. That is why, they are called polyarch. In monocot root, there is a large sized pith made up of parenchyma tissues.
44. (c) Phellem is a suberized cell formed outside the region of stem followed by endodermis, inside the phloem.
    Pericycle lies immediately inside the endodermis of the dicotyledonous stem phloem. Phloem in characterized by the presence of sieve tubes.
45. (c) Tunica corpus theory was proposed by Schmidt. According to this theory, two tissue zones occur at the apex.

DPP /CB07

1. (d) Squamous epithelium is formed of thin discoidal and polygonal cells that fit like tiles in a floor, so is also called pavement epithelium. It is found in the walls of blood vessels, in the alveoli of lungs for exchange of gas, and in Bowman's capsule of nephron for ultra filtration.
2. (b) The cell junctions called tight, adhering and gap junctions are found in epithelial tissue. Epithelial tissue covers the whole surface of the body. It is made up of cells closely packed and arranged in one or more layers.
3. (b) 4. (d)
5. (d) Myoglobin is a single-chain globular protein of 153 amino acids, containing a heme (iron-containing porphyrin) prosthetic group in the center around which the remaining apoprotein folds. It has a molecular weight of 16,700 daltons, and is the primary oxygen-carrying pigment of muscle tissues. Unlike the blood-borne hemoglobin, to which it is structurally related, this protein does not exhibit cooperative binding of oxygen, since positive cooperativity is a property of multimeric/oligomeric proteins only.
6. (a)
7. (c) Silver fish (Insecta), scorpion (Arachnida); crab (Crustacea) and honey bee (Insecta) all belong to phylum Arthropoda which is characterized by the presence of jointed appendages. Poison glands occur only in scorpion. Compound eyes and metamorphosis are the characteristic of insects.
8. (c)
9. (a) Lymph differs from blood in that it contains less protein and no red blood cells. Lymph contains white blood cells.
10. (b)
11. (c) Dolphin is an aquatic mammal and breathes through lungs.
12. (c) 13. (b) 14. (a) 15. (a) 16. (c)
17. (a) Sexual dimorphism is found in cockroach. A pair of anal style are present in male cockroach only. Female has broader abdomenal segments.
18. (a) Presence of air passages in the bones (pneumatic bones) is a characteristic of birds.
19. (a) 20. (c)
21. (b) Nephridia in earthworm, malpighian tubules in cockroach and urinary tubules in the rat are excretory in function.
22. (c) 23. (a) 24. (c) 25. (b) 26. (d)
27. (d) 28. (b) 29. (c) 30. (c)
31. (c) Spiracles are 20 in number
32. (c)
33. (d) Propelling mucus is the function of simple epithelium.
34. (c) Imitability is a property only shown by cardiac muscles
35. (b) 36. (b) 37. (c)
38. (c) Myocardium is the muscular part. Pericardium and endocardium are the linings of epithelium.
39. (c) This cartilage is found in intervertebral discs.
40. (d) These granules are composed of RNA and protein.
41. (b) The GI tract is lined with simple columnar epithelium, which allows a maximum number of cells to contact food particles.
42. (a) Most epithelia have a basement membrane between the epithelial cells and the underlying connective tissue.
43. (a)
44. (b) Bone is surrounded by dense, white fibrous sheath called periosteum.
45. (b) A - Ocellus; B - Compound eye; C - Mandible; D - Maxilla; E - Labrum; F - Labium.

DPP /CB08

1. (c) Glycosylation is the process or result of addition of saccharides to proteins and lipids. The process is one of the four principal co-translational and post-translational modification steps in the synthesis of membrane and secreted proteins and the majority of proteins synthesized in the rough ER undergo glycosylation.
2. (b) Vacuoles are present mainly in the plant cells. Each vacuole is surrounded by cytoplasmic membrane called as tonoplast which is similar to plasma membrane.
3. (b) The membrane surrounding a lysosome allows the digestive enzymes to work at the 4.5 pH they require. They are created by the addition of hydrolytic enzymes to early endosomes from the Golgi apparatus.
4. (c)
5. (c) Nucleolus was discovered by Fontana (1781) and given name by Bowman (1840). It does not have membrane and is attached to chromatin at nucleolar organiser region (NOR). Nucleolus is the site for elaboration of r-RNA and synthesis of ribsomes, hence called ribosomal factory.
6. (b)
7. (a) The fluidity of membranes in a plant in cold weather may be maintained by increasing the number of phospholipids with unsaturated hydrocarbon tails.
8. (c) 9. (a) 10. (b)
11. (b) Fimbriae are hair like structures present in large number in bacteria. They help in attaching bacteria to solid surfaces or host tissues.
12. (d) Proteins have very specific shapes which make them ideal as receptor molecules for chemical signalling between cells. Branching side chain glycolipids on the outer surface of cell membranes are also involved in cell-cell recognition.
13. (a) The SER provides surface for the synthesis of lipids, including phospholipids, cholesterol, steroid hormones (sex hormones, adrenal corticoid hormones), ascorbic acid and visual pigments.
14. (a)
15. (b) Phospholipids in the lipid bilayer limit the permeability of the membrane.
16. (c) Vacuole is a non - living reservoir, bounded by a selectively permeable membrane, the tonoplast. It is not a air filled cavity but it is filled with a highly concentrated solution called vascular sap or cell sap. pH of vacuolar cell sap is acidic and hypertonic.
17. (c) Plastids are double membranous, semi-autonomous organelles which store and synthesise various types of organic compounds. They develop from colourless precursor proplastids. Proplastids have the ability to divide and differentiate into various types of plastids.
18. (c) Only statements B and E are incorrect as plastids have circular DNA and are found to be present in higher plants.
19. (b) 20. (a)
21. (d) Prokaryotes possess gas vacuole.
22. (d) Membrane-less organelles.
23. (b)
24. (d) All of these
25. (b) 2n = 34
26. (d) 27. (a) 28. (b) 29. (d) 30. (b)
31. (d) 32. (c)
33. (b) In 1935, Danielli and Davson proposed that cell membrane is made of a double layer of phospolipid molecules sandwiched between two single layers of proteins. The three layers are held together by electrostatic forces while phospholipid layers are kept adhered by vander Waal's forces.
34. (a)
35. (d) The cell membrane is asymmetric and has different properties, and functions of the cytoplasmic side versus the extracellular side. These properties arise from differences in the constituents of the membrane.

36. (b) Both glycolipids and glycoproteins serve as recognition signals.
37. (a) The charged or polar lining of the channel proteins allows passage of polar and charged molecules.
38. (d) All cells have a plasma membrane. The other structures listed are organelles and therefore are present only in eukaryotes.
39. (d)
40. (c) The cytoskeleton supports the cell and allows for movement of the entire cell and listed are part of the cytoskeleton.
41. (d) 42. (c) 43. (b)
44. (c) Plants are autotrophs and synthesize their food in the process of photosynthesis with the help of chloroplast (plastid).
45. (d) Peroxisomes contain glycolic acid and oxidase, which oxidises glycolic acid (a product of photosynthesis) to glyoxylic acid.

DPP /CB09

1. (c) The nucleic acid (DNA and RNA) are repeating units of nucleotides i.e. polynucleotides. Each nucleotide comprises of nitrogenous heterocyclic bases viz. purines or pyrimidines, pentose sugar and phosphoric acid.
2. (b) Glycogen is an animal starch, stored in the liver and is polysaccharide of α – Glucose. About 30,000 glucose molecules joined by 1-4 α -glycosidic bonds and its branches by 1-6 glycosidic bonds. It gives red colour with iodine solution.
   - Galactose is a monosaccharide from milk.
   - Fructose is a monosaccharide from fruit.
   - Sucrose is a disaccharide (Glucose + Fructose).
3. (b) Cellulose is the most abundant organic compound, most abundant polysaccharide and most abundant biopolymer found on earth.
4. (c) In the cytoplasm, protein synthesis is restricted to the ribosomes. Mitochondria being a semiautonomous organelle, has its own genetic machinery to synthesize its proteins.
5. (d) RNA is a single chain of ribonucleotides. Each ribonucleotide consists of nitrogenous bases (adenine, guanine, cytosine and uracil), ribose sugar and phosphoric acid.
6. (a) Reducing sugar is any sugar that has an aldehyde group or is capable of forming one in solution through isomerisation. Reducing monosaccharides include glucose, fructose, glyceraldehyde and galactose. Many disaccharides, like lactose and maltose also have a reducing form, as one of the two units may have an open- chain form with an aldehyde group.
   However, sucrose and trihalose in which the anomeric carbons of the two units are linked together, are non reducing disaccharides since neither of the rings is capable of opening.
7. (d) α-amylase enzyme stimulates the germination of barley seed.
8. (a) Massive ions like Na^+ facilitate the absorption of substances like amino acid and glucose through cotransport.
9. (d) Enzymes accelerate chemical reaction by lowering the activation energy.
10. (c) Competitive inhibition is the reversible inhibition of enzyme activity due to presence of substrate analogues. Non competitive inhibition is the reduction of enzyme activity by a factor that has no real structural similarity with the substrate. Allosteric inhibition is reversible non-competitive inhibition occurring in case of allosteric enzymes . Herein, the inhibitors are the products or intermediates of reactions catalyzed by the enzymes. Hence, it is also called end product inhibition or feedback inhibition.
11. (c) There are only 20 amino acids . These can be arranged in different order in the polypeptide chain to form a diverse array of proteins.
12. (a) Length of one turn of DNA helix is 34Å.
   $1Å = 10^{-1}$ nm
13. (c) K_m (Michealis Menten constant) is defined as that substrate concentration at which under optimum conditions the rate of an enzyme catalysed reaction reaches half the maximum rate. K_m is inversely proportional to affinity of enzyme for its substrate.
14. (b) 15. (a) 16. (d) 17. (a) 18. (a)
19. (d) 20. (b) 21. (d) 22. (c) 23. (b)
24. (c) 25. (b)
26. (b) Phospholipids are triglyceride lipids, in which one fatty acid is replaced by phosphoric acid residue often linked to additional nitrogenous groups like choline (in lecithin), ethanolamine (in cephalin) etc.

$$\begin{array}{l} CH_2\text{—O—CO—}R_1 \\ | \\ CH\text{—O—CO—}R_2 \\ |\qquad\qquad O \\ |\qquad\qquad | \\ CH_2\text{—O—}\overset{}{P}\text{—}(CH_2)_2\text{—}\overset{+}{N}(CH_3)_3 \\ \qquad\qquad \| \\ \qquad\qquad O \end{array}$$

Phosphate — Choline

Fig. : Structure of lecithin.

27. (b) Polysaccharides, monosaccharides are linked together by glycosidic bond (C–O–C). Water has hydrogen bonding. Proteins and polymers of different amino acids, which are linked by peptide (CONH) bond formed by dehydration (polymerisation) between COOH group of one amino acid and NH_2 group of next amino acid with the removal of H_2O. The three hydroxyl groups of a glycerol molecule join with the the carboxyl groups of three fatty acid molecules to form a fat or triglyceride molecule. The chemical linkage between the glycerol and fatty acids is called the ester bond.
28. (a) Chitin is the second most abundant polysaccharide (after cellulose). Chitin is not a storage polysaccharide rather it is a structural homopolysaccharide, formed of N-acetylglucosamine (NAG) monomers joined together by β-1, 4 linkages. Also, it forms the structural component of fungal walls and exoskeleton of arthropods.
29. (b)

Some Secondary Metabolites	
Pigments	Carotenoids, anthocyanins, etc.
Alkaloids	Morphine, Codeine, etc.
Terpenoids	Monoterpenes, Diterpenes, etc.
Essential oils	Lemon grass oil, etc.
Toxins	Abrin, Ricin
Lectins	Concanavalin A
Drugs	Vinblastin, Curcumin, etc.
Polymeric substances	Rubber, Gums, Cellulose.

Vinblastin is used as an anticancer drug, whereas curcumin is a component of turmeric and also used as drug.

30. (b) Lyases enzyme catalyze the breakage of specific covalent bonds and removal of groups without hydrolysis producing double bonds or removal of double bonds by adding groups e.g., histidine decarboxylase that splits C–C bond of histidine, forming CO_2 and histamine.

$$\begin{array}{cc} X & Y \\ | & | \\ C & \text{—}C \end{array} \xrightarrow{\text{lyase}} X-Y+C=C$$

Fructose-1, 6 diphosphate $\xrightarrow{\text{Aldolase}}$ Dihydroxyacetone-3-phosphate + Glyceraldehyde-3-phosphate.

31. (c) Successive glucose units are linked together by $\alpha\ 1 \rightarrow 4$ linkage and at branchings $\alpha\ 1 \rightarrow 6$ linkage is present.

32. (c) 33. (c)

34. (b) Tertiary structure of proteius is devoid of phosphodiester bonds. There bonds are present in nudeic acids.

35. (b)

36. (a) A high k_m indicates low affinity of an enzyme for its substrate.

37. (d) The multiple molecular forms of an enzyme occuring in the same organisms and having a similar substrate activity are called isoenzymes or isozymes. They have similar properties but different molecular weights and location. Over 100 enzymes are known to have isoenzymes. α-amylase of wheat endosperm has 16 isozymes, lactate dehydrogenase has 5 isozymes.

38. (d) Essential amino acids are the amino acids which cannot be synthesized by organisms in the body and are obtained from plants e.g., valine, leucine, isoleucine, phenylalanine, threonin, lysine, tryptophan, methionine.
Non-essential amino acids can be synthesized by the organism and may not be the requisite components of diet e.g., serine, tryosine, aspartate, cysteine, glutamate, proline, glycrine, alanine, asparagine, glutamine and tyrosine.

39. (a) Acidic amino acids have an extra carboxylic group (mono-amino dicarboxylic), e.g., glutamate (glutamic acid, Glu), asparate (aspartic acid, Asp).
Basic amino acids have an additional amino group without forming amides (diamino monocarboxylic), e.g., arginine (Arg), lysine (Lys).
Neutral amino acids have one amino group and one carboxylic group (mono-amino mnocarboxylic) with noncyclic hydrocarbon chain, e.g., glycine (Gly), alanine (Ala), valine (Val), Leucine (Leu), isoleucine (lle). Essential amino acids are the amino acids which cannot be synthesized by organisms in the body and are obtained from plants e.g., valine, leucine, isoleucine, phenylalanine, threonin, lysine, tryptophan, methionine.
Non-essential amino acids can be synthesized by the organism and may not be the requisite components of diet e.g., serine, tryosine, aspartate, cysteine, glutamate, proline, glycrine, alanine, asparagine, glutamine and tyrosine.

40. (d)

41. (b) Trehalose is a non-reducing sugar. It is made up to 2 molecules of glucose

42. (c) 43. (a) 44. (d)

45. (a) Palmitic acid is one of the most common saturated fatty acids found in animals and plants. It has 16 carbons including the carboxyl carbon.

DPP /CB10

1. (d) G_1 phase, also called Gap I phase is characterized by increase in cell size. In the S phase or synthetic phase DNA molecules replicate. G_2 is the second growth phase or Gap II where in there is intensive formation of RNAs and proteins. In the mitotic metaphase, the chromosomes are arranged at the equatorial plate.

2. (b) Kinetochore is the proteinaceous covering of centriole, to which spindle fibers attach.

3. (b) Mitosis is the process in which eukaryotic cell separates the chromosomes in its cell nucleus, into two identical sets in two daughter nuclei. It is generally followed immediately by cytokinesis, which divides the nuclei, cytoplasm, organelles and cell membrane into two daughter cells containing roughly equal shares of these cellular components. Mitosis and cytokinesis together define the mitotic (M) phase of the cell cycle - the division of the mother cell into two daughter cells, genetically identical to each other and to their parent cell.

4. (d) In bivalent formation of chromosomes during meiosis, the homologous chromosomes are arranged in pairs. The phenomenon is called synapsis and it occurs during zygotene stage. DNA replication occurs during S phase or synthetic phase which is the second phase of interphase.

5. (a) 6. (b) 7. (b) 8. (b) 9. (b)

10. (c) 11. (b)

12. (d) During telophase, the individual chromosomes are no longer seen and chromatin material tends to collect in a mass at the two poles. Chromosomes cluster at opposite spindle poles and their identity is lost as discrete elements. Nuclear envelope assembles around the chromosome clusters. Nucleolus, Golgi complex and ER reform.

13. (a) 14. (c)

15. (a) Condensation initiates in prophase.

16. (c) Interkinesis is the stage that occurs in between meiosis-I and meosis-II

17. (a) During anaphase, centromeres split resulting in the separation of sister chromatids towards opposite poles.

18. (c) 19. (d) 20. (a)

21. (d) A = Diplotene B = Dissolution
C = Pachytene D = Anaphase - II

22. (d) The lampbrush chromosomes occuring in prophase of meiosis II are highly elongated special kind of synapsed mid-prophase or diplotene chromosome bivalents which have already undergone crossing over. Lampbrush chromosomes occur in pairs. Each chromosome of a pair has a double main axis due to presence of two elongated chromatids. Both the adjacent chromatids bear rows of large number of chromomeres. Two adjacent chromomeres are separated by interchromomeric stretches. Many of the chromomeres give out lateral projections or loops. The lateral loops provide a test tube or lampbrush-like appearance to the chromosome pair. Lateral loops take part in rapid transcription of DNA to mRNA meant for synthesis of yolk and other substances required for growth and development of meiocytes.

23. (a) The mitotic spindle forms and the nuclear membrane disperses during prophase.

24. (b) If at G_1 cells are given the signal to divide, they are unlikely to be stopped at subsequent checkpoints.

25. (a) Sister chromatids are the two identical strands of a duplicated chromosome.

26. (d) Meiosis reduces the chromosome count from diploid to haploid and halves the amount of genetic material.

27. (d) Kinetochores are assembled at the centromere of each chromosome and are the sites where microtubules attach to segregate the chromosomes. In meiosis there is only one fused kinetochore per chromosome; in mitosis there are two kinetochores per chromosome.

28. (d) Chromosomes are attached tot he microtubules at their kinetochores. There are molecular motors at the kinetochores which help move the chromosomes to the poles by the shortening of the kinetochore microtubules.

29. (c) Programmed cell death occurs during the development of many organisms (for instance, tadpoles lose their tails to become adult frogs). One of the stimuli for programmed cell death. Necrosis (cell death that is not programmed) occurs when cells have been deprived of cell cycle, in which cells reproduce, and is not a step in programmed cell death.

30. (b) After fertilization, the DNA content in the cell increases because of the fusion of genetic material of the parents. Mitosis then occurs and is followed by cytokinesis at x, which is the division of the cytoplasm into 2 compartments, i.e. 2 cells.

31. (a) Although no centrioles are visible, there is spindle formation and the chromosomes do exhibit equatorial arrangement. Spindle formation is probably by another organelle, unknown as yet.
32. (c) At prophase I, DNA replication has already occurred, and the original amount of DNA has been doubled to x. At anaphase I, the amount of DNA in the cell remains the same because no cytokinesis has occurred yet to separate the cytoplasm.
33. (d) G_0 represents a stage in G_1 in which cells are supposed to be withdrawn from division.
34. (c) Restriction point represents a stage in G_1 phase If the cell has passed restriction point, it would divide.
35. (c) Pairing can be procentric proterminal pairing or intermediate condition (also called as random synapsis which may occur simultaneously at all chromomeres).
36. (b) Stock is the one that receives the graft which has 2n = 48. This would produce the root which will have 2n = 48. Scion has 2n = 24, would produce microspore (n = 2)
37. (c) The number of chromosome will be same (50), but each chromosome will have 2 chromatids.
38. (a) No. of required meiosis $= n + \frac{n}{4} = \frac{5n}{4}$

where n = no. of seeds.

$= \frac{5 \times 10}{4} = \frac{50}{4} = 12.5 = 13$

10 Megaspore + 10 Microspore = 10 seeds

To produce 10 Megaspores, 10 meiotic division would be needed

13 – 10 = 3 meiosis will produce 12 microspore

So, wastage of pollen grain = 2
39. (d) The correct order of mitotic events which occur during meiosis is: Formation of synaptonemal complex, recombination, separation of homologous chromosomes, separation of sister chromatids.
40. (b)
41. (c) Any agent that stimulates cell division is called mitogen e.g., temperature, cytokinin, auxin, gibberellin, insulin and steroids.
42. (c)
43. (d) During the S-period, the centrioles separate and undergo duplication which produces two pairs of centrioles still contained within the radiating masses of microtubules.
44. (a)
45. (c) The given figure shows anaphase I of meiosis. In anaphase I, the homologous chromosomes break apart while sister chromatids remains associated at their centromere. At the end of anaphase I, two groups of chromosomes are produced at two poles, having half the number of parental chromosomes.

DPP/CB11

1. (d)
2. (c) Osmosis is the movement of solvent particles from a region of low solute concentration to a region of high solute concentration through a selectively permeable membrane.
3. (b) 4. (c)
5. (d) The factors which affect the rate of transpiration are light – stomatal opening, temperature – increases rate of transpiration, humidity – low humidity increases the rate of transpiration whereas high humidity exerts an opposite effect *i.e.* decreases the rate, wind – high wind causes stomatal closure, thereby, reducing the rate of transpiration in plants.
6. (a) Osmotic potential is the potential of a solution to cause water movement into it across a semi- permeable membrane. Water potential is the tendency of water to leave a system.
7. (d) Turgor pressure is the pressure that develops in a cell due to osmotic diffusion of water inside it and is responsible for pushing the membrane against cell wall. Stomata open under conditions of increased turgor pressure of guard cells and stomata gets closed under conditions of decreased turgor pressure of guard cells. When turgid, they swell and bend outward. As a result, the stomatal aperture opens. When they are flaccid, the tension from the wall is released and stomatal aperture closes.
8. (b) Active transport involves movement of ions against concentration gradient involves the expenditure of energy. Diffusion involves the movement of solute particles from region of higher concentration to a region of lower concentration. Pinocytosis is cell drinking. Brownian movement is the random to and fro movement of atoms and molecules.
9. (c)
10. (d) A solution having more solutes as compared to solution having less number of solutes : former is known as hypertonic and latter hypotonic. Hence, solution outside the cytoplasm has more solutes (higher concentration) and are hypertonic. If two solutions have same solute concentration then they are isotonic.
11. (b)
12. (a) During sodium-potassium pump, the concentration of sodium ions will be about 14 times more in extra cellular fluid (outside) and concentration of potassium ions will be about 28-30 times more in axoplasm (inside). Thus, $3Na^+$ and $2K^+$ are transported during the process.
13. (a) In rainy season, door gets swelled due to the phenomenon of imbibition. It is the process of absorption of water without forming a solution.
14. (c) Cell placed in a hypotonic solution becomes turgid due to endosmosis. Water exerts a pressure on the walls of the cell called turgor pressure. Cell wall being a rigid and hard structure also exerts a pressure on Cytoplasm in response to Turgor pressure when TP = WP, DPD = 0 (No net exchange of water).
15. (a) The pure water, at atmospheric pressure has zero water potential. The addition of any solute particles reduces the free energy of water. Thus, the water potential will be negative.
16. (b)
17. (b) Water logging of soil makes it physiologically dry because this condition does not allow oxygen to enter the soil.
18. (c) 19. (b) 20. (b)
21. (b) The absorption of water still occurs when concentration of outer soil water is more than root hair cell sap.
22. (d)
23. (b) With increasing humidity, the rate of transpiration decreases linearly, because the high saturation of water vapour in the atmosphere prevents the evaporation of more water from the leaf interior to the exterior.
24. (b) 25. (b)
26. (d) If we cut a small soft-stemmed plant on a day, when there is plenty of atmospheric moisture, early in the morning we will soon see drops of solution ooze out of the cut stem; this comes out due to the positive root pressure. If we fix a rubber tube to the cut stem as a sleeve we can actually collect and measure the rate of exudation, and also determine the composition of the exudates. Root pressure is a positive pressure that develops in the xylem sap of the root of some plants. It is a manifestation of active water absorption. Root pressure is observed in certain seasons which favours optimum metabolic activity and reduce transpiration. It is more during rainy season in the tropical countries and during spring in temperate habitats.

27. (c) Cohesion tension theory (Cohesion-tension and transpiration pull theory) explains the ascent of sap best. It was put forward by Dixon and Jolly in 1894. It was further improved by Dixon in 1914. Therefore, the theory is also named after him as Dixon's theory of ascent of sap. Today most of the workers believe in this theory.
28. (a) Sap will flow from high to low pressure areas in the plant. Adding sugars to the sap causes water to enter the phloem, thus increasing pressure. Removing sugars causes water to leave the phloem, thereby reducing pressure.
29. (c) The water potential is the sum of the osmotic potential (usually negative) and the pressure potential (usually positive), so WP = –0.24 + 0.16 = –0.08 mPa.
30. (a) Guttation is caused by root pressure. It is only important to the movement of water in quite small plants.
31. (d) As leaf cells lose water, they release abscisic acid, which causes potassium ions to leave guard cells, decreasing their pressure potential and closing stomata.
32. (b) As carbon dioxide is fixed into organic acids at night the pH decreases; metabolism of these acids during the day (with the carbon dioxide going into the citric acid cycle) causes the pH to increase.
33. (a) Choices b, c, and d are true for phloem transport only.
34. (c) Sugar molecules diffuse through the symplast of mesophyll cells, are pumped into the apoplast near a vein, re-enter the symplast of a companion cell, and then move into a sieve tube element.
35. (a) The intercellular spaces and cell walls of the plant constitute the apoplast.
36. (a) Transpiration causes tension.
37. (d) Abscisic acid, light, and carbon dioxide levels all regulate stomatal opening and closing.
38. (a) Stomatal regulators work by activating and deactivating the proton pump in guard cells.
39. (d) 40. (c) 41. (c) 42. (a)
43. (b) The tension created by the evaporation of water from leaf surfaces generates a pull on the water column that moves water and nutrients up the xylem from the roots to the leaves.
44. (c) 45. (d)

DPP /CB12

1. (b) 2. (c)
3. (a) Molybdenum is a component of nitrogenase and nitrate reductase involved during N_2 fixation. Copper is essential for photosynthesis and respiration. Zinc is involved in synthesis of auxin. Manganese is activator of enzymes involved in nitrogen metabolism.
4. (b) The root nodules of leguminous plants contain a symbiotic nitrogen fixing bacteria *Rhizobium*. Root nodules are small irregular outgrowth of the roots which are pinkish internally due to presence of a pigment called leghaemoglobin. It is related to blood pigment haemoglobin. The cells of root nodules are tetraploid and contain polyhedral bacteria called bacteroids. Leghaemoglobin is an oxygen scavenger and protects the nitrogen fixing enzyme nitrogenase of bacteroids.
5. (c) 'Whip-tail' disease in cauliflower is noted due to deficiency of molybdenum. Cauliflower is the most sensitive of crops to molybdenum (Mb) deficiency. Light sandy soils with pH levels close to or below pH 6 under conditions of high fertility can show severe Mb deficiency. Molybdenum is the only plant essential micronutrient that becomes less available as the soil acidifies–the opposite of copper, zinc and iron.
6. (b) The essential elements are divided into macroelements and microelements based on the quantity in which they are required by the plants. Mn, Fe, Cu, Mo, Zn, B, and Cl are the micronutrients needed in very small quantities by the plants. C, H, O, N, P, S, K, Ca, Mg, Fe are the macronutrients required in more quantity.
7. (d) Heterotrophic plants are less dependent on nitrogen obtained from nitrification since they receive some nitrite and nitrate through their parasitic or carnivorous nutritional modes.
8. (d) 9. (b)
10. (d) Root hairs absorb water, minerals and oxygen but they do not absorb CO_2. CO_2 intake takes place in leaves.
11. (c) The ash that is left after burning any dry part of the plant contains only mineral elements and is called plant ash.
12. (b) The atmospheric nitrogen is fixed in the soil in the form of nitrate. Nitrate and ammonium can be taken up by plants or used by other soil organisms. Any nitrate or ammonium that is not used for growth is added to the pool of inorganic nitrogen in the soil.
13. (c) Cultivation of plants by placing the roots in the nutrient solution without any soil is called hydroponics. It is also known as soilless culture/ water culture/ solution culture. It is used to determine which elements are essential for plant growth and what symptoms are produced by the absence or deficiency of essential elements.
14. (c) 15. (c)
16. (a) Cobalt stimulates growth of legumes such as beans, clover and alfalfa. This stimulation of growth by cobalt is due to its use, not by the plant itself, but by nitrogen-fixing bacteria that live in roots of plants.
17. (b) 18. (a) 19. (a) 20. (b) 21. (b)
22. (d)
23. (b) Magnesium activates the enzymes of respiration, photosynthesis and is involved in the synthesis of DNA and RNA. Manganese activates many enzymes which are involved in photosynthesis, respiration and nitrogen metabolism.
24. (a) NO_3^-, is negatively charged and not tightly bound to soil particles.
25. (d) Legume roots have swellings called nodules that contain nitrogen-fixing bacteria of the genus Rhizobium.
26. (b) The main difference between micronutrients and macronutrients is in how much of them a plant needs to survive.
27. (c) Symptoms caused by deficiency of essential mineral elements on plants are called Hunger signs.
28. (a) The deficiency of calcium results in death of leaf, stem and root apices.
29. (c) Hydrogen sulphide has little to do with ion exchange.
30. (b) Due to immobile property of sulphur, chlorosis occurs first in young leaves.
31. (c) All other statements are false.
32. (a) The characteristic symptom of iron deficiency is the interveinal chlorosis. The symptom may be general or localised to a single leaf or a single branch because of the limited mobility of iron in the tissues. The deficiency symptom first appears in young leaves.
33. (d) An iron deficiency is suspected in a plant. If the older leaves are yellow, a nitrogen deficiency would be suspected because nitrogen is easily relocated in the plant while iron is not.
34. (d) 35. (d)
36. (d) The bulk of the organic material of a plant is derived from carbon dioxide assimilated from the atmosphere.
37. (b) For chlorophyll formation, a plant needs Fe, Mg & light
38. (c) 39. (a) 40. (b) 41. (c) 42. (a)
43. (b) "Reclamation" and "Little leaf" disease is caused by deficiency of Cu and Zn.
44. (a) 45. (a)

DPP /CB13

1. (a)
2. (c) The C_4 pathway allows photosynthesis to occur at very low concentrations of carbon dioxide as PEP carboxylase has an extremely high affinity for carbon dioxide. This pathway also works well at high temperatures and light intensity, enabling efficient photosynthesis in tropical plants.
3. (a) Photorespiratory loss of CO_2 occurs when RuBisCo starts functioning as an oxygenase instead of carboxylase under conditions of high O_2 and low CO_2. It involves three organelles chloroplast, mitochondria and peroxisomes. Half of the photosynthetically fixed carbon (in the form of RuBP) may be lost into the atmosphere through this process and no ATP formation occurs.
4. (d) 5. (a)
6. (c) The process of photorespiration requires 3 cell organelles-mitochondria, chloroplasts and peroxisomes. In chloroplasts glycolate is formed from ribulose biphosphate which passes into peroxisomes to be changed into glyoxylate. In this reaction H_2O_2 is evolved. Glyoxylate is changed to glycine which now enters mitochondria. Two molecules of glycine combine to form one molecule of serine with the evolution of CO_2.
7. (b) Grana are present inside the chloroplasts. Each granum may contain 10 – 100 thylakoids. Thylakoids or baggy trousers are membrane lined flattened sacs. Thylokoids (or granna) contain chloroplast proteins, photosynthetic pigments and other factors required for photosynthesis.
8. (b) Photosynthesis takes place only in the visible part (400 - 700 nm wavelength) of electromagnetic radiations. Hence, this component comprises the photo-synthetically active radiation.
9. (a) 10. (a) 11. (a)
12. (b) RuBisCo is the enzyme involved in Calvin cycle. Nitrogenase catalyses nitrogenation. Invertase catalyses breaking of sucrose to glucose and fructose.
13. (a) The primary acceptor of CO_2 in C_4 plants is phosphoenol pyruvate or PEP. PEP in mesophyll cells combine with CO_2 and is converted into 4 carbon compound oxaloacetic acid by PEP carboxylase. In C_3 plant, Ribulose 1, 5–diphosphate is primary acceptor of CO_2.
14. (a) Photolysis is catalysed by the protein-bound inorganic complex containing manganese ions (oxygen evolving complex) of photosystem II.
15. (c) 16. (c) 17. (d) 18. (d) 19. (d) 20. (a)
21. (b) 22. (c) 23. (c) 24. (a) 25. (d) 26. (b)
27. (c) 28. (c) 29. (d)
30. (a) Oxygen is a product of photosynthesis. A small quantity of O_2 is essential for photosynthesis to take place. But as O_2 concentration rises, rate of photosynthesis decreases. It may be because (i) Oxygen takes part in oxidation of photosynthetic pigments, intermediates and enzymes in the presnce of strong light (photo-oxidation), (ii) Oxygen is a strong quencher of excited state of chlorophyll. Oxygen competes with CO_2 for reducing power. It converts RuBP-carboxylase to RuBP-oxygenase. At very high oxygen concentration, the rate of photosynthesis begins to decline in all plants. This phenomenon is referred to as Warburg effect.
31. (c) PS I is involved in both cyclic and non-cyclic photophosphorylation. PS II is involved only in non-cyclic photophosphorylation. PS II is present in the appressed (linner) part of grana thylakoids. PS I is located in the non-appresed (outer) part of grana thylakoids as well as stroma thylakoids.
32. (d) In cyclic photophosphorylation, only PS–I is involved.
33. (a) In Calvin cycle, three molecules of CO_2 will require 3 x 3 = 9 ATP and 2 x 3 = 6 $NADPH_2$.
34. (c) In photosynthesis, light energy trapped by chlorophyll is used to excite electrons in the chlorophyll. (An electron extracted from water by photolysis fills the electron hole.) The excited electrons are then transferred through the electron transport chain in the thylakoid membrane to $NADP^+$, forming NADPH in the stroma.
35. (b) The key compound of C-3 cycle or Calvin cycle is 3-PGAL as it is starting point for many other metabolic pathways in the plant.
36. (d) Photosynthesis is the process by which plants manufacture food. This occurs when CO_2 combines with ribulose 1,5-bisphosphate (RuBP) to form the products of photosynthesis.
37. (d) 38. (d) 39. (c) 40. (b) 41. (b)
42. (a) 43. (c) 44. (a) 45. (a)

DPP /CB14

1. (b) During aerobic respiration, oxygen is consumed and CO_2 is released. R.Q is defined as ratio of volume of CO_2 released and O_2 consumed by the respiratory substrate.

$$R.Q. = \frac{\text{Volume of } CO_2}{\text{Volume of } O_2}$$

2. (b) As a result of glycolysis, one molecule of glucose is broken down to two molecules of pyruvate. During glycolysis, 4 molecules of ATP are produced but 2 molecules of ATP are consumed in the activation of glucose molecule. Therefore, there is a net gain of 2 molecules of ATP.
3. (d) The end product of glycolysis is pyruvate. It enters mitochondria and is oxidatively decarboxylated to acetyl CoA before entering into Krebs cycle.
4. (c) Mesosomes are the invaginations of the plasma membrane that can form into vesicles. They are found to be present in both gram-positive and gram-negative bacteria. Mesosomes may play a role in cell wall formation during cell division and/or chromosome replication and distribution and/or electron transfer systems of respiration.
5. (d) Terminal cytochorome is cyt a_3. cyt a_3 posseses two copper centers. It helps in transfer of electrons to oxygen.
6. (c)
7. (d) Glycolysis is a reductive process in which glucose is broken down in the absence of oxygen within the cytoplasm of the cell. It forms 2 pyruvate molecules, 2 net ATP molecules, 2 $NADH_2$ molecules.
8. (d) In Kreb's cycle, the FAD participates as electron acceptor during the conversion of succinic acid to fumaric acid.
9. (d) 10. (a) 11. (a) 12. (a) 13. (d)
14. (d) 15. (d) 16. (b) 17. (c) 18. (b)
19. (b) 20. (a) 21. (b) 22. (b) 23. (d)
24. (c) Muscle fatigue is the reduction in force of contraction of a muscle after prolonged stimulation. In the absence of oxygen, skeletal muscle of human beings can contract for a short time, but it gets fatigued soon. This is due to the fact that in the absence of oxygen, products of glycolysis mainly lactic acid is not disposed off and accumulates in the muscles. This leads to muscle fatigue and pain in the muscles. A muscle gets fatigued sooner after a strenuous exercise than after a mild exercise. Faster breathing for sometime after a strenuous exercise supplies extra oxygen, disposes off excess lactic acid and muscle fatigue disappears.
25. (b)

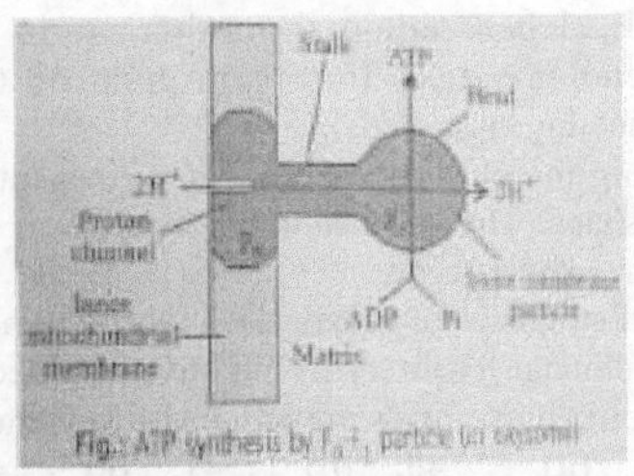

Fig.: ATP synthesis by F_0-F_1 particle (in oxysome)

26. **(a)** Under normal circumstances, the cell must have sodium pumps (Na^+/K^+ ATPase) which actively transport ATP into the cell. When KCN is added, the carriers are destroyed, thus ATP can enter freely.

27. **(c)** Since oxygen will not participate, the Pasteur effect will not take place.

28. **(d)** During Krebs cycle fumarate is formed from succinate and hence, synthesis of fumarate is blocked.

29. **(a)** Substrate-level-phosphorylation means synthesis of ATP by utilizing energy released directly by the substrates. In glycolysis, there are two steps in which ATP is synthesized. In Krebs cycle, there is only one step where GTP is synthesized when succinyl CoA is converted to succinic acid.

30. **(c)** Mitochondria oxidize not only carbohy drates but also the products of fats i.e. fatty-acids through acetyl CoA and glycerol through phosphoglyceraldehyde.

31. **(b)** Yeast causes alcoholic fermentation without using oxygen and hence depletion of oxygen has no effect on the metabolism or growth of yeast cells.

32. **(c)** When oxygen is available $NADH_2$ produced in EMP-Krebs cycle pathway is released in the form of NAD. When there is shortage of oxygen NAD is not available and NADP takes up its place.

33. **(d)**

34. **(d)** ATP molecules from ADP are generated maximum in electron transport chain.

35. **(b)** 12 kcal of energy present in one molecule of ATP & on oxidation of one mole of glucose into CO_2 and H_2O energy released in 686 kcal. So no. of ATP which can store this energy would be = 57.1 = 57 ATPs.

36. **(c)** Succinate:ubiquinone oxidoreductase, also known in mitochondria as Complex II, provides a link between the citric acid cycle and the membrane-bound electron-transport system. The membrane extrinsic, water-soluble domain, known as succinate dehydrogenase (SDH), contains the fumarate/succinate active site with a covalently bound FAD group and three iron-sulfur clusters: $[2Fe\text{-}2S]^{2+/1+}$, $[4Fe\text{-}4S]^{2+/1+}$, and $[3Fe\text{-}4S]^{1+/0}$. The enzyme catalyzes the interconversion of fumarate and succinate, and is closely related to fumarate reductase.

37. **(d)** Glycolysis takes place in all body cells and is of two types-
(a) Anaerobic glycolysis-From glycogen or glucose to lactic acid in muscles.
(b) Aerobic glycolysis- From glycogen or glucose to pyruvic acid (all cells of body).

38. **(d)**

39. **(b)** One of the functions of the electron transport chain and the conversion of pyruvate to lactate is the regeneration of NAD from reduced NAD. NAD is needed in glycolysis.

40. **(d)** ATP formation occurs both at 2, where glyceraldehyde-3-phosphate is converted to pyruvate; and at 5, where the six-carbon compound, citrate, releases water and CO_2 in the Krebs cycle.

41. **(a)** $FADH_2$ is produced during the formation of fumaric acid from succinic acid in the kreb's cycle of aerobic respiration.

42. **(d)** The oxygen obtained from cellular respiration combines with the hydrogen obtained from the oxidation of organic molecules to form water.

43. **(d)** Fats make more electrons available to the electron transport system and more hydrogen ions available for chemiosmosis.

44. **(c)**

45. **(c)** Primary respiratory substrate is carbohydrate and secondary respiratory substrate is fat.

DPP /CB15

1. **(d)** Long day plants undergo flowering after receiving light above a critical day length. *e.g.* spinach, sugarbeet. Short day plants flower only when they receive light below a critical day length, *e.g.* *Glycine max,* tobacco.

2. **(d)** The hormone, which controls cell division and cell differentiation, is cytokinin. Both the auxin and gibberellins cause elongation of cell and plants.

3. **(d)** Long-day plants usually flower in the spring or early summer; they flower only if the light periods are longer than a critical length, which is usually 9-16 hours. For example, wheat plants flower only when light periods exceed fourteen hours.

4. **(c)**

5. **(c)** Abscisic acid (ABA), also known as abscisin II and dormin, is a plant hormone. It functions as many plant developmental processes, including bud dormancy. Abscisic acid is a derivative of carotenoids. It was called "abscisin II" originally because it was thought to play a major role in abscission of fruits. At about the same time another group was calling it "dormin" because they thought it had a major role in bud dormancy. The name abscisic acid (ABA) was coined by a compromise between the two groups.

6. **(c)** Differentiation of root is controlled by high auxin concentration. in tissue culture auxin concentration is made high to promote rooting.

7. **(c)** **8.** **(c)** **9.** **(b)** **10.** **(a)** **11.** **(c)**

12. **(b)**

13. **(d)** Parthenocarpy refers the development of fruits without fertilization. Such fruits are seedless but otherwise normal in appearance, e.g. banana, pineapple etc. The auxins (IAA, IBA, 2, 4-D) are applied in low concentration in a lanolin paste to the stigma of flower to induce parthenocarpy.

14. **(c)** **15.** **(a)**

16. **(b)** Cytokinin are usually amino purine so they resemble nucleic acids in structural aspects.

17. **(a)** Natural auxin are naturally occurring auxin plant hormones which are called phytohormones. Eg. IAA and IBB. These are synthesized in shoot apices, leaf primordial and developing seeds from tryptophan. Some of the important synthetic auxins are 2, 4-D; 2, 4, 5-T (2, 4, 5- trichlorophenoxy acetic acid).

18. **(c)** 2, 4-Dichlorophenoxy acetic acid is used for causing defoliation of forest trees.
2, 4-D or 2, 4 dichlorophenoxy acetic acid is an auxin hormone. It stimulates the growth activities of the cells of the root due to which roots get destroyed and thus plants finally destroys. 2, 4-D is used as a defoliant for broad leaved dicots.

19. **(d)** **20.** **(a)** **21.** **(c)** **22.** **(d)** **23.** **(b)**

24. **(c)** Geometric growth is quite common in unicellular organisms when growth is nutrient rich medium. Here, every cell divides. The daughter cells grow and divide, the grand daughter cells repeat the process and so on. Number of cells in initially small so that initial growth is slow. Later on, there is rapid growth at exponential rate (log or exponential growth).

Sigmoid Growth Curve : Geometric growth cannot be sustained for long. Limited nutrient availability slows down the growth. It leads to stationary phase. Plotting the growth against time gives a typical sigmoid or S-curve. The exponential growth can be expressed as
$W1 = W_0e^{rt}$
where W_1 = final size (weight, height, number etc.)
W_0 = initial size at the beginning of the period
r = growth rate
t = time of growth
e = base of natural logarithms

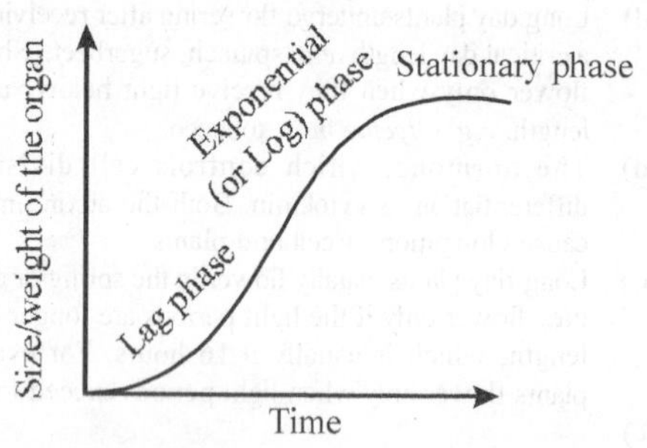

25. (a) Low temperature required for vernalization is usually 0—5°C. In nature, the plants requiring vernalization are commonly biennials (e.g., cabbage, sugarbeet, carrot), which complete their life cycle in two years. They germinate and grow vegetatively in the first year and produce flowers in the second year of growth. These plants fulfill their cold requirement during winters. However, such biennial plants can be made to flower in one growing season by providing low temperature treatment (i.e., 0—5°C temperature) to young plants or moistened seeds.

26. (c) The pulse of red light converts P_r to P_{fr}. Subsequent pulses of far-red light stimulate changes that lead to chlorophyll synthesis.

27. (d) Mechanical abrasion, leaching of inhibitors by water, and exposure to fire may all trigger germination, but actual germination cannot begin until the seed imbibes water.

28. (b) Ethylene gas promotes senescence and is one of the by products of burning your gas grill. You should move your grill or your camellia bush.

29. (b) Less of chlorophyll and degeneration of proteins are the two important symptoms of senescence. Cytokinins delay these processes and, thus the senescence is also delayed. This effect of cytokinin is known as Richmond-Lang effect.

30. (d) Receptors receive environmental cues and trigger hormone release or a signal transduction pathway that involves hormones. The signal alters gene expression by altering genes which are transcribed and translated. Cellular response depends on expression of those genes.

31. (b) The ratios of cytokinins to auxins control cell differentiation in plant tissue culture. When both are present in relatively equal quantities, cells divide but do not differentiate. If there is more of cytokinin than auxin, shoot buds develop from a callus. If there is relatively more auxin than cytokinins, roots develop. Thus proportion of these two hormones control organ formation in callus tissues. Abscissic acid (ABA) is a naturally occurring growth inhibitor.

32. (c) Cryptochromes respond to blue and ultraviolet light.

33. (d) The curvature induced in plant organs in response to the unidirectional light is called photoropism.

34. (a) The first naturally occuring auxin was isolated by Kogl and Haagen-Smit (1931) from human urine. It was identified as auxin-a (auxentriolic acid). Later, in 1934 Kogl, Haagen-Smit and Erxleben obtained another auxin called auxin-b (auxenolonic acid) from corn germ oil (extracted from germinating corn seeds).

35. (b) Altering the plasticity allows for permanent changes in cell wall shape. The cell wall must increase in size in order for cell growth to occur.

36. (b) Dwarf strains were genetically unable to produce gibberellin, so additional fertilizers would have no effect.

37. (a) Paratonic movements are induced by external stimuli.

38. (d) 39. (a) 40. (d) 41. (c) 42. (a)

43. (c) Prunning of plants promotes branching, because the axillary buds get sensitized to cytokinin.

44. (c) 45. (d)

DPP /CB16

1. (b) Spleen (also known as blood bank and graveyard of RBCs) is a dark, purple gland present in the left side of abdomen against the stomach. It is internally made up of white pulp and red pulp. Its main functions are storage of blood and break down of old and senescent RBCs.

2. (a) Vitamins are accessory food factors required in small amount for growth and metabolism. Vitamins are either water soluble → B. complex and vit.C or fat soluble → vit. A, D and K.

3. (c) 4. (d)

5. (b) Premolar and molar are lophodont teeths. Lophodont teeth with the cusps elongated to form narrow ridges. The molars in elephants and horses have cusps fused by means of intermediate masses of dentine to form ridges or lophs.

6. (d)

7. (c) Continued consumption of fat rich diet causes hypercholesterolemia. Hypercholesterolemia is the presence of high levels of cholesterol in the blood. High cholesterol raises risk for heart disease, heart attack, and stroke. Kidney stones are solid mass made up of tiny crystals. There are different types of kidney stones. The exact cause depends on the type of stone like, calcium stones, uric acid stone *etc.* Vitamin A toxicity or hypervitaminosis A is having too much of vitamin A in the body. Ketonuria is condition in which ketone bodies are present in urine. Body productes excess ketone bodies as an alternate source of energy during starvation or diabetes mellitus (type 1).

8. (a) Fat is mainly digested by pancreatic lipase while protein is digested by enzymes in pancreatic juice, intestinal juice and stomach.

9. (c) Beri-beri is a deficiency disease caused by the lack of vitamin B_1 (Thiamine) in the diet. The deficiency of riboflavin (Vit.B_2) causes ariboflavinosis. Scurvy is the result of lack of vitamin C in the diet. Deficiency of calciferol (vitamin D) in the diet causes rickets in children and osteomalacia in adults.

10. (a) Oxyntic cells or Parietal cells, are the stomach epithelium cells that secrete gastric acid and intrinsic factor. These cells secrete hydrochloric acid (HCl) which makes the gastric juice acidic (pH = 2.0-3.0).
Alpha cells of islets of Langerhans secretes glucagon hormone which increase the glucose level in the blood by converting glycogen to glucose in liver cells. Kupffer's cells are specialized cells in the liver that destroy bacteria, foreign proteins, and worn-out blood cells. Sebaceous glands and microscopic glands in the skin that secrete an oily/ waxy matter (called sebum) to lubricate the skin and hair of mammals.

11. (c) Kwashiorkor and Beri-Beri are deficiency diseases which occur mostly in children. Kwashiorkar occurs due to deficiency of protein and Beri-Beri due to deficiency of vit-B_1 (Thiamine).

12. (d) Colostrum (also known as beesting or first milk) is a form of milk produced by the mammary glands in late pregnancy and the few days after giving birth. Human and bovine colostrums are thick, sticky and yellowish. In humans, it has high concentrations of nutrients and antibodies, but it is small in quantity.

Colostrum is high in carbohydrates, high in protein, high in antibodies, and low in fat (as human new borns may find fat difficult to digest). Newborns have very small digestive systems, and colostrum delivers its nutrients in a very concentrated low-volume form. It has a mild laxative effect, encouraging the passing of the baby's first stool, which is called meconium. This clears excess bilirubin, a waste product of dead red blood cells which is produced in large quantities at birth due to blood volume reduction, from the infant's body and helps to prevent jaundice.

13. (b) Long term intake of alcohol causes damage to liver which is known as cirrhosis of liver with continued alcohol intake, there is destruction of hepatocytes and fibroblasts (cell which form fibres) and stimulates of collagen protein formation.

14. (d) **15. (d)** **16. (b)** **17. (b)** **18. (d)** **19. (a)**
20. (d) **21. (b)**

22. (a) The chylomicrons are formed inside enterocytes and are absorbed in lacteals.

23. (a) The caecum is a pouch-like portion of the large intestine which hosts some symbiotic micro-organism. The caecum absorbs water and salts from undigested foods before they continue on to the large intestine.

24. (b) **25. (a)**

26. (d) In the small intestine, pancreatic juice containing an enzyme, α-amylase acts on starch and converts it into dissacharide, maltose and isomaltose and 'limit' dextrins.

Starch Pancreatic . . . . . 'Limit' dextrins

27. (a) Fructose is generally absorbed by facilitated transport. The digestive wastes, solidified into coherent faeces in the rectum initiate a neural reflex causing an urge or desire for its removal.

28. (c) Many of the water-soluble vitamins play a vital role in enzyme function.

29. (d) The pancreas produces a mumber of important digestive enzymes, without which digestion and nutrient absorption are greatly hampered.

30. (c) A gastrovascular cavity has a single opening.

31. (d) Enzymes that digest proteins, lipids, and carbohydrates all function in the small intestine.

32. (c) Essential amino acids must be acquired through diet because an animal cannot directly synthesize all of the amino acids needed for protein

33. (d) The stomach is protected from digestive enzymes and low pH by the neutralizing, buffering, and coating mucus secreted over its inner surface.

34. (a) The small intestines are the site for the majority of digestion in humans.

35. (c) The large intestine is the site of water and ion absorption. The large population of bacteria in the large intestine contributes vitamins that are useful to the host.

36. (b) The membranes of the digestive tract are, from the inside to the outside: mucosa, submucosa, circular and longitudinal muscles, serosa.

37. (c) the stomach stores food (and performs some digestion too) before passing it on to the intestines. The small intestine (midgut) finishes tint digestion and carries out most of the nutrient absorption, while the large intestine (hindgut) reabsorbs water and ions.

38. (d) Insulin increases the rate of glucose uptake from the blood into the muscle cells by increasing the number of glucose transporters in the plasma membrane.

39. (a) The liver produces bile salts and add to the bile pigments bilirubin from the breakdown of red blood cells. The bile pigments are purely excretory.

40. (c)

41. (d) Generally, fatty acids upto a chain length of 10 carbon atoms are primarily absorbed through the blood caillaries, but those with higher chain length through lymphatic route (lymph vessels).

42. (c) Mammals may drink water and also get it from oxidation of glucose.

43. (b) Because saliva converts starch into maltose.

44. (a) Anxiety and eating spicy food together in normal healthy man can lead to indigestion which is difficulty in digestion.

45. (a) Young infant may be feeding entirely on mother's milk which is white in colour but the stools which the infant passes out is quite yellowish because bile pigments passed through bile juice. Bile pigments are any of several coloured compounds derived from porphyrin that are found in bile; principally bilirubin and biliverdin. Bile pigment is produced regularly when old red blood cells are broken down, mainly by the spleen. In some blood-disorders where the red cells are destroyed, more bile pigment is produced.

DPP /CB17

1. (a) 70% to 75% CO_2 is transported as primary buffer of the blood bicarbonate ion (HCO_3) in blood plasma. When CO_2 diffuses from tissues into blood then it is acted upon by the enzyme carbonic anhydrase.

2. (a) A large portion of oxygen is left unused in the human blood even after its uptake by the body tissues. This O_2 acts as a reserve during muscular exercise.

3. (a) It is the relationship between the percentage saturation of haemoglobin (by volume) in the blood and the oxygen tension (in partial tension) pO_2 of the blood.

It is usually a sigmoid plot. Haemoglobin molecules can bind up to four oxygen molecules in a reversible way. The shape of the curve results from the interaction of bound oxygen molecules with incoming molecules. The binding of the first molecule is difficult. However, this facilitates the binding of the second and third molecules, and it is only when the fourth molecule is to be bound that the difficulty increases, partly as a result of crowding of the haemoglobin molecule, partly as a natural tendency of oxygen to dissociate.

4. (d) Brain is the most vital organ. It stops functioning in the absence of O_2.

5. (d) When the CO_2 concentration in blood increases, breathing becomes faster and deeper. As CO_2 levels increase, patients exhibit a reduction in overall level of consciousness as well as respiratory effort. Severe increase in CO_2 levels can lead to respiratory arrest.

6. (c) CO_2 enters RBC and reacts with water to form carbonic acid. Carbonic acid dissociates to form bicarbonate and hydrogen ions. Some bicarbonate ions are transported in erythrocytes while some diffuse into the blood plasma. Exit of bicarbonate ions change the ionic balance between the plasma and erythrocytes. To restore this balance chloride ions diffuse from plasma into erythrocytes. Due to this, the pH of blood is maintained.

7. (a)
8. (c) The maximum volume of air a person can breath in after a forced expiration or the maximum volume of air a person can breathe out after a forced inspiration is called vital capacity (VC). This includes tidal volume, inspiratory reserve volume and expiratory reserve volume. (TV + IRV + ERV). VC varies from 3400 ml – 4800 mL, depending upon age, sex and height of the individual.
9. (b) 10. (d)
11. (c) Emphysema is lung condition that causes shortness of breath due to damage of air sacs in the lungs (alveoli). The main cause of emphysema is a long-term exposure to air-borne irritants including tobacco smoke, air pollution, dust.
12. (b) 13. (d) 14. (b) 15. (d) 16. (d)
17. (a) 18. (a)
19. (c) When the external intercostal muscles and diaphragm relax, the ribs move downward and inward and diaphragm becomes convex (dome shaped), thus decreasing the volume of thoracic cavity and increasing the pressure inside as compared to the atmospheric pressure outside. This will cause the air to move out (expiration).
20. (d) Pulmonary, CO_2 is found to be lower than cutaneous O_2 at all temperatures. Statement A is a wrong answer. This is because at 5°C, pulmonary O_2 is lower than cutaneous O_2, although as the temperature increases, the reverse becomes true.
21. (c) Myoglobin has an affinity for oxygen. It is present in the muscles. As oxygen levels reduce, myoglobin would release the stored oxygen to keep the muscles working.
22. (a) This decreases the size of the thoracic cavity, which increases the pressure within the thoracic cavity, forcing air out of the lungs.
23. (d) Debris normally gets trapped in mucus lining the respiratory passage ways. This mucus is then swept up and away from the lungs by the action of cilia.
24. (d) Myoglobin has the greatest affinity for oxygen and does not unload the oxygen until very low partial pressures of oxygen (below 20 mmHg). Foetal haemoglobin has a higher affinity for oxygen than adult haemoglobin, thus facilitating the transfer of oxygen from maternal haemoglobin to fetal, haemoglobin at the placenta. When respiration is occurring a fast rate, more oxygen is needed. High levels of respiration are associated with high levels of carbon dioxide, hence the affinity of haemoglobin for oxygen decreases at higher carbon dioxide partial pressures so that oxygen can be released to respiring tissues.
25. (a) Inspiration in the mammalian lung occurs by means of negative pressure produced by contraction of the diaphragm. Thus, the pressure in the lungs falls below atmospheric pressure.
26. (d) The Bohr effect describes the action of pH on the oxygen-binding curve. Decreases in pH result in a net right shift of the entire oxygen equilibrium curve.
27. (d) Movement of O_2 and CO_2 from the blood to the tissues always occurs by diffusion of O_2 and CO_2 down their partial pressure gradients.
28. (d) This mechanism serves a very important cleansing role for the mammalian respiratory system.
29. (d) A left shift in the oxygen-binding curve means that the pigment operates at lower PO_2 values.
30. (d) Actively respiring tissues require more oxygen. They are also associated with high partial pressures of carbon dioxide since it is a product of respiration. The curves show that oxygen would be more readily released at higher partial pressures of carbon dioxide, i.e. at active tissues.
31. (c) When the chest expands, stretch receptors in the lungs reflexly excite the expiratory center in the medulla which tends to produce expiration. Partial exhalation lessens the intensity of this reflex, allowing breath to be held for a longer time. Exhalation does not change the blood CO_2 concentration unless fresh air (low in CO_2) is brought into the lungs, allowing more CO_2 to diffuse out of the blood.
32. (b) Thin layer of water on the inner surface of alveoli serves to contract the alveoli and makes them behave like unfilled balloons that resist inflation. This would mean tremendous effort to inhale and expand the alveoli and therefore to overcome this, cells of the alveoli secrete pulmonary surfactant to decrease the surface tension of water lining the alveoli.
33. (d) Gas is brought into the lungs when the negative pressure exists between the lungs and the rest of the chest. This negative intrapleural pressure is aided to occur by all the choices mentioned.
34. (c) 35. (d)
36. (c) Changes in pH due to the accumulation of acidic metabolites resulting from increased cell activity are the basis of the Bohr effect.
37. (d)
38. (d) The hyoid is a bone of the skeletal system that supports the larynx and tongue.
39. (c) The alveoli do not contain "air" with 20.9% O_2 because incoming air is mixed with air left in the "dead space" of the trachea and bronchi that has had some of the O_2 removed by the lungs.
40. (c) Water breathing is more difficult than air breathing because the higher density of water makes it more expensive to move across the respiratory surfaces than air, it has less oxygen than air, and the oxygen content is very dependent on the temperature of the water. CO_2 however, dissolves easily into water and is easy to get rid of.
41. (d) All of the listed choices have larger volume than the resting tidal volume.
42. (a) Myoglobin has only one haem group and there can be no co-operative binding, hence the shape of the dissociation curve is hyperbolic rather than sigmoidal. Also, myoglobin has a very high affinity for oxygen and does not release oxygen until a very low oxygen tension.
43. (b)
44. (a) During the transport of CO_2 through the blood, bicarbonate ions diffuse out of RBCs while chloride ions from plasma enter the RBCs to maintain ionic equilibrium. This is called chloride shift.
45. (b) The total volume of air that can be expelled from the lungs after maximum inspiration and then expiring to the maximum is known as the vital capacity. The volume of air that remains inside lungs at the end of maximum forceful expiration is the residual volume. Expiratory reserve volume is the maximum extra volume of air that can be expired by forceful expiration after a normal tidal expiration. Total lung capacity is the maximum volume of air that can be contained in the lungs after maximum inspiration.

DPP/CB18

1. (a)
2. (c)
3. (a) Lubb sound is caused partly by the closure of the bicuspid and tricuspid valves and partly by the contraction of the muscles in the ventricles. Lubb is the first heart sound.
4. (b) Rh factor was discovered by Karl Landsteiner. A child of Rh^+ man will be Rh^+ whether the mother is Rh^+ or Rh^-. If the mother is Rh^+ then there will be no problem but if mother is Rh^- so when the blood of Rh^+ child (in womb)

mixes with the blood of Rh^- mother then some antibodies in mother's blood are formed against Rh^+ factor which coagulate the womb blood causing death. If birth takes place then there is a possibility of child death in early years. This in known as erythroblastosis foetalis. In most cases the Ist pregnancy may succeed but after that it fails.

5. (c)
6. (c) Peak P-causes diastolic phase in ventricle while R-Peak causes systole in ventricle means diastolic and systolic phases represented by P & R.
7. (c) Sino-Auricular node (SA node) present in the walls of right auricle has a myogenic initiation of heartbeat in a regular fashion and controls the pace of heartbeat called pacemaker.
8. (c)
9. (d) Haemoglobin has 250 times more affinity for CO as compared to oxygen. Hb readily combines with CO, forming carboxyhaemoglobin (COHb), COHb interferes with the transport of oxygen.
10. (c) An adult healthy human has average systolic and diastolic blood pressure as given below :

$$\text{B.P.} = \frac{\text{Systolic}}{\text{Diastolic}} = \frac{120 \text{ mm Hg}}{80 \text{ mm Hg}}$$

11. (b) Blood group A – Antigen-A & Antibody 'b' or Anti A or a.
Blood group B – Antigen-B & antibody a or Anti B or b.
Blood group AB – Antigen-A & B and no antibody.
Blood group O – No antigen & Antibodies a & b both.
12. (d) The number of blood platelets per cubic mm in human blood is 3 lacs while WBCs are 5000/cubic mm of blood. Veins are as complex as the arteries. Veins and arteries both are types of blood vessels. Arteries carry blood from heart to different organs while vein carries blood from different organs to heart. At any given time in a healthy human, the blood amount is same in both, as the circulation of blood never stops.
Blood consists of two parts:
The **plasma** (water, proteins, inorganic salts and other elements) constitutes 55-60% of blood while **cellular** part constitutes 40-45% of total blood. WBC are produced in red bone marrow, lymph nodes and sometimes even in liver and spleen.
13. (a) High proportion of cholesterol in blood leads to deposition of cholesterol on the walls of blood vessels. This causes the arteries to lose their elasticity and get stiffened. This is called arteriosclerosis or hardening of arteries.
14. (d)
15. (b) 'Bundle of His' are a typical cardiac muscle fibres, connecting the atria with ventricle.
16. (d) 17. (d) 18. (a) 19. (c) 20. (d)
21. (d) 22. (c) 23. (a) 24. (a) 25. (a)
26. (d) 27. (b) 28. (a)
29. (a) The degree of oxygenation of blood markedly affects the amount of CO_2 transported in blood. The lower the pO_2 and the haemoglobin saturation with O_2, the more the CO_2 that can be carried in the blood. This phenomenon, is called the Haldane effect. It depicts the greater ability of reduced haemoglobin to form carbaminohaemoglobin and to buffer H^+ by combining with it. In the pulmonary circulation, uptake of O_2 facilitates the release of CO_2. As haemoglobin becomes saturated with O_2, the hydrogen ions released combine with HCO_3^-, helping to unload CO_2 from the pulmonary blood. The Haldane effect is quantitatively more important in promoting O_2 transport than the Bohr effect in promoting O_2 transport. It results from the simple fact that combination of O_2 with haemoglobin causes the haemoglobin (oxyhaemoglobin) to become a stronger acid. This in turn displaces CO_2 from the blood.
30. (d) Christmas factor (factor IX) is a plasma thromboplastin component i.e. a thromboplastin activator present in blood plasma. Its deficiency causes congenital disease called hemophilia B (also called Christmas disease).
31. (c) Veins bring blood from different body parts to the heart. The flow of blood in veins is not so fast because the blood in veins is under low pressure. Veins possess valves which prevent backward flow of blood.
32. (d) Cardiac output = Stroke volume × heart beats per minute

$$\therefore \text{ Stroke volume} = \frac{\text{Cardiac output}}{\text{Heart beats per minute}}$$

$$= \frac{5250}{75} = 70 \text{ ml}$$

33. (a) Heart valves only permit unidirectional flow of blood.
34. (b)
35. (c) The left ventricle pumps blood into the aorta.
36. (a) Contraction of the ventricle produces pressure in the human circulatory system. Blood leaves the ventricles and then immediately enters the arteries moving away from the heart. Blood pressure will have its highest recording in these arteries.
37. (c) The small diameter of capillaries offers great resistance to blood flow. This slows the blood to a speed at which nutrient and gas exchange with neighboring cells is more efficient.
38. (b) The atrium has thinner walls and generates lower pressures than the ventricles.
39. (a)
40. (c) The left ventricle generates a greater pressure in the blood flowing to the systemic circuit than the right ventricle with blood flowing to the pulmonary circuit.
41. (c) The AV node delays the ventricular depolarization relative to atrial depolarization, so atrial contraction occurs before ventricular contraction. The Purkinje fibers transmit the cardiac impulse to very small localized groups of ventricular fibers.
42. (d)
43. (a) Presence of calcium will remove heparin - blood anticoagulant and will promote blood clotting. Sodium oxalate and heparin containing test tubes will not allow the blood to clot.
44. (a) Blood serum is liquid minus clotting elements of pale yellow colour. It does not have fibrinogen and other clotting materials. It does not take part in blood clotting.
45. (a) Folic Acid (Cyanocobalamine vit. B_{12}) works in the formation and maturation of RBCs. In the deficiency of this RBCs formation decreases and the formed RBCs will not mature i.e. they will not loose the nucleus and hence remain nucleated and lack haemoglobin. Such RBCs cannot carry oxygen & person suffers from anaemia. Supplementing his diet with folic acid and cobalamine will lead to the increased formation and proper maturation of enucleated RBCs with haemoglobin and eliminate anaemia.

DPP /CB19

1. (c) Liver is the primary site of detoxification and elimination of body wastes and poisons. Liver detoxifies endotoxins, *e.g.* toxic NH_3 combined with CO_2 to form less toxic urea. It also detoxifies alcohol and convert them to acetaldehyde and then harmless acetyl CoA.
2. (d) Renin (also called rennet or chymosin) is an coagulating enzyme produced from stomach of human body. It catalyzes the coagulation of milk by converting milk with

soluble protein caesin into insoluble semi fluid calcium paracaesinate. This is called curdling of milk. Rennin produced in the infants immediately after birth. As the child grows, rennin production goes down and is replaced by pepsin digestive enzymes.
Renin is an enzyme which acts as hormone secreted by juxtaglomerular cells. It converts angiotensinogen into angiotensin.

3. (b) Nephron is the excretory unit of human excretory system. Each nephron has a Bowman's Capsule, a Proximal Convoluted Tubule (PCT), Loop of Henle (Descending & Ascending limbs) and Distal Convoluted Tubule (DCT) which then enter into collecting duct.

4. (d)

5. (b) Thermoregulation is the ability of an organism to keep its body temperature within certain boundaries, even when temperature surrounding is very different. In humans, sweating is primarily a means of thermoregulation.

6. (c) Due to a long hunger strike and survival on water, a person will have less urea in his urine because urea comes to kidney as a waste product from liver which is formed after the breakdown of protein fat, carbohydrate during hunger. It is not synthesised but the synthesised ones are catabolised.

7. (c)

8. (b) It is named after its discoverer, F. G. J. Henle. In the kidney, the loop of Henle is the portion of the nephron that leads from the proximal convoluted tubule to the distal convoluted tubule. The loop has a hairpin bend in the renal medulla. The main function of this structure is to reabsorb water and ions from the urine. To do this, it uses a countercurrent multiplier mechanism in the medulla.

9. (a) **10. (d)** **11. (c)** **12. (b)** **13. (d)**

14. (a) Glomerular podocytes are highly specialized cells with a complex cytoarchitecture plays a major role in establishing the selective permeability of glomerular filtration barrier.

15. (c) Haematuria is the presence of blood cells in urine.

16. (c) **17. (c)**

18. (a) ANF stimulates the loss of sodium in urine while aldosterone absorbs sodium from glomerular filtrate.

19. (b) **20. (d)** **21 (d)** **22. (b)**

23 (a) Kangaroo rat lives in desert, therefore, it needs to conserve water in the body. In kangaroo rat, loop of Henle is significantly longer as it descends further into the medulla and produces a higher concentration gradient in the surrounding tissue so as to produce urine which is 18 times concentrated than that of their blood. This gradient allows more water to diffuse into the surrounding tissues to be reabsorbed.

24. (b) Dialysis is of two types : haemodialysis and peritoneal dialysis. Peritoneal dialysis uses a natural filter inside the body i.e. peritoneal membrane to remove wastes and extra fluid from the body. The dialysis fluid fills the belly and pulls out wastes and extra fluid from the body. Blood is not removed from the body as in the case of haemodialysis.

25. (b) If the walls of the collecting duct are water-permeable, water leaves the ducts to pass into the hyperosmotic surrounding and concentrated urine is produced. Thus, when there is insufficient ADH, less water is reabsorbed and more dilute urine is produced in copious amounts.

26. (c) Ectotherms on poikilothermic animals are cold-blooded animals whose body temperature is dependent upon their environment.

27. (c) A and B involve ultrafiltration.
In the proximal convoluted tubules, the mitochondria provide energy for active transport. The cells here are adapted for reabsorption. The amino acids diffuse into the cells and are actively transported to the intercellular spaces, where they diffuse into the surrounding capillaries.

28. (a) The removal of urea and excess ions occurs by diffusion from the blood in the abdominal capillaries to the dialysis fluid, down a concentration gradient.

29. (d) This is the correct flow of urine after it leaves the kidney.

30. (b) About 1 percent of the filtrate is excreted as urine, so about 1.5 litres of the original 150 litres would be urinated.

31. (c) Antidiuretic hormone acts on the collecting ducts by increasing the permeability to water. Antidiuretic hormone secretion is stimulated by a decrease in blood pressure.

32. (b) Reabsorption returns substances to the blood, secretion moves substances from blood into the kidney tubules.

33. (a) The route of water and solutes through the nephron from the glomerulus, to Bowman's capsule, to proximal tubule, to loop of Henle, to distal tubule, to collecting ducts.

34. (d) Glucose and amino acids found in the filtrate are reabsorbed.

35. (a) The proximal tubule is the site of active transport of Na^+ out of the tubule. Na^+ also moves out of the tubule at the ascending limb of the loop of Henle, but this is a passive transport with Cl^- being actively transported out.

36. (a) Red blood cells are too large to be filtered out of the blood at the glomerulus and thus will not be found in the filtrate.

37. (b) Glucose is 100% reabsorbed and thus a healthy person will excrete no glucose in the urine. Sodium and water are usually over 99% reabsorbed. Urea is the main excretory product formed as a result of protein breakdown.

38. (b) The sole mechanism for water reabsorption in the renal tubules is by osmosis.

39. (d) A – Adrenal gland – located at the anterior part of kidney, secrete Catecholamines which stimulate glycogen breakdown.

40. (d)

41. (c) Salamander (Amphibia; Caudata) excrete urea by help of kidneys.

42. (c) Normally albumin can't pass out through the filtering pores due to their large size but in some pathological conditions viz. anoxia or heart failure, the filtration membrane becomes damaged and filtering pores enlarge in size, so that serum albumin passes out in the largest amount and appears in the urine.

43. (b) **44. (b)** **45. (c)**

DPP /CB20

1. (c) Movement of Ca^{2+} out in sarcoplasmic reticulum controls the making and breaking of actin and myosin complex actomyosin due to which muscle contraction and relaxation takes place. Albert Szent Gyorgyi worked out biochemical events of muscle contraction.

2. (a) Elbow joint is an example of hinge joint. The elbow is a hinge joint; it can open and close like a door. Hinge joint is a form of diarthrosis (freely movable joint) that allows angular movement in one plane only, increasing or decreasing the angle between the bones e.g. elbow joint, knee joint etc.

3. (b) MRI machine does not show face image of bone and calcium, *e.g.* scapula, canine. It is also not suitable for patients with cardiac pacemakers.

4. (b) Hyoid is a horse shoe shaped bone present in neck between lower law and sound box (larynx). It is not articulated to any bone, but is simply suspended, from temporal bones by means of ligaments. Hyoid provides surface for the attachment of tongue muscles.

5. (b)

6. (d) Ball and socket joint is a type of synovial joint in which two bones are articulated. Shoulder joint is an example of

ball and socket joint in which humerus is joined with pectoral girdle.

7. (a) Actin and tropomyosin are part of thin filaments of skeletal muscle. Tubulin is presents in microtubules. Myosin is muscle protein.

8. (d)

9. (a) The pelvic girdle is formed by two innominate bones consists of three separate bones ilium, ischium and the pubis.

10. (d) The last two pairs *i.e.* 11th and 12th pairs ribs remain free anteriorly, hence, they are called as floating ribs.

11. (d)

12. (d) The sensation of fatigue in the muscles after prolonged strenuous physical work is caused by the accumulation of lactic acid.

13. (d) 14. (c) 15. (c) 16. (c) 17. (a)

18. (c) Troponin is a protein which is found on actin filament and myosin protein is found in myosin filament. Both actin and myosin are complex proteins in striated muscles. Thymosin is a hormone secreted by the thymus that stimulates development of T-cells. Prolactin is a hormone released by the pituitary gland that stimulates breast development and milk production in women. Rhodopsin, also known as visual purple, is not a hormone. It is a biological pigment in photoreceptor cells of the retina that is responsible for the first events in the perception of light.

19. (a) 20. (d) 21. (a) 22. (a) 23. (a)

24. (a) Smooth muscles are non-straited, unbranched and spindle shaped. Skeletal muscles are unbranched. Cardiac muscles fibres are uni-nucleated.

25. (d) Tarsals, femur, metatarsals and tibia are boned of the legs which are involved in running during chasing the ball by cricket player.

26. (b) Bone is made up of 60&70% inorganic matter and 30&40% organic matter.

27. (a) Hyaline cartilage is most abundant cartilage in body.

28. (b) The ER releases calcium ions only if stimulated by an action potential, which requires acetylcholine to diffuse across the neuromuscular junction.

29. (c) Fine motor control is accomplished by the presence of smaller, more numerous motor units. Each motor unit requires an individual motor neuron.

30. (c) Smooth muscle is found mainly in the walls of hollow organs, such as digestive tract organs and blood vessels. Smooth muscles propel substances through the hollow organ by alternately contracting and relaxing.

31. (b) A hydrostatic skeleton consists of fluid held under pressure in a closed body compartment.

32. (a) Mandible is a large bone constituting the lower jaw.

33. (b) Stapes is one of the three ear ossicles in the middle ear resembling a tiny stirrup. It transmits sound vibrations from the incus to the internal ear.

34. (b) ATP provides energy that is used to detach myosin from actin.

35. (b) Endoskeletons (such as those of mammals) are found inside the body, and exoskeletons (such as those of insects) are found outside the body.

36. (c) In gout, there occurs a defect in uric acid metabolism resulting into its elevated level in blood (hyperuricemia). This is followed by precipitation of excessive uric acid which gets deposited in the joint spaces. These deposited crystals of uric acid causes pain in different bony joints.

37. (b) 38. (c) 39. (d)

40. (a) Glenoid cavity is a shallow concavity on the lateral side of pectoral girdle in which the head of humerus fits making the shoulder joint.

41. (c) 42. (c) 43. (b) 44. (b) 45. (a)

DPP /CB21

1. (b) Afferent nerve fibres – sensory nerve fibres which carry senses from receptors to brain or spinal cord (CNS). Efferent nerve fibres – motor nerve fibres that carry orders or responses from CNS to effectors (muscles and glands).

2. (a) Hypothalamus contains important nerve centres that controls the body temperature, thirst, hunger, eating, water balance and sexual function.

3. (c) The rods and cones are exceptionally specialized bipolar neurons which have developed some structural features to carry out the transduction of light energy into neuronal signals.

4. (c) The transmission of nerve impulse through synapse requires a chemical neuro transmitter. The most common neurotransmitter is acetylcholine.

5. (b) Alzheimer disease in humans is associated with the deficiency of acetylcholine. Acetylcholine is the neurotransmitter produced by neurons (referred to as cholinergic neurons). Acetylcholine plays a role in skeletal muscle movement, as well as in the regulation of smooth muscle and cardiac muscle. Acetylcholine is synthesized from choline and acetyl coenzyme-A through the action of the enzyme choline acetyltransferase and becomes packaged into membrane-bound vesicles.

6. (c) Conditioned reflexes are acquired reflexes and are dependent on past experiences, training and learning. I.R. Pavlov demonstrated conditioned reflexes in a hungry dog. He called food and salivation in response to it as unconditioned stimulus and sound of bell and salivation in response to bell as conditioned reflexes.

7. (a) The inner layer of the posterior two-thirds of the eyeball consists of a light sensitive layer, called retina that possesses two types of photoreceptors called the rods and the cone cells. Retina reduces the internal reflection, so any damage to it leads to greater internal reflection of light often causing an increase in light sensitivity.

8. (c) Epinephrine or adrenaline, norepinephrine or noradrenaline and acetylchloline are the neurotransmitters. These are released by the nerve fibres to transmit the impulse to the next neurone. Cortisone is not a neurotransmitter.

9. (d) Total sum of physio-electrochemical changes that takes place along the length of nerve fibre is known as nerve impulse. Change in potential due to stimulation of nerve fibre is called action potential. During propagation of nerve impulse, Na^+ enters inside so (+ve) change is formed inside the membrane. K^+ ions come out.

10. (c) 11. (c) 12. (b)

13. (c) The intensity of movement of basilar fibres regulates the loudness or amplitude of sound.

14. (d) 15. (a) 16. (b)

17. (a) Epinephrine and norepinephrine are secreted by adrenalmedulla in response to stress of any kind and during emergency situations and are called emergency hormones or hormones of flight, or fight.

18. (c) 19. (a) 20. (d) 21. (d) 22. (b)

23. (a)

24. (a) During stress condition, stimulation of the sympathetic nerves to adrenal medulla causes large quantities of adrenaline to be released into the blood circulation and then this hormone is carried to the specific tissues of the body where it produces its effect e.g., increase in heart beat.

25. (b) Tectorial membrane is a gelatinous membrane covering the sensory hair cells in the scala media of cochlea.

26. (b) Once differentiated, nerve cells do not divide as they lack mitotic activity. Lack of mitotic activity indicates that they do not have centrosome.
27. (d) Reflex arcs allow for rapid processing and response.
28. (d) These drugs alter perception of stimuli and work mainly at the axon or the synapse
29. (b) An action potential is a localized electrical event—a membrane depolarization at a specific point of stimulation.
30. (d) An influx of sodium depolarizes the membrane and opens more sodium channels, causing still more depolarization, an example of positive feedback.
31. (c) The embryonic forebrain gives rise to the telencephalon and the diencephalon. The thalamus and the hypothalamus develop from the diencephalon; the cerebral cortex develops from the telencephalon.
32. (c) Since the action potential started in the middle of the axon, rather than at the axon hillock as typically occurs, there is no previous patch of membrane currently undergoing a refractory period. The action potential is thus free to propagate in both directions.
33. (b) Hyperpolarizing neuron B by 15 mV would make its membrane potential –65mV, actually taking it further away from the threshold voltage needed to fire an action potential.
34. (d) The medulla oblongata contains centers that control several visceral functions, such as breathing, heart and blood vessel activity, swallowing, vomiting, and digestion.
35. (d) For an action potential to occur in an axon, the membrane must be depolarized above a certain level. This level is known as the threshold.
36. (d) Electrical synapses join two cells with protein tunnels known as connexons. The junctions that are formed provide for very fast transmission between cells.
37. (c) Atropine is a substance that blocks the inhibitory effects of muscarinic receptors in muscle tissue, especially the heart.
38. (b)
39. (c) Arbor vitae is composed of white matter.
40. (a)
41. (c) The white matter in the CNS consists of tracts that convey sensations from one structure or region to another.
42. (c)
43. (c) The olfactory epithelium lining the superior border of the nasal cavity is innervated by the olfactory (first cranial) nerve.
44. (a) These are called the limits of audibility. Actual value is 20 to 20000 cycles/sec.
45. (c)

DPP /CB22

1. (a) Cortisol is secreted by the middle region of adrenal cortex. It increases the blood glucose level (which is anti-insulin effect) by converting proteins & fats into glucose.
2. (b) Pancreas is a gland which is both exocrine and endocrine. Cells of acini are exocrine which secrete enzymes and islet of Langerhans part is endocrine which secretes hormones like insulin, glucagon, *etc.*
3. (b) Hormones are chemical messengers formed by endocrine cells. Chemically hormones are of the following types: Amines– composed of amino group e.g., Melatonin.
   Amino acids – eg. thyroxine
   Peptides –eg. insulin
   Steroids –eg. aldosterone
4. (a) Parathamone is a peptide hormone secreted by the parathyroid gland in response to low levels of calcium in the blood. It acts to maintain normal blood levels of calcium by increasing the number of osteoclasts, which break down the bone matrix and release calcium into the blood. It also increases the reabsorption of calcium and magnesium ions in the kidney tubules, so that their concentration is maintained in the blood.
5. (d) Electroencephalogram is the recording of electric potentials originating from different parts of the brain in the form of waves. Fractionation is the technique of rupturing cells and separating their components especially cell organelles for studying their chemistry and functions. Electrophoresis is the technique of separation of charged solutes on the basis of their differential migration in an applied electric field.
6. (b)
7. (a) Luteinizing hormone (LH) stimulates ovulation. Deficiency of insulin causes diabetes mellitus. Deficiency of ADH or vasopressin causes diabetes insipidus. Deficiency of parathormone causes tetany. Deficiency of thyroxine causes cretinism in infants and myxoedema in adults.
8. (b) In heart cells AMP acts as secondary messenger which is made in the response of adrenaline and it stimulates Ca^{2+} ions to come out from the sarcoplasmic reticulum of muscle fibres which causes muscle contraction.
   cGMP (Cyclic Guanosine Monophosphate) which acts as secondary messenger works on the action of acetylcholine, increase in flow of Ca^{2+} into muscle fibres & hence causes muscle relaxation. There is no role of sodium in hormonal action.
9. (c) Pancreas is a heterocrine gland comprising both endocrine and exocrine parts. Its endocrine part consists of small masses of hormone secreting cells called islets of Langerhans. The α-cells of latter secrete glucagons and its β-cells secrete insulin. These two hormones have antagonistic effects on the glucose level in the blood which means that insulin decreases the blood glucose level while glucagon increases blood glucose level.
10. (a) Neurons regulate endocrine activity, but not *vice-versa*.Neurons in the hypothalamus secrete thyroid releasing hormone (TRH), which stimulates cells in the anterior pituitary to secrete thyroid-stimulating hormone (TSH).
11. (b) Posterior pituitary releases vasopressin which stimulates reabsorption of water in the distal tubules in nephron.
12. (d) Osmotic pressure of body fluids is mainly maintained by the plasma proteins (albumins, globulins) and electrolyte ions (Na^+, K^+ etc). Phosphorus has nothing to do with the osmotic pressure of body fluids.
13. (c) Sertoli cells are found in the walls of seminiferous tubules of the testes. They anchor and provide nutrition to the developing germ cells especially the spermatids.
14. (b) In menstrual cycle, menstrual phase lasts for 4 days, proliferating/ovulating phase for about 10 days and secretory phase for 14 days.
15. (c) 16.(a) 17. (a) 18. (b) 19. (b)
20. (c) 21.(d) 22. (a) 23. (b) 24. (d)
25. (c) Calcium plays an important role in blood clotting. Parathormone, a hormone released by parathyroid glands, increases calcium level in the blood. Therefore, deficiency of this hormone will decrease Ca^{2+} level in the blood, thus leading to delay in blood clotting and increase in bleeding time.
26. (d) Estrogen and testosterone being steroid hormones are soluble in lipids, therefore they can cross the plasma membrane and bind to the cytoplasmic receptors to trigger their action.
27. (b)
28. (d) Hydrophobic molecules, such as steroid based hormones, can move easily through the plasma membrane.
29. (d) Insulin acts to lower blood sugar levels.
30. (d) Receptors on target cells react specifically to chemical messengers. The way they react will determine the cellular response.

31. (d) The adrenal glands secrete hormones in response to stress.
32. (b) Hormones typically act on a receptor in or on target cells to stimulate the production of proteins.
33. (d) The internal transport system may be the cardiovascular system in animals or the xylem and phloem in plants.
34. (c) Oxytocin stimulates muscle contractions during childbirth and milk letdown.
35. (c) In diabetics, the rise in blood glucose above the normal level of 80–90 mg/100ml fails to stimulate the production of insulin from the pancreatic β-cells of the islets of Langerhans. As a result, the glucose is not removed from the blood to be stored in muscle cells as glycogen, resulting in a high blood glucose level for a long period of time.
36. (c) Paracrine hormones act on local cells that are hormones act on the very same cells that secrete them.
37. (a) "Upregulation" of hormone receptors on a cell is the production of more receptors when the hormone is present over time in the blood or other fluids surrounding the cell.
38. (b) Thymic hormones stimulate T cell development.
39. (c) In many cases, hormones trigger cyclic AMP when they bind with the target cells. Cyclic AMP then carries out the action of the hormone within the cell as a second messenger.
40. (d) The pituitary gland is considered the master gland because it produces hormones that control endocrine glands throughout the body.
41. (c) Inhibited release of a releasing hormone will lead to a reduction in the secretion of a specific anterior pituitary hormone.
42. (d) 43. (c) 44. (b)
45. (d) Iodine is essential for the normal rate of hormone synthesis in the thyroid. Deficiency of iodine in our diet results in **hypothyroidism** and enlargement of the thyroid gland, commonly called **goitre.**
Hypothyroidism during pregnancy causes defective development and maturation of the growing baby leading to stunted growth (cretinism), mental retardation, low intelligence quotient, abnormal skin, deaf-mutism, etc.

DPP /CB23

1. (c)
2. (b) Cloning is a technique by which genetically same individuals can be produced without including any sexual reproduction eg. Dolly sheep.
3. (d) 4. (d)
5. (b) The parents that have survived to reproduce asexually are able to survive in the current stable environment. Therefore, the offspring should be preadapted for this stable environment.
6. (a) Species that exhibit parthenogenesis develop from unfertilized eggs produced by the mother. Therefore, the genetic make-up should be 100 percent the same as the mother.
7. (a) Since external fertilization can only take place in an aquatic habitat, there are no terrestrial animals that use it.
8. (a) Many animals reproduce both by asexual and sexual means.
9. (b) All these animals can reproduce sexually. However, only the chicken lays an external egg.
10. (b) All of these statements concerning asexual reproduction are correct, except that asexual reproduction is best in favorable, stable environments, ones that don't change rapidly. The reason for this is that asexual reproduction, in contrast to its sexual counterpart, results in the formation of identical offspring. Although asexual organisms can often produce many more offspring in a single reproductive event than sexual organisms, these asexually produced young do not usually have the genetic variation caused by meiosis and crossing-over to be able to survive a rapidly changing environment or times of environmental stress.
11. (a) Buds in 'eyes' form new plants.
12. (b) Bud giving rise to new plant is present towards base.
13. (b) New plants in cane-sugar are formed from nodes which are absent.
14. (b) Water absorption & gaseous exchange stop due to presence of wax
15. (c) Parthenocarpy is the development of a fruit without the formation of seeds as a result of lack of pollination, lack of fertilization and lack of development. This condition can be artificially induced by application of hormones.
16. (a) 17. (a) 18. (a) 19. (d)
20. (c) Most of banana varieties are triploid and triploidy is associated with seedlessness.
21. (d) 22. (b) 23. (b) 24. (a) 25. (d)
26. (a) Homothallic and monoecious are terms used to denote bisexual condition. The example indudes fungi and plants. Heterothallic and dioecious are terms used to denote unisexual condition.
27. (b) Morphallaxis is a mechanism of regeneration involving reorganization of body cells. In epimorphosis, new cells proliferate from the surface of the wound to form the missing structure. In accretionary growth some specialized cells retain the ability to divide and produce new cells to replace the worn-out.
28. (a) Cells become variable in shape, size & getting their specialization for the formation of particular tissue or organ in future foetus. They place themselves at some specific regions in embryo for further organogeny.
29. (d) Earthworm, sponges, tapeworms are bisexual animals and hermaphrodites as they prossess both male and female reproductive organs.
30. (c)
31. (c) Sowing is related with sexual reproduction.
32. (c) Vegetative propagation in mint occurs through sucker. Vegetative reproduction is a type of asexual reproduction for plants, and is also called vegetative propagation, vegetative multiplication, or vegetative cloning. It is a process by which new plant "individuals" arise or are obtained without production of seeds or spores. It is a natural process in many plant species (as well as non-plant organisms such as bacteria and fungi) and used or encouraged by horticulturists to obtain quantities of economically valuable plants. A related technique used in cultivation is tissue culture, which involves vegetative reproduction under sterile conditions.
33. (d) Vegetative reproduction and apomixis, both are asexual methods of reproduction, which gives the progeny genetically similar to parent.
34. (c) 35. (c) 36. (b) 37. (d) 38. (d)
39. (c) 40. (d)
41. (b) In oomycetes female gamete is large and non motile while male gamete is small & motile.
42. (c) Male sex organ is called antheridium or globule while female sex organ is called oogonium. They develop on the same branchlet in the same plant in *chara.*
43. (c)
44. (c) In most aquatic organisms, such as a majority of algae and fishes as well as amphibians, syngamy occurs in the external medium (water), *i.e.*, outside the body of the organism. This type of gametic fusion is called external fertilisation.
45. (d)

DPP /CB24

1. (c) In porogamy the tip of pollen tube enters the micropyle, pushes through the nucellar tissue & finally pierces the egg-apparatus end of the embryo sac. If pollen tube enters through the chalazal side it is called chalazogamy & if it enters laterally it is called mesogamy.

2. (b) Dry indehiscent single-seeded fruit formed from bicarpellary syncarpous inferior ovary is cypsela. Cypsela is also called inferior, false or pseudocarpic achene, the thin fruit wall (developed from pericarp and thalamus)is attached to the seed at one point but the fruits develops from an inferior, unilocular and uniovuled ovary, *e.g.,* sunflower, marigold. Some cypsela develop pappus for dispersal *e.g. Sonchus, Taraxacum.*

3. (d) Each pollen has two layered wall. The outer layer is thick, tough, cuticularised called exine which is composed of a material called "sporopollenin". It is highly resistant to biological and physical decomposition, due to which pollens are preserved for a long time in fossils.

4. (b) 5. (d)

6. (c) Embryo sac is 7-celled structure. There is a large central cell with two polar nuclei, egg apparatus with egg cell and 2 synergids present at micropylar end and its chalazal end, 3 antipodal cells are present.

7. (b) Nucellus - 2n, antipodal cells - n
Antipodal cells - n, egg cell - n
Antipodal cells - n, megaspore mother cell - 2n
Nucellus - 2n primary endosperm nucleus - 3n
Antipodal and egg cell are the product of meiotic division and rest are not.

8. (a) Polyembryony refers to the formation of more than one embryo within a seed of a flowering plant. *e.g.* Citrus.

9. (c)

10. (b) In oogamy male and female gametes are morphologically as well as physiologically different. Female gametes are large and non-motile. Male gametes are small but motile.

11. (d) Cotyledons and testa respectively are edible parts in groundnut and pomegranate. A cotyledon is a significant part of the embryo within the seed of a plant. Upon germination, the cotyledon may become the embryonic first leaves of a seedling. Testa is often thick or hard outer coat of a seed.

12. (a) 13. (c)

14. (a) Some fleshy fruits such as mango, plum etc. usually have a single hard stone that encloses a seed, called drupe.

15. (a) Herkogamy – when there is some physical barrier present between the stamens and carpels avoiding any chance of self pollination.
Cleistogamy – when the flowers remain closed, self-pollination is the rule.
Dichogamy – when the two sexes mature at different times.

16. (b)

17. (a) Unisexuality of flowers prevents autogamy, but not geitonogamy. In self fertilisation, the male and female gametes are derived from the same individual. Among plants, self fertilization also called autogamy is common in many cultivated species, *eg*., wheat and oats. However, self fertilization is a form of inbreeding and does not allow for the mixing of genetic material; if it occurs over a number of generations it will result in offspring being less vigorous and productive than those resulting from cross fertilization.

18. (a) Amphitropous : Both body of ovule and embryo sac are curved. The embryo sac assumes horse-shoe shape. *e.g. Papaveraceae.*

19. (d) The male gamete will be haploid (n). 2 polar nuclei will be diploid (2n). Endosperm formed by fusion of male gamete with two polar nuclei will be pentaploid.

$$\text{Male gamete} + 2\text{ polar nuclei} \rightarrow \text{Endosperm}$$
$$(n) \qquad (2n) + (2n) \qquad (5n)$$

20. (a) Monocarpic plants are the plants in which flowers and fruits are formed only once after vegetative growth of several years e.g., some bamboos, *Agave*, etc.

21. (c) In angiosperms, microsporangium is equivalent to pollen sac, megasporangium is equivalent to nucellus and embryo sac is equivalent to female gametophyte. Pollen grain or microspore represents immature male gametophyte.

22. (a) Ovule is an integumented megasporagium found in spermatophytes which develops into seed after fertilization. An angiospermic ovule is typically an ovoid and whitish structure. It occurs inside ovary where it is attached to a parenchymatous cushion called placenta either singly or in a cluster. The ovule is stalked. The stalk is called funiculus or funicle. The point of attachment of the body of the ovule with the funiculus is known as hilum.

23. (d) Autogamy (Gk. autos-self, gamos-marriage) is a type of self pollination in which an intersexual or perfect flower is pollinated by its own pollen.
Geitonogamy (Gk. geiton-neighbour, gamos-marriage) is a type of pollination in which pollen grains of one flower are transferred to the stigma of another flower belonging to either the same plant or genetically similar plant. In geitonogamy, the flowers often show modifications similar to ones found in xenogamy or cross pollunation.
Xenogamy (Gk. xenos-strange, gamos-marriage) or cross pollination is the transfer of pollen grains from anther of one flower to the stigma of a genetically different flower.

24. (d) *Cannabis* is wind pollinated plant.

25. (b) The typical and the most common type of embryo sac, found in 80% flowering plants is called *Polygonum* type of embryosac. It contains 8 nuclei but 7 cells – 3 micropylar, 3 chalazal and one central. It is formed by one meiosis and three mitosis.

26. (a) The pollen tube eats its way through the solid part of the stigma and style by secreting pectinases and hydrolytic enzymes. Pollen tube travels intercellularly and chemotropically along the concentration gradient of calcium – boron -inositol sugar complex.

27. (a) Stigma is the terminal receptive part of the pistil which functions as landing platform for the pollen grains. It also determines the compatibility of the pollen grains.

28. (c) The tapetal layer is of great physiological significance as all the food material entering into the sporogenous tissue diffuses through this layer. Ultimately, the cells of tapetal layer disorganise. Thus, tapetum makes a nutritive layer for the developing microspores.

29. (b) The typical and the most common type of embryo sac, found in 80% flowering plants is called *Polygonum* type of embryosac. It contains 8 nuclei but 7 cells – 3 micropylar, 3 chalazal and one central. It is formed by one meiosis and three mitosis.

30. (b) Outer layer (exine) of pollen grain is made of a highly resistant substance called sporopollenin (Zelisch, 1932). Sporopollenin is not degraded by any enzyme. It is not affected by high temperature, strong acid or strong alkali. Because of sporopollenin, pollen grains are well preserved as fossils.

31. (c) One or more thin areas present in the exine of pollen grain are known as germ pores. The germ pores are apertures in the exine layer of the pollen grain where the sporopollenin is absent. The germ pore helps in the formation of the pollen tube and the release of the male gametes during fertilization. There are usually three germ pores in dicots (tricolpate) and one in monocots (monocolpate).

32. (a) Megaspore mother cell (MMC) is the sporogenous cell that divides to give 4 haploid megaspores. Out of these 4 megaspores, only one is functional and other three degenerate. Hence 100 meiotic divisions are required for the formation of 100 functional megaspores.

33. (c)

34. **(c)** Autogamy is possible only when anther and stigma are close together and there is synchrony in pollen release and stigma receptivity. As in case of cleistogamous flowers, the flowers remain closed so that anthers and stigmas are never exposed. The flowers undergo only self pollination. No external pollinating agency is required.

35. **(b)** Apogamy is the reproduction without the fusion of gametes, and usually without meiosis. The term may include any form of vegetative reproduction. The production of a diploid gametophyte from the sporophyte due to the absence of meiosis is known as apospory. Parthenogenesis is the development of an unfertilized egg into a complete individual without fertilization. Production and development of seedless fruits is called parthenocarpy.

36. **(a)** Endosperm represents the food storing tissue of a seed. It is produced as a result of double fertilization in angiosperms. In most monocots and some dicot seeds, the food reserve remains in the endosperm. They are called endospermic or albuminous seeds, e.g., cereals, castor, coconut, rubber. However, in the majority of dicot seeds (e.g., pea, gram, bean, mustard, groundnut) and some monocot seeds (e.g., orchids, *Sagittaria*), the endosperm is consumed during seed development and the food is stored in cotyledons and other regions. They are called nonendospermic or exalbuminous seeds.

37. **(b)** Part of embryo axis between the plumule and cotyledonary node is epicotyl (above the level of cotyledons) while the part between radicle and cotyledonary node is called hypocotyl (below the level of cotyledons).

38. **(b)** Pollination will not occur

39. **(b)** There will not be insect pollination

40. **(a)** Germination of pollens will not occur in ovule.

41. **(d)**

42. **(c)** Embryo can not grow without food stored in cotyledons.

43. **(b)** Pollens are formed by meiosis.

44. **(b)**

45. **(c)** Obligate symbiotic relationship is present between *Yucca* flowers and moth, *Tageticula*.

DPP /CB25

1. **(c)** During breast feeding, prolactin hormone (required for the release of milk) present in the blood inhibits the release of LH from pituitary. This will counter the effect of LH on the ovarian follicles and therefore no ovulation will occur. Hence, no pregnancy occurs during this period.

2. **(a)** **3.** **(d)**

4. **(b)** The corpus luteum is essential for establishing and maintaining pregnancy in females. In the ovary, the corpus luteum secretes estrogens and progesterone, which are steroid hormones responsible for the thickening of the endometrium and its development and maintenance, respectively.

5. **(b)**

6. **(b)** Foetal ejection reflex in human female is induced by fully developed foetus and placenta. When a woman is in a lithotomy or semi-sitting position, the foetal ejection reflex is impaired and the increased pain caused by the sacrum's inability to move as the baby descends can be intolerable.

7. **(b)** Acrosome a small pointed structure at the tip of nucleus. It breaks down just before fertilization, releasing hydrolytic enzymes that assist penetration between follicle cells that surrounds the ovum, thus facilitating fertilization.

8. **(c)** Sertoli cells are found in the walls of seminiferous tubules of the testes. They anchor and provide nutrition to the developing germ cells especially the spermatids.

9. **(b)** In menstrual cycle, menstrual phase lasts for 4 days, proliferating/ovulating phase for about 10 days and secretory phase for 14 days.

10. **(a)** Menopause is the period when ovulation and menstrual cycle stop in human females. The period of menopause is between 45-55 years.

11. **(d)**

12. **(c)** The gland corresponding to this in female is Bertholin.

13. **(a)** **14.** **(a)** **15.** **(b)** **16.** **(b)**

17. **(b)** The part of fallopian tube closest to the ovary is infundibulum. Infundibulum possess finger-like projections called fimbriae that help in collection of ovum after ovulation. It leads to wider part of oviduct called ampulla. The last part of oviduct is isthmus that has a narrow lumen and joins the uterus.

18. **(c)** **19.** **(b)** **20.** **(b)**

21. **(a)** Vasa efferentia are fine ciliated ductules that arise from the seminiferous tubules of testis (where sperms are formed) and open into epididymis which is a mass of long, narrow, closely coiled tubule lying along the inner side of testis. Epididymis stores the sperms. Thus, if vasa efferentia get blocked, sperms will not be transported from testes to epididymis.

22. **(c)** Wall of each seminiferous tubule is formed of a single layered germinal epithelium. Majority of cells in this epithelium are male germ cells and at certain places, there are present tall Sertoli cells. These cells act as nurse cells providing nutrition to the developing sperms.

23. **(d)** Secretion of seminal vesicle, prostrate gland and bulbourethral gland constitute seminal plasma which is rich in fructose, calcium and certain enzymes.

24. **(c)** The prostate gland is a single large gland that surrounds the urethra. It produces a slightly alkaline, milky secretion which forms 25% of the volume of semen. It possesses citric acid, enzymes (acid phosphatase, amylase, pepsinogen) and prostaglandins. Secretion of the prostate gland nourishes and activates the spermatozoa to swim.

25. **(c)** The epididymis is a mass of long narrow closely coiled tubule which lies along the inner side of each testis. Coiling forms three parts - upper caput epididymis or head, middle corpus epididymis or body and lower cauda epididymis or tail.

26. **(d)** Release of ovum from the ovary is called ovulation. The graafian follicle rises to the surface, sends out a protuberance or stigma and everts to release the ovum into peritoneal cavity. The empty Graafian follicle contains a blood clot which is called corpus haemorrhagic. Its granulosa cells continue to proliferate, develop yellow carotene pigment or lutein and get converted into lutein cells. This converts the ruptured a temporary endocrine gland secreting progesterone with small quantity of estrogen.

27. **(c)** The female external genitalia include mons pubis, labia majora, labia minora, hymen and clitoris.

28. **(a)** Secondary spermatocyte contains half the number of chromosomes i.e., 8 Each chromosome has 2 chromatids, therefore, 8 chromosomes will have 16 chromatids in all.

29. **(d)** During spermatogenesis, at the end of first maturation division (reductional division or meiosis), the primary spermatocyte divides into two haploid daughter cells called secondary spermatocytes.

30. **(c)** In the head of the epididymis, the sperms undergo physiological maturation, acquiring increased motility and fertilizing capacity.

31. **(d)** Oogenesis is the process of formation of functional haploid ova from the diploid germinal cells in the ovary. Oogenesis begins during embryonic development but is completed only after fertilization of the secondary oocyte with the sperm.

32. **(b)** The middle piece of sperm contains mitochondria coiled around the axial filament. They provide energy for the movement of the sperm.

33. (a) Release of ovum from the ovary is called ovulation. The Graafian follicle rises to the surface, sends out a protuberance or stigma and everts to release the ovum into peritoneal cavity. The empty Graafian follicle contains a blood clot which is called corpus haemorrhagic. Its granulosa cells continue to proliferate, develop yellow carotene pigment or lutein and get converted into lutein cells. This converts the ruptured a temporary endocrine gland secreting progesterone with small quantity of estrogen.

34. (a) FSH acts on the Sertoli cells and stimulate secretion of some factors which help in the process of spermiogenesis.

35. (a)

36. (d) TSH or thyroid stimulating hormone has no role in menstruation.

37. (c) On the 14th day of menstrual cycle, ovulation occurs, so chances of conception are optimum.

38. (c) Estrogen secreted from ovarian follicles under the influence of FSH, causes proliferation of the endometrium of the uterine wall.

39. (a)

40. (a)

41. (d) Acrosome, a cap like structure present at the tip of the sperm, is a lysosome like organelle derived from golgi apparatus.

42. (b) Prolactin is secreted by the anterior lobe of pituitary gland. After parturition, secretion and storage of milk in mammary glands is under the influence of this hormone.

43. (c) After birth, the first milk released by mammary glands is called colostrum. It is released for 2-3 days. It is thin, yellowish fluid containing cells from the alveoli of glandular tissue of mammary glands and is rich in protein, antibodies, but low in fat.

44. (a) Fusion of male and female gamete produces a zygote. Repeated division of the zygote is called cleavage forming a solid morula. After further division and rearrangement a fluid filled cavity surrounded by blastomeres - blastula is formed. The appearance of germ layers mark the gastrula.

45. (b) The type of reproduction shown in the given figure of yeast is budding. In budding, a daughter individual is formed from a small projection, the bud, arising from the parent body. In yeast, the division is unequal and a small bud is produced that remains attached initially to the parent body. Later on the bud gets separated and matures into a new yeast organism.

DPP /CB26

1. (c) Statements (i) and (iii) are correct.
   - Medical Termination of Pregnancy (MTP) during first trimester is generally safe.
   - Intrauterine device like copper-T are effective contraceptives.

2. (b) Amniocentesis is the most widely used method for prenatal detection of many genetic disorders. It is also a technique used for determining the sex of the foetus.

3. (d)

4. (a) Copper T is an intrauterine device which prevents the fertilized egg becoming implanted in the wall of the womb.

5. (b) 6. (a)

7. (a) Gamete intrafallopian transfer (GIFT) is recommended for those females who cannot produce an ovum. In this process, the eggs of the donor woman are removed and in a form of mixture with sperm, transferred into fallopian tube of another woman who cannot produce ovum, but can provide suitable environment for fertilization. Thus in GIFT, site of fertilization is fallopian tube, not laboratory.

8. (a)

9. (b) Condoms are barriers made of thin rubber/latex sheath that are used to cover the penis in the male or vagina and cervix in the female, just before coitus so that the ejaculated semen would not enter into the female reproductive tract. This can prevent conception.

10. (c) 11. (a) 12. (d) 13. (a) 14. (b) 15. (b)

16. (c) 17. (c) 18. (b) 19. (d)

20. (b) The birth control pill interferes with the maturation of the follicles and the ova, inhibiting release of an egg.

21. (d) Sterilization techniques can be considered as the safest birth control measures. It provides a permanent and sure birth control. It is called vasectomy in male and tubectomy in female.

22. (c)

23. (a) Copper 'T' is an intrauterine device which prevents the fertilized egg from becoming implanted in the wall of the womb.

24. (c) Cu ions released by copper releasing intra uterine devices suppresses sperm motility.
   Intra-uterine devices are inserted by doctors in the uterus through vagina. They are available as the non-medicated IUDs, copper releasing IUDs and hormonereleasing IUDs.

25. (d) Intra uterine device (IUD) is a method of contraception in India. The IUD is inserted in the woman's uterus through the cervix.

26. (d) Test-tube baby programme employs zygote intrafallopion transfer (ZIFT) technique. In this technique fusion of ovum and sperm is done outside the body of woman to form zygote which is allowed to divide forming 8 blastomeres, then it is transfered the fallopion tube of the woman.

27. (d)

28. (c) Jaundice cannot be detected in a developing foetus by amniocentesis. It is a foetal sex determination test based on chromosomal pattern in the amniotic fluid surrounding the developing embryo. It is now legally ban in India.

29. (d) Medical termination of pregnancy (MTP) or induced abortion became legal in India from 1971 with some strict conditions to avoid its misuse. Such restrictions are important to check indiscriminate and illegal female foeticides which are reported to be high in India. All the other options are of natural methods of contraception.

30. (c) Oxytocin is birth hormone and acts on the smooth muscles of our body and stimulates their contraction. Vasopressin acts mainly at the kidney and stimulates resorption of water and electrolytes and reduces loss of water through urine. Hence, it is also called as anti-diuretic hormone (ADH).

31. (a) STDs are more common in 15-24 years age group. Gonorrhoea, syphilis, genital herpes, chlamydiasis, genital warts, trichomoniasis, hepatitis-B and AIDS are some of the common STDs. Haemophilia is sex-linked recessive disease, which show its transmission from unaffected carrier female to some of male progeny. Sickle cell anaemia is an autoimmune linked recessive trait in which both the partners are carriers for the gene.

32. (d) In Intra-Uterine Transfer (IUT), embryo with more than 8-blastomeres stage (morula) is used for transfer into the uterus.

33. (c) Reproductive health in society can be improved by creating awareness among people about various reproduction related aspects and providing facilities and support for building up a reproductively healthy society.

34. (b) 35. (b) 36. (b) 37. (a)

38. (d) Amniocentesis involves prenatal diagnosis of metabolic error and other genetic abnormalities.

39. (a) The programmes called 'family planning' for reproductive health awareness were initiated in 1951.

40. (d) Reproduction-related areas are currently in operation in India come under the popular name 'Reproductive and Child Health Care (RCH) programmes.

41. (b) 'Saheli'–an oral contraceptive for the females was developed at Central Drug Research Institute (CDRI) in Lucknow, India.

42. (b) In artificial insemination (AI) technique, the semen collected either from the husband or a healthy donor is artificially introduced either into the vagina or into the uterus (IUI – intrauterine insemination) of the female.

43. (c) An ideal contraceptive should be user-friendly, easily available, effective and reversible with no or least side-effects.

44. (c) Reproductive health is a crucial part of general health and a central feature of human development. It is a reflection of health during childhood, and crucial during adolescence and adulthood, Reproductively healthy societies have normal sex related emotional and behavioural interactions.

45. (a) To get rid of unwanted pregnancies and to prevent the fatality or harmfulness to the mother or to foetus or both due to the continuation of pregnancy are the reasons on the basis of which pregnancy can be terminated

DPP /CB27

1. (a) AaBB × aaBB on crossing gives 50% individuals having genotype AaBB and 50% individuals having genotype aaBB.

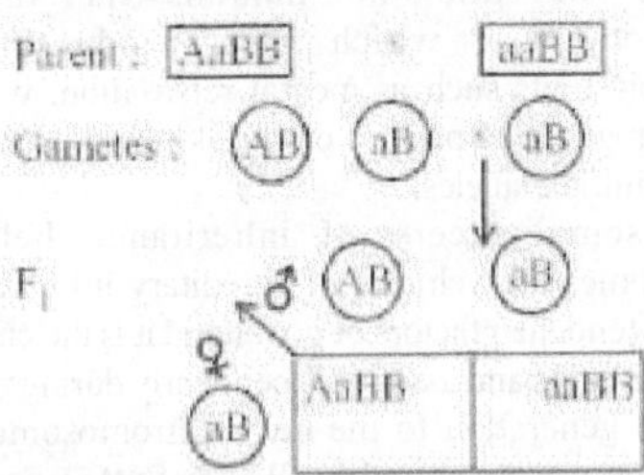

2. (b) Incomplete dominance is the phenomenon where none of the alleles are dominant, with the effect that the hybrid produced by crossing two pure individuals is a mixture between the parents.

3. (d) When a pair of contrasting characters are studied during hybridization, it is called dihybrid cross. In F_2-generation, its ratio is 9 : 3 : 3 : 1.
It means when parents having –
Yellow, round Crossed with green wrinkled
seeds seeds
(IIRR) × (iirr)
will give 9 – Yellow round seeds.
3 – Yellow wrinkled seed.
3 – Green round seeds.
1 – Green wrinkled seeds

4. (b) Genetic mapping help in studying the architecture of the chromosomes and of entire genome.

5. (a) Epistasis is the phenomenon of masking or suppressing the phenotypic impression of a gene pair by a non allelic gene pair which impresses its own effect.

6. (b) Independent assortment of genes takes place only when they are located on separate non-homologus chromosomes. Where two or more than two genes are located on same chromosome, independent assortment will not be possible.

7. (b) In eukaryotic cells, two cytoplasmic organelles, mitochondria and chloroplast of green plants, contain their own genetic materials.

8. (b) A cross of F_1 hybrid with its recessive homozygous parent is called the test cross. It is done to determine the genotype of a given plant. If the given plant has homozygous dominant traits then on test cross it gives all dominant trait plants but if it is heterozygous dominant than it gives dominant and recessive phenotypes in 1 : 1 ratio.

9. (b) Pleiotropy is the condition in which a single gene influences more than one trait.
Polyploidy is a condition in which individuals have more than two complete sets of chromosomes.
Apomixis is a reproductive process in plants that superficially resembles normal sexual reproduction but there is no fusion of gametes.

10. (c) A gene may have more than two alternative forms occupying the same locus on a chromosome, such alleles are known as multiple alleles and the phenomenon is turned as multiple allelism. ABO blood group has 3 alleles. I^O, I^A or A, I^B or B.

11. (b) Aneuploidy is the numerical change in the chromosome number of the genome. Euploidy is the phenomenon of having multiple or more than twice the number of genome.

12. (d) Haemophilia is a sex linked recessive trait carried by X chromosome (also known as bleeder's disease).

13. (d) Sickle cell anaemia (in which nucleotide triplet CTC is changed to CAC) affects the β-globin chain of haemoglobin. Since these changes occur at a particular locus or point of a chromosome where specific gene is located, they are called as point mutation.

14. (b) Down's syndrome is due to trisomy of 21st chromosome and is an autosomal abnormality.

15. (d) A barr body is one of the X-chromosomes in somatic cells of females. If there are more X-chromosomes, then there will be more barr bodies.

16. (b) Down's syndrome develops due to trisomy of chromosome number 21. In Turner's syndrome, the effect appears due to fusion of a gamete without sex chromosome and a gamete with one X-chromosome (44 + X). Klinefelter individuals are phenotypically males. The defect appears due to fusion of egg having unreduced sex complement (A + X X) with a gamete carrying Y chromosome (44 + XXY). Gynandromorphism occurs among *Drosophila.* In such individuals one half of the body shows male characters and the other half shows female characters.

17. (c) Sex in *Drosophila* is a function of the ratio of the number of X chromosomes to the number of autosomal sets. Therefore a *Drosophila* with a X/A =1.0 will be a female whereas the one with a X/A ratio = 0.5 will be male. However, in humans the presence or absence of the Y chromosome determines sex.

18. (c) The lack of independent assortment in sweet pea and *Drosophila* is due to linkage.

19. (c) Baldness is a sex influenced trait. The dominance of alleles may differ in heterozygotes of the two sexes.

20. (a) Cri-du-chat/cat cry syndrome is due to the deletion of a large part of the small or one of the 5th chromosome.

21. (d) Biometric genetics is the mathematical or statistical study of genetic phenomenon. In this branch, data of various genetic traits are analysed by applying the principles of statistics. It helps in the investigation of various genetic principles and checks their correctness and probability.

22. (b) In grasshopper the males lack a Y-sex chromosome and have only an X-chromosome. They produce sperm cells that contain either an X chromosome or no sex chromosome, which is designated as O.

23. (d) Linkage is the inheritance of genes of same chromosome together and capacity of these genes to retain their parental combination in subsequent generation. The strength of linkage between two genes is inversely proportional to the distance between the two. This means, two linked genes show higher frequency of recombination if the distance between them is higher and lower frequency if the distance is smaller.

24. (a)

25. (c) Genotype is the genetic make up of an individual irrespective of the mendelian characters or genes impressing.

Parent : AABB × aabb

Gametes : (AB) (ab)

F_1 : AaBb

	AB	Ab	aB	ab
AB	AABB	AABb	AaBB	AaBb
Ab	AABb	AAbb	AaBb	Aabb
aB	AaBB	AaBb	aaBB	aaBb
ab	AaBb	Aabb	aaBb	aabb

26. (d)

27. (c)

28. (d) Phenylketonuria is a human genetic disorder in which the body does not contain the enzyme phenylalanine hydroxylase, necessary to metabolize phenylalanine to tyrosine, and converts phenylalanine instead to phenylpyruvic acid. As PKU is an autosomal recessive genetic disorder each parent must have at least one defective allele of the gene for PAH, and the child must inherit a defective allele from each parent. As such, it is possible for a parent with a PKU phenotype to have a child without PKU if the other parent possesses at least one functional allele of the gene for PAH. A child of two parents with the PKU phenotype will always receive two defective alleles so will always have PKU. The gene for PAH is located on chromosome 12.

29. (a) Chromosomes occur in homologous pairs. Somatic cells have diploid number of chromosomes. Humans have 23 pairs *i.e.* 46 chromosomes. 22 pairs of autosomes and 1 pair of sex chromosomes i.e. XX in females and XY in males.

30. (a)

31. (d) Given that recombinant percentage is 7% and 5% therefore, total recombinants would be 7 + 5 = 12%. It is known that one map unit is the distance that yields 1% recombinant chromosomes. Hence distance between two non-allelic genes = 12 map units.

32. (b) In the given cross, passing of disease is from carrier female to male progeny (criss-cross inheritance). Any trait that shows criss-cross inheritance is located on the sex chromosome. Presence of a single recessive gene i.e. X^c in carrier individuals (XX^c) does not cause the disease, thus the trait is recessive.

33. (b)

34. (b) A cross between heterozygous long-winged flies and (homozygous) vestigial winged flies represents an example of test cross, in which the exact Mendelian ratio of 1 : 1 is obtained. i.e., 96 long-winged flies and 96 vestigial winged flies.

35. (b) Incomplete dominance is the phenomenon of neither of the two alleles being dominant so that expression in the hybrid is intermediate between the expressions of the two alleles in homozygous state. F_2 phenotypic ratio is 1 : 2 : 1, similar to genotypic ratio.

36. (a) To determine the genotype of a tall plant of F_2 generation, Mendel crossed the tall plant from F_2 generation with a dwarf plant. He called this a test cross. In a typical test cross an organism (pea plants) showing a dominant phenotype whose genotype is to be determined is crossed with the recessive parent instead of self-crossing. The progenies of such a cross can easily be analysed to predict the genotype of the test organism. Normal test cross ratio for a monohybrid cross is 1 : 1 and for a dihybrid cross is 1 : 1 : 1 : 1.

37. (a) Quantitative inheritance (polygenic inheritance) is a type of inheritance controlled by one or more genes in which the dominant alleles have cumulative effect with each dominant allele expressing a part or unit of the trait, the full trait being shown only when all the dominant alleles are present. e.g., kernel colour in wheat, skin colour in human beings, human intelligence, height in human beings and several plants, etc.

The ability of a gene to have multiple phenotypic effects because it influences a number of characters simultaneously is known as pleitropy. The genes involved are called pleiotropic genes. It is not essential that all the traits are equally influenced. Sometimes the effect of a pleiotropic gene is more evident in case of one trait (major effect) and less evident in case of others (secondary effect). An example of this in humans is the disease phenylketonuria, which also produces abnormal phenotypic traits such as mental retardation, widely placed incisors, pigmented patches on the skin and excessive sweating for multiple alleles.

38. (b) Chromosomal theory of inheritance believes that chromosomes are vehicles of hereditary information which possess Mendelian factors or genes and it is the chromosomes which segregate and assort independently during transmission from one generation to the next. Chromosomal theory of inheritance was proposed by Walter Sutton and Theodore Boveri independently in 1902. But it was later modified and expanded by Morgan, Sturtevant and Bridges.

39. (c) Linked genes are those genes which do not show independent assortment but remain together because they are present on the same chromosome. In linkage, there is a tendency to maintain the parental gene combination except for occasional crossovers.

40. (b) In ZW-ZZ type of sex determination, the male has two homomorphic sex chromosomes (ZZ) and is homogametic, and the female has two heteromorphic sex chromosomes (ZW) and is heterogametic. There are, thus, two types of eggs: Z and W, and only one type of sperms, i.e., each with Z. Fertilization of an egg with Z chromosome by a sperm with Z chromosome gives a zygote with ZZ chromosomes (male). Fertilization of an egg with W chromosome by a sperm with Z chromosome yields a zygote with ZW chromosomes (female). This mechanism operates in some vertebrates (fishes, reptiles and birds).

41. (c) Since colourblindness is a sex-linked recessive trait and males just have one X chromosome, they can never be the carriers. Males will always express the disease/phenotype.

42. (d) In a stable population, for a gene with two alleles, 'A' (dominant) and 'a' (recessive), if the frequency of 'A' is p and the frequency of 'a' is q, then the frequencies of the three possible genotypes (AA, Aa and aa) can be expressed by the Hardy-Weinberg equation:

$p^2 + 2pq + q^2 = 1$

where p^2 = Frequency of AA (homozygous dominant) individuals q^2 = Frequency of aa (homozygous recessive) individuals

2pq = Frequency of Aa (heterozygous) individuals

so, p = 0.6 and q = 0.4 (given)

$\therefore$ 2pq (frequency of heterozygote) = $2 \times 0.6 \times 0.4 = 0.48$.

43. **(b)** The distance between genes is measured by map unit. 1% crossing over between two linked genes is known as 1 map unit or centi Morgan (cM). 100% crossing over is termed as Morgan (M) and 10% crossing over as deci Morgan (dm).

44. **(d)** Dominant autosomal traits are caused by dominance autosomal genes. Some of the dominantly autosomal inherited disorder in human beings are : Polydactyly– presence of extra fingers and toes , Huntington's disease or Huntington's chorea — a disorder in which muscle and mental deterioration occurs and there is gradual loss of motor control resulting in uncontrollable shaking and dance like movement (chorea), phenylthiocarbamide (PTC) tasting, etc.

45. **(a)** Most of the gene mutations involve a change in only a single nucleotide or nitrogen base of the cistron. These gene mutations are called point mutations. e.g. sickle cell anaemia in which polypeptide chain coding for hemoglobin contains valine, instead of glutanic acid due to substitution of T by A in second position of triplet codon.

DPP /CB28

1. **(d)** rRNA occurs inside ribosomes. m RNA brings information from DNA to polypeptides. hnRNA are heterogenous nuclear RNA.

2. **(d)** (a) **Helicase** also known as unwindase, these enzymes separate the two strands of DNA.
(b) **Topoisomerases :** Tension produced by unwinding of DNA strands is reduced by these enzymes.
(c) **RNA Primase :** This is an RNA polymerase, which helps in the synthesis of a small fragment of RNA called primer.

3. **(a)** DNA ligase is the enzyme which helps in joining two fragments of DNA. The enzyme is used in DNA replication as it joints the Okazaki segments (also in proof reading). It also finds its use in genetic engineering as it can join two or more desired nucleotide sequences of DNA.

4. **(a)** In 1953 Wilkins obtained very fine X-ray crystallographic pictures of DNA from which Watson and Crick developed the double helix model of DNA.

5. **(c)** Okazaki fragments in DNA are linked up by the enzyme DNA *ligase*. Replication always occurs in 5' - 3' direction. Okazaki fragments synthesized on 3' - 5' DNA template, join to form lagging strand which grows in 3' - 5' direction.

6. **(d)** The phenomenon of making DNA over RNA genome through enzyme reverse transcriptase is called reverse transcription or teminism.

7. **(c)** Telomerase is a ribonucleoprotein which synthesize the rich strand of telomers in DNA. **Telomerase** is an enzyme that adds specific DNA sequence repeats ("TTAGGG" in all vertebrates) to the 3' ("three prime") end of DNA strands in the telomere regions, which are found at the ends of eukaryotic chromosomes. The telomeres contain condensed DNA material, giving stability to the chromosomes. The enzyme is a reverse transcriptase that carries its own RNA molecule, which is used as a template when it elongates telomeres, which are shortened after each replication cycle. Telomerase was discovered by Carol W. Greider in 1984.

8. **(b)** During transcription, from the DNA template, complementary mRNA is formed and thymine is replaced by uracil.

9. **(d)** Replication begins at the *Ori* - origin of replication and proceeds on both sides from the Ori. Unidirectional replication is rare. RNA primers are involved in both prokaryotes and eukaryotes.

10. **(c)** During the process of replication the enzyme that helps to cut one strand of DNA duplex to release the tension of coiling of two strands is topoisomerase.

11. **(c)**

12. **(d)** It was given by Geneticists George W. Beadle and E. L. Tatum which states that each gene in an organism controls the production of a specific enzyme. It is these enzymes that catalyze the reactions that lead to the phenotype of the organism.

13. **(b)** The transforming chemical discovered by Griffith in his experiments with *Pneumococcus*, was confirmed as DNA by Avery, McLeod and McCarty.

14. **(a)**

15. **(a)** Statement (i) and (iv) are / is correct
Regulation of *lac* operon by repressor is referred to as negative regulation. In negative regulation, a repressor molecule binds to the operator of an operon and terminates transcription. In positive regulation, an activator interacts with the RNA polymerase in the promoter region to initiate transcription. Human genome contains some 20,000-25,000 billion genes bases.

16. **(a)** In negative (repressible) operon, the repressor co-repressor complex binds with the operator. The free repressor cannot bind to the operator.

17. **(b)**

18. **(a)**

19. **(c)** In eukaryotes, DNA organization is complex. There is a set of positively charged, basic proteins called histones. Histones are rich in the basic amino acid residues lysine and arginine. There are five types of histone proteins – H_1, H_2A, H_2B, H_3 and H_4. Four of them (H_2A, H_2B, H_3 and H_4) produce histone octamer called nu body or core of nucleosome. The negatively charged DNA is wrapped around the positively charged histone octamer to form nucleosome. DNA connecting two adjacent nucleosomes is called linker DNA which bears H_1 histone proteins.

20. **(c)** Transformation is a phenomenon by which DNA isolated from one type of cell when introduced into another type is able to give some of its properties to the latter. In 1944, Avery, McCarty and MacLeod discovered that protein-digesting enzymes (proteases) and RNA-digesting enzymes (RNase) did not affect transformation, so the transforming substance was not a protien or RNA. Digestion with DNase did inhibit transformation, suggestion that the DNA caused did inhibit transformation, suggesting that the DNA caused the transformation. They concluded that DNA is the hereditary material.

21. **(d)** Synthesis in DNA by DNA polymerases occurs only in 5' $\rightarrow$ 3' direction. One strand called leading strand, is copied in the same direction as the unwinding helix. The other strand is known as the lagging strand. Replication of lagging strand is in a discontinuous way, synthesizing short segments of DNA which are always in the 5' $\rightarrow$ 3' direction. These short segments are called Okazaki fragments joined together by the action of DNA ligase.

22. **(b)** In tRNA, there is a TΨC loop which contains pseudouridine and ribothymidine. The loop is the site for attaching to ribosomes. Another loop, DHU loop contains dihydrouridine. It is binding site for aminoacyl synthetase enzyme. tRNA molecules have unpaired (single stranded) CCA–OH sequence at the 3' end. This is called amino acid attachment site, because the amino acid becomes covalently attached to adenylic acid or A of CCA sequence during polypeptide synthesis. Anticodon loop is made up of three nitrogen bases for recongnising and attaching to the codon of mRNA.

23. (d) According to Chargaff's rules (1950), purine and pyrimidine base pairs are present in equal amounts in a DNA molecule. i.e., [A + G] = [T + C]

or $\frac{[A \quad G]}{[T \quad C]}$ 1.

24. (c) 2'–OH group present in ribose sugar of every nucleotide of RNA is a reactive group. Uracil present in RNA is less stable as compared to thymine (= methyl uracil) of DNA. Being unstable, RNA mutates at a much faster rate and there is no repairing system. That is why RNA viruses have shorter life span. They mutate and evolve very fast.

25. (c) The segment of DNA that takes part in transcription is called transcription unit. It has three components (i) a promoter, (ii) the structural gene and (iii) a terminator. Besides a promoter, eukaryotes also require an enhancer. Promoter is located upstream of structural gene. It is called 5' end of the coding strand which is 3' end of template strand. Terminator region is present downstream of structural gene at the 3' end of coding strand which if actually 5' end of the template strand. Promoter has different parts for attachment to various transcription factors. In many cases, the promoter has an AT rich region called TATA box. The area has a groove to which specific protein components can combine. TATA containing region is also called Pribnow box.

26. (b) Eukaryotes have three RNA polymerases. RNA polymerase I is located in the nucleolus and transcribes for rRNAs (28s, 18s and 5.8S), RNA polymerase II is localized in the nucleoplasm and used for hnRNA, mRNA and RNA polymerase III is localized in the nucleus, possibly the nucleolar-nucleoplasm interface and transcribes for tRNA, 5s tRNA and snRNAs.

27. (a) UAA is a nonsense codon. It signals for polypeptide chain termination. Hence, only 24 amino acids chain will be formed.

28. (d) Regulation of gene expression can be exerted at four levels:
(i) Transcriptional level during formation of primary transcript.
(ii) Processing like splicing, terminal additions or modifications
(iii) Transport of RNAs from nucleus to cytoplasm and
(iv) Translational level

29. (c) A typical nucleosome contains 200 bp of DNA helix. Nucleosomes constitute the repeating unitof a structure in nucleus called chromatin. Chromatin is held over a scaffold of non histone chromosomal (or NHC) proteins. At some places chromatin is densely packed to form darkly stainned heterochromatin. At other places chromatin is loosely packed. It is called euchromatin. It is transcriptionally active chromatin whereas heterochromatin is transcriptionally inactive and late replicating or heteropycnotic.

30. (d) Process of DNA synthesis where by a parent DNA molecule is faithfully copied, giving rise to two identical daughter molecules is called DNA replication. In DNA synthesis, DNA polymerase plays important role having the capability to elongate an existing DNA strand but cannot initiate the synthesis. So, the synthesis is initiated with the help of RNA primer formed by RNA primase. RNA primase synthesizes the short primer RNA of about 10 nucleotides that is elongated by DNA polymerase to form an okazaki fragment of DNA during DNA replication. Helicase unzips the two strands of DNA and topoisomerase reduces the coiling tension developed due to the unwinding of the two strands.

31. (b) Lactose operon in *E.coli* is a catabolic pathway in which the structural genes remain switched off unless the inducer (Lactose) is present in the medium.

32. (d)

33. (a) DNA fingerprinting is the technique of determining nucleotide sequences of certain areas of DNA which are unique to each individual. DNA contains non cistronic hypervariable repeat sequences called VNTR. DNA fingerprinting involves the identification of these VNTRs.

34. (a) In negative (repressible) operon, the repressor co-repressor complex binds with the operator. The free repressor cannot bind to the operator.

35. (a) Operator gene allows the functioning of the operon.

36. (c) If the *lac* repressor is non functional, it cannot bind the operator site and transcription of the *lac* operon will occur at all times, whether or not lactose is present.

37. (d) Option *a* refers to the *lac* and *trp* repressors, option *b* to the CRP protein, and option *c* refers to promoter that have different transcriptional efficiencies.

38. (a) **39. (b) 40. (b) 41. (a)**

42. (b) The presence or absence of tryptophan determines whether the genes that code the necessary enzymes in tryptophan synthesis will even be transcribed.

43. (c) Option (a) describes transcription in eukaryotic cells; Option (b) describes translation.

44. (c) DNA replication is an energy-consuming process that must have an input of energy to proceed. Energy is provided in the breaking of the triphosphate tails of each nucleotide.

45. (a) AUG is initiating codon. UCG codes for serine, UUU codes for phenylalanine, UGU codes for cysteine.

DPP/CB29

1. (a) Genetic drift can operate only on smaller populations where its fluctuation can be observed in the proportion of allele distribution in the presence of external disturbances. Genetic drift is the random change in allele number and frequency in a gene pool.

2. (a)

3. (d) Flipper of seal and wing of birds are modified forelimbs, thus, have same fundamental structure but have different functions. Flippers are meant for swimming and wings are meant for flying. Therefore, these organs are homologous organs.

4. (c) Prevalence of pesticide resistant insects supports natural selection theory e.g., DDT was thought to be an effective insecticide against household pests (like mosquitoes house-flies, body lice) in 1945. But, within 2 to 3 years of its introduction, new DDT resistant mosquitoes appeared in the population. These mutant strains, soon became well established in the population by natural selection and thus replaced the original DDT-sensitive mosquitoes.

5. (b) Paleozoic era is the era of ancient life. The correct arrangement of periods of this era is Cambrian → Ordovician (age of invertebrates) → Silurian → Devonian (age of fishes) → Carboniferous (age of Amphibians) → Permian.

6. (b) *Homo erectus* appeared about 1.7 million years ago in middle pleistocene. Its brain capacity was about 900 cc. He was about 1.5-1.8 m tall having erect posture.

7. (b) Stabilizing selection favours average sized individuals while eliminates small sized individuals. It reduces variation and thus, does not promote evolutionary change. But it maintains the mean value from generation to generation. In directional selection, the population changes towards one particular direction. It favours small or large sized individuals and more individuals of that type will be present in next generation. The mean size of the population changes. Disruptive selection favours both small-sized and large-sized individuals, It eliminates most of members with mean expression, so produces two peaks in the distribution of the trait that may lead to development of two different individuals. It is important in bringing about evolutionary change.

8. (b) It is presumed that the first living organisms were chemoheterotrophs that obtained energy by the fermentation of complex organic substances available to them from the sea broth. They were anaerobes.

9. (d) The cranial capacity of Neanderthal man was 1400 cc, of Peking man was 850 - 1100 cc, of java ape man was 800 - 1000 cc and of African man was 500 cc.

10. (b) Hardy-Weinberg principle describes a theoretical situation in which a population is undergoing no evolutionary change. It states that allele frequencies in a population are stable and constant from generation to generation. There are five factors that affect Hardy-Weinberg Principle. These are - mutation, gene flow, genetic drift, genetic recombination and natural selection pressure.

11. (c) In allopatric speciation, a part of the population becomes geographically isolated from the main population. The population becomes entirely separated and finally constitutes a new species in Galapagos Islands is an example of allopatric speciation.

12. (d) In a stable population, for a gene with two alleles, 'A' (dominant) and 'a' (recessive), if the frequency of 'A' is p and the frequency of 'a' is q, then the frequencies of the three possible genotypes (AA, Aa and aa) can be expressed by the Hardy-Weinberg equation:

$p^2 + 2pq + q^2 = 1$

where p^2 = Frequency of AA (homozygous dominant) individuals

q^2 = Frequency of aa (homozygous recessive) individuals

2pq = Frequency of Aa (heterozygous) individuals

so, p = 0.6 and q = 0.4 (given)

∴ 2pq (frequency of heterozygote) = 2 × 0.6 × 0.4 = 0.48.

13. (d)

14. (b) According to Lamarckism (proposed by Lamarck), whatever characters an individual acquires in its life time due to internal vital force, effect of environment, new needs, use/disuse of organs, they are inherited to next generations. This process continues and after several generations, the variations are accumulated upto such an extent that they give rise to new species. Melanization in peppered moth favours Drawin's theory of natural selection.

15. (c)

16. (a) In 1953, Stanley Miller and Harold Urey created, in the laboratory, conditions comparable to those of early Earth, with water vapour, hydrogen, methane, and ammonia. The Miller-Urey apparatus produced a variety of amino acids and other organic compounds found in living organisms today.

17. (b) This is the definition of fitness.

18. (b)

19. (d) Fitness is a measure of an organism's genetic contribution to the next generation.

20. (b) 21. (b)

22. (b) The analogous organs show convergent evolution due to similar adaptation. They do not support organic evolution. Whale, seal and shark shows convergent evolution due to similar habitats.

23. (b) Speciation takes place via reproductive isolation which is the most important consequence of geographical isolation.

24. (d) Scales on their hind limbs and eggs with calcareous shell, indicates in birds about their reptilian ancestry.

25. (d) Jurassic period of meoszoic era was about 19-20 crore years ago & lasted for about 5.5-6 crore years. The climate was hot and damp. It is called the age of dinosaurs. Ist primitive bird *Archaeopteryx* evolved from reptiles. I[st] angiosperm appeared as a dicotyledon but gymnosperms were dominant.

26. (a) The theory of the continuity of the germplasm was published by August Weismann (1834-1914) in 1886. It proposes that the contents of the reproductive cells (sperms and ova) are passed on unchanged from one generation to the next, unaffected by any changes undergone by the rest of the body. It thus, rules out any possibility of the inheritance of acquired characteristics, and has become fundamental to Neo-Darwinian theory.

27. (b) The eye of octopus and eye of cat show different patterns of structure, yet they perform similar functions. This is an example of analogous organs.

Analogous organs have evolved due to convergent evolution . Analogous organs have developed in the evolutionary process through adaptation of quite different organisms to similar mode of life.

28. (a) Genetic drift is variation in gene frequencies within populations that can occur by chance rather than by natural selection. Sometimes the change in allele frequency is so different in the new sample of population that they become a different species. The original drifted population becomes founder and the effect is called founder effect.

29. (b) Darwin observed an amazing diversity of creatures on Galapagos islands. He realised that there were many varieties of finches in the same island like seed-eating, with altered beaks insectivorous and vegetarian finches. Carrion are dead bodies. No finches feed on carrion.

30. (b) The term Evolution was given by Herbert Spencer that is "Descent with modifications". Evolution helps us to understand the history of life.

Evolution is a process in which something changes into a different and usually more complete or better form over time and in response to environment. This results in descendents becoming different from ancestors.

31. (c) Organs that are similar in fundamental structure but different in functions are "Homologous organs", Richard Owen, introduced the term homologous. Pectoral fins of fish and fore limbs ofhorse similar in structure but different in functions are homologous organs.

Rest of the organs compared in the question are analogous organs.

32. (b) One of the first attempts to explain the mechanism of evolution was made by Jean Baptiste de Lamarck. His theory was Inheritance of Acquired Characters. The theory states that the characters acquired during life time are passed on to the progeny and then to subsequent generations and new species are produced.

33. (b) Progressive evolution is development of organisms with more elaborate and specialized structures from those having less elaborate features e.g. amphibians from reptiles .Retrogressive or degenerative evolution is development of simpler forms from more complex ones. Such evolution has occurred in case of vestigial organs, parasitic forms, and in reduction of overspecialized structures such as wings in flightless birds. Parallel evolution is formation of similar traits in related groups of organisms independently due to similar requirement e.g. running of two toed deer and one toed Horse. Evolution of wings in insects and birds serve as example of convergent evolution.

34. (b) In the embryos of all vertebrates, the presence of gill slits supports the theory of recapitulation (repeating the early stages of embryogenesis in earlier evolved animals).

35. (a) The change in the colour of peppered moth is due to the mutation of single mendelian gene for the survival in the smoke-laden industrial environment. It is called Industrial melanism.

36. (c) Chimpanzees and gorillas are our closest relatives among the living primates.

37. (c) Electron Spin Resonance method is the most accurate method for dating of fossils.

38. (a) Adaptive radiation refers to evolution of different species from a common ancestor. The mammals are adapted for different mode of life i.e. they show adaptive radiation. They can be aerial (bat), aquatic (whale and dolphins), burrowing or fossorial (rat), cursorial (horse), scantorial (squirrel) or arboreal (monkey). The adaptive radiation, the term by Osborn, is also known as Divergent evolution.

39. (d) The most significant trend in evolution of modern man (*Homo sapiens*) from his ancestors is development of brain capacity.

40. (b) According to Oparin, the atmosphere of primitive earth was reducing because its atoms were most numerous and most reactive. Free oxygen was not present in significant amount. Large quantities of H_2, N_2, water vapour, CO_2, CH_4 and NH_3 were present.

41. (b) There is no life on moon due to absence of water. There are many other reasons for being no life on moon, like insufficient oxygen, unsuitable temperature etc.

42. (b) *Cro-Magnon* fossils have been found in Europe. Neanderthal man is a transitional stage. *Australopithecus* appeared in early Pleistocene.

43. (c) The primitive atmosphere was reducing due to lack of free molecular oxygen. The early atmosphere contained ammonia (NH_3), water vapour (H_2O), hydrogen (H_2), methane (CH_4).

44. (a) Marsupials in Australia and placental mammals in North America show convergent evolution. These two subclasses of mammal have adapted in similar ways to a particular food supply, locomotor skill or climate.

45. (b) *Australopithecus* is one of the longest-lived and best-known early human species whose remains were found between 3.85 and 2.95 million years ago in Eastern Africa (Ethiopia, Kenya, and Tanzania). Evidences shows that they hunted with stone weapons but essentially ate fruits.

DPP /CB30

1. (c) A special defence system works specially against viral infection. It has no effect on micro-organism. Cells invaded by a virus produce an antiviral protein called interferon (IFN). The latter is released from the infected cell and on reaching the nearby non-infected cells it makes them resistant to the virus infection.

2. (c) Syphilis is caused by bacteria *Treponema pallidum*.

3. (c) The lymphocytes which differentiate in the thymus are known as the T - lymphocytes. T-lymphocytes are responsible for the cellular immune response.

4. (b) The major obstacle in transplantation of organs is that the recipient body does not accept the donor's organ. The body defence mechanism rejects & treats the transplanted organ as a foreign particle and reacts actively.

5. (b) Tranquilizers bring about depression of brain activity and feeling of calmness. Stimulants stimulate the nervous system, and increase alertness and activity. Hallucinogens (*Cannabis* products) alters thoughts, perceptions and feelings.

6. (b) *Rauwolfia serpentina* belong to the family *Apocynaceae*, its roots yield a chemical useful for high blood pressure.

7. (d) Tetanus and typhoid are caused by bacteria. Whooping cough is caused by bacteria. Sleeping sickness is caused by parasitic flees. Syphilis is caused by bacteria while AIDS is the final stage of HIV viral disease. Measles and Rabies are caused by virus.

8. (a) Allergy means all kinds of untoward reactions manifested in the body due to hypersensitivity of substances called allergens. Allergy reactions occur due to liberation of histamine by tissue cells.

9. (c) Carcinomas are malignant growths of the epithelial tissue that cover or line the body organs.

10. (c) HIV is a spherical, enveloped virus of about 90-120 nm diameter. Its nucleocapsid is icosahedral and its genome consists of a single-stranded RNA filament segmented into two identical filaments and associated with a "reverse transcriptase enzyme".

11. (c)

12. (d) AIDS is caused by HIV virus and Gonorrhoea is caused by *Neisseria gonorrhoea*. Urethritis is inflammation of the urethra by bacteria. Syphilis is caused by *Treponema pallidum.*

13. (c) The saliva in the mouth and the tears from the eyes belong to physiological barriers.

14. (d) Widal test is used for the diagnosis of typhoid. It is an agglutination test for the presence of antibodies against the *Salmonella* organisms, which cause typhoid fever.

15. (b) Amphetamines bring about increased alertness and sleeplessness. Hashish is a hallucinogen. Valium is a tranquilizer. Valium depresses brain activity and produces feeling of calmness, relaxation and drowsiness. Morphine is an opiate narcotic.

16. (c) Whenver our body gets attacked by some foreign invadors, our body's immune system produces some chemicals to kill or to react against the invader. These chemicals are actually made up of carbohydrates & proteins *i.e.* glycoproteins called antibodies.

17. (a) B and T-lymphocytes produce antibodies against pathogen in the body, if due to some reason B and T-lymphocytes are damaged the body will not produce antibodies against a pathogen. Each B cell and T cell is specific for a particular antigen.

18. (b) Pollen–grains of many species are responsible for some of the severe allergies and bronchial affliction in some people often lead to chronic respiratory disorders - asthma, bronchitis.

19. (c) Tetanus toxoid is a vaccine consisting of growth products of *Clostridium tetani* treated with formaldehyde serving as an active immunising agent. Hence it is weakened germs. The first tetanus toxoid (inactivated toxin) was produced in 1924 and was used successfully to prevent tetanus in the armed services during World War II. In the mid-1940s, tetanus vaccine was combined with diphtheria toxoid and inactivated pertussis vaccine to make the combination DTP vaccine for routine childhood immunization.

20. (d) AIDS (Acquired Immune Deficiency Syndrome) is caused by HIV retrovirus. The virus destroys the helper T lymphocytes thus reducing their numbers.

21. (d) **22. (a) 23. (a) 24. (b)**

25. (a) Preformed antibodies need to be injected to treat the bite by a viper snake. It is also a type of immunization which is called as passive immunization.
Antibodies, produced by B-cells, are typically made of basic structural units—each with two large heavy chains and two small light chains. B cells differentiate into plasma cells that secrete antibodies. Antibodies are proteins that bind to specific antigens and mark them for destruction by, for example, making them more recognizable to phagocytic cells. Rejection of a kidney graft is not a function of B lymphocyte.

26. (b) Infectious diseases are those diseases that can spread from one person to another e.g., AIDS. Non-infectious diseases are those diseases that cannot spread from one person to another e.g., cancer.

27. (a) Common cold is caused by some 100 types of Rhino viruses. Symptoms of common cold are irritation of nasal tract, nasal congestion, flow of mucus, sore throat, sneezing, cough, headache and slight fever.

28. (c) Elephantiasis or filariasis is caused by a number of worms. However, in India only two types of worms are responsible for this disease, *Wuchereria bancrofti* and *W. malayi*. This disease is transmitted by female *Culex* mosquitoes. Elephantiasis affects lymphatic vessels of the lower limbs.

29. (d)

30. (d) Hepatitis B is a viral disease, transmitted through both blood and sexual intercourse.

31. (d) Plague is a bacterial disease caused by *Yersinia pestis*. Symptoms of this disease are high fever and a bubo in the groin or armpit. It is primarily a disease of rodents but accidently affects man. The vector of this disease is rat fleas that carries germs of plague from rat to rat. When the rats dies of plague, the fleas leave the dead rat and if any man is nearby, they bite him and inject some plague germs into his blood. Shigellosis or bacillary dysentery, is a diarrheal illness resulting from an acute inflammatory reaction caused by *Shigella*.

32. (a) Skin amd mucous membrane are the physical barriers that provide the first line of defence or the external defence. These barriers prevent the enrty of micro-organisms into the body. Outer tough layer of the skin, the stratum corneum prevents the enrty of bacteria and viruses. Mucus secreted by mucous membrane traps the microorganisms and immobilises them.

33. (a) The substances to which an exaggerated response of the immune system occurs are called allergens. The antibodies produced to these are of IgE type.

34. (d) If the immune system fails to recognize 'self' from 'non-self' and starts destroying the body's own cells, this leads to some malfunctions, which are termed as autoimmune diseases. Both rheumatoid arthritus and myasthenia gravis are autoimmune diseases. In rheumatoid arthritis, inflammation of the synovial membrane in synovial joints occurs. When this membrane, which is the source of synovial fluid, becomes inflamed, it produces too much fluid. Thus, the joints swell and become extremely painful. Myashtenia gravis is a chronic disease marked by abnormal fatigability and weakness of selected muscles. The degree of fatigue is so extreme that these muscles are temporarily paralysed. In this disease, antibodies bind to cholinergic receptors on muscle cells, which impairs the ability of the neurotransmitter acetylcholine to induce muscle contraction.

35. (b) 'A'-lymph nodes, 'B'-Thymus, 'C'-spleen, 'D'-Bone marrow. Thymus and bone marrow are the primary lymphoid organs where maturation of T-cells and B-cells take place respectively. Lymph nodes and spleen are the secondary lymphoid organs where T-cells and B-cells undergo proliferation and differentiation.

36. (c) When a host is exposed to antigens, which may be in the form of living or dead microbes or other proteins, antibodies are produced in the host body. This type of immunity is called active immunity. Infecting the microbes deliberately during immunization/vaccination induces active immunity e.g., polio vaccine.

37. (c) 38. (d)

39. (d) The given figure is of cannabinoid molecule. Cannabinoids are hallucinogenic chemicals obtained from leaves, resin and inflorescence of hemp plant, Cannabis sativa. Marijuana, hashish and ganja and cannabinoids.

40. (c) Cannabinoids bring about euphoria, excitement, sometimes uncontrolled laughter and dilation of pupil of eyes. These days cannabinoids are taken by some sports persons.

41. (d) Rabies, Influenza and AIDS are viral diseases, Amoebiasis, Ascariasis and Trypanosomiasis are caused by Protozoa; Taeniasis, Ascariasis and Elephantasis are the diseases caused by Helminths but Cancer, Tuberculosis and Tetanus are not related diseases. Tuberculosis and Tetanus are bacterial diseases while cancer is not.

42. (c) Lipoproteins are conjugated proteins having polypeptides in association with lipids. Immunoglobulins are the constituent of antibodies. Interferons (INFs) are a group of three vertebrate glycoproteins.
Interferon induces among adjacent cells, as antiviral state by inducing synthesis of the enzymes which inhibit the viral production cycle.

43. (d) 44. (b) 45. (a)

DPP /CB31

1. (a) Parthenocarpy is the development of fruits without prior fertilization which results in the formation of seedless fruits. In some plant species, parthenocarpic fruits are produced naturally or they may be induced by treatment of the unpollinated flowers with auxin.
Removal of androecium before pollen release is called emasculation which is helpful in preventing unwanted pollination. Vernalized seeds are the chilled treated seeds for breaking dormancy. Phenyl Mercuric Acetate is an antitranspirant.
Gibberellins and Auxins are known to induce parthenocarpy in plants. If a tomato plant is treated with a low concentration of auxin and gibberellic acid it wll produce fruits without fertilization i.e. parthenocarpic fruits.

2. (a)

3. (b) Breeding of crops with high levels of minerals, vitamin and minerals is called biofortification. This is most practical aspect to improve the health of people.

4. (b) Jaya and Ratna are two rice varieties developed for green revolution in India.
   - The scientific name of Jaya is IET-723. This paddy variety takes about 130 days to grow and the grain is long, bold and white. Its yield is 50-60 quintals per hectare.
   - The scientific name of 'Ratna' is IET-1411. It takes about 130-135 days to grow. The grain is long, slender and white. Its yield is 45-50 quintal/hectare.

5. (a) DDT was subsequently banned for agricultural use worldwide under the Stockholm Convention, but its limited use in disease vector control continues to these days in certain parts of the world and remains controversial. Along with the passage of the endangered Species Act, the US ban on DDT is cited by scientists as a major factor in the comeback of the bald eagle in the contiguous US.

6. (c) *Catla catla* and *Labeo rohita* are the two Indian major carps whereas *Heteropneustes* is a catfish. *Cyprinus* is the exotic breed.

7. (a) Food chain is the transfer of energy from green plants (primary producers), through a sequence of organisms in a food chain occupying different trophic level. Therefore, statements (ii) and (iii) are correct.

8. (a) Azadirachtin is a chemical compound belonging to the limonoids. It is a secondary metabolite present in the Neem tree seeds. The molecular formula is $C_{35}H_{44}O_{16}$. Azadirachtin is a highly oxidised tetranortriterpenoid which boosts a plethora of oxygen functionality, comprising an enol ether, acetal, hemiacetal, and tetra-substituted oxirane as well as a variety of carboxylic esters. It is classified among the plant secondary metabolites.

9. (c) 10. (d) 11. (c) 12. (d)

13. (d) In superovulation, a high milk yielding cow is induced to shed 4–6 eggs (instead of one) every 6–8 weeks (instead of 20–21 days). The superovulated donor is artificially inseminated with semen from a quality bull. The embryos developing from the eggs so fertilised are flushed out. These good quality embryos are now transferred to surrogate mother for delivery.

14. (d) Karl Von Frisch, carried out many experiments and determined that when a foraging bee returns to the hive, it performs a waggle dance. Honey is predominantly glucose and fructose. *Dorsata* is a bigger bee than *Apis indica* (a medium sized bee). Bees wax is secreted by special wax glands to make compartments.

15. (a) 16. (a) 17. (a) 18. (b) 19. (b) 20. (a)

21. (a) White leghorn is a mediterranean breed.

22. (a)

23. (c) Golden rice is vitamin A rich variety developed by rDNA technology and used in the treatment of vitamin A deficiency.

24. (d) Golden rice is a transgenic crop of the future with high Vit. A content. Millions of people suffer from Vit. A deficiency which leads to vision impairment. Transgenic rice has been developed which is capable of synthesizing beta carotene, the precursor of Vitamin A. The rice variety is now being crossed into adapted varieties with field tests possible in an year or two.

25. (a)

26. (d) Main objective of production/use of herbicide resistant GM crops is to reduce herbicide accumulation in food articles for health safety. GM plants has been useful in many ways. Genetic modifications has made crops more tolerant to abiotic stresses, reduced reliance on chemical pesticides, enhanced nutritional value of food.

27. (d) In order to obtain virus-free plants through tissue culture, the best method is meristem culture. Meristem tip culture is used successfully to remove viruses, bacteria, in order to produce the greatest number of plants. Meristem culture is used to produce healthy propagation stock for crops and ornamentals.

28. (c) Gene library contains DNA fragments representing the entire genome of an organism. So collection of alleles of the genes of a crop is called gene library.

29. (b) Vector DNA are the DNA molecules that can carry a foreign DNA segment and replicate inside the host cells. Vector DNA may be plasmids, a bacteriophage, cosmids, yeast artificial chromosomes.

30. (c)

31. (a) Three crops that contribute maximum to global food grain production are Wheat, rice and maize, which belong to the family *Poaceae* (*Graminae*).

32. (d) 'Himgiri' developed by hybridisation and selection for disease resistance against rust pathogens is a variety of wheat. It is resistant to leaf/stripe rust and hill bunt.

33. (b)

34. (b) Controlled breeding experiments are carried out using artificial insemination. The semen is collected from the male and is injected either deep into the cervix or at the beginning of the body of the uterus of the selected female by the breeder. The semen may be used immediately or can be frozen and used later. It can also be transported in a frozen form.

35. (b) Breed is a group of animals related by descent and similar in most characters like general appearance, features, size, configuration, etc.

36. (b) Inbreeding strategies allow the desirable qualities of more closely related individuals to be continued within the same breed for 4-6 generations. It increases homozygosity and thus, is necessary for evolving a pure line. Continued inbreeding, especially close inbreeding usually leads to reduces fertility and even productivity. This is called inbreeding depression.

37. (b) Products of honey bee are honey, bee wax, bee venom and royal jelly. Medicinal importance includes its uses as laxative, antiseptic, sedative, etc. It is also used against digestive disorders. Bee venom is used to cure gout and arthritis. Bee was is used is producing cosmetics, paints polishes, etc. It is not a labour intensive process and is source of additional income to the farmers.

38. (c) Germplasm is the sum total of all the alleles of the genes present in a crop and its related species. The germplasm of any crop species consists of the following types of materials: cultivated improved varieties, improved varieties that are no more in cultivation, old local or 'desi' varieties, pure lines produced by plant breeders, and wild species related to the crop species.

39. (d) Tobacco mosaic is caused by Tobacco Mosaic Virus. The infection causes characteristic patterns, such as "mosaic"-like mottling and discolouration on the leaves. Turnip Mosaic Virus causes turnip mosaic. Chlorotic lesions, mosaic and mottling are the common symptoms of this disease.

40. (d) The cell from micro-organisms such as bacteria, fungi, filamentous algae, treated in various ways and used as food, are called single cell protein (SCP). The biomass is not only obtained from unicellular microorganisms but also from multicellular microorganisms.

41. (d) Meristem is a localized group of cells, which are actively dividing and undifferentiated but ultimately give rise to permanent tissue. Cultivation of axillary or apical shoot meristems is called meristem culture. Meristem culture is carried out in potato, banana, cardamom, orchids (protocorm stage), sugarcane, strawberry, sweet potato, etc.

42. (b) The plant cell without cell wall are called protoplasts.

43. (b) The above figures of maize, wheat and garden pea are some Indian hybrid crops. The term "hybrid" refers to a plant variety which is developed through the interbreeding of two or more varieties, genera or species. Though hybrids contain the best properties of the parent plants, they usually do not breed true and often revert to one of the parent plants. Hybrids are favoured for greater disease resistance, more vigorous growth, earlier maturity, higher quality of vegetables, better uniformity and improved flavour.

44. (b)

45. (c) Because iron promotes the formation of chlorophyll and magnesium is an integral part of chlorophyll molecule so in the absence of these nutrients plants show chlorosis & leaves get yellowing. Application of iron and magnesium to promote the synthesis of chlorophyll can cure the disease.

DPP /CB32

1. (d) Activated sludge is a process for treating sewage and industrial wastewaters using air and a biological floc composed of bacteria and protozoans. During the process, the primary effluent is taken to aeration tank that contain large number of aerobic heterotrophic microbes. They form flocs that digest a lot of organic matter. As the biological oxygen demand of waste water is reduced, it is passed into settling tank to undergo sedimentation. The sediment of the settling tank is called activated sludge that is a rich source of aerobic bacteria. Hence, the statement (d) is correct.
Biogas is produced by anaerobic breakdown of biomass with the help of methanogenic bacteria. It is made up of methane, carbon dioxide with traces of nitrogen, hydrogen sulphide and hydrogen.
Methanobacterium is an anaerobic bacterium that is found in rumen of cattle and is helpful in the breakdown of cellulose.

2. (c) 'Each antibiotic is effective only against one particular kind of germ' is not correct.

3. (d) *Monascus purpureus* is a yeast used in the production of statins which are used in lowering blood cholestrol.

4. (d) A common biocontrol agent for control of plant diseases is *Trichoderma. Trichoderma* is a tree living fungus that exerts biocontrol over several plant pathogens for the control of plant diseases. It is the natural method of pest and pathogen control.
5. (c) A fed batch is a biotechnological batch process which is based on feeding of a growth limiting nutrient substrate to culture. It is done for purifying enzymes.
6. (a) 2100 antibiotics have been reported so far from actinomycetes alone. Of these maximum antibiotics have been reported from streptomyces alone. Waksman isolated streptomycin from *Streptomyces griseus.*
7. (c) 8. (a)
9. (c) *Aspergillus niger* is used for production of citric acid in industries.
10. (d)
11. (a) Human insulin is being commercially produced from a transgenic species of *Escherichia coli. E. coli* is a bacterium that is commonly found in the lower intestine of warm blooded animals. The bacteria can also be grown easily and its genetics are comparatively simple and easily manipulated, making it one of the best studied prokaryotic model organisms, and an important species in biotechnology.
12. (d) Organic farming involves use of organic wastes and other biological material along with beneficial microbes to release nutrients to crop to increase the soil fertility in an ecofriendly, and pollution free environment. *Glomus,* earthworm and *Oscillatoria* can be used in organic farming while snail cannot.
13. (d)
14. (c) *Clostridium butylicum* industrially produces butyric acid.
15. (b) 16. (b) 17. (c) 18. (c)
19. (b) 20. (a) 21. (c)
22. (a) Farmers have reported over 50% higher yields of rice by using the biofertilizer *Azolla pinnata*.
23. (d) Microbes are omnipresent, found in soil, water, air, ice, iside bodies of human beings, animals and plants. Some are found in hot springs (upto 80°–100°C) and even in geysers (thermal vents).
24. (b) Prions are highly resistant glycoportein particles which function as infectious agents. Prions can also act as catalyst converting nomal protein into prion state. Prions are not affected by proteases, nucleases, temperature upto 800°C, UV radiations and formaldehyde.
25. (c) Probiotics are live microrganisms (bacteria in most cases) that are similar to beheficial micriorganisms found in the human gut. They are also called "friendly bacteria" or " good bacteria". Probiotic microorganisms consist mostly of strains of *Lactobacillius, Bifidobcterium* and *Streptococcus*. Probiotics are taken as food supplement and energy drinks (e.g. Yakult).
26. (c)
27. (b) Flocs are masses of bacteria held together by slime and fungal filaments to form mesh like structures.
28. (a) Secondary treatment of sewage (or biological treatment) depletes 90-95% of the BOD and many pathogens are removed. Reduction of BOD by 90% is achieved through mineralization of small fraction of organic matter and conversion of large proportion to removable solids.
29. (d) Methanogens are microorganisms that produce methane as a metabolic byproduct in anoxic condition. They include *Methanobcterium, Methanobrevibacter* and *Methanococcus*.
30. (d) 31. (b)
32. (b) Organic farming includes serval methods to enhance soil fertility. In such farming, methods of biological origin are used e.g., biopesticides biofertilizers, IPM (Integrated Pest Management) green manure, bioherbicides etc. Chemical fertilizers are not used in organic farming.
33. (b)
34. (c) *Oscillatoria* and *Nostoc* are nitrogen fixing cyanobacteria. They add orgnaic matter as well as extra nitrogen to the soil. Cyanobacteria are very important and low-cost biofertilizers.
35. (d) Commercial ethanol or ethyl alcohol is produced by yeast *Saccharomyces cerevisiae.*
36. (d) Lactic acid bacteria help in souring milk. Ripening of cheese is done by bacteria (*Propionibacterium shermanii*) or moulds (*Penicillium roqueforti*).
37. (a) Brewer's yeast lack sufficient diastase and amylase therefore if complex carbohydrates have to be acted upon by them, 1% malt or inoculation with fungus like *Rhizopus* is done to degrade sugars.
38. (b) Baggasse is crushed sugarcane from which sugar has been extracted. It is used for fuel in sugar refineries and in making of fibre board.
39. (b) The bacteria *Clostridium acetobutylicum* is used to produce butanol and acetone from starch. This bacteria was first used by Chaim Weizmann in 1920.
40. (a) 41. (d)
42. (c) *Aspergillus niger* is related with production of gallic acid.
43. (b) The label A represents sludge, label B represents methane and carbon dioxide, and the label C represents dung and water.
44. (a)
45. (a) BOD is the method of determining the amount of oxygen required by microorganisms to decompose the waste present in the water supply. It is a measure of organic matter present in the water. If the quantity of organic wastes in the water supply is high then the number of decomposing bacteria present in the water will also be high. As a result, BOD value will increase.

DPP /CB33

1. (a) The linking of antibiotic resistance gene with the plasmid vector became possible with DNA ligase. DNA ligase is an enzyme that is able to join together two portions of DNA and therefore plays an important role in DNA repair. DNA ligase is also used in recombinant DNA technology as it ensures that the foreign DNA is bound to the plasmid into which it is incorporated.
2. (b) DNA or RNA segment tagged with a radioactive molecule is called probe. They are used to detect the presence of complementary sequences in nucleic acid samples. Probes are used for identification and isolation of DNA and RNA.
3. (a) Restriction endonucleases are enzymes that makes cuts at specific positions within the DNA molecule. They acts as molecular scissors. They recognise specific base sequence at palindromic sites in DNA duplex and cut its strands.
4. (d) In gel electrophoresis, agarose extracted from sea weed used as gel agarose, made of 0.7% gel show good resolution of large DNA and 2% gel will show good resolution of small fragments.
5. (d)
6. (b) Cloning is the production of an organism with exactly similar genetic make up as in the mother individual. Cloning is done to preserve genotype of an individual. This is achieved by cell culture, tissue culture or genetic engineering.
7. (d) Retrovirus has the ability to transform normal cells into cancerous cells. Hence, it can used as a vector for cloning desirable genes into animal cells.
8. (d) In pBR 322;
-ori-represents site of origin of replication
-rop-represents those proteins that take part in replication of plasmid.

-*Hind* III, *Eco*RI- Recoginition sites of Restriction endonucleases
-amp^R and tet^R - They are antibiotic resistant gene part.

9. (c) Electroporation is the method of making cell membrane permeable for the entry of recombinant DNA into the bacteria.

10. (c) The Southern blot is used to detect and identify certain DNA sequences in a sample of bodily fluid. It uses single-stranded DNA to search out their complementary strands. When a Southern blot is performed on DNA, the first step is digestion of DNA with restriction enzymes. Restriction enzymes cut DNA at known sequences, and produces DNA fragments of a certain length. Once the DNA is cut into pieces, scientists conduct electrophoresis to separate them by size.

11. (a) PCR is now a common and often indispensable technique, developed in 1984 by Kary Mullis, used in medical and biological research labs for a variety of applications. These include DNA cloning for sequencing, DNA-based phylogeny, or functional analysis of genes; the diagnosis of hereditary diseases; the identification of genetic fingerprints (used in forensic sciences and paternity testing); and the detection and diagnosis of infectious diseases. In 1993, Mullis won the Nobel Prize in Chemistry for his work on PCR.

12. (d) For gene transfer into the host cell without using vector microparticles made of tungsten and gold coated with foreign DNA are bombarded into target cells at a very high velocity.

13. (b) The first fecombinant DNA was constructed by Stanley Cohen and Herbert Boyer in 1972. They cut the piece of DNA from a plasmid carryign antibotic-resistance gene in the bacterium *Salmonella typhimurium* and linked it to the plasmid of *Escherichia coli*. The vector transfers the piece of DNA attached to it.

14. (a) Genetic engineering is the artificial synthesis, isolation, modification, combination, addition and repair of the genetic material (DNA) to alter the phenotype of the host organims to suit human needs. It is the manipulation of genes by man in vitro. Restriction endonucleases play major role in genetic engineering as they can cut DNA at specific sites.

15. (c) Electrophoresis is a technique of separation of molecules such as DNA, RNA or protein, under the influence of an electrical field, so that they migrate in the direction of electrode bearing the opposite charge, viz, positively charged molecule move towards cathode (–ve electrode) and negatively charged molecues travel towards anode (+ve electrode) under an electric filed through a matrix of agarose gel. The DNA fragments separate according to their size through the agarose gel, with smaller fragments moving farther away as compared to larger ones. The DNA fragments can be visualized by staining them with ethidium bromide followed by exposure to UV radiations. Bright orange coloured bands of DNA can be observed. The separated DNA bands are then cut out from the agarose gel and extracted from the gel piece. This step is known as elution. The DNA fragments purified in this manner are used in constructing recombinant DNA by joining them with cloning vectors.

16. (c) The DNA used as a carrier for transferring a fragment of foreign DNA into a suitable hosts is called vehicle DNA of cloning vector or gene carrier. When desired gene is introduced into a vector, recombinant DNA is formed. Vectors may be plasmids, bacteriophages, cosmids, phagemids, Yeast Artificial Chromosomes (YACs) Bacterial Artificial Chromosomes (BACs), transposons, viruses, etc.

17. (d) **18. (b)**

19. (d) *Escherichia coli* is a bacterium found in human colon. On this bacterium scientists have made extensive genetic experiments to make some vital chemicals like insulin. Another bacterium is *Agrobacterium tumefaciens* which causes crown gall in plants is extensively used for genetic experiments.

20. (c) PCR is a technique for enzymatically replicating DNA without using a living organism such as *E. coli* or yeast. It is commonly used in medical and biological research labs for a variety of tasks like detection of hereditary diseases, identification of genetic fingerprints etc.
The correct steps shown in the given figure are:
A – Denaturation at a temperature of about 94° to 98°C. During the denaturation, the double strand melts open to single stranded DNA, and all enzymatic reactions stop.
B – Annealing (binding of DNA primer to the separated strands. Occurs at 50° to 65°Celsius, which is lower than the optimal temperature of the DNA polymerases)
C – Extension or elongation of the strands using the DNA primer with heat-stable DNA polymerases, most frequently Taq (*Thermus aquaticus*) at 72°C.

21. (d) **22. (c) 23. (a) 24. (b)**

25. (c) Transformation is a process by which a cell takes up naked DNA fragment from the environment, incorporates it into its own chromosomal DNA and finally expresses the trait controlled by the incoming DNA. Since DNA is a hydrophilic molecule, it can not pass through membranes, so that bacterial cells must be made competent to take up DNA. This is done by treating them with a specific concentration of a divalent cation, such as calcium (Ca^{2+}) which increases the efficiency with which DNA enters the bacterium through pores in its cell wall. Recombination DNA (rDNA) can then be forced into such cells by incubating them briefly at 42° C (heat shock), and then putting them back on ice. This enables the bacteria to take up the recombination DNA.

26. (c) Polymerase chain reaction is a technique used to replicate a fragment of DNA so as to produce many copies of a particular DNA sequence. A single PCR amplification cycle involves three basic steps: denaturation, annealing and extension (polymerisation).

27. (a) Eukaryotic genes do not function properly when transferred into bacterial cell because introns are present in eukaryotic cells but are absent in prokaryotic cells. Hence, when bacterial cell is transformed with recombinant DNA generated using human gene, it could not process it. As a result desired protein will not be produced.

28. (b) **29. (b) 30. (a) 31. (c)**

32. (d) Direct gene transfer is the transfer of naked. DNA into plant cells but the presence of rigid plant cell wall acts as a barrier to uptake. Therefore, protoplasts are the favoured target for direct gene transfer. Polyethylene glycol mediated DNA uptake is a direct gene transfer method that utilizes the interaction between polyethylene glycol, naked DNA, salts and the protoplast membrane to effect transport of the DNA into the cytoplasm.

33. (d) **34. (c)**

35. (d) The fact that DNA is structured the same way in all known organisms means that similar methods can be used to study the hereditary material.

36. (a) pBR 322 is an artificially constructed vector plasmid. It is widely used in gene cloning experiments.

37. (d) *Eco*RI is an endonuclease enzyme isolated from strains of *E.coli* and a part of restriction modified system. So, .co part stands for *coli*.

38. (a) Restriction endonuclease-*Hind* II, always cuts DNA molecules at a particular point by recognizing a specific sequence of six base pairs. This specific base sequence is known as the recognition sequence for *Hind* II.

39. (c) Chitinase is an enzyme that cleaves the glycosidic bonds in chitin, thereby breaking down the structural polysaccharide component of the hard outer covering of many animals and of the cell wall of fungi.

40. (d)

41. (c) Palindromic sequences in DNA molecule are group of bases that forms the same sequence when read in both forward and backward direction. In the given question, only option (c) represents a palindromic sequence.

42. (d) During heat shock to the bacterium, the temperature used for giving thermal shock is 42° C. This enables the bacteria to take up the recombinant DNA.

43. (d) A stirred-tank bioreactors is usually cylindrical or with a curved base to facilitate the mixing of the reaction contents. The stirrer facilitates even mixing and oxygen availability throughout the bioreactor. Alternatively air can be bubbled through the reactor.

44. (b) Downstream processing refers to the recovery and purification of biosynthetic products, particularly pharmaceuticals. It is an essential step in the manufacture of pharmaceuticals such as antibiotics, hormones (e.g. insulin and human growth hormone), antibodies and vaccines; antibodies and enzymes used in diagnostics; industrial enzymes; and natural fragrance and flavour compounds.

45. (d) Polymerase chain reaction (PCR) is a technique of synthesizing multiple copies of the desired gene (or DNA) *in vitro*. At the start of PCR, the DNA from which a segment is to be amplified, an excess of the two primer molecules, the four deoxynucleoside triphosphates and the DNA polymerase are mixed together in reaction mixture that has appropriate quantities of Mg^{2+}. The PCR operation is followed in a sequence where denaturation, primer extention occurs.

DPP /CB34

1. (b) In this technique nematode specific genes are introduced in the host plant in such a way that it produces both sense and antisense RNA. The two RNA's being complementary to each other form a double stranded RNA (dsRNA) which is also called interfering RNA responsible for initiating RNA interference (RNA i). This (dsRNA) bind to and prevent translation of specific mRNA of nematode (gene silencing). Thus transgenic plants based on RNAi technology are resistant to nematode.

2. (a)

3. (a) RNA interference technique, sense & antisense RNA fused to form dsRNA that silent the expression of m- RNA of nematode. RNA interference is a novel strategy adopted to prevent infestation of nematode *Meloidegyne incognitia* in roots of tobacco plants. .

4. (a) Recombinant DNA technology is the process joining together two DNA molecules from two different species that are inserted into a host organism to produce new genetic combination.

5. (d) Transgenic Rosie is actually cow. Restriction enzymes cut the DNA at specific sites.

6. (b)

7. (c) Common activities in bioinformatics include mapping and analysing DNA and protein sequences, aligning different DNA *etc.* are the part of human genome project.

8. (a) Dolly sheep was obtained by cloning the udder cell (somatic cell) fused with enucleated oocyte.

9. (a)

10. (a)

11. (b) Biowar or biological war or bioterrorism is the development of biological weapons against people, their crops and animals.

12. (d) The transgenic food, containing enzyme produced by antibiotic resistance gene, can cause allergies because it is a foregin protein. The bacteria present in alimentary cancal can take up antibiotic resistance gene and become resistant to that antibiotic. Transgenes can endanger native species if Bt toxin is expressed in pollens which are transported by pollinator honey bees. GM crop prodution causes changes in natural environment which may be harmful for all organisms.

13. (c) Two cry genes, cry IAc and cry IIAb have been incorporated in cotton. The genetically modified crop is called Bt cotton as it contains Bt toxin genes. The genes cry I Ac and cry II Ab control cotton bollworms.

14. (c) Hirudin is a protein that stops blood clotting. The gene encoding hirudin was chemically synthesized. This gene was then transferred ito *Brassica napus*, where hirudin accumulaes in seeds. The hirudin is purified and used as medicine.

15. (d) Using conventional methods of diagnosis (serum and urine analysis, etc.) early detection is not possible. Recombinant DNA technology, Polymerase Chain Reaction (PCR) and Enzyme Linked Immunosorbent Assay (ELISA) are some of the techniques that serve the purpose of early diagnosis.

16. (b) Gene therapy is a collection of methods that allows correction of gene defect that has been diagonsed in a child/embryo. Correcting of a genetic defect involves delivery of a normal gene into the individual or embryo to take over the function of and compensate for the non-functional gene.

17. (a) Bioethics may be viewed as the set of standards that may be used to regulate various activites based on their effects on the biological world. This is because biotehncology has aroused social as well as political concerns, which have ranged from biotechnology being unnatural to detrimental to biodiversity.

18. (c) 19. (d) 20. (b) 21. (b) 22. (a)
23. (b) 24. (d) 25. (d) 26. (b) 27. (c)
28. (c) 29. (a)

30. (d) The first commercial example of enzyme modification of a protein for human use is the conversion of pig insulin to human insulin called "humulin".

31. (a)

32. (b) These are all good examples of the end products of various r-DNA technologies.

33. (c) The use of BGH does not appear to have serious health or environmental drawbacks, but it could offer such a competitive edge to corporate farming operations that smaller, more traditional family farms might be forced out of business. This scenario is a good example of how genetic engineering can have unintended social consequences.

34. (d) DNA vaccines are used as injection of pure DNA or RNA into skeleton muscles leading to expression of DNA in the muscle cell.

35. (c) 36. (c)

37. (d) Golden rice is a transgenic food crop which may help in solving the problem of night blindness in developing countries. Golden rice or miracle rice is rich in vitamin A or β-carotene and iron and decaffeinated coffee are also valuable achievements of gene transfer technology.

38. (d) Bt toxin genes were isolated from *Bacillus thuringiensis* and incorporated into cotton plant to form a genetically modified crop called Bt cotton. Bt cotton has high yield and resistance to bollworms.

39. (c) Only mammalian cells can correctly attach these sugars to proteins.

40. (b) *Agrobacterium tumefaciens* is the causal agent of crown gall disease (the formation of tumours) in over 140 species of dicot. This disease caused by a DNA plasmid (T_i plasmid) carried by bacterium and transferred to the plant cells. T_i plasmid has been widely used in plant engineering as a vector in order to inject gene in host plant to form transgenic plant.

41. (d) 42. (b) 43. (b) 44. (b) 45. (a)

DPP /CB35

1. (d) Tropical rain forests occur in equatorial and subequatorial regions. The forests receive all the external inputs for optimum plant growth. Due to abundant plant growth, a large number of animals live in tropical rain forests. In grasslands fire occur periodically which prevent tree growth. In savannahs, periods of drought are common. In deciduous forests broad leaved hard wood deciduous trees are found predominantly.
2. (b) Competition is most severe between the members of a population belonging to same habitat.
3. (a) Symbiosis or mutualism is a mutually beneficial association necessary for the survival of both the partners.
4. (c) Praying mantis is a good example of camouflage. Camouflage is the natural colouring of an animal which enables it to blend in with its surroundings.
   A praying mantis, or praying mantid, is the common name for an insect of the order Mantodea. Often mistakenly spelled **preying mantis** (a tempting mistake, as they are notoriously predatory) they are in fact named for the typical "prayer-like" stance. The word *mantis* derives from the Greek word mantis for prophet or fortune teller.

A praying mantis from **India**

5. (a) Population is an aggregation or grouping of individuals of the same species at the same time in a particular area. Species is a grouping of individuals of one or more populations resembling each other in important morphological, anatomical and biochemical characters and can potentially interbreed. Biotic community is an assemblage of interdependent, and ecosystem is a self sufficient and self regulating segment of nature comprising of a biotic community and its physical environment, both interacting and exchanging materials. The basic level of ecological organisation starts with the individual, the next more complex levels are populations, species, community and then ecosystem.
6. (d) The keystone species in an ecosystem are those who are the main contributors to the ecosystem.
7. (b) Niche indicate the habitat of a particular species and the interaction of that species with the resources present in the habitat. Niche overlap means that two or more species sharing the resources present in a particular niche.
8. (a) **Stenohaline :** Steno meaning narrow and haline meaning salt. A fish cannot handle a wide fluctuation of salt content in water. Many fresh water fish tend to be stenohaline and die in environments of high salinity such as the ocean. Fish living in coastal estuaries and tide pools are often euryhaline (tolerant to changes in salinity) as are many species which have life cycle requiring tolerance to both fresh water and sea water environments such as *Salmon*.
9. (d) Human population growth in India can be regulated by following the national programme of family planning.
10. (b) Some species make permanent burrows deep into the soil to escape high temperature or sunlight and some cold blooded animals often like to bask in the sun to warm up their body.
11. (b) **12. (b) 13. (c)**
14. (c) This age pyramid represents the declining population of any organism. Population decline is the reduction over time in region's census. It can be caused for several reasons that includes heavy immigration disease, famine or sub-replacement fertility.
15. (b) The logistic population growth is expressed by the equation $\frac{dN}{dt} = rN\frac{K - N}{K}$ where N is population density at time t, r is the Malthusian parameter (rate of maximum population growth) and K is the so called carrying capacity (*i.e.* maximum sustainable population). It is a type of population growth when resources are limiting.
16. (b) The best pH of soil for cultivation of plants is 6.5 – 7.5.
17. (d) A self regulatory mechanism of maintaining favourable internal conditions for uninterrupted chemical reactions in the living system despite changes in the external environment is homeostasis. Total energy stored within system is enthalpy.
18. (b) Gause's hypothesis states that two different competing species cannot coexist together for a long in the same niche, one has to get eliminated by the other competing species, if they do coexist, then they do so as a result of niche differentiation, i.e. differentiation of then realized niches.
19. (b) Speciation takes place via reproductive isolation which is the most important consequence of geographical isolation.
20. (d) Hotspots are the geographical area where biodiversity is maximum. Two hotspots in India are Western ghats and North eastern himalayan region.
21. (b) **22. (c)**
23. (d) A natural association of interdependent populations of different species inhabiting a common environment or habitat as a visible, self-contained unit is called a biotic community or biocoenosis.
24. (b) A biome is a large regional unit characterized by a major vegetation type and associated fauna found in a specific climatic zone some examples of terrestrial biomes are desert, temperate deciduous forest, tropical rain forest, etc.
25. (d) Protoplasm is the material comprising the living content of a cell. It consists of 90% water alongwith the minerals, sugars, amino acids, proteins, enzymes, etc. The translocation of nutrients inside the body occurs via blood which also majorly consists of water. In an environment with excessive heat, water loss in the form of sweating and transpiration keeps the body cool. Thus, water maintains the body temperature.
26. (a) The red colour of red algae is due to abundant formationof phycoerythrin which is able to absorb blue green wavelengths of light. Being shorter, these wavelengths are able to reach the maximum depth in water. Therefore, red algae can be found in deepest ocean waters.
27. (c) The seasonal movement of complete populations of animals to a more favourable environment is called migration. It is usually in response to uneven precipitation and lower temperatures resulting in a reduced food supply and is often triggered by a change in day length. Migration is common in mammals (e.g. porpoises), fish (e.g. eels and salmon) and some insects but is most marked in birds.
28. (d) Competition is defined as the active demand by two or more individuals of the same species or members of two or more species at same trophic levels for a common resource. Intraspecific competition is competition amongst members of the same species for a common resource such as for food, space and mate. Interspecific competition amongst members of different species for a common resource such as for food, space and mate. Interspecific competition is rivalry amongst members of different species. The severity of competition depends upon similarity in the requirement of food and shelter.
29. (b) Population density is the number of individuals present per unit area of volume at a given time. If the total number of individuals is represented by letter N and the number of

units of space (area for land organisms and volume for water organisms) by letter S, then the population density D can be represented as D = N/S. For instance, number of animals per square kilometer, number of trees per acre in a forest etc.

30. (a) $\text{Date rate} = \frac{\text{Number of deaths}}{\text{Initial number of individuals}}$

$= \frac{200}{800} = \frac{1}{4} = 0.25$

31. (d) Tropical rain forest shows the maximum diversity and also the maximum productivity. More solar energy and resources are available in tropics which promotes higher productivity. On land the maximum primary production rate is found in tropical rain forests, followed by tropical deciduous forests, temperate forests, savannah, temperate grasslands and desert shrubs.

32. (c) 33. (a) 34. (c) 35. (c) 36. (c) 37. (a)

38. (c) Permafrost, a permanently frozen subsoil, is a distinctive characteristic of the tundra.

39. (a) Grasslands are typically maintained by grazing herbivores and by periodic fires. Overgrazing, such as may occur if agricultural livestock are introduced, which can lead to destruction of the grassland.

40. (c) Deserts have less than 25 centimeters of rainfall per year.

41. (c) Aestivation is the cessation or slowing of metabolic activity during the summer period to avoid problems of heat and desiccation etc.

42. (d) Blubber is the fatty layer present between the skin and muscle of whales and other cetaceans. It acts as an insulator and protects the animal from heat loss and serves as a food reserve.

43. (a) Predation is a relation between two organisms in which one organism captures and feeds on other. Commensalism is a relation between two organisms in which one benefits and the other neither derives benefit nor harms. Parasitism is a relation between organisms in which one lives as a parasite on another and harms the host. Competition is process in which the fitness and survival ability of one species is significantly lower in the presence of another species.

44. (a) Population density varies due to changes in the following basic process - A: Natality + Immigration; B: Mortality + Emigration
Natality is the proportion of births to the total population in a place in a given time. Immigration is the number of individuals of the same species that have come into the habitat from elsewhere during the time period under consideration. Mortality is the number of deaths in the population during a given period of time. Emigration is the number of individuals of the population who left the habitat and have gone elsewhere during a given period of time.

45. (c)

DPP /CB36

1. (c) 2. (b)

3. (a) A blanket of air surrounding the earth is called as atmosphere and it is divided into different strata (zone). The zone of atmosphere that lies near the ground is troposphere.

4. (b) Detritus food chain begins with dead organic matter of dead plant parts, animals and their excretory product which is being acted upon by decomposers such as saprophytes to obtain energy needed for their survival.

5. (b) Biogeochemical cycles : Two types :
(i) Gaseous : Biogenetic materials involved in circulation are gases *e.g.* N_2, O_2, CO_2 *etc.*
(ii) Sedimentary : Biogenetic materials involved in circulation are non-gaseous *eg.* P, Ca, S etc. form rocks.

6. (a) A standing crop is the quantity or total weight or energy content of the organism, which are in a particular location at a particular time.

7. (c) The decomposers act on the dead organic matter and break them down into simpler compounds and minerals.

8. (d) In soil science, humus refers to any organic matter which has reached a point of stability where it will break down no further and might, if conditions do not change, remain essentially as it is for centuries, if not millennia. In agriculture, humus is sometimes also used to describe mature compost, or natural compost extracted from a forest or other spontaneous source for use to amend soil. It is also used to describe a topsoil horizon that contains organic matter (humus type, humus form, humus profile).

9. (b) The rate at which organic molecules are formed in a green plant is called gross productivity.

10. (a) An ecosystem having low stability can be easily damaged. An ecosystem having high resilience will take less time to recover.

11. (a) Hydrarch succession takes place in wetter areas and xerarch succession respectively, which takes place in dry areas. So, both hydrarch and xerarch successions leads to medium water conditions.

12. (c) Coral reefs are often called "rainforest of the sea". It forms some of the most diverse ecosystem on earth. Coral reef ecosystem have the highest gross primary productivity in the sea.

13. (c) Polar ice caps & glaciers have the largest amount of pure and fresh water without any dissolved salt or impurity.

14. (d) Four important functional aspects of the ecocystem are (i) Productivity (ii) Decomposition, (iii) Energy flow and (iv) Nutrient cycling.

15. (a) 16. (a) 17. (a)

18. (b) The rate at which food energy is assimilated at the trophic level of consumers is called secondary productivity. In other words, the rate of synthesis of organic matter by consumers is called as secondary productivity. The primary producers have produced the food for consumers, and secondary productivity reflects only the utilization of this food for the production of consumer biomass.

19. (d) Humus constitutes the organic component of soil and is rich in lignin and cellulose. Humus is formed from organic remains through the activity of decomposer microorganisms. Process of formation of humus from raw organic remains is called humification. Humus is quite resistant to microbial action.

20. (c) Less than 50% of the solar energy incident over earth is present in PAR (photosynthetically active radiation).

21. (a) A single species may occupy more than one trophic level. Sparrow can be a primary consumer if it feeds on seeds, fruits and peas or a secondary consumer if it feeds on insects and worms.

22. (b) Secondary sucession (= subsere) is the biotic succession that occurs in an area which become secondarily bare due to the destruction of community previously present there. Secondary succession starts from previously built up substrata with already existing living matter. The action of any external force, such as sudden change in climatic factors, biotic intervention, fire, etc. had resulted in the destruction of previous community. Thus, area become devoid of living matter but its substratum, instead of primitive, is built up. It has organic matter so is biologically fertile and thus the successions are comparatively more rapid.

23. The cycling pool of carbon consists of 6×10^{14} kg (29%) of free CO_2 in the atmosphere, 1.45×10^{15} kg (71%) of dissolved CO_2 occurs in the oceans. Oceans also regulate the amount of CO_2 in the atmosphere.

24. (d) 25. (b) 26. (a) 27. (d) 28. (c)
29. (c) 30. (a) 31. (b) 32. (b) 33. (b)

34. (b) Phosphorus is mostly used as phosphate. Its reservoir pool is phosphate rocks while cycling pool is soil for terrestrial ecosystems and water for aquatic ecosystems.

35. (b) Only 10% of the mass is flown from one trophic level to another in the form of energy.

36. (c) Lichens are the early settlers on a barren area because they can tolerate desiccation, heating during summer noon or excessive cooling during winter nights. They secrete lichen acids and carbonic acid. The acids slowly corrode rock surface and release minerals required for proper growth of lichens.

37. (a) The first biotic community which develops in a bare area is called pioneer community. It has very little diversity. This stage takes the longest time to change the environment for invasion of the next community.

38. (d) Climax community is the stable, self perpetuating and final biotic community that develops at the end of biotic succession and is in perfect harmony with the physical environment.

39. (c) Identification and enumeration of plants and animal species of an ecosystem gives its species composition.

40. (d) Communities is an assemblage of interacting populations occupying a given area. Climate, species interaction, feeding relationship among organisms and succession are the factors which influence communities.

41. (d) Grassland can support greater grazing rates by herbivores than forests because grassland produces less woody plant tissue.

42. (a)

43. (c) Arrow III is incorrect.

44. (d) A : Presence of 3 - 4 storeys of plant grown in a forest is called stratification.
B : A biome having grasses with scattered trees is called savannah.
C : Man made ecosystem is dam. Man - made ecosystem are the artificial ecosystems which rely on the human efforts to sustain.
D : Pioneer in hydrosere is blue green algae.

45. (d) Decomposition involves breakdown of complex organic matter by decomposer to inorganic raw materials like CO_2, water & various nutrients. It consists of the following processes:
Fragmentation : It is the formation of smaller pieces of dead organic matter or detritus by detritivores.
Catabolism : Chemical conversion of detritus into simpler inorganic substances with the help of bacterial and fungal enzymes is called catabolism.
Leaching : Water soluble substances (formed as a result of decomposition) are leached to deeper layers of soil.
Humification : If decomposition leads to the formation of colloidal organic matter (humus), the process is called humification.
Mineralization : Formation of simpler inorganic substances (like CO_2, water and minerals) is termed mineralization.

DPP /CB37

1. (a) In situ means keeping endangered species of animals or plants into natural environment and not in the environment that looks like natural but is man made, like zoological & botanical gardens. *In situ* includes, national parks, sanctuaries and biosphere reserve.

2. (d) Biodiversity Act of India was passed by the parliament in the year 2002.

3. (b) Biodiversity is the number of variety of organism found within a specified geographic region.

4. (d) The 'Cheetah' has become extinct from India. The endangered subspecies *Acinonyx jubatus venaticus* lives in Asia (Iran). In 1990, there were reports in the Times of India of a cheetah sighting in eastern India. There is a chance some cheetahs remain in India, though it is doubtful. There have also been reports of Asiatic cheetahs in the Balochistan Province of Pakistan, though these continue to be unverified. The cheetah prefers to live in an open biotope, such as semi-desert, prairie, and thick bushes.

5. (d) Hot spots are those areas which were rich in biodiversity but now are under threat due to direct or indirect interference of human activites. These regions are on the edge to get some of their species extinct due to humans. Western Ghats in India are under threat due to continuous developmental activities and Doon valley is under threat due to continuous mining activities.

6. (b)

7. (d) *Podophyllum* is one of the endangered species of Indian medicinal plants. It is used to remove benign (not cancer) growths, such as certain kinds of warts. It works by destroying the tissue

8. (a) Kaziranga National Park in Assam is famous as a protection environmental park for rhinoceros.

9. (c) 10. (b)

11. (c) Lesser inter-specific competition is not observed in biodiversity hotspots.

12. (b) In India maximum genetic diversity is seen in rice. Total of 23 variety of irrigated ecology and 8 varieties of rainfed ecology has been invented in India.

13. (d) The statements (i) and (ii) are correct. The statement (iii) is incorrect because predators cannot lead to the extinction of prey species. Predator and prey evolve together. The statement (iv) is incorrect because chemical such as nicotine and strychnine produced by the plants are not metabolic disorders but are metabolic wastes.

14. (b) Diversity is due to the long term evolutionary changes. Adaptability to continuous changes in environmental conditions is important for natural selection of variants and variations generation after generation leading to emergence of diverse descendants.

15. (d) Higher diversity in tropical areas is because (i) Speciation is a function of time. Temperate areas have undergone frequent glaciation in the past. It killed most of the species. No such disturbances occured in tropics where species continuted to flourish and evolve undisturbed for millions of years (ii) There are no unfavouable seasons in tropics. Continued favourable environment has helped tropical organisms to gain more niche specialisation and increased diversity. (iii) More solar energy is available in tropics. This promotes higher productivity and increased biodivesity. (iv) Resoruce availability is higher in tropics (v) There is reduced competition in tropics due to favourable environment. (vi) Rate of extinction is low in tropics.

16. (b) *Lantana*, *Eichhornia* and Arican catfish are exotic species. Non-native or exotic species are often introduced inadvertently for their economic and other uses. They often become invasive and drive away the local species. These species are considered to be second major cause of extinction of species. Exotic species have proved harmful to both aquatic and terrestrial ecosystems.

17. (d) A stable community is self-prepetuating and final biotic community that is in perfect harmony with the physical

environment. It does not show much of changes and is usually resilient to natural or man made disturbances, it is also resistant to evasion by alien species.

18. (b) The number of species of aves, reptiles, amphibians and mammals are 1232, 456, 209 and 300 respectively.

19. **(a)** IUCN is International Union of Conservation of Nature and Natural Resoruces, which is now called World Conservation Union (WCU).It has its headquarters at Morges, Switzerland. It maintains a red data book or red list which is a catalogue of taxa facing risk of extinction.

20. **(a)** The relationship between species richness and area is rectangular hyperbola for a wide variety of taxa whether they are birds, bats, fresh water fishes or flowering plants and the equation can be given as $S = CA^Z$. On a logarithimic scale, it is straight line described by the eqution $\log S = \log C + z \log A$. Where S is species richness, Z is slope of line or regression coefficient, C is Y-intercept while A is area.

21. **(b)** Genetic diversity is the diverty in the numbers and types of genes as well as chromosomes present in different species and the variations in the genes and their allels in the same species. Genetic variation in *Rauwolfia vomitoria* indicates genetic diversity. Another example is the presence of more than 50,000 genetically different strains of rice,and 1000 varieties of mango in India.

22. **(d)** Keystone species are those species which have significant and disproportinonaely large influence on the community structure and characteristics. It often has considerably low abundance and biomass as compared to dominant species. Removal of such species causes serious disruption in structure and function of community.

23. **(b)** National park is a reserved area used for conservation purposes. It is maintained by the government. Cultivation grazing, forestry and habitat manipulation are not allowed. Protection is provided to the entire ecosystem.

24. **(d)** Bishnois of Rajasthan protect *Prosopis cineraria* and Black buck religiously.

25. **(b)** Exotic species tend to cause a the reat to biodivesity. They are not found in hot spots which are regions of high biodiversity. In India, 3 hotspots ae found, i.e., Western Ghats and Sri Lanka, Indo-Burma (NE India) and Himalays.

26. (c) Ex situ (off site) conservation is conservation of selected rare plans/animals in places outside their natural homes. Ex situ conservation includes offsite collections, gene banks in vitro fertilization, cryopreservation techniques and tissue culture.

27. (a) Jim Corbett National Park (district Nainita Uttrakhand) is the first national park of Inida, established in 1936.

28. (b) Earth Summiti at Rio de Janeiro (1992), Brazil, promoted Convention on Biological Diversity (CBD) which was signed by 152 nations, Its recommendations came into effect on 29th Dec, 1993. India became a party to this Convention on Biological Diversity in May, 1994.

29. **(a)**

30. **(a)** Alpha diversity is present within community.

31. **(d)** **32. (b) 33. (c) 34. (d)**

35. **(b)** Those species whose populations have been seriously depleted and whose ultimate security is not assured are known as Endangered species.

36. **(b)** **37. (c)**

38. **(d)** Chiru is Himalayan Ibex, also called Tibetan Antelope. Shahtoosh is obtained from its neck hair. It is banned under the law.

39. **(c)**

40. **(d)** Nature selects those set of characters that are best adapted to the environment. This has resulted in the great diversity seen in the population of finches.

41. **(d)** **42. (b) 43. (b) 44. (d) 45. (b)**

DPP/CB38

1. **(a)** NO_2 and SO_2 after reacting with water in atmosphere produce HNO_3 & H_2SO_4 which fall down as acid rain.

2. **(d)** DDT residues are rapidly passed through food chain causing biomagnification because DDT is liposoluble. Biomagnification, also known as bioamplification or biological magnification, is the increase in concentration of a substance, such as the pesticide DDT, that occurs in a food chain as a consequence of:
-Persistence (slow to be broken down by environmental processes)
-Food chain energetics
-Low (or non-existent) rate of internal degradation/excretion of the substance
(often due to water-insolubility).

3. **(d)** Silicosis is caused by exposure to dust containing silica.

4. **(d)** In secondary treatment mainly settled sewage flows to an aerobic biological treatment stage where it comes into contact with micro-organisms which remove and oxidise most of the remaining organic pollutants.

5. **(c)** Biochemical Oxygen Demand (BOD) in river water increases when sewage gets mixed with river water. Biochemical oxygen demand or biological oxygen demand (BOD) is a chemical procedure for determining how fast biological organisms use up oxygen in a body of water. It is used in water quality management and assessment, ecology and environmental science. BOD is considered as an indication of the quality of a water source.

6. **(d)** **7. (a) 8. (a)**

9. **(b)** Ozone layer is present in the stratosphere in the atmosphere of the earth. It protects the earth from UV rays (less than 300 nm). Hole in the ozone layer means thining of ozone layer that allows harmful UV rays to reach earth which cause skin diseases.

10. **(a)** Lots of urea and phosphate fertilizer were used in the crops in the vicinity and the lake water turned green and stinky. Due to this, lake near a village suffered heavy mortality of fishes within a few days.

11. **(c)** If natural habitat is destroyed, wild life is affected worstly.

12. **(c)**

13. **(d)** The correct acronym is IUCN, International Union for Conservation of Nature and Natural resources.

14. **(b)** Photochemical smog is made by the deposition of dust & carbon particles on the pollutant gases and water vapours. The gases found in photochemical smog are NO_2, PAN, O_3 etc.

15. **(a)** Due to dumping of huge amount of sewage, the oxygen levels are depleted, which are reflected in terms of BOD values of water. The number of microbes also increases tremendously and these also consume most of the oxygen. Thus BOD of the river water will increase.

16. **(c)** Primary pollutants perist in the form they are released in the environment. Secondary pollutants are formed by reaction amongst the primary pollutants. Secondary pollutants are more toxic than primary pollutants. So, DDT is a primary pollutant.

17. **(c)** Bhopal gas tragedy, the world's worst industrial catastrophes was not due to radioactive fallout, but it was because of leakage of phosgene and methyl isocyanate gas.

18. **(b)** Lichens are indicators of air polution, especially of the concentration of sulphur dioxide in the atmosphere. Air pollutants dissovled in rain water, mainly sulphur dioxide can damage lichens and prevent them from growing. So, lichens are natural indicators of air pollution.

19. **(b)** Noise pollution causes anxiety, stress and irritability. Adernaline hormone is released when body is under stress. So noise pollution leads to secretion of adernaline hormone, which provides added energy to the body to deal the stressful situation.

20. **(b)**

21. **(a)** In the given pie diagram, greenhouse gases are shown. 'A' is CO_2 which is produced due to deforestation and large scale combustion of fossil fules. 'B' is CH_4 which is produced by incomplete biogas combustion and is the major constituent of biogas. 'C' is water vapour.

22. **(b)** Ultraviolet radiations are of three types - UV - C (100) - 280 nm), UV - B (280 - 320 nm) and UB - A (320 - 390 nm). UV radiation of wavelengths shorter than UV-B, are almost completely absorbed by earth's atmosphere, given that the ozone layer is intact. But, UV-B damages DNA and mutation may occur. It causes aging of skin, damage to skin cells and various types of skin cancers.

23. **(b)** In Jhum cultivation crops are cultivated for 2-3 years without manuring. It results in nutrient depletion, reduced moisture retention of soil and increased soil erosion.

24. **(b)** **25.** **(c)** **26.** **(d)** **27.** **(b)**

28. **(c)** Yellowing and blackening of Taj Mahal at Agra is due to SO_2 and other pollutants released by Mathura refinery.

29. **(b)** **30.** **(b)** **31.** **(c)** **32.** **(a)** **33.** **(d)** **34.** **(a)**

35. **(a)** The 'Jhum cultivation' is also known as 'Slash and burn agriculture'.

36. **(c)**

37. **(a)** Unlike domestic sewage, waste water from industries like petroleum, paper manufacturing, metal extraction and processing, chemical manufacturing, etc. often contain toxic substances, notably heavy metals (defined as elements with density > $5 g/cm^3$); such as mercury, cadmium, copper, lead etc.) and a variety of organic compounds.

38. **(b)** DDT causes egg shell thinning in birds because it inhibits calcium ATPase.

39. **(b)** Friends of the Arcata Marsh (FOAM) is a non-profit organization which for the last twenty years has advanced knowledge and educated the public about treatment and reuse of wastewater. The purpose of the FOAM is to stimulate understanding of the Arcata Marsh and Wildlife Sanctuary, its relationship with Arcata's integrated wastewater treatment system, the surrounding watersheds and bay, and their link with the earth's water cycle.

40. **(d)** The green house effect is a naturally occuring phenomenon that is responsible for heating of Earth's surface and atmosphere. Without a green house effect, the average temperature at earth's surface would have been around –18°C, rather than the present average of +15°C.

41. **(a)** The presence of a large amount of nutrients in water causes excessive growth of planktons or free-floating algae to produce algal bloom. Such algal blooms deteriorate the quality of water, cause the mortality of fishes and impart distinct colour to water bodies.

42. **(b)** A : Catalytic converter is a device incorporated in the exhaust system of a motor vehicle, containing a catalyst for converting pollutant gases into less harmful ones. As the exhaust passes through the converter, unburnt hydrocarbons are converted into carbon dioxide, water, carbon monoxide and nitric oxide are changed to carbon dioxide and nitrogen gas respectively.

B : Electrostatic precipitator is a device that removes suspended dust particles from a gas or exhaust by applying a high-voltage electrostatic charge and collect the particles on charged plates.

C : Earmuffs is a pair of soft fabric coverings, connected by a band across the top of the head, that are worn over the ears to protect them from cold or high noise level.

D : Land - fill is a site for the disposal of solid waste in which refuse is buried between layers of dirt so as to fill in or reclaim low-lying ground.

43. **(a)** Electrostatic precipitator is a device that removes suspended dust particles from a gas or exhaust by applying a high-voltage electrostatic charge and collecting the particles on charged plates. In the given diagram of electrostatic precipitator, the parts marked as A, B and C are respectively discharge corona, negatively charged ions, and collection plate grounded.

44. **(d)**

45. **(a)** An algal bloom is a rapid increase in the population of phytoplankton algae in an aquatic system. Typically only one or a few species are involved and the bloom is recognized by discoloration of the water resulting from the high density of pigmented cells.